6th Digital Earth Summit 2016

IOP Conference Series: Earth and Environmental Science
Volume 46

Beijing, China
7 – 8 July 2016

Editors:

Fang Chen
Yuan Zeng
Linlin Guan

ISBN: 978-1-5108-3271-8
ISSN: 1755-1307

Printed from e-media with permission by:

Curran Associates, Inc.
57 Morehouse Lane
Red Hook, NY 12571

Some format issues inherent in the e-media version may also appear in this print version.

Printed by Curran Associates, Inc. (2017)

For permission requests, please contact the Institute of Physics at the address below.

Institute of Physics
Dirac House, Temple Back
Bristol BS1 6BE UK

Phone: 44 1 17 929 7481
Fax: 44 1 17 920 0979

techtracking@iop.org

Additional copies of this publication are available from:

Curran Associates, Inc.
57 Morehouse Lane
Red Hook, NY 12571 USA
Phone: 845-758-0400
Fax: 845-758-2633
Email: curran@proceedings.com
Web: www.proceedings.com

Table of contents

Volume 46

6th Digital Earth Summit
7–8 July 2016, Beijing, China

Accepted papers received: 3 November 2016
Published online: 21 November 2016

Preface

The papers in this proceeding are contributions written by participants at the 6th Digital Earth Summit held 7-8 July, 2016 in Beijing, China. Under the theme "Digital Earth in the Era of Big Data", the 6th Digital Earth Summit aims to provide an international platform to focus on the application and technologies of Digital Earth for new developments and advances in the context of big data. In total, over 184 abstracts were selected for presentations in "Digital Earth Application in the Context of Big Data", "Big Data and Mountain Surface Process", and "The Role of Open Standards in the Era of Big Data" at 13 scientific sessions and 2 poster sessions. The main topics of the Summit are well represented by the papers in this proceeding. The Co-Chairs of the Summit was Huadong GUO (Institute of Remote Sensing and Digital Earth, Chinese Academy of Sciences) and Alessandro ANNONI (Joint Research Center, European Commission). The Organizing Committee for the Summit was chaired by Yongwei LIU (Chinese National Committee of International Society for Digital Earth) and Xinyuan WANG (Key Laboratory of Digital Earth Sciences, Chinese Academy of Sciences).

We would like to thank the International Society for Digital Earth (ISDE), the Institute of Remote Sensing and Digital Earth (RADI) of CAS, the International Centre on Space Technologies for Natural and Cultural Heritage (HIST), under the Auspices of UNESCO, the CAS-TWAS Centre of Excellence on Space Technology for Disaster Mitigation (SDIM), Integrated Research on Disaster Risk (IRDR), Chinese National Committee of ISDE, Key Laboratory of Digital Earth Sciences of CAS, without whose generous support the Summit could not have taken place. Special thanks also go to the staff at IOP Publishing Ltd., UK.

Fang Chen, Yuan Zeng, Linlin Guan
Institute of Remote Sensing and Digital Earth
Chinese Academy of Sciences, China

Peer review statement

All papers published in this volume of *IOP Conference Series: Earth and Environmental Science* have been peer reviewed through processes administered by the proceedings Editors. Reviews were conducted by expert referees to the professional and scientific standards expected of a proceedings journal published by IOP Publishing.

Monitor key parameters of winter wheat using Crop model

Yue Jibo[1,2], Feng Haikuan[2] and Qi Xiudong[1]

[1] School of Surveying and land Information Engineering, Henan Polytechnic University, Jiaozuo, China
[2] National Engineering Research Center for Information Technology in Agriculture, Beijing, China

E-mail: fenghaikuan123@163.com

Abstract. Estimation of biomass, canopy cover and yield is very important to agricultural decision Precision Farming. During the winter wheat growing season of 2013/2014, field measurements were conducted at Yangling District, Shaanxi Province at the jointing stage, heading stage and filling stage. AquaCrop model and Particle swarm optimization algorithm was used to find the global optimal simulation when the intermediate variable was the biomass. Through the simulation for each of the experimental data, biomass, canopy coverage and soil moisture were verification by ground measurements. Based on 8 sets of data, the simulation accuracy was calculated. The RMSE, nRMSE, MAE and R2 between simulation and measured biomass were 1.06 ton/ha, 11.92%, 0.90 ton/ha and 0.92. The RMSE, nRMSE, MAE and R2 between simulation and measured canopy cover were 8.92%, 9.84%, 7.84% and 0.66, respectively. The simulation results show that the AquaCrop model can help the decision making of winter wheat field in arid areas.

1. Introduction

Winter wheat biomass, crop coverage and soil moisture are frequently used variables in precision agriculture. Those parameters play an important role in monitoring the growth of agriculture, the accurate and continuous parameter estimation can provide guidance for farming production scientifically. Biomass refers to the total amount of organic matter (dry weight) in the unit area of a certain time (including the weight of the food stored in the body), crop coverage refers to the ratio of the vertical projection area to the area of a plant, those parameters can be intuitive to use to monitor the growth of crops. Soil moisture refers to the proportion of water in a certain volume (weight) of soil, it is a parameter to characterize the water in the soil during crop growth, in many crop growing areas, the growth of crops is affected by water stress. In agricultural production, the water use in irrigated agriculture was about 72% of the fresh water resources in the whole world, but drought in arid regions is still a widespread phenomenon for the shortage of water resources [1]. Therefore, monitor the parameters of crops, take the drought in arid and semi-arid regions effectively is very important for agricultural management department.

Crop growth simulation model has become one of the most powerful tool in the agricultural scientific research and management of crop planting, sustainable agriculture and the role of the government decision has been gradually known by people, is also expanding its application field. Because the attention to the developing countries, the United Nations food and agriculture organization (FAO) developed the AquaCrop to help make the irrigation plan, and improve the utilization efficiency of water [2][3]. Many studies were carried out, such as maize, wheat, soybean, Rice [4][5][6].

AquaCrop crop model is widely used in various types of crop growth simulation and irrigation analysis. By calculating the crop transpiration, water productivity was used to calculate the biomass, harvest index in AquaCrop crop model, then complete the simulation of crop growth. The aim of this paper is to explore the performances of AquaCrop model by monitoring key parameters of winter wheat.

2. Methodology

2.1. Study area and Data Collection

The study site is located in Yangling district (34°2′25″ to 34°7′23″ N, 107°5′10″ to 108°9′23″ E) of Shaanxi, China. The campaigns were carried out during the winter wheat growing season of 2014. During the winter wheat growing season of 2013/2014, field measurements were conducted at Yangling District, Shaanxi Province.8 fields were employed to addition in the experiment with Biomass, canopy cover and soil moisture at the jointing stage, heading stage and filling stage. The ground was measured in 8 field, winter wheat LAI, biomass and soil moisture were collect. Canopy cover were calculated using equation 1.

$$CC=1\text{-}\exp\,(-0.65\times LAI) \tag{1}$$

2.2. Aquacrop and model data

Aquacrop model is developed by the Food and Agriculture Organization of the United Nations (FAO, http://www.fao.org/nr/water/aquacrop.html), AquaCrop uses the first Doorenbos and Kassam [7] equation for the biomass calculation. The most important part of AquaCrop model was the crop response to water, using the experience of first Doorenbos and Kassam equation.

$$\frac{Y_x\text{-}Y_a}{Y_x} = K_y \frac{ET_x - ET_a}{ET_x} \tag{2}$$

In equation 2, Yx is the maximum output of winter wheat, ETx is the maximum evapotranspiration, Ya is the actual output of winter wheat, ETa is the actual evapotranspiration and Ky is the output of winter wheat experience response factor between evapotranspiration and yield. The core growth mechanism of AquaCrop model are as follows:

$$B = WP \times \Sigma Tr \tag{3}$$

$$Y = B \times HI \tag{4}$$

Equation 3 describes how the AquaCrop model simulated of crops growth, Equation 4 gives the method of biomass convert to the output. Where, WP is the water productivity, the unit is kg/m^3, it means how many crop output can be obtained by a certain amount of water consumption. Different varieties of crops have different water productivity. The Tr is crop transpiration, B means the biomass of crop. HI is Harvest index, HI is the response factor between biomass and output of the crops.

Before the simulation, sensitivity analysis was performed, the following model sensitive crop parameters were select [8][9]. Canopy decline coefficient, Canopy growth coefficient, Maximum canopy cover, From sowing to maximum rooting depth, Number of plants per hectare, From sowing to emergence, Water productivity, Soil water depletion factor for canopy senescence and Shape factor for water stress coefficient for stomatal control. Soil investigation has been performed, there are four types of soil in the study area, they are Silty clay loam, Silty loam, Silty clay loam, Silty loam.

2.3. Particle Swarm optimization

Particle swarm optimization (PSO) was originally proposed by Kennedy and Eberhart [10], by simulating the flock foraging behaviour and developed a kind of random search algorithm based on group collaboration. It is a kind of Swarm intelligence (Swarm intelligence, SI). Particle swarm optimization iteratively trying to improve a candidate solution with regard to a given measure of quality. PSO initialized to a group of random particles (random solutions), and then find the optimal solution, through iteration in each iterative, particles by tracking two extreme solutions to update themselves. In particle swarm optimization initial value, Max value, Min value, V Max, V Min of optimize parameters were needed for optimization. The initial value was set as the starting value, the Max value and Min value were the max and min parameters value, V Max, V Min were the maximum and minimum values of each change.

2.4. Optimization method

When optimization was proceeding, 9 crop parameters and 4 soil parameters were adjusted and optimized using Particle swarm optimization. The initial value, Max value, Min value, V Max, V Min of optimize parameters were list in table 1.

Table 1. Optimize parameters.

Crop Parameters	initial value	Max value	Min value	vMax	vMin
Canopy decline coefficient	0.05	0.03	0.07	0.001	-0.001
Canopy growth coefficient	0.05	0.02	0.07	0.001	-0.001
Maximum canopy cover	0.75	0.82	0.99	0.01	-0.01
From sowing to maximum rooting depth	1500	1200	1800	15	-15
Number of plants per hectare	4250000	2000000	5500000	125	-125
From sowing to emergence	170	120	250	4	-4
Soil water depletion factor for canopy senescence	0.65	0.55	0.75	0.01	-0.01
Shape factor for water stress coefficient for stomatal control	2.5	1.5	3.5	0.1	-0.1
WP	15	13	18	0.2	-0.2
Soil Parameters					
Silty clay loam	0.10	0.05	0.15	0.01	-0.01
Silty clay	0.30	0.20	0.40	0.02	-0.02
Silty clay loam	0.10	0.05	0.15	0.01	-0.01
Silty loam	0.30	0.20	0.40	0.02	-0.02

In the AquaCrop model, soil parameters and crop parameters which were sensitive were selected to assist model simulation optimization. Particle swarm optimization algorithm was used to adjust the above parameters, using 30 particle populations, the 80 iteration to find the global optimal simulation with intermediate variable was the biomass. Through the simulation for each of the experimental data, biomass, canopy coverage and soil moisture were verification by ground measurements. Finally, the final biomass and yield were simulated and predicted.

2.5. Optimal relationship

Each run of AquaCrop model will simulate the crop growth, biomass, canopy coverage and soil moisture. Then, the relationship between measured value and simulated value could calculated by equation 4. Assisted by particle swarm optimization algorithm, simulation of the growth process is recorded when the Minimum f was obtained.

$$\mathrm{f} = \sqrt{\frac{\sum_{1}^{n}(Bsim_i - Bm_i)^2}{n}} \tag{4}$$

In equation 4, $Bsim_i$ is the estimated values, Bm_i is the observed values, n means the sampling number (this paper for 3 times). In fact, each spot had simulated 2400 times, gBest is the parameters when f gets minimum value in 2400 times simulation. The gBest was accepted as the winter wheat growth simulation.

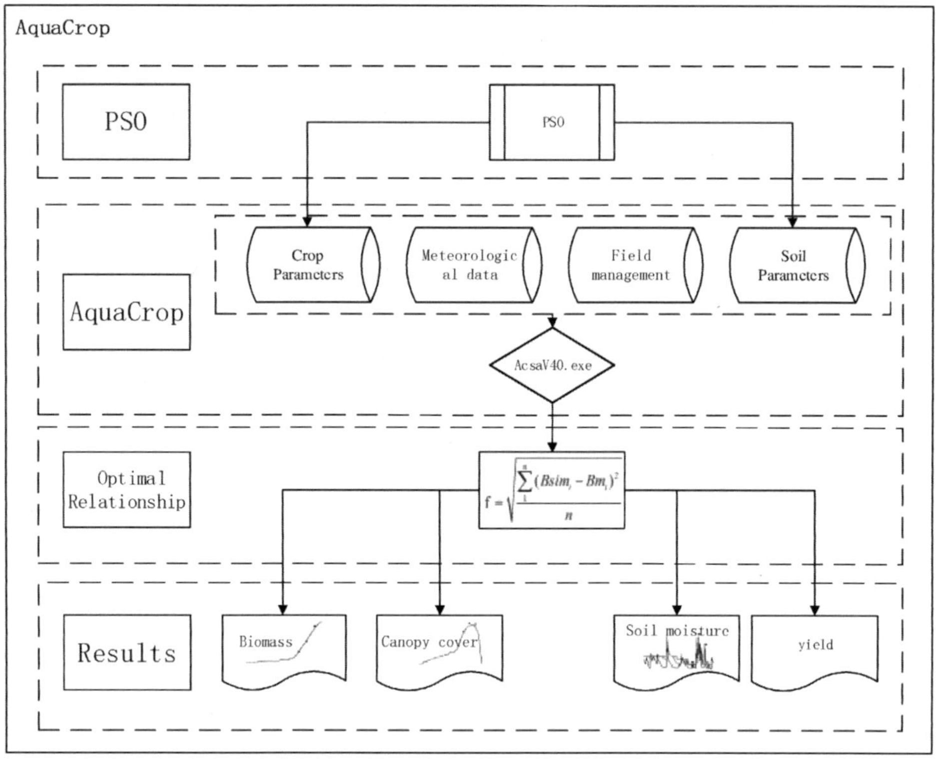

Figure 1. The flow chart.

3. Results

3.1. Biomass

In simulation experiments, the role of biomass data was intermediate variable, therefore, simulation accuracy can be show from observed biomass and AquaCrop model estimated biomass directly. In figure 2, the curve means the estimated biomass from planting to harvest, observed values means the ground measurement biomass.

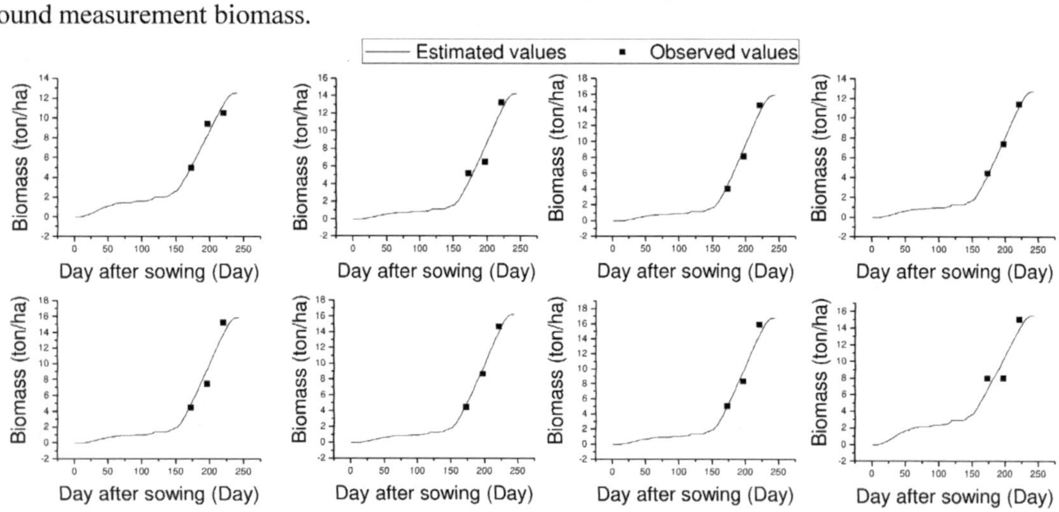

Figure 2. Biomass simulation results.

In figure 2, the estimated biomass was simulated by Aquacrop in whole growth period, figure 2(a), 2(b), 2(c), 2(d), 2(e), 2(f), 2(g), 2(h) were the observed values and the results of simulation each sampling field. The observed values and the estimated are very close. The model was obtaining a precision results by using PSO algorithm. The RMSE, nRMSE, MAE and R2 between simulation and measured biomass were 1.06 ton/ha, 11.92%, 0.90 ton/ha and 0.92.

3.2. Canopy cover

In Figure 2, canopy cover and soil moisture were saved. Figure 2(a), 2(b), 2(c), 2(d), 2(e), 2(f), 2(g), 2(h) were the canopy cover of simulation each sampling field. The RMSE, nRMSE, MAE and R2 between simulation and measured canopy cover were 8.92%, 9.84%, 7.84% and 0.66, respectively.

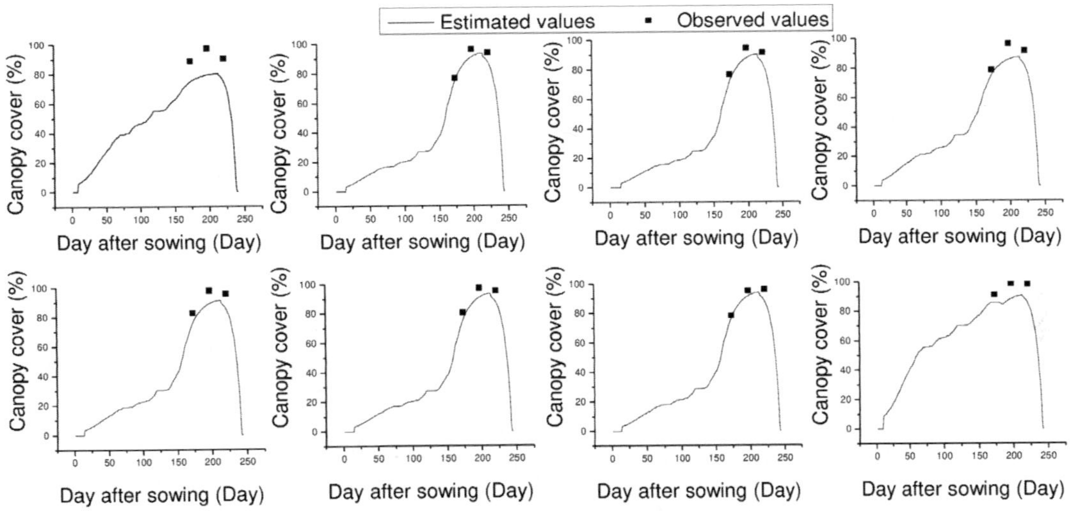

Figure 3. Canopy cover simulation results.

3.3. Soil moisture

AquaCrop model can simulate and gain the soil moisture in 0-0.05cm, 0.05-0.15 cm, 0.15-0.25 cm, 0.25-0.35 cm, 0.35-0.45 cm, 0.45-0.55 cm, 0.55-0.67 cm, 0.67-0.82 cm, 0.82-0.97 cm, 0.97-1.13 cm, 0.13-1.27 cm and 1.27-1.45 cm. When the ground measurements, only the 0-20 cm of soil moisture were obtained. In Figure 4, soil moisture simulated by AquaCrop in 0-0.05cm, 0.05-0.15cm and 0.15-0.25cm were used to contrast measurements.

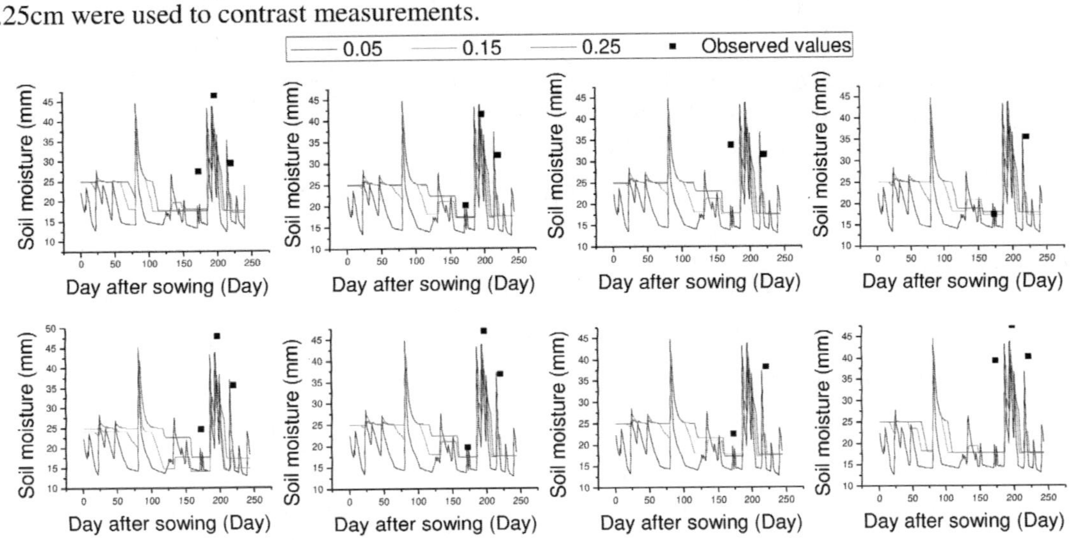

Figure 4. Soil moisture simulation results.

4. Discussion

In this paper, using the AquaCrop model and ground survey data, crop parameters and environmental conditions were estimated, such as biomass, canopy cover and soil moisture which have an important influence on crop yield, and the growth state of the crops was analysed, these calculations can be done on the computer. The growth of winter wheat was simulated, and the data of the three items of the growth cycle, canopy cover and soil moisture were obtained. Result shows that the potential of AquaCrop in farmland management.

References

[1] Andarzian B, Bannayan M and Steduto P 2011 Validation and testing of the AquaCrop model under full and deficit irrigated wheat production in Iran. *Fuel & Energy Abstracts* **100** 1-8.

[2] Trombetta A, Iacobellis V and Tarantino E 2016 Calibration of the AquaCrop model for winter wheat using MODIS LAI images. *Agricultural Water Management* **164** 304-316.

[3] Soddu A, Deidda R and Marrocu M 2013 Climate variability and durum wheat adaptation using the AquaCrop model in southern Sardinia. *Procedia Environmental Sciences* **19** 830-835.

[4] Mkhabela M S and Bullock P R 2012 Performance of the FAO AquaCrop model for wheat grain yield and soil moisture simulation in Western Canada. *Agricultural Water Management* **110** 16-24.

[5] Nyakudya I W, and Stroosnijder L 2014 Effect of rooting depth, plant density and planting date on maize (Zea mays, L.) yield and water use efficiency in semi-arid Zimbabwe: Modelling with AquaCrop. *Agricultural Water Management* **146** 280-296.

[6] Montoya F, Camargo D and Ortega J F 2016 Evaluation of Aquacrop model for a potato crop under different irrigation conditions. *Agricultural Water Management* **164** 267-280.

[7] Doorenbos J, Kassam A H and Bentvelsen C1980 Yield Response to Water. *Irrigation & Agricultural Development* **14** 257-280.

[8] Jin X L, Feng H K and Zhu X K 2014 Assessment of the AquaCrop model for use in simulation of irrigated winter wheat canopy cover, biomass, and grain yield in the North China Plain. *Plos One* **9** 86938-86938.

[9] Vanuytrecht E, Raes D and Willems P 2014 Global sensitivity analysis of yield output from the water productivity model. *Environmental Modelling & Software* **51** 323-332.

[10] Kennedy J and Eberhart R 1995 Particle swarm optimization. *IEEE International Conference on Neural Networks* **4** 1942-1948.

Acknowledgments

This work was supported in part by the Beijing Natural Science Foundation (No.6144025) and National Natural Science Foundation of China(No.41301475).

Fractal Dimension Change Point Model for Hydrothermal Alteration Anomalies in Silk Road Economic Belt, the Beishan Area, Gansu, China

H H Han, Y L Wang, G L Ren, J Q LI, T Gao, M Yang and J L Yang

Xi'an Center of Geological Survey, Geological Survey of China, Xi'an, China

E-mail:hanhh06@hotmail.com

Abstract. Remote sensing plays an important role in mineral exploration of "One Belt One Road" plan. One of its applications is extracting and locating hydrothermal alteration zones that are related to mines. At present, the extracting method for alteration anomalies from principal component image mainly relies on the data's normal distribution, without considering the nonlinear characteristics of geological anomaly. In this study, a Fractal Dimension Change Point Model (FDCPM), calculated by the self-similarity and mutability of alteration anomalies, is employed to quantitatively acquire the critical threshold of alteration anomalies. The realization theory and access mechanism of the model are elaborated by an experiment with ASTER data in Beishan mineralization belt, also the results are compared with traditional method (De-Interfered Anomalous Principal Component Thresholding Technique, DIAPCTT). The results show that the findings produced by FDCPM are agree with well with a mounting body of evidence from different perspectives, with the extracting accuracy over 80%, indicating that FDCPM is an effective extracting method for remote sensing alteration anomalies, and could be used as an useful tool for mineral exploration in similar areas in Silk Road Economic Belt.

1. Introduction

Mineral exploration is a key building content for "One Belt One Road" plan. Remote sensing techniques plays an important role in extracting mineralization alteration areas, which is a key in exploration of mineral deposits [1]. Over the past years, multispectral and hyperspectral digital imaging have been successfully used for providing detailed information about the different altered minerals which were exposed on the earth's surface, such as clay minerals (kaolinite, illite), sulfate minerals (alunite), carbonate minerals (calcite, dolomite), iron oxides (hematite, goethite), and silica (quartz) [2]. The studies have shown that the band ratioing, principal component analysis (PCA) and the spectral angle mapper (SAM) image processing techniques could be employed to identify hydrothermally altered rocks and associated mineralization effectively [3,4]. However, these researches mainly focused on the identification of principal component image that containing alteration anomalies and ignores the discrimination for threshold of anomalies on principal component image. Based on the normal distribution of mineralization anomalies, Zhang proposed "De -Interfered Anomalous Principal Component Thresholding Technique" (DIAPCTT) [5], and the 'N' times standard deviation was used as the measure for anomaly slicing. Chen has reported the obvious nonlinear characteristics (self-similarity and mutability) of geological anomalies [6]. However, whether these nonlinear characteristics can be used to quantitatively calculate the critical threshold of anomalies on principal component image hasn't been documented yet.

In this study, a fractal dimension-change point model (FDCPM) which is based on calculating two independent attribute that are respectively self-similarity and mutability is proposed to extract geological anomaly's nonlinear characteristics. To test the model, the reflectance data of ASTER imagery in Beishan, an important ore concentration area in Silk Road Economic Belt, were evaluated as an example for discriminating the critical threshold of alteration anomalies.

2. Geological setting and data preparation

2.1. Geology

Beishan mineralization belt located at the boundary between Gansu and Xinjiang of China goes along NEE-EW-NWW trending beishan rift-orogeny. It is an important metal metallogenic belt for Au, Ag, Cu, Zn and others, in which there occur a number of middle -large scale mineral deposits [7]. Fangshankou area is located in the southwest of the Liuyuan ore concentration area in the beishan mineralization belt. The emergence stratum is the Devonian Three Wells Formations, and the main lithology includes feldspar quartz sandstone, feldspathic litharenite, feldspar sandstone with siltstone and pelitic siltstone. Moreover, the rift is growing, and the magma activity is intense. The metallotect characteristics of magmetic hydrothermal deposits which were formed in the Hercynian epoch is strictly controlled by faults, distributed spatially along the faults. The main types of the wall rock alteration include quartzitification, sericitization, chloritization, pyritization, limonitization, and carbonatization.

2.2. Remote sensing data preparation

The Advanced Spaceborne Thermal Emission and Reflection Radiometer (ASTER) data are collected in 14 bands, three in the visible and near-infrared portion of the electromagnetic spectrum with 15 m spatial resolution, six in the short-wave infrared with 30 m spatial resolution, and five in the thermal infrared with 90 m spatial resolution. The relatively large number of bands (14 bands) and the wider spectrum (VIR 0.520–0.860 μm, SWIR 1.600–2.430 μm and TIR 8.125–11.650 μm) covered by the ASTER data allow discrimination between a wide range of rock compositions. Because of this advantage, ASTER data have been used increasingly for extracting mineralization alteration [2].

Data set ASTL1B 00101604572311060600600 was used for this research. To remove atmospheric effects and Crosstalk from ASTER SWIR data, the ENVI 4.8 software contain FLaash model and CrossTalk3.0 software were used. The resulting data can be assumed to be more representative of the soils or lithologies of the exposed areas than the unprocessed data. Hence, the spectrum will be more closely comparable to its corresponding library spectrum. Moreover, river, vegetation and other interferents were removed by mask technique.

3. Fractal dimension change point model

3.1. Self - similarity and Fractal Dimension Model

The essence of fractal dimension is self-similarity [8], and the size and number of geological anomalies have self-similarity characteristics [9]. Suppose there is a data set $\{X_i, i=1,2, \ldots,N\}$, the fractional model can be shown as below [10]:

$$N(r) = \sum_{i=1}^{N} Xi = Kr^{-D} \quad (1)$$

where N(r) = Number of scale which is greater than or equal to X_i
r= Characteristic scale
K = Constant
D= Fractal dimension

$$\lg N(r) = \lg K - D \lg r \quad (2)$$

Equation 2 is a deformation formula by evaluating the logarithm of equation 1. It is observed that there is an exact linear relation between lg*r* and lgN（*r*）.

3.2. Mutability and change point mode

Han has reported the optimal number and position of change point could effectively be detected by the change point model, and its model as shown below [11].

1）When i=2，…，N, then the sample will divided into two sections (X_1, X_2, …, X_{i-1} and X_i, X_{i+1}, …, X_N) for each i. Nest ,the arithmetical mean（$\overline{X_{i1}}$ and $\overline{X_{i2}}$）and the square deviation （S_i）are calculated for each section.

$$S_i = \sum_{t=1}^{i-1}(X_t - \overline{X_{i1}})^2 + \sum_{t=i}^{N}(X_t - \overline{X_{i2}})^2 \quad (3)$$

2）Calculate the arithmetical mean（$\overline{X}$）and the squared deviation for the sample（S）.

$$\overline{X} = \sum_{t=1}^{N} X_t / N \quad (4)$$

$$S = \sum_{t=i}^{N}(X_t - \overline{X})^2 \quad (5)$$

3）Calculate the expected value (E（S-S_i）).

$$E（S-S_i）= E[N^{-1}（i-1）（N-i+1）（\overline{X_{i1}} - \overline{X_{i2}}）2]，\quad i = 2, 3, \ldots N \quad (6)$$

In a word, we can see that the change point can increase the difference between S and S_i. As shown in the formula 5, the S is a fixed value, and when the value of S_i reaches its minimum, the change point can be detected easily. The sample series is established by taking logarithmic operation for lgN（*r*）and lg*r*.

4. Extracting of hydrothermal minerals and its analysis

4.1. Extracting of hydrothermal minerals

Three principal component images containing Limonitization, Sericitization, and Chloritization are extracted by the Adaptive Coherence Estimator (ACE) method based on USGS mineral spectrum library for ASTER data. Then the fractal dimension model was used to analyze the self-similarity of different geological anomalies. Nest, the change point method was used to analyze the mutabiliyt of different geological anomalies, and when Si reaches its minimum, we got the precise location of the change point for different classification.

The results show that when the value of *i* is 59, Si reaches its minimum, and the value of *r* is 60, so we can consider that the value(*r*=60) is the critical threshold to distinguish the geological background and the grade 3 Limonitization. On the same principle, the value(*r*=111) was gotten to distinguish the grade 3 Limonitization and the grade 2 Limonitization, last the value(*r*=119) were got to distinguish the grade 2 Limonitization and the grade 1 Limonitization. In addition, the three values of r for Sericitization are 74, 125 and 132 respectively, and the three values of r for Chloritization are 170, 221 and 229 respectively.

Figure 1-a shows the self-similarity and the optimum imitative straight line for geological background and limonitization, with a fitting coefficient of 0.7553, which indicate that self-similarity can effectively distinguish the alteration anomalies. Figure 1-b shows the three critical thresholds and imitative straight line for different limonitization alteration grade. Then, the alteration anomaly map can be made. Similarly, figure 2 and figure 3 describes the fractal characteristic for Sericitization and Chloritization.

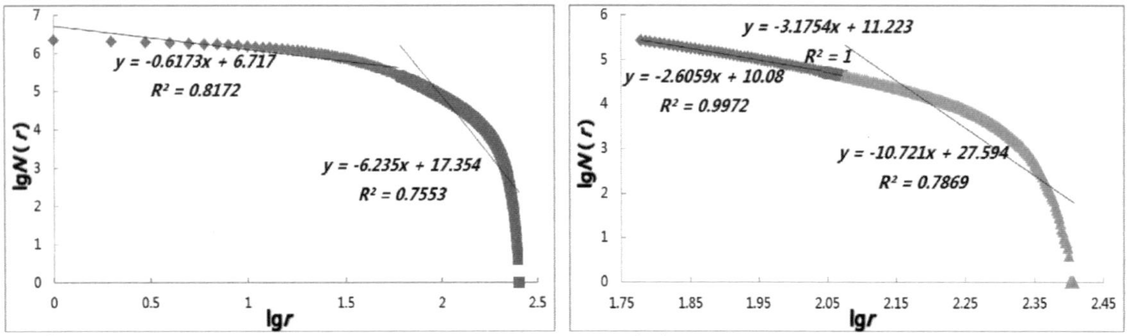

（a）log-log plot of anomalous principal component （b）log-log plot of anomalous classification

(blue represents geological background and red represents Alteration Anomalies) （intersection of different colors represents a different critical threshold for classification）

Fig.1 Fractal characteristic analysis diagram of limonitization

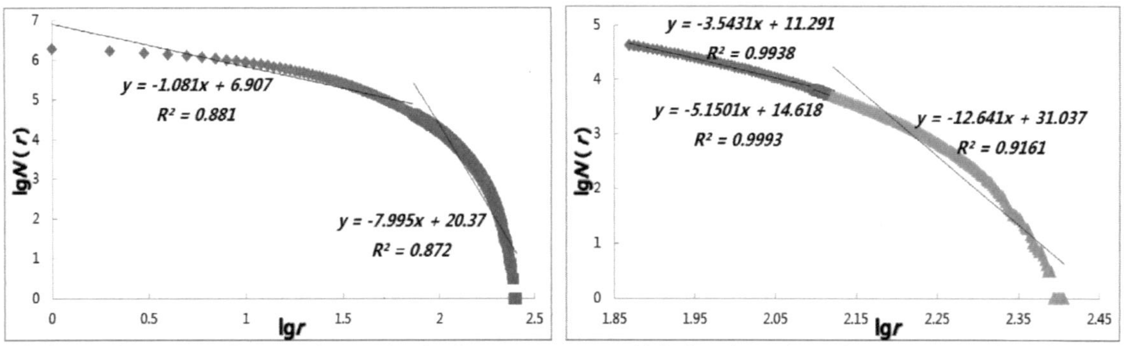

（a）log-log plot of anomalous principal component （b）log-log plot of anomalous classification

(blue represents geological background and red represents Alteration Anomalies) （intersection of different colors represents a different critical threshold for classification）

Fig.2 Fractal characteristic analysis diagram of sericitization

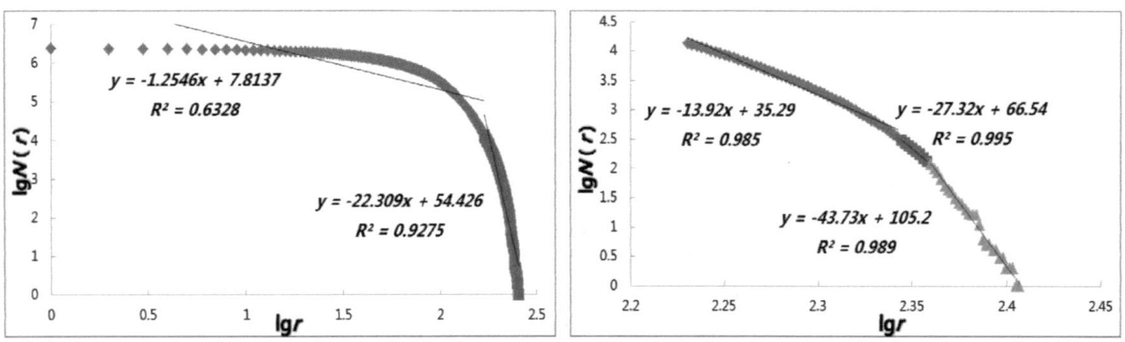

（a）log-log plot of anomalous principal component （b）log-log plot of anomalous classification

(blue represents geological background and red represents Alteration Anomalies) （intersection of different colors represents a different critical threshold for classification）

Fig.3 Fractal characteristic analysis diagram of chloritization

4.2. Spatial distribution of alteration minerals

Figure 4 is a remote sensing alteration anomalies map in the testing area. As can be seen from the map, the alteration anomalies show a linear distribution along nearly EW trending which is consistent with regional structures. The known ore and mineralized points are almost lying in the anomaly area or its boundary, moreover the remote sensing alteration anomalies are agreed well with a mounting body of evidence from heavy mineral anomaly and geochemical anomaly. What's more important is the field verification also confirmed the above phenomenon.

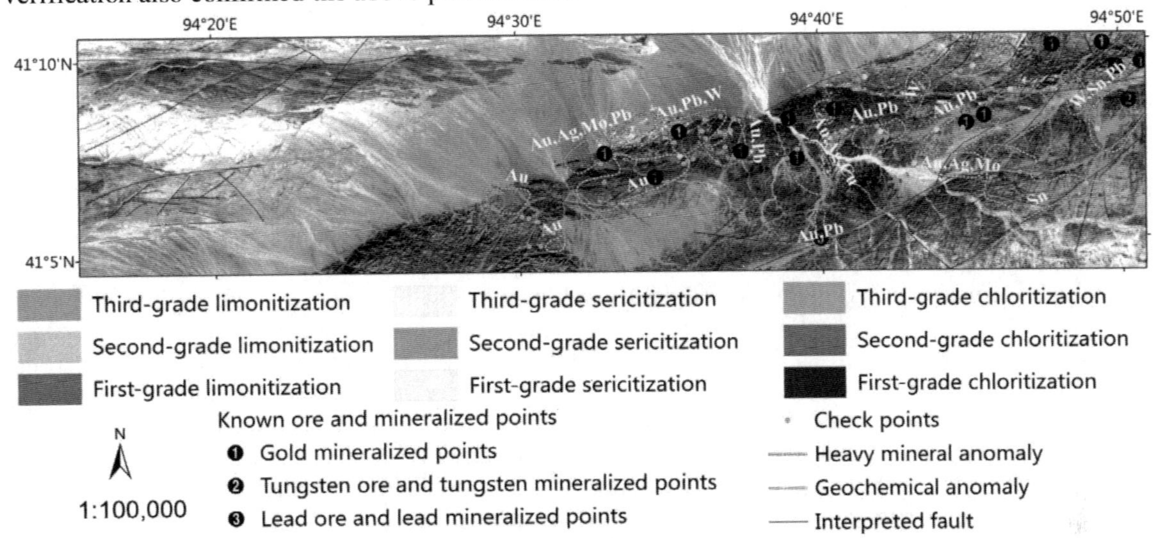

Fig.4 Remote sensing alteration anomalies map in the testing area

64 points were randomly selected to verify the accuracy of remote sensing alteration anomalies. The results show that the fractal dimension-change point model is an effective extracting method for remote sensing alteration anomalies, and its accuracy could over 80% (see table 1).

Table 1. Confusion matrix of alteration minerals.

Extracting / Checking	Limonitization	Sericitization	Chloritization
Limonitization	28	3	2
Sericitization	1	17	
Chloritization	3	1	8
None	1		
Total	33	21	10
Accuracy（%）	84.85	80.96	80

4.3. Compare with DIAPCTT

Table 2 shows the different results of critical threshold by DIAPCTT and FDCPM, also the corresponding 'N' value for DIAPCTT were given. It can be seen that there is a great difference for threshold by different methods. The proportion of limonitization information which is obtained by DIAPCTT is 7.48%, and the proportion which is obtained by FDCPM is 10.48%, this means that we have got more limonitization information. For sericitization and Chloritization, the proportion of alteration anomalies information got by DIAPCTT much larger than the result obtained by FDCPM, this may reflect that the threshold range got by previous method is too wide. On the other hand, the N values in DIAPCTT can directly affect the amount of anomalies information, but it can't be quantitative selected until now. Through the above comparison, we can sure that the FDCPM be more objective and effective for people who is deficient in geological experience or had no information about the geological background of the study area.

Table 2. Statistical information of remote sensing alteration anomalies.

AA	Method	Average	SD	Third-grade			Second-grade			First-grade		
				TH	PP	'N'	TH	PP	'N'	TH	PP	'N'
LI	DIAPCTT	25.94	29.36	69.98	7.48	1.5	84.65	4.61	2	99.33	2.93	2.5
	FDCPM			60	10.48	1.16	111	2.14	2.9	119	1.72	3.17

SE	DIAPCTT	12.29	18.38	39.86	6.76	1.5	19.06	4.37	2	58.25	3.02	2.5
	FDCPM			74	1.70	3.36	125	0.26	6.13	132	0.20	6.51
Ch	DIAPCTT	54.03	38.37	111.58	8.15	1.5	130.76	3.71	2	149.94	1.52	2.5
	FDCPM			170	0.53	3.02	221	0.01	4.35	229	0.01	4.56

(Abbreviations: AA-alteration anomalies, LI-Limonitization, SE-Sericitization, CH-Chloritization, DIAPCTT-De-Interfered Anomalous Principal Component Thresholding Technique, FDCPM- Fractal Dimension-Change Point Model, SD-Standard deviation, TH- Threshold, PP- Pixel proportion)

5. Conclusions

This study proposed a new method which was named fractal dimension-change point model (FDCPM). Then, we explore the performance characteristics of the model for quantitative calculating the critical threshold of remote sensing alteration anomalies and locating the hydrothermal mineralized alteration zones in Beishan, which is an important mineralization area in the silk road economic belt. This approach is based on the self-similarity and mutability of alteration anomalies. The experiment result indicates that FDCPM is an effective method and its extracting accuracy could over 80%. What's more important is its findings agree well with a mounting body of evidence from different perspectives. These results indicate that our method is promising for identifying alteration zones and is a useful tool for mineral exploration in similar areas in the silk road economic belt. Also, this model will be a useful addition for DIAPCTT to choose the thresholding value quantificationally.

6. References

[1] Gabr S, Ghulam A and Kusky T 2010 Detecting areas of high-potential gold mineralization using ASTER data *Ore Geology Reviews* **38** 59-69

[2] Meer F D V D, Werff H M A V D, Ruitenbeek F J A V, Hecker C A, Bakker W H B, Noomen, M F, Meijde M V D, Carranza E J M, Smeth J B D and Woldai T 2012 Multi- and hyperspectral geologic remote sensing: A review *International Journal of Applied Earth Observation and Geoinformation* **14** 112-128

[3] Pour A B and Hashim M 2011. Identification of Hydrothermal Alteration Minerals for Exploring of Porphyry Copper Deposit Using ASTER Data, SE Iran *Journal of Asian Earth Sciences* **42** 1309–1323

[5] Amer R, Kusky T and Mezayen A E 2012 Remote Sensing Detection of Gold Related Alteration Zones in Um Rus Area, Central Eastern Desert of Egypt *Advances in Space Research* **49** 121–134

[6] Zhang Y J, Zeng Z M, Chen W 2003 The Methods for Extraction of Alteration Anomalies from the ETM+（TM） Dada and Their Application: Method Selection and Technological Flow Chart *Remote Sensing for Land and Resources* **56**(2) pp 44-49

[7] Zeng C H, Wu D J, Xia W B, Wang W G and Cao W 2002. Metallogenic Regularity and Prospect of Gold Deposits of Beishan Mineralization Belt *Xinjiang Geology* **20**(3) pp 219-223

[8] Li W Y, Dong F C, Jiang H B, Tan W J and Wang Y H 2006 Metallogenetic Characteristics and Prospecting Potential of Major Metallic Minerals in Northwest China *Northwestern Geology* **39**(2) 1-16

[9] Cheng Q M, Zhang S Y, Zuo R G, Chen Z J, Xie S y, Xia Q L, Xu D Y and Yao L Q 2009. Progress of Multifractal Filtering Techniques and Their Applications in Geochemical Information Extraction *Earth Science Frontiers* **16**(2) 185-198

[10] Cheng Q M 2008 Singularity of Mineralization and Multifractal Distribution of Mineral Deposits *Bulletin of Mineralogy, Petrology and Geochemistry* **27**(3) 298-305

[11] Liang Y Q, Wang G W, Zhu Y Y and Qu J N 2011 Alteration from ETM+ Data Rating based on Fractal Technologies *Remote Sensing Technology and Application* **26**(2) 508-511

[12] Han H H, Gao T, Yi H, Yang M, Yan X J, Ren G L and Yang J L 2012 Extraction of Relief Amplitude Based on Change Point Method: A Case Study on the Tibetan Plateau *Scientia Geographica Sinica* **32**(1) 101-104

The Application of GeoRSC Based on Domestic Satellite in Field Remote Sensing Anomaly Verification

Ting Gao and Min Yang and Haihui Han and Jianqiang Li and Huan Yi

Xi'an Center of Geological Survey, Geological Survey of China, Xi'an, China

E-mail: gt8212@163.com

Abstract. The Geo REC is the digital remote sensing survey system which based on domestic satellites, and by means of it, the thesis carriedy out a remote sensing anomaly verification field application test in Nachitai area of Qinghai. Field test checks the system installation, the stability of the system operation, the efficiency of reading and show the romoate image or vector data, the security of the data management system and the accuracy of BeiDou navigation; through the test data, the author indicated that the hardware and software system could satisfy the remote sensing anomaly verification work in field, which could also could make it convenient forconvenient the workflow of remote sense survey and , improve the work efficiency, . Aat the same time, in the course of the experiment, we also found some shortcomings of the system, and give some suggestions for improvement combineding with the practical work for the system.

1. Introduction

In recent years, with the gradual development of prospecting breakthrough strategic action, China Geological Survey has carried out exploratory researches trials on new techniques and methods of geological survey. Not only remote sense technology is widely popularized to improve the efficiency and quality of geological survey, but also the complete flow informatization is strengthened. It has developed regional geological survey digital acquisition system, realizing the main process informatization regional geological survey field data acquisition, indoor settlement and result output.

In order to satisfy the needs that geological survey can continuously advance to the tough and complicated areas in the west and the needs of developing field geological work in arid areas and areas without the electronic signals, China Geological Survey has organized the project "Construction and Application of Field Geological and Mineral Survey Service and Management System", which has also carried out the applied research and promotion experiment of domestic satellite technology with the financial support of NDRC's "Demonstration Projects of Applying High Technology to Production". It has researched and developed GeoRSC with vast essential data, remote sensing data processing, distribution and service, geological information identification and extraction technology, space information bearing and display platform based on moveable equipment. The main purpose is satisfying the requirements of modernization, elaboration and hommization for field geological investigation and establishing Chinese field geological survey business operation and integrated management system based on domestic satellites, thus maintaining the safety of operating personnel on remote sensing geology in their field work and improving the overall efficiency.

2. Introduction to the Field Terminal of GeoRSC

The field terminal of GeoRSC belongs to the moveable residence version of remote sensing satellite data service system. According to "Remote Sensing Satellite Data Service System Construction Plan" (Table 1), it mainly realize remote sensing geological survey functions including remote sensing anomaly verification, control spot film collection and spectral curve collection, which also realize operation management combined with Beidou Satellite. There are specifically remote-sensing image and vector data demonstration, GPS navigation and location, input and output of remote sensing data and vector data, adding and editing the property of observation points, control spot film collection, Beidou communication, spectral curve collection and result data output(Fig 1).

Table 1 Functions of Different Versions of Remote Sensing Satellite Data Service Systems[6]

Version	Function Module	Technical Index
Central Version of Remote Sensing Data Service	Routine data processing	Realizing remote sensing data process under regular state, including cutting, joining, fusion and thematic information extraction
	Remote sensing data resource catalog	Under regular state, the center nodes of remote sensing data service can only provide remote sensing data resource catalog under private geological survey network, existing data can be referred to but it doesn't provide data download
	Immediate emergent data processing	Realizing immediate remote data processing in emergency, including immediate geometric correction, cutting, joining and fusion of field control spot film as well as interpretation of emergent special information.
	Emergent data service	Realizing emergent data transmission from remote sensing data service centre node to regional center nod and professional applied center node inside the private geological survey network, covering an area of 10 km × 10 km, offering high-mark data, the product data is 2 G and time 30 min.
	Control spot film management	Connecting to the backstage field control spot film database, realizing control spot film management
	Remote sensing anomaly verification result management	Connecting the backstage result database, realizing the remote sensing anomaly verification result management
	Spectral curve management	Connecting the backstage spectral curve database, realizing spectral curve management
Regional Central Version of Professional Application Central Version	Routine data query	Refer to the released remote sensing data resources through the remote sensing data resource catalog under normal state
	Emergent data acquisition	Obtaining the emergent remote sensing data sent by the node of remote sensing data service center through point-to-point method under emergent state
	Control spot film management	Managing the control spot film data gathered in moveable stations or field terminals, entering the local field control spot film database
	Remote sensing anomaly verification result management	Managing the remote sensing anomaly verification result data gathered in moveable stations or field terminals, entering the local result database
	Spectral curve management	Managing the spectral curve data gathered in moveable stations or field terminals, entering the local spectral database
Moveable Residence	Control spot film management	Managing the control spot film data gathered in field terminals, entering the local field control spot film database

Version	Function Module	Technical Index
Version	Remote sensing anomaly verification result	Managing the remote sensing anomaly verification result data gathered in field terminals, entering the local result database
	Spectral curve management	Managing the spectral curve data gathered in field terminals, entering the local spectral database

3. Experiment on Anomaly Verification of GeoRSC Terminal

3.1. Natural conditions of experimental site

The thesis selects Nachitai area of Qinghai to carry out the field test of GeoRSC. Within the experimental site, the natural, economic and geographical conditions are quite severe. The average height above sea level is 4500m and the highest is 5600mm; the mountains are high and precipitous, whose relative height difference is 500 to 1600 m. Apart from the highway lines, most places are dead zone of communication. Within the area, it is very cold and it has longer winters and shorter summers; it belongs to arid or semiarid areas. Its summer is arid and the highest temperature can reach 25℃; while its winter is long, with strong wind and sands, less snow and extremely low temperature, whose lowest temperature can reach -30℃. Thus, the field working time is very short, which can only last from middle May to late September. Within the area, vegetations don't grow, most of which are herbages, generally growing along mountains or rivers valleys, particularly centering in Dong Da Tan and Xueshuihe drainage basin. The basement is bare and its soil is mainly high and cold desert soil. Most places here are no man's land, whose economic development is very unbalanced.

3.2. Experimental objective and content

According to the needs of field work environment, the experiment carried out a test aimed at the stability, accuracy and data security of the software platform of GeoRSC Terminal; it mainly focuses on the stability of the new software system, fluency of reading and showing remote sensing data and vector data, accuracy of Beidou Satellite positioning as well as security of data usage.
Based on this, we select two types of remote sensing data respectively with medium-low resolution ratio and high resolution ratio, Shape-type and MapGIS-type vector data, 20 layers and three abnormal vector data with different spectral resolution to have a test. Meanwhile, 30 outliers are selected to test according to the process of anomaly verification. Besides, two different types of locations, maintain environment and plain and hill environment are selected to test the accuracy of Beidou positioning.

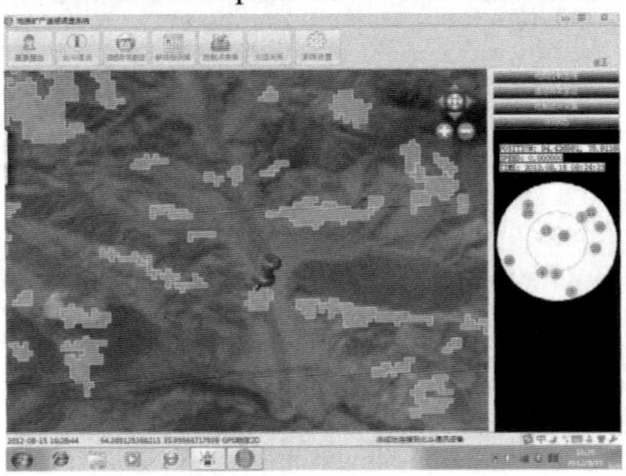

Fig 1 Main Interface of GeoRSC

Fig 2 Field Remote Sensing Anomaly Verification Screenshot

3.3. Experiment workflow

System installation is done indoors to make preparations for field experiment and test. Firstly, because of the updating of software, it is necessary to install the latest GeoRSC for tablet PC and PDA and complete online registration. At the same time, connect and install the drive programs of access equipment like GPS. Secondly, based on ETM remote sensing data, Aster remote sensing data and TASI/CASI hyperspectral data, extract abnormal information of different spectral resolution in the test area used for verifying the system's applicability to different data. Use the aerial photo data at the height of 50 m and 0.5 m (data with medium-low resolution) and at the height of 300 m (data with high resolution) from China-Pakistan satellite CBERS-2 to make the operation base map of the experimental site, which is used for testing the system's load capacity. Besides, respectively load them into the system according to the usage procedures.

According to the designed verification spots, field verification is carried out one by one.The system will record the geological and alternation phenomenon, which will is recorded in the database (Fig 2) .

In order to verify and analyze the positioning ability and stability of Beidou-1 bluetooth portable terminal, it carries out positional accuracy measurement of static single-point long-term sequence of Beidou, and two schemes are designed. The first is to choose positional accuracy measurement of static single-point long-term sequence, which is used for simulating mountainous environment. This experiment chooses field stations for experiment. Beidou terminal is put in southward one-way windows with good intervisibility, having a long-term sequence monitoring of the terminal's locating data so as to confirm the positional accuracy and stability. With 7 hours' continuous positioning (with an internal of 1 minute), about 400 locating data is obtained. The second is the positional accuracy measurement of single-point long-term sequence under multiway sighting conditions, which is used for simulating plain and hilly environment, conducted in the base station of Golmud Seismostation, where the view is open and there are no high buildings around, which can be viewed as intervisibility. The base station has a triangulation point (XZ01), and according to the data provided by State Bureau of Surveying and Mapping, its Xi'an 80 geographical coordinates is 36°25' 57.854" in latitude and 94°52'28.11539" in longitude, and elevation: 2749.660 m. On the site, ongoing continuous base station PT measurement is adopted, whose continuous measurement duration is up to 8 hours, which totally measures 438 locating points.

3.4 Field measurement result of GeoRSC terminal

System installation and debugging test: the system installation process is successful, there is no report error or circumstance where the connection fails influenced by internet connection in networking activation; Beidou Navigation Satellite System and the communication system needs to connect with the tablet PC through Bluetooth. Influenced by Bluetooth connection, there are occasional conditions where the equipment can't be found; GPS Module needs to connect with the tablet PC through USB data line, and the device attachment is normal. **Data loading test:** By loading the measurement of 2 image layer and 20 vector layers, the system and normally load single vector and remote sensing data (documents whose data size is less than 150M), and the batch of loaded documents may likely to fail; it doesn't support the loading and display of remote-sensing images of single document more than 150 M, and it can only be loaded normally after cutting the single image; when the loaded amount is more than 300M, the image dragging and zooming will have delayed display (the total data in this test is about 770M); the software system of the current version doesn't support the customized line styles, styles and colors of vector layers, and the vector files are of relatively lower recognition.**Data security test:** The system of the new version has good stability and there is no forced termination in the operation process, which ensures the security of field work data. Aimed at the stability of the system data storage, we design 30 verification spots in the database, whose data storage is complete and the query and derivation are normal. Aimed at the system's performance of mass data management, in terns if the 30 verification spots, we collect them in the database respectively in five days in different orders. There is no confusion in the data storage and the query and derivation are normal.**Beidou positioning accuracy test:** Through the test, in areas with good sighting conditions, after deducting system errors (southing offset), the positioning circular error probable (CEP) of Beidou 1 Bluetooth portable terminal is about 3.83 m, and the mean square error (standard deviation) is about 5.09 m. the better the sighting condition is, the smaller the circular probable error and mean square error is. According the stipulations of mapping 1:50000 topographic map error, Beidou 1 Bluetooth portable terminal can satisfy the working needs of simple measurement, positioning and navigation of 1:50000 or smaller measuring scales.

4. Major Advantages and Improvement Suggestions

4.1 Major advantages

This system can better realize the function of field positioning accuracy and real-time information passing; besides, tablet PC is applied in the field verification of remote sensing geological survey, which brings remote sensing survey from a traditional way to a digital age, realizing traditional pin-points in actual material maps and recording of paper books, thus effectively improving the work efficiency. It makes it convenient for field display and referring to image and remote sensing anomaly of different measuring scales; it also realizes database management of real-time logging field verification materials in the field.

4.2 Improvement suggestions

In this field test process, the author finds some defects of the system, including: in the process of connecting GPS positioning module to Beidou Bluetooth communication module, it is restricted by the system stability, hoping to integrate these two modules with the system hardware and improve the usability and portability of the whole system; it is in need of adding the function of adjusting layer display order, which is thus convenient for the leading-in and display of newly added survey data. Besides, it is necessary to increase the display of customized line types, symbols and colors of vector data so as to enrich the drawing expression.

5. Conclusion

As to traditional remote sensing geological survey, the GeoRSC is a revolutionary leap, helping the survey staff change from original manual recording to all-digital information age, thus realizing intelligence and process of remote sensing geological survey; this system also applies Beidou communication satellite to have an online management of the working staff, thus realizing real-time and accurate monitoring of field staff, maintaining the security of working staff in the tough areas, realizing real-time information communication, solving problems in the field work on the site and greatly improving the efficiency of geological survey. It marks that the geological industry has entered a universe integrated information age. Though the system still needs continuous improvement in some hardware and software, this system will be accepted and adopted by field geological survey staff, which will have a wide application prospect. The wide popularization of the system will greatly increase the application level of Chinese proprietary technology, strengthening the working security of field working staff, making remote sense survey more convenient and better applying remote sensing technology to Chinese geological mineral course.

References:

[1]. Li Chao ling. Current situation and development trend of chinese digital mapping technology [J]. *Sinkiang Geology*. 2003.**21**(supplement):1-6.
[2]. Yu Qingwen, Li Chaoling, Zhang Kexin. Current situation and development trend of digital mapping [J]. *Earth Science-Journal of China University of Geosciences*. 2003. **28**(4) pp:370-376.

[3]. Guo Fusheng, Wu Zhichun, Xie Caifu. Several improvement suggestions and practical techniques of digital geological mapping system [J]. *Geology of China.* 2012. **39**(1) pp :252-259.
[4]. Wang Runsheng, Xiong Shengqing, Nie Hongfeng. A study on remote sensing geological exploration technology and application [J]. *Geological Journal.* 2011, **85** (11) pp: 1699-1743.
[5]. Liu Dechang, Li Zhizhong, Wang Junhu. Scientific progress and development prospect of chinese remote sensing geological mining [J]. *Geo-Information Science.* 2011. **13** (4) pp:431- 438.
[6]. He Kaitao, Li Zhizhong, Wang Daming. Overview of Field Geological Survey Service and Management System Design Based on Remote Sensing Satellite and Beidou Satellite [J]. Geomechanics Journal. 2012, 18 (3) : 203 -212.
[7]. Xu Chong, Dai Fuchu, Chen Jian. Detailed interpretation of remote sensing of secondary geological disaster in severely afflicted area of wenchuan Ms 8. 0 [J]. *Remote Sensing Journal.* 2009, **13** (4) pp :754 -762.
[8]. Fan Yida, Zhang Baojun. Overview and prospect of application of chinese beidou satellite in disaster reduction [J]. *China Aerospace*. **2010**, (2) pp : 7-9.
[9]. Gao Ting, Li Jianqiang, Yang Min, Han Haihui. A discussion on application of digital survey terminal in remote sensing anomaly verification [J]. *Technological Management of Territorial Resources.* 2013, **30** (6) pp : 63- 67.

Using ASD data to identify the altered minerals for exploring of gold deposit in the Beishan area, North China

G L Ren, H Yi, M Yang, N Liang, J Q Li and J L Yang

Key Laboratory for the Study of Focused Magmatism and Giant Ore Deposits, MLR/

Xi'an Center of Geological Survey, Geological Survey of China, Xi'an, China

E-mail:155219321@qq.com

Abstract. Hyperspectral information of altered minerals plays an important role in the identifications of mineralized zones. In this study, the altered minerals of two gold deposits from Fangshankou-Laojinchang regions of Beishan metallogenic belt were measured by ASD field Spectrometer. Many gold deposits would have a close relationship with Variscan magma intrusion, which have been found in study region. The alteration minerals have been divided six types by the spectral results, i.e. sericite, limonite, dolomite, chlorite, epidote and calcite. The distribution characteristics and formations of altered minerals were discussed here. By the ASD, the spectral curve of different geological units in the Jintanzi and Fangshankou gold deposits were analysed and summarized. The results show that the sericite and limonite are mainly related with the gold mineralization and widely occurred in the gold deposits. Therefore, we proposed that the sericite and limonite are the iconic alteration mineral assemblages for gold mineralization and the models of altered minerals for gold deposits could be established in this region.

1. Introduction

Hyperspectral remote sensing technology are widely used in mineral identification, mineral mapping, lithological mapping, mineral resources exploration, and so on, for its wide spectral bands and high spatial resolution system [1, 2]. The Hyperspectral technology system was used by researchers to recognize altered minerals information, which plays an important role to position mineralized zones and trace mineralized centre of hydrothermal alteration [3-10]. In this study, ASD field Spectrometer was used to identify the altered minerals of two gold deposits from Fangshankou-Laojinchang regions of Beishan metallogenic belt, where belongs to the Gobi hilly topography and the rocks are well bareness. So it is a favourable geological conditions and good natural environment to carry out the hyperspectral measurement.

The deposits in Beishan metallogenic belt are closely associated with magmatism [11, 12]. Different types of metal ore bodies have been documented in Shuangyingshan - Huaniushan arc Unit [13], which had developed an Au-W-Cu-Fe-Ni magmatic hydrothermal system. The central and southern Beishan region is one of the most important gold metallogenic provinces in northwest China. Five groups of gold deposits can be distinguished according to the difference of host rocks, including (a) volcanic and volcaniclastic rocks, (b) granitoid intrusions, (c) skarn type, (d) Precambrain schists, and (e) sedimentary rocks. Gold mineralization in study area is almost confined to the gold-bearing beresite veins and the intrusions were related to gold deposits [14-16]. The gold veins mainly hosted in the lenticular alteration zones. The hydrothermal alteration minerals contain quartz, sericite, limonite, chlorite, carbonate, epidote, iron oxides, and clay minerals [15-20]. Thus the alteration zones that have high gold concentrations can use Hyperspectral technology to identify gold ore body.

The main purpose of this study is application of ASD spectral data in determining the distribution pattern of these alteration minerals, and try to locate the potential alteration belts from these minerals.

2. Geological Setting

The study area (40°52′30″N-41°12′00″N and 94°16′00″E-95°04′30″ E) is located approximately 75 km at west of side the Liuyuan town, in the Beishan mountains, showing in figure 1(a).

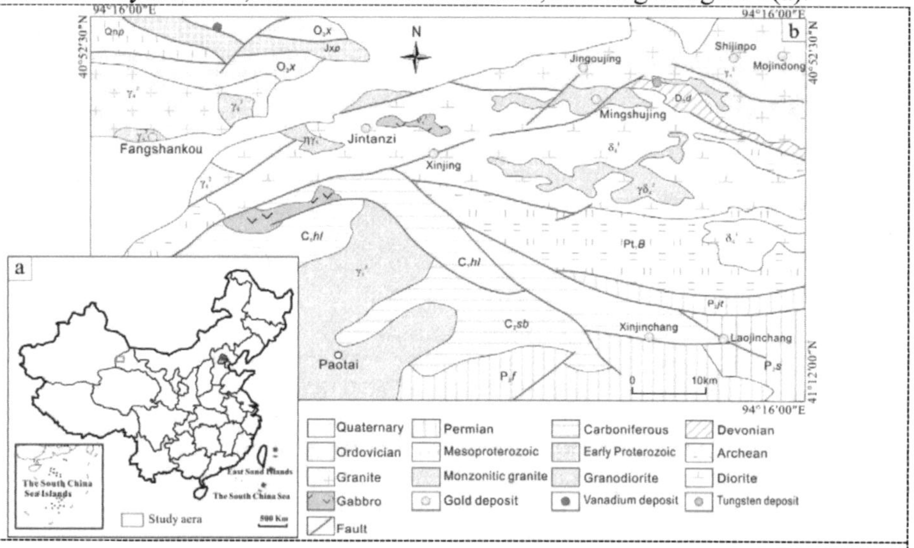

Figure 1. Geological map of Fangshankou - Laojinchang area in Beishan, (a) Location map of the Jintanzi - Laojinchang gold Mine in northwestern of China, (b) Geological setting of the Jintanzi-Laojinchang region.

The Shuangyingshan - Huaniushan arc Unit, is located in the southern margin of the Beishan epicontinental active belt, and between the tarim and sino-korean plates. The unit is a polymetallic metallogenic belt of copper, gold, silver, iron, lead, and zinc [12, 19, 20]. The study area has undergone a very complex tectonic evolution history, all stratigraphic sequences from the Ordovician to the Quaternary thus have been deformed. The mineralization and alteration are widely distributed in this region, and the main types of alterations are carbonation, chloritization, epidotization, ferritization, sericitization, and silicification [13].

The strata in the study area are mainly composed of Early Proterozoic metamorphic basement, Late Proterozoic- and Early Paleozoic- carbonates and clastic rocks, Late Paleozoic volcanic - clastic rocks and carbonates. The Paleoproterozoic rocks contain migmatite, gneiss, quartzite, and granulite. The Neoproterozoic meta-sedimentary rocks, which are marble, linestone, and meta-sandstone, emerged in the area [11]. The Early Paleozoic rocks and Late Paleozoic rocks are composed by dolostone and clastic rocks, and volcanic - clastic rocks, lava and carbonates, respectively (figure. 1(b)).

It is characterized by development of brittle fracture with active fault structures and magmatic activity. The intermediate acidic intrusive rock grows, which contains with adamellite, biotite granite, granodiorite and other acidic intrusive body, minor gabbro in Variscan period and monzogranite in Indosinian period.

Systematic geological investigations indicated that there are different types of host rocks for the gold deposits and most of them showed a close genetic relation with the granitoid intrusions or Paleozoic felsic volcanic rocks [15, 18]. The forming processes of the gold deposits are genetically associated with Hercynian magmatic activities [14]. The tectonic movement and the granitoid emplacement could provide a pass for the ascent and unloading of the deep-seated ore fluids. The Fangshankou and Jintanzi gold deposits are related with intrusion rocks.

Fangshankou gold deposit is located in the northwest part of the study area. Gold mineralization occured in the late intrusive alteration granite, which is situated the contact zone between the granite porphyry and monzonitic granite (figure 2(a)). The gold mineralization hosted in quartz veins within the beresitization granite. Ore bodies contain three gold-bearing quartz veins, which were nearly parallel outputs, with the attitude of 230°∠81°.

Jintanzi gold deposit is located in the north part of the study area, which is situated the interchange of NE and NW trending secondary faults. Gold mineralization produced in the late intrusive cataclastis - alteration granite, which is in the contact zone between the Middle Variscan moyites and Early Variscan quartz diorite (figure. 2(b)). The gold mineralization hosted in quartz fragmentation zone within the altered granite. Ore bodies contain two gold-bearing broken quartz veins, which were nearly parallel outputs, with the attitude of 345°∠76°.

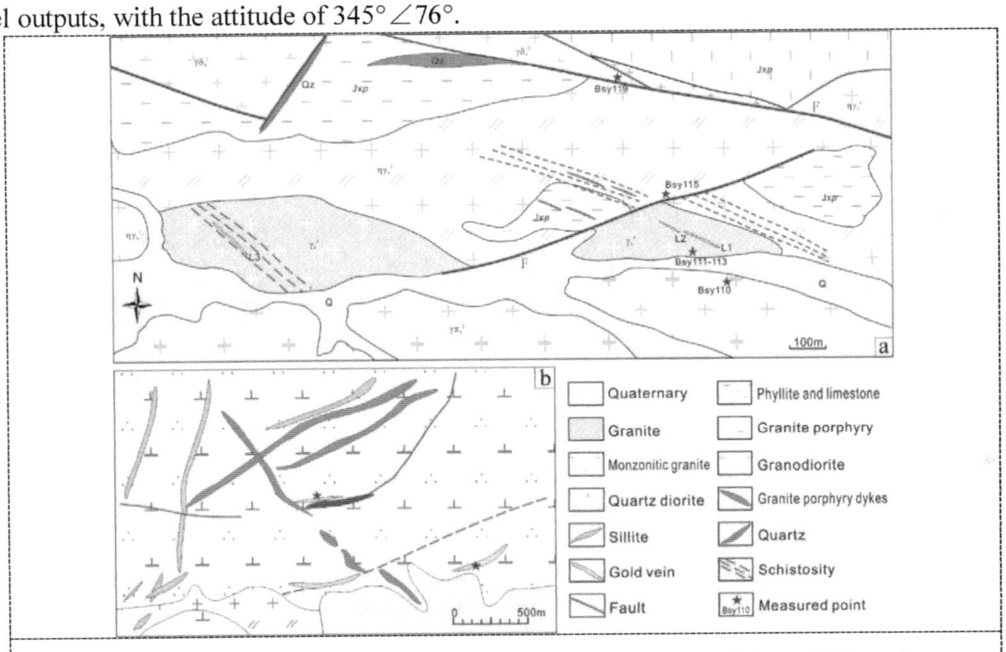

Figure 2. Geological map of Fangshankou (a) and Jintanzi (b) gold deposits.

3. Data and methods

The measurement of ground spectra were carried out under cloudless conditions from 12:00 pm to 16:00 pm in study area, by using an ASD Field Spec Pro spectrometer (Analytical Spectral Devices, Boulder, Co., USA). This spectrometer was fitted with 10° field of view fiber optics, operated in the 350 - 2500 nm spectral regions with sampling intervals of 1 nm. Surface reflectance measurements were taken at 20 sites over each plot, and for each measurement, the ASD spectrometer was kept stationary and the average of 100 consecutively acquired spectra was recorded to reduce the noise level. We used the average of 20 sites acquired spectra as the measured spectra.

In order to analyze the alteration zoning, a series of situ spectral measurements were carried out in Fangshankou and Jintanzi gold deposits for different geological units with ASD instrument.

3.1. The Spectral measurements in Jintanzi gold deposit

ASD ground spectral measured different rocks in Jintanzi gold deposits. Spectral characteristics of different rocks in this area are as follows:

3.1.1. Quartz diorite. There are six absorption bands for altered quartz diorite in Jintanzi gold deposit, showing in Fig. 3(a). The Fe^{3+} and Fe^{2+} absorption was at wavelengths between 600 to 750 nm, 750 to 1000 nm, 1000 to 1200 nm, and the peak was gentle. Al-OH absorption, with shallow peak, was occurred at wavelengths between 2190 to 2220 nm. Fe-OH bands are located at wavelengths of 2250 nm. Mg-OH bands are shown at wavelengths between 2320 to 2360 nm. All these absorption peaks are associated with the alteration of surrounding rocks of the chloritezation, epidotization, weak sericitezation.

3.1.2. Sericite crushed rock. As showing in figure 3(b), there are also six absorption peaks for gold mineralization rock in mylonitized fracture zone. Its original rock is quartz diorite and it has similar alterations with quartz diorite, which contains chlorite, epidote, and sericite absorption peaks. The Al-OH absorption, which was at the wavelengths between 2190 to 2220 nm, was obvious deep. It displayed strong sericitezation.

3.1.3. Beresitization. The beresitization has four absorption bands (figure 3(c)). The wavelengths between 600 to 750 nm, 750 to 1000 nm, are the Fe^{3+} absorption and showing in tensive limonitezation. Al-OH and -OH has absorption at wavelengths of 2205 to 2345 nm, respectively.

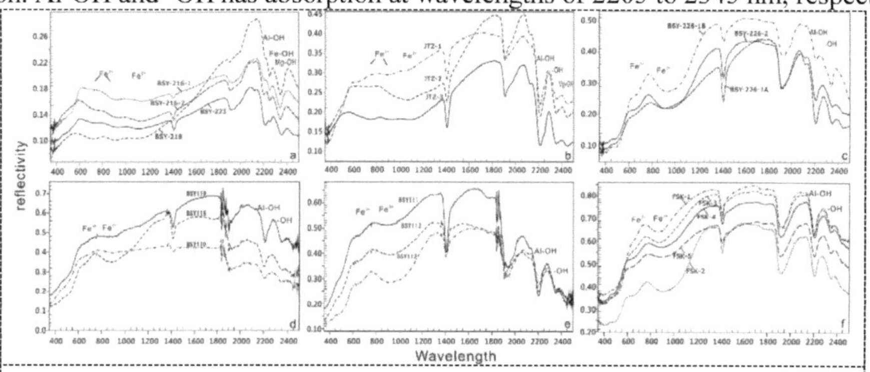

Figure 3. Spectral of typical rocks and minerals in Jintanzi and fangshankou gold deposits, a - Quartz diorite, b - Sericite crushed rock, c - Beresitization; d - Granite; e - Phyllic granite; f - Beresitization.

3.2. The Spectral measurements in Fangshankou gold deposit

Spectral characteristics of different rocks from Fangshankou gold deposit are as follows:

3.2.1. Granitoid. They include including porphyry (BSY-109), monzogranite (BSY115) and granodiorite (BSY119). Spectral analysis showed that sharp absorption peaks of the Al-OH were appeared at 2207 nm, and -OH absorption peaks at 2345 nm, suggesting the attendance of the sericite minerals. At 600 to 750 nm and 750 to 1000 nm, it appeared weak Fe^{3+} absorption (figure 3 (d)).

3.2.2. Phyllic altered granite. In addition to its Al-OH and -OH absorption peaks, 600 to 750nm and 750 to 1000 nm absorption bands of Fe^{3+} (figure. 3 (e)), they also showed strong limonitization and obvious combination of limonite and high alumina sericite anomaly. Microscopic identification indicated phyllic occurred due to gas-liquid rock metasomatism. In the development process of faults, the rock appears a lot of pyritized phyllic vein, and both sides of the surrounding rocks were breaking contacts.

3.2.3. Beresitization. Spectral analysis of pyritized phyllic veins showed that there were two absorption peaks due to Fe^{3+} ions located at 600 to750 nm and 750 to 1000 nm, sericite characteristic absorption peaks were occured at 2204 nm, -OH absorption peaks appeared at 2345 nm (figure 3 (f)).

Its anomaly types, geological features and spectral peaks are similar to the Mingjingou gold deposits [7] within the pyritized phyllic gold ore. Altogether, it showed relatively good mineralization potential of gold ore.

The cross sections in Fangshankou gold deposit were studied for which spectral analysis of the rocks. Its gold ore bodies occur in the altered crush zone, which were hosted in the cataclastis - alteration granite. The geological cross section shows, since the ore body to the periphery development silicification, limonite (pyrite), chlorite, epidote, minor carbonate, sequentially.

As a result, prospecting models of hyperspectral technology were established on the basis of the gold deposits in this region. The pyrite phyllic gold ore are the impartant gold minerazation type in this area and limonitezation and sericite are the important mineral alteration anomalies for gold minerazation prospecting. The typical alteration zoning section analysis showed that zoning characteristics of gold minerazation zone, representing centre limonite + sericite, which change to chlorite + epidote + weak sericite (Fig. 4).Therefore the selected iconic alteration mineral anomaly combined by the CASI/SASI was effective for the mineral prospecting. its conclude that hyperspectral remote sensing can provide precise and reliable information for mineral exploration.

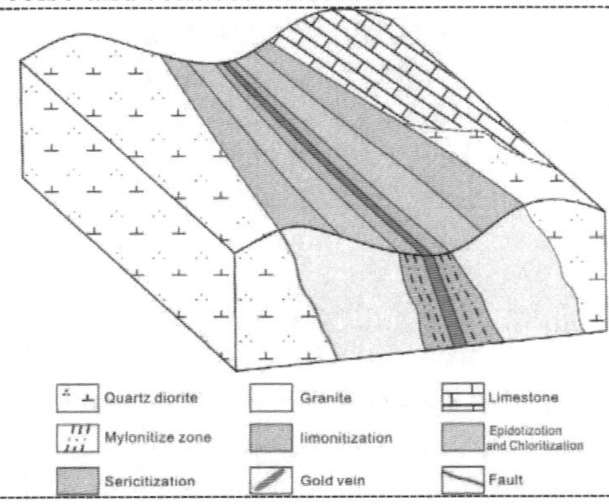

Figure 4. Model of altered minerals anomaly by hyperspectral remote sensing in gold deposit, Beishan.

4. Conclusions

The ASD field Spectrometer was used to analyse and summarize the spectral curve characteristics of different geological units at Jintanzi and Fangshankou gold deposits. Prospecting models of hyperspectral information are established here. It investigated that mineralized bodies were related with pyrite phyllic and mylonitized alteration zone. Hyperspectral data demonstrated that limonite + sericite are the typical alteration minerals for identification of golden mineralization. Chlorite + epidote abnormal are possibly caused by the alteration of the host rocks.

References

[1] Li Z Z, Yang R H, Dang F X, Zhang X F, Tan B X and Zhao H J 2009 The hyperspectral remote sensing technology and its application *Geological Bulletin of China*. 28(2-3) 270-77

[2] Du X D, Tang Y, Liu D C, Li Z and Wang H D 2015 The application of aerial hyperspectral technology to the geological survey for oil and gas in the East Junggar Basin *Geology in China*. 42(1) 275-87

[3] Bierwirth P, Huston D and Blewett R 2002 Hyperspectral mapping of mineral assemblages associated with gold mineralization in the central Pilbara, Western Australia *Economic Geology*. 97(4) 819-26

[4] Berger B R, King V V T, Morath L C and Phillips J D 2003 Utility of high-altitude infrared spectral data in mineral exploration: Application to northern Patagonia Mountains, Arizona. *Economic Geology*. 98(5) 1003-18

[5] Wang R S, Gan F P, Yan B K, Yang S M and Wang Q H 2010 Hyperspectral mineral mapping

and its application *Remote Sensing for Land and Resources*. 22(1) 1-13

[6] Wang R S, Xiong S Q, Nie H F and Liang S N 2011 Remote sensing technology and its application in geological exploration *Acta Geologica Sinica*. 85(11) 1699-1743

[7] Ren G L, Yang J L, Yang M, Li J Q, Gao T, Yi H, Han H H and Zhao Y J 2013 Application of hyperspectral remote sensing anomaly information on metallogenic prediction in the Jintanzi – Mingjingou area of Beishan, Gansu *Geotectonica et Metallogenia*. 37(4) 765-76

[8] Wang R J, Li M S, Wang B, Deng G W, Sun Y B, Wang S S and Niu H W 2014 Study on spectrogram of altered mineral assemblages and its application in Xiaogangou gold deposit, Qinghai Province *Gold*. 35(5) 10-15

[9] Liu D C, Qiu J T,Tian F and Sun Y 2015 Application of airborne hyperspectrum remote sensing to mapping of ore-control faults： A case study of the Heishishan - Huaniushan fault *Geology and Exploration*. 51(2) 366-75

[10] Sun Y, Zhao Y J, Li H B, Lu D H, Tian F, Qin Kai, Yang G F, Zhou J J and Liu P F 2015 HySpex hyperspectral mineral mapping of Asiha gold ore district in Dulan county, Qinghai province and its prospecting implications *Acta geologica sinica*. 89(1) 195-203

[11] Yang H Q, Li Y, Li W M, Yang J G, Zhao G B, Sun N Y, Wang X H and Tan W J 2008 General discussion on metallogenitic tectonic setting of Beishan Mountain, Northwestern China *Northwestern Geology*. 41(1) 22-28

[12] Yang J G, Yang H Q, Yang L H, Li W M and Li Y 2004 Preliminary discussion on the controlling effect of the northeast - trending structure on the mineralization of Au-W-Sn-Mo (REE) deposits in Beishan, Gansu and inner Mongolia *Geotectonica et Metallogenia*, 28(4) 404-12.

[13] Xiao W J, Mao Q G, Windley B F, Han C M, Qu J F, Zhang J E, Ao S J, Guo Q Q, Cleven N R, Lin S F, Shan Y H and Li J L 2010 Paleozoic multiple accretionary and collisional processes if the Beishan orogenic collage *American Journal of science*. 310 1553-94.

[14] Nie F J, Jiang S H, Bai D M, Liu Y, Zhang Y, Zhao Y M, An C J, Wang X L and Su X X 2003 Temporal-spatial distribution and metallogenic processes of gold and copper deposits in Inner Mongolia—Gansu Xinjiang Border (Beishan) region *Mineral deposits*. 22(3): 234-245.

[15] Hu P 2008 Tectonomagmatic evolution and gold metallogeny in south Beishan Mountain, Northwest China *Beijing: Chinese Academy of Geological Sciences*. 10-60

[16] Jiang S H, Nie F J, Hu P and Lai X R 2006 Discussion on the relationship of magmatism and gold metallogeny in Beishan mountain area *Mineral Deposits*. 25(s1) 269-72

[17] Jiang S H 2004 Magmatism and gold metallogeny in Beishan Mt, Northwestern China *Beijing: Chinese Academy of Geological Science 12-62*

[18] Liu W and Pan X F 2006 Evolution of ore-forming fluilds and formational mechanism for gold deposits in the southern Beishan, Xinjiang-Gansu border area of China *Acta Petrologica Sinica*. 22(1) 171-88

[19] Wang Z Z and An G B 2007 Ore controlling factors and minerogenic prognosis of western metallogenic belt of south belt of Beishan, Gansu *Xinjiang Geology*. 25(3) 307-12

[20] Tian Z L and Wu X D 2001 Distribution regularities and prospecting guideline for gold deposits prospects in Beishan ore-forming belt *Xinjiang Geology*. 19(2) 127-29

IOP Conf. Series: Earth and Environmental Science **46** (2016) 012005 doi:10.1088/1755-1315/46/1/012005

An Evaluation of Spatial Distribution of Public Parking Facilities in Huizhou Downtown

Jiasheng Chen[1], **Yang Bai**[1] **and Ying Chen**[1]

[1] Environmental evaluation and GIS development center, Huizhou University, Huizhou Guangdong 516007, China

E-mail: baiyang@hzu.edu.cn

Abstract. The survey and evaluation of existing public parking facilities were carried out, which had important practical significance to resolve conflicts over demand and supply of parking facilities. Taking Huizhou downtown as a study area, we surveyed parking facilities mainly by daily observing and recording. Parking facilities supply, characteristics, and demand were analysed by calculating parking utilization and turnover rate. Based on GIS, the distance-based and time-based accessibility of parking facilities were analysed to evaluate spatial distribution. The results indicated that a large spatial difference in supply and characteristics of parking was shown in public parking facilities of Huizhou downtown and that the parking demand was large. Furthermore, there existed imbalance in spatial distribution of parking facilities in Huizhou downtown area. Our study suggested that it was imbalanced and irrational between parking facilities supply and parking demand, that the planning of parking facilities was inadequate and that management system was incomplete.

1. Introduction

The spatial distribution of parking facilities, as a part of traffic problems in urban areas, has been extensively studied in recent years. Since the implementation of China's reform and opening-up policy in 1980s, China has witnessed enormous social changes. Not only has the economy boomed, but also the quality of people's live has improved sharply. At the same time, urbanization process has been accelerating. With the increasing number of private cars, it is more convenient and efficient for people to travel. However, the parking problem has becoming one of the most outstanding problems in many cities, especially big cities. A series of contradictions were produced between demand and supply of parking facilities, which has had a deep impact on traffic operation and hindered the sustainable development of the city

Researchers have discussed the parking demand, parking behaviors, spatial distribution, parking facilities planning and evaluation. These have been studied by researchers abroad since 1940s. The prediction model of parking demand was built by large-scale surveys to revise indicators of parking facilities in the USA [1]. Location models of parking facilities have been set up by many researchers based on various indexes about their distribution. For example, a location model of minimum walking distance to parking facilities was proposed in Japan, under the circumstance that the demand and the supply of parking facilities were constant [2]. The related studies were launched late in China, which just focused on parking facilities surveys at the beginning, and the parking location planning was studied in 1990s [3]. What's more, the Maximum Entropy (Max Ent) model for the distribution of parking facilities was established so as to coordinate the parking location with road networks [4]. The evaluation index and decision for parking planning were discussed and the index system was set up to evaluate the planning as well as socioeconomic benefits ,by studying the decision making system of parking planning in the urban business district. Additionally, some researchers designed another system of evaluating planning of parking facilities [5-7]. Although a number of studies have been carried out to analyze the demand of parking and to evaluate distribution of parking facilities in the big cities, especially in downtown area, less attention has been paid to the medium and small cities. However, in the future, most of them are likely to become big cities, so it is also uneatable and urgent to focus on the public parking problem in big or small cities and solve it.

Our study was designed to evaluate the spatial distribution of parking facilities by taking downtown of Huizhou as a study area. The parking facilities were surveyed mainly by daily observing and recording. Parking facilities supply, demand, and characteristics were analyzed by calculating parking utilization and turnover rate. Based on GIS, the distance-based and time-based accessibility of parking facilities were analyzed to evaluate spatial distribution.

2. Dataset and Method

Huizhou is located in the northeast of the Pearl River Delta, the southern coast of Guangdong province, China. Taken into account the typicality of the spatial distribution of public parking facilities, Huizhou downtown, with a large population and a high level of economy, ranging from the 3rd Ring Road South to the East River, was divided into 3 regions including He Nan'an, Qiaoxi and Qiaodong as shown in Figure 1. To achieve the purpose of the paper, the data of public parking facilities in Huizhou downtown, including location, types and parking spaces etc., were collected by parking survey from July to August in 2015. They were combined with Huizhou City Planning and Construction Bureau statistics and other network search results like Baidu Map etc. The basic road network and other data were obtained by registration and vectorization based on the map of the related area.

For the purpose of collecting the data to analyse the parking characteristics and calculating parking utilization rate and turnover rate in different time, the method of field survey was used by means of daily observing and recording on weekdays (from Monday to Friday) and weekends (from Saturday to Sunday) in each divided region of the study area. Based on the data above, the evaluation indicator system of parking facilities was established as shown in Figure 2.

2.1. The indicators of parking supply and demand

The parking supply refers to the ability to meet the demand of the parking facilities [8]. Generally, parking space is the most important and direct indicator to reflect the parking supply, which is the sum of actual parking spaces. Meanwhile, there are two major types of parking spaces including Off-street Parking and On-street Parking.

The parking characteristics refer to the use of parking facilities. In our study, two representative indicators including Parking utilization rate and Parking turnover rate were selected to analyse the parking characteristics of Huizhou downtown [9-10]. Parking utilization rate is the foundation of assessing the utilization with many factors, such as the location of parking lots, the development of surrounding economy and population status, etc. [11]. Parking turnover rate refers to the average frequency of parking in a unit parking space in a certain period of time, reflecting the parking capacity. The higher the turnover rate is, the stronger the ability of parking service is with more vehicles accommodated [12].

2.2. The indicators of spatial distribution analysis

Accessibility refers to the traffic condition from one place to others or from other places to one. It is also an important factor in the evaluation of spatial distribution. There are a variety of evaluation methods of accessibility. The common model were used to evaluate the accessibility based on the minimal traffic impedance in the ArcGIS when multiple choices of parking, the distance to destination and the time spent in walking were taken into account. The two models, including distance-based accessibility model concerning shortest distance and time-based accessibility model that focuses on time spent, were selected for the analysis of the minimum impedance. Network Analyst Extension in the ArcGIS was used to set up the traffic network, analyse and obtain the results.

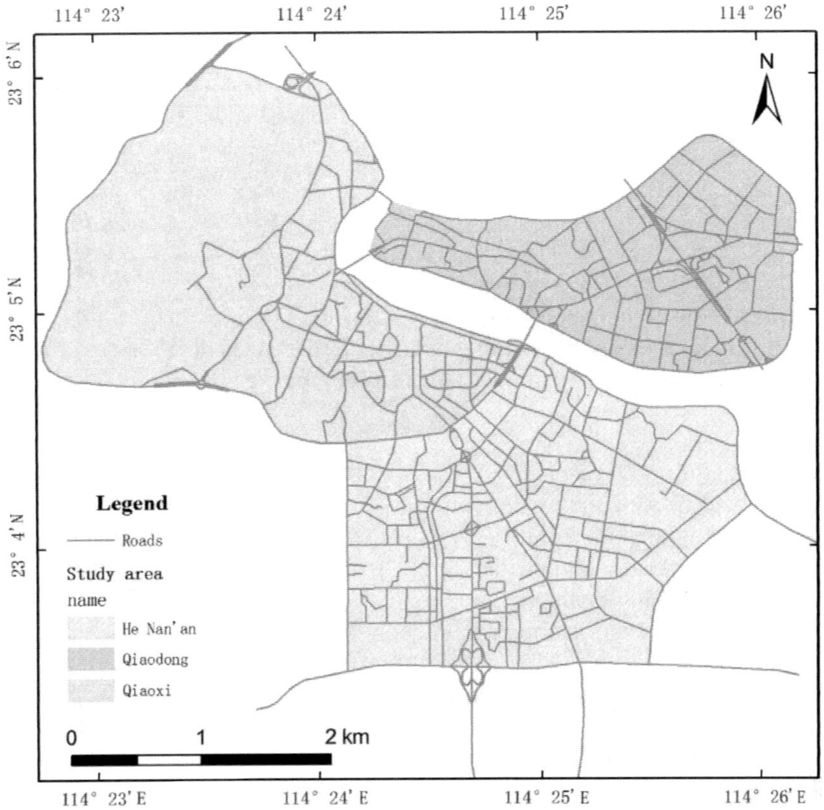

Figure 1. Geographical location of the study area

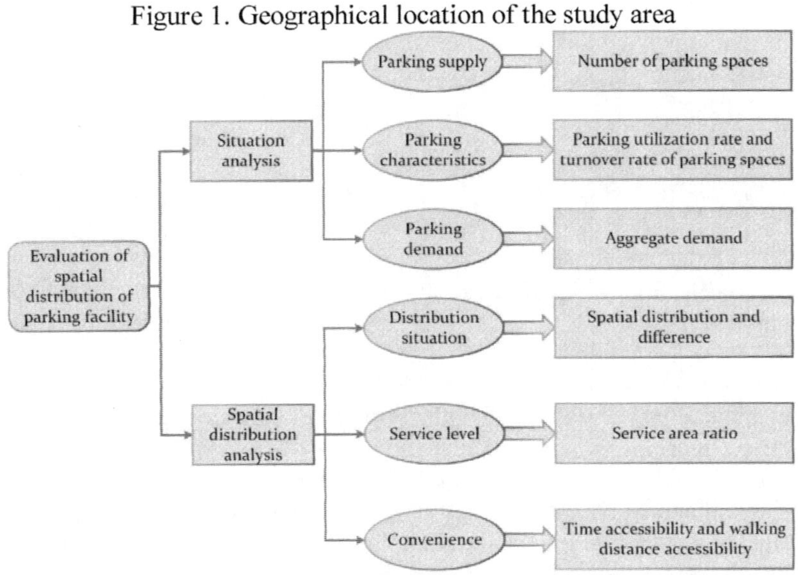

Figure 2. the evaluation indicator system of parking facilities

3. Results and analysis

3.1. Public parking situation analysis

3.1.1. Supply situation analysis
According to the methods above, the situation of parking spaces for the supply of parking facilities in Huizhou downtown is shown in Table 1.

Table 1. The current supply of public parking area of Huizhou city downtown

Region	The number of parking facilities			The number of parking spaces		
	Off-street Parking	On-street Parking	Total	Off-street Parking	On-street Parking	Total
He Nan'an	25	6	31	3358	336	3694
Qiaoxi	16	4	20	1930	850	3240
Qiaodong	9	13	22	790	1065	1855
Total	50	23	73	6078	2251	8789

As can be seen from Table 1, there are altogether 73 parking facilities in the study area, including 50 off-street parking lots and 23 on-street parking lots, which provide 8789 standard parking spaces, including 6078 off-street parking spaces and 2251 on-street parking spaces, accounting for 25% of the total.

3.1.2. Parking characteristics analysis
According to the collected data of the three regions, the parking utilization rates in each interval on weekdays and weekends are shown in Figure 3. It indicates that the utilization rates of parking facilities in Huizhou downtown are generally higher than 80%. And on weekends, they are almost over 100 % , which means that the supply of parking spaces may fall short of the demand.

Parking turnover rates are shown in figure 4. It shows that the rates of weekends are less than the weekdays between 8: 00 and 15: 00. After 15: 00, the turnover rates of weekends are higher than weekdays. But trends of parking turnover rates are substantially the same at all times, and it reaches maximum between 17: 00 and 18: 00. It also can be seen from the figure 4 that parking turnover rates have two peaks between 9: 00 and 11: 00 and between 17: 00 and 18: 00, when residents travel frequently.

.

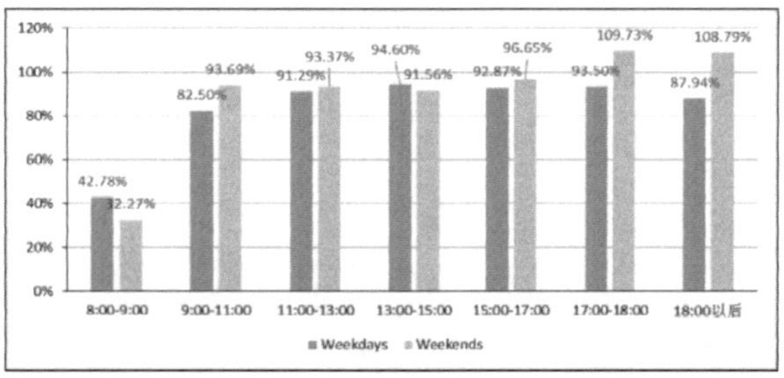

Figure 3. Parking utilization rate of parking facilities in Huizhou downtown.

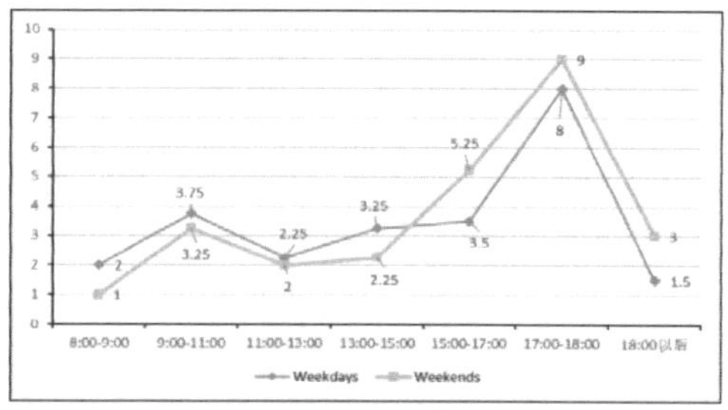

Figure 4. Parking turnover rate of public parking facilities in Huizhou downtown.

3.1.3. Demand situation analysis
As mentioned above, the results reveal that the utilization rate of parking facilities in Huizhou downtown is generally high. In the morning, when residents rush to work between 8: 00 and 9: 00, most of vehicles are mobile, so the parking utilization rate is below 50%; Between 9: 00 and 18: 00, the parking utilization rates exceed 80%, and the turnover rates wavily rise, which suggests that parking demand and fluidity of parking spaces increase; Many activities like shopping, dining, etc., mainly distribute in public service facilities after 17:00 on weekends, which leads to the fact that the parking utilization rates are over 100% and the demand is greater than the supply.

3.2. Spatial distribution analysis of public parking facilities based on GIS

3.2.1. Spatial distribution situation analysis

Based on the location of public parking facilities, the map of spatial distribution was obtained in ArcGIS. In Figure 5, it is obvious that public parking facilities mainly distribute in the northwest of He Nan'an district, southeastern of Qiaoxi district and south-central of Qiaodong district on both sides of the main roads, and the quantities of facilities in the east of He Nan'an district, the east of Qiaoxi district and the west of Qiaodong district turn out to be less, where the old town used to be located.

3.2.2. Service level analysis

The service area ratio is one of the indicators to measure differences in public services and facilities, which equals to the ratio of the services and facilities area to the total area. The ratio of He Nan'an district, Qiaoxi district and Qiaodong district were calculated as shown in Table 2, from which we can know that the ratio of service area in Qiaodong district is highest and Qiaoxi district is the lowest.

Table 2. Statistics of rate of service area of parking facilities in Huizhou city downtown.

Region	Number of facilities	Service area	Service area ratio
Henanan	31	2.218	39.33%
Qiaoxi	20	1.397	22.60%
Qiaodong	22	1.598	46.45%

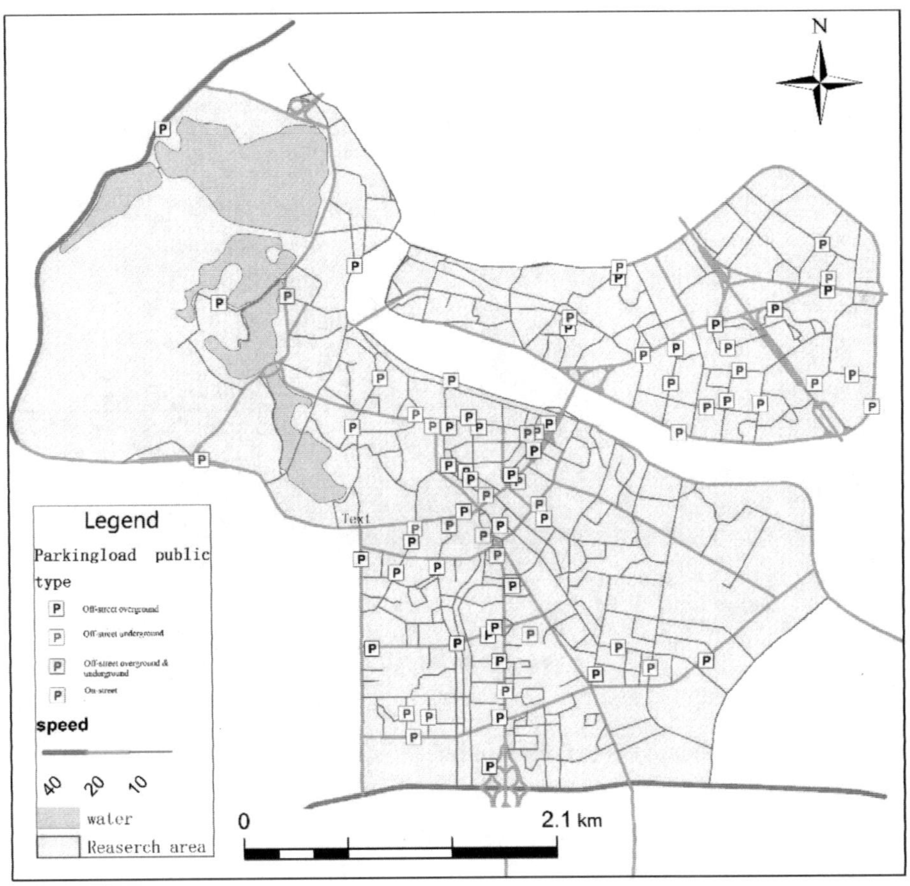

Figure 5. Spatial distribution of parking facilities in Huizhou downtown.

3.2.3. Public parking accessibility assessment

It is shown in Figure 6 that the accessibility of Maidi block in He Nan'an district, Dongping block in Qiaodong district are less than 50 or range from 50 to 100; The accessibility of the West Lake in Qiaoxi district, the old town in Qiaodong district, and both sides of Maidi Road and so on range from 100 to 200 and even exceeds 200. Similarly, as shown in Figure 7, the district with small traffic impedance and great accessibility has a better time-based accessibility, and vice versa.

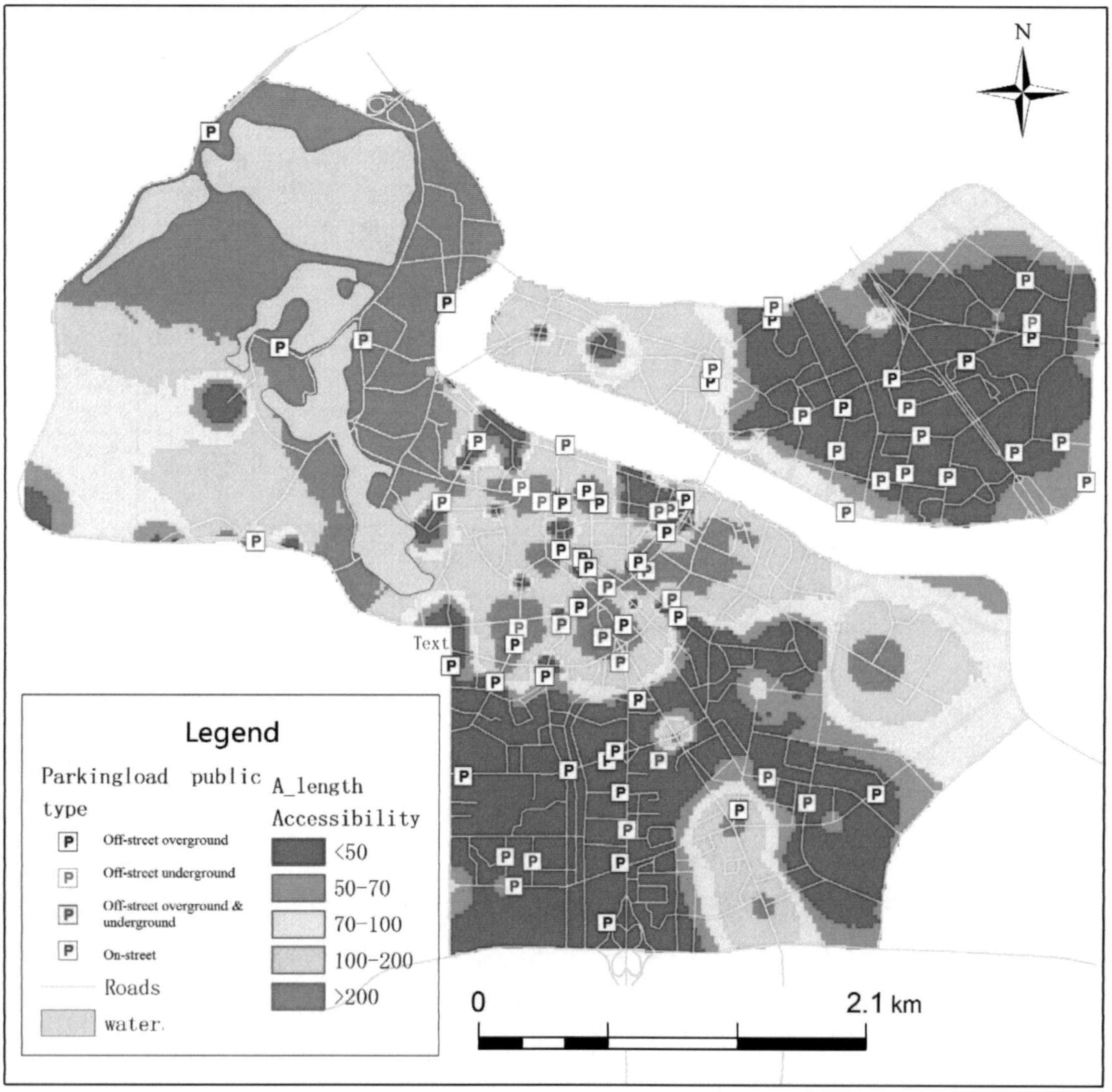

Figure 6. Distance-based accessibility of public parking in Huizhou downtown.

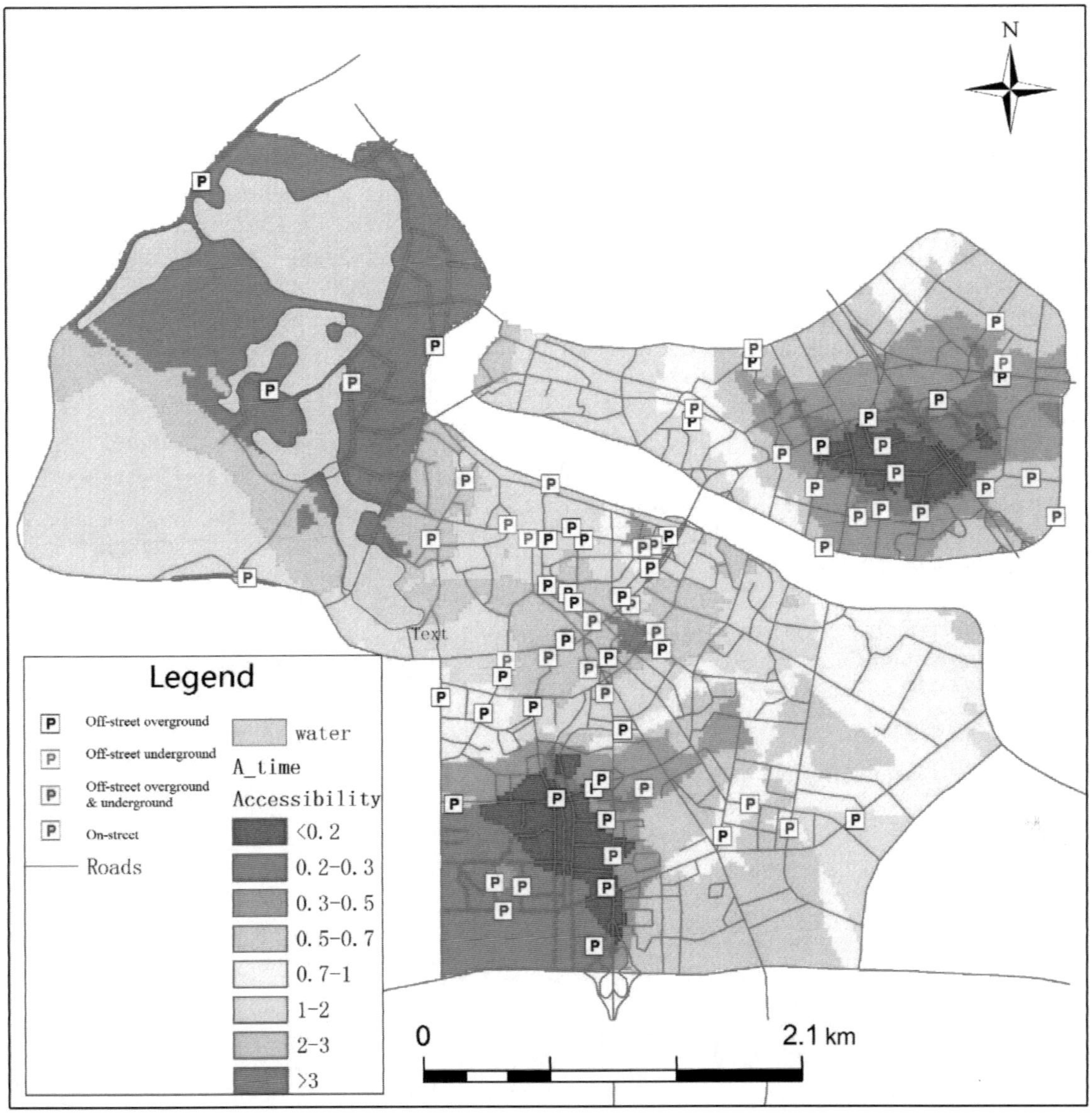

Figure 7. Time-based accessibility of public parking in Huizhou downtown.

4. Conclusion

From the analysis above, we can conclude that the quantities of parking spaces and public parking facilities in Huizhou downtown are apparently inadequate. As far as the supply of public parking facilities is concerned, the gap between new urban areas and the old areas is wide, so it is with the core areas and the edge areas. Thus, there exists an imbalance and irrational structure between the supply of parking facilities and the demand of parking in Huizhou downtown, the planning of parking facilities and the management system are incomplete

References

[1] Lijuan W, Xi C and Shifeng L 2011 Investigation and analysis of present situation of public parking fascility in central area of Shijiazhuang City *J Journal of Shijiazhuang Tiedao University* **5** 24-27.

[2] Linwei L 2014 Some Problems of Zhengzhou city downtown public parking Construction Management *J Henan Science and Technology* **3** 213-215.
[3] Xiaohua X, Ruizhang W, Donghong W and Zhiyong Z 2015 Evaluating the Medical facilities spatial distribution based on GIS: an application of Xiang'an District J Journal of Geo-information Sciense **17** 317-328.
[4] Feng Z, Guangliang X and Xiao Q 2015 Smart city planning and construction based on geographic perspectives: some theoretical thinking *J Progress in Geography* **34** 402-409.
[5] Xiaoshu C, Wenyue Y and Xiaoyan H 2015 Aaccessibility and CO2 emissions from travel of smart transportation: theory and empirical studies *J Progress in Geography* **34** 418-429.
[6] Chunguang S, Xintman Y and Ji W 2013 Study on the optimization of social parking distribution in Central District of Ji'nan City *J Urban Roads Bridges and Flood Contro* **27** 24-26
[7] Jinqi Z 2005 Parking Planning and Construction of Small and Medium Cities *J China Science and Technology Information* **22** 153
[8] Xiaolan Z 2005 Study on the Relationship Between the Supply of Parking Facilities and the Capacity of Railway Network in Urban Central District *J Traffic and Transportation* **1** 77-79
[9] Zhongping T, Huahu L and Jifeng D 2004 Parking Feature Survey Method and Data Processing Research J Central South Highway Engineering **29** 10-11
[10] Jun G and Yuli Z 2005 The Study of City Parking Planning Survey Method *J City Bridge and Flood Contro* **19** 23-25
[11] Sizhen G and Yuanxiang Y 2012 Study on the Model of Parking Utilization *J Journal of Chongqing Jiaotong University* **31** 568-570
[12] Xiaodan W and Jiaxin W 2015 A Microscopic Model of Parking Demand Forecasting Based on the Actual Turnover Rate *J Shandong Traffic Science and Technology* **29** 10-12

High resolution remote sensing and potential analysis of iron ore prospecting——Taking Datong Township,West KunLun Area for example

J Q LI, H YI, G L REN, T GAO, M YANG, H H Han, J L YANG

Xi'an Center of Geological Survey, Geological Survey of China, Xi'an, China

E-mail: 405739307@qq.com

Abstract. By the high resolution remote sensing anomaly verification, two magnetite ore belt was fist found in the paleoproterozoic Kulangnagu group complex(Pt_1K) at Datong township area. Which located in the southeast of West KunLu Taxkorgan iron deposit prospect area. The magnetite presents ribbon, lenticular, hosted in quartz schist with marble stratum. Based on the regional stratigraphic paleogeographic environment, primary rock formation, ore controlling environment geological background analysis, think that the Kulangnagu group complex have the potential to find metamorphosed sedimentary iron ore. Based on the analysis of high resolution remote sensing, combined with geochemical and geophysical data, We discussed the ore prospecting area and put forward two magnetite favorable prospecting area, one at the south of Datong township area, another at the southeast of Bulunmusha township, given the direction For the next step of prospecting.

In the previous work, contact metasomatic or hydrothermal jade deposit, hydrothermal molybdenum deposit have been found in Kulangnagu group complex [1][2], But, few rich magnetite were reported. Only in the 1:25 million regional geological survey, some laminated magnetite quartzite are recorded in the measured geological profile, but no more detailed expositions were carried out. In 2010-2011, by the remote sensing anomaly inspection, one banded iron formation (BIF) belt ware found in the southern part of Datong town area by the authors, then two magnetite ore belt was fist found in the BIF belt. Based on the analysis of the high resolution remote sensing, starting from the ancient geographical environment, primary rock formation, ore controlling environment and remote sensing anomaly characteristics, combined with the geological characteristics of magnetite ore, this paper discussed the origin and spatial distribution of the magnetite ore in Datong Township, West Kunlun.to provide basis for the next prospecting direction.

1 Regional Geology

Kulangnagu group complex belonging to the Western Kunlun stratigraphic division, Widely distributed in Northern Kunlun, Qiaerlong Township, Bulunmusha Township, Datong Township, Xihexiu Township and other places. Especially, in the south of Datong Twonship, Kulanglagu River area, North of Kudi and North of Saitula， it's well outcropped and look like a long belt [1][3]. The area between the Kegang fault zone and Kang Xiwa fault zone is the study area in this paper (Figure 1).

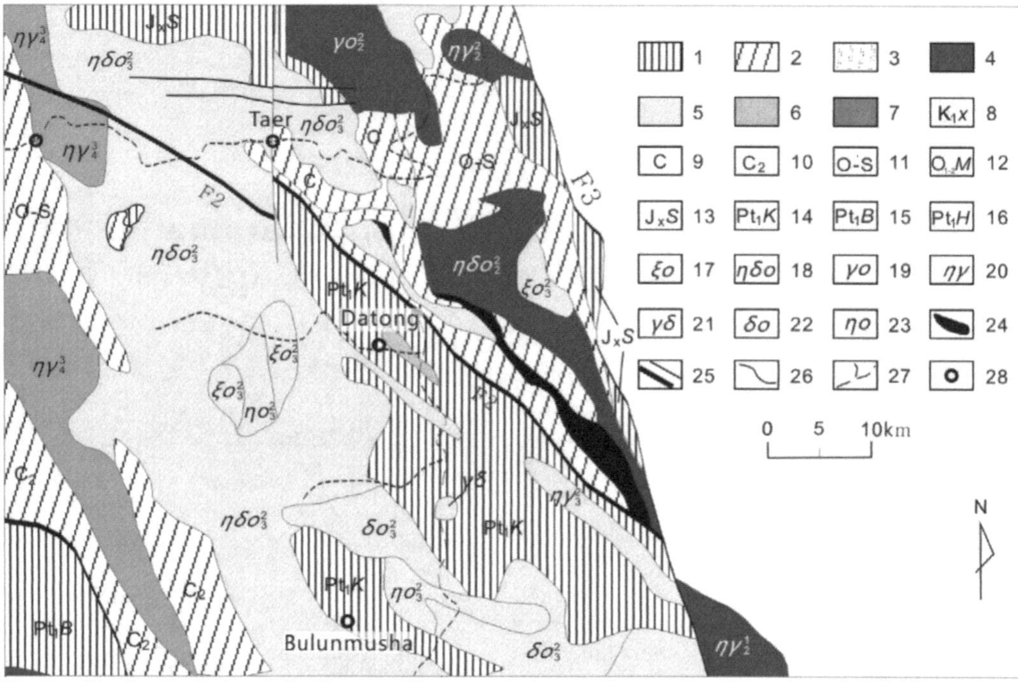

Fig.1 Geological sketch in Datong Town area (according to Liu Jianping, 1:25 000 regional geological data, modified)

1-Precambrian strata, 2-palaeozoic stratum, 3-Mesozoic strata, 4-Proterozoic granitoids, 5-Caledonian granitoids, 6-Hercynian granite, 7-Yanshan period granite, 8-Xiadilafu group,9-Carboniferous, 10-Middle Carboniferous, 11-Ordovician – Silurian, 12-Maliezikcn Group, 13-Sangzhutage Group, 14-Kulangnagu Group Complex, 15-Bulunkuole Group Complex, 16-Heluositan Group Complex, 17- Quartz syenite, 18-Quartz monzodiorite, 19-Tonalite, 20-Monzonitic granite, 21-Granodiorite, 22-Quartz diorite, 23-Adamellite, 24-Ultrabasic rock, 25-Fault, 26-Geological boundary, 27-River, 28-Township, F1-Kongxiwa fault zone, F2-Kegang fault zone, F3-Kusilafu fault zone.

In the High resolution image (True color, composited by bands 5-3-2), the magnetite ore body's color is light gray, texture clear. Along the extension of texture direction (North to West) can see the dark, white and brown strips obviously. And NW parallel faults are well developed. According to result of remote sensing image interpretation, the lithology unit of this area has the image characteristics of quartz schist and marble, the ore belts were located in the area of ductile brittle shear zone. A intrusive rock appear on the south side of the shear zone.

Through the field verification, the rocks has well-developed joints, strongly crushing and alteration. their deep parts has been mylonization. Caused by the tensile shear,the dark mineral in the invasion rocks have been oriented arrangement, part of the intrusive rocks have become felsic mylonite. There have some later granite intrusion in the rocks, and the vein's alterations are well-developed. This shows that the late hydrothermal activities were strong.

In Datong area, Kulangnagu group complex's lithology is mainly crystalline schist, such as dark gray-black biotite quartz schist, gray mica quartz schist, grey-black quartzite, laminated magnetic quartz, white marble. The magnetite quartzite laminae is obvious, and the fold, crumpled structure are obviously on hand specimen scale.

The original rock of the BIF and its surrounding rocks in Kulangnagu group complex is a set of clastic rocks (mainly argillaceous sandstone) - carbonate rocks intercalated with basic volcano rocks. The Datong magnetite ore produced in the banded iron quartzite section along the formation strike, has the feature of syngenetic sedimentary mineralization, By the late (Caledonian) magmatic activity, the magnetite ore bodies reformed by hydrothermal alternation and further enrichment [1][4].

2 Geological Characteristics of Iron Deposits

2.1 The Output Characteristics of Magnetite Ore Seam

The Datong Magnetite Deposits located at about 1.5 kilometers south of the township government, Taxkorgan Tajik Autonomous County, two magnetite ore outcrop was found, one is the southern Datong magnet mineralized bodies,the other is the Yilifa magnet mineralized bodies. The surrounding rock of the mineralized bodies are grey - black mica quartz schist intercalated with marble, the strike of the strata is relatively stable, the occurrence of the strata is 40∠70. at the Part of the quartzite intercalated with mica quartz schist near the marble, where developed some lamellar magnetite quartzite and magnetite ore. The thickness of the laminated magnetite quartzite is about 300 meters, The thickness of the magnetite ore body is 10-20cm, and the maximum thickness is up to 100cm, Presenting single layer or multiple layers output. In local area, the magnet mineralized bodies shaped as lenticular, pod, cystic. In the ore belt, the mineralization is not continuous, intercalated with several layers of different thickness of iron quartzite. The mineralization of the upper part of is strong, the iron ore layer has large thickness, stable occurrence and good continuity. The average grade of iron is 18%.

The ore bodies are controlled clearly by strata and interlayer fault (figure 2). The magnetite veins are banded, its width varies greatly, mainly in 2-20mm. the main gangue minerals are quartz. In overall, the ore body shows a continuous stripe, but the light and dark stripes are interspersed with each other and the theirs thickness changes greatly, this may reflect the more turbulent sedimentary environment at that time.

Fig.2 Magnetite ore belt and its Geological section in Datong Town

a- Irregular banded and layered magnetite ore bodies (Yilifa magnet mineralized points), b- Banded magnetite ore bodies(Yilifa magnet mineralized points), c- Contact relationship between magnetite ore bodies and mica quartz schist (southern Datong Town), d- Section of magnet mineralized points, Mt- Magnetite, ibr- Itabirit, qs- Quartz schist, 1- Quartz monzodiorite, 2- Quartzite, 3- Quartz schist, 4- Biotite schist, 5- Marble, 6- Magnetite ore bodies, 036 and 037- Sample number.

2.2 Texture, structure, mineralogical composition of the magnetite ore

Ore mineralogical composition: ore minerals are mainly magnetite and a small amount of hematite. The main gangue mineral are quartz and a small amount of plagioclase, mica, hornblende, diopside, tremolite, calcite, chlorite, pyrite etc.

Ore texture: mainly allotriomorphic - hypidiomorphic granular texture, a small amount of Granoblastic texture, hypidiomorphic granular and allotriomorphic granoblastic texture.
Ore structure: The structure of the ore is mainly banded (or lamellar), dense block, local disseminated, vein etc. Iron minerals and quartz chequered with black and white, with obvious banded, striped, fine lines structure characteristics, form the Banded Iron Formation (BIF). The compact massive rich ore are mainly produced in the stratum of high silica content. The magnetite is mainly concentrated in the transition zone that from the pelitic rocks to the sandy rocks. The ore structure and enrichment characteristics are similar to Tashikuergan "Pamir type iron ore". Show the characteristics of the marine sedimentary deposit that affected by the change of sedimentary environment [5][6].
Wall rock alteration: skarn, chloritization, sericitization, limonitization, pyritization.

3 Remote sensing anomaly characteristics

The color of the high resolution remote sensing image (worldview-2 true color image, band5,3,2) of Datong magnetite ore area is light gray, texture clear, in the ore belt, there have dark gray, white and brown strips. NW trending parallel joint fissure development, indicates that the area has experienced brittle ductile shear tectonic activity.
By using principal component analysis method to extracted the -OH and Fe oxide anomaly from Landsat ETM dat (a middle resolution multi spectral data), only weak -OH anomaly displayed in this region. Field investigation has proved that the -OH anomaly Mainly caused by the marble belt. By using ratio method (B5/B2) to extracted the Fe oxide anomaly from Worldview-2 data (a high resolution multi spectral data) the result shows that the Fe anomaly is obviously in this area (Figure 3). There have strong Fe oxide anomaly development in the magnetite belts. Shown that the high resolution Fe oxide anomaly and the iron ore belts have a good agreement.
In the Fe oxide anomaly map, the Fe oxide anomaly have a zonal distribution of north west direction. From the north to the south, there have 3 Fe oxide anomaly zones. a few meters to tens of meters wide, extends several hundred meters to several kilometers. Field verification shows that the Fe oxide anomaly zone is mainly caused by the banded magnetite quartzite.

4 Mineralization potential analysis

In western Kunlun, a large number of BIF type magnetite were found in Paleoproterozoic bulunkuole group complex and Changcheng Period Saitula group complex[8].
The rock assemblage characteristics and metamorphic degree of palaeoproterozoic Kulangnagu group complex same as palaeoproterozoic Bulunkuole group complex and Changcheng Period Saitula group complex. They are the primary Precambrian metamorphic basement strata in Tashikuergan area. Therefore, the found banded iron formation (BIF) and magnetite strips indicate that Kulangnagu group complex has metallogenic conditions of magnetite, with important prospecting indicator significance.
By high resolution remote sensing interpretation, the writer thinks that，there are several stable banded magnetite quartzite stratum in Kuruke Langan, Datong and Bulunmusha area. They magnetite quartzite stratum have obviously image characteristics, its thickness of 200-1000m, extending the length of about 10-25 km. in Datong area, anomaly investigation has confirmed that there have magnetite quartz formation and have found 1m thickness of magnetite ore. It can be concluded that, Kulangnagu group complex have a certain magnetite ore prospecting potential.

5 Conclusions

High resolution remote sensing technology provides powerful help for the geological survey in the Western hard and dangerous areas [7]. Worldview-2 images have high spatial resolution and high spectral resolution in the visible and short wave infrared region (V-NIR), and more accurate iron containing altered minerals information could be extracted and identified, such as limonite, jarosite, goethite, chlorite and etc[8].
Some BIF in Kulangnagu group complex, its iron ore body has experienced several stages, such as the original volcanic sedimentary mineralization, the further enrichment of the regional metamorphism, the

IOP Conf. Series: Earth and Environmental Science **46** (2016) 012006 doi:10.1088/1755-1315/46/1/012006

reenrichment of the late magmatic hydrothermal activity and the transformation of ductile brittle deformation.

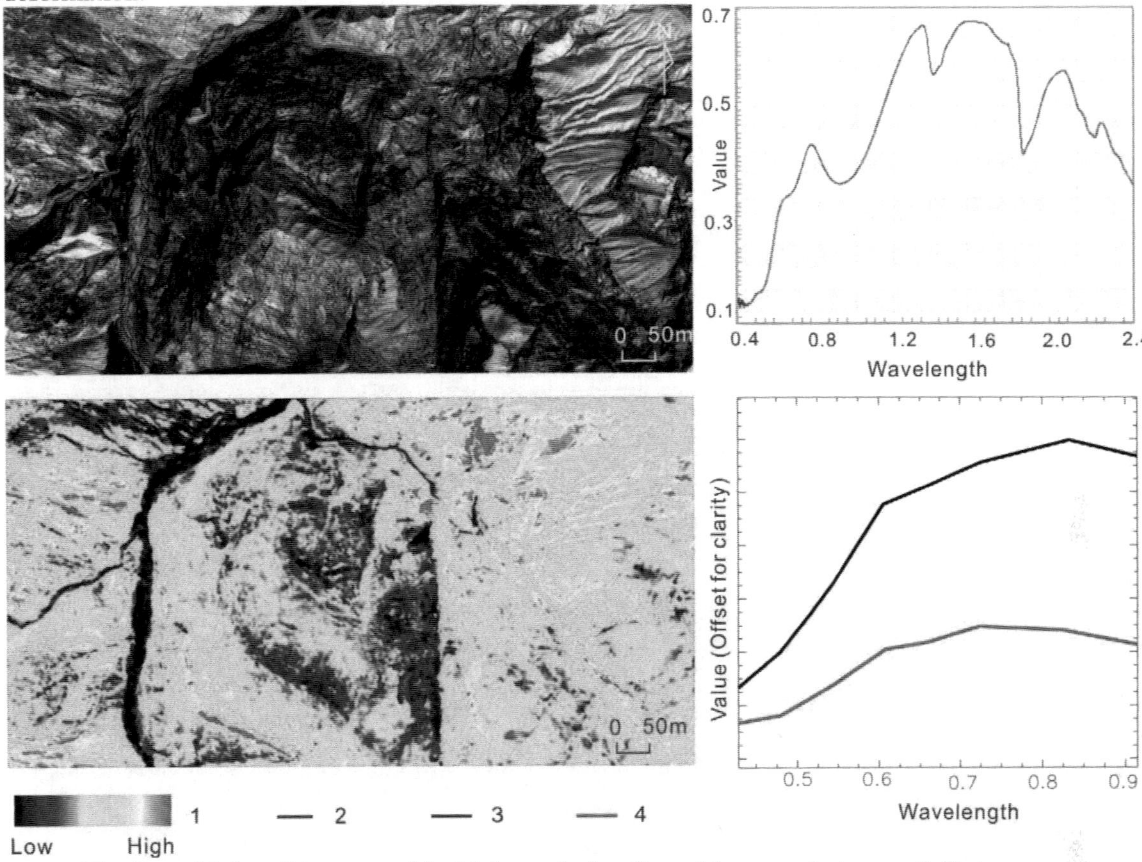

Fig. 3 worldview-2 image and its high resolution Fe oxide anomaly in sourth Dotong area

1- Fe anomaly, 2- Limonite spectral (field), 3- Worldview-2 image spectral,4- Limonite spectral(USGS)

The discovery of banded iron formations in Kulangnagu group complex, has provided a new prospecting direction for the iron ore prospecting work in the western Kunlun area. Recommendations would carry out large scale aeromagnetic surveys or geochemical exploration works, increasing the degree of geological work, further evaluated of magnetite ore prospecting potential for Kulangnagu group complex in Datong area.

References:

[1] Li R S, Ji W H, Yang Y C *et al* 2008 *Kunlun Mountains and Its Adjacent Area Geology* (Beijing: Geological publishing house) p 18.

[2] Liu J P, Wang H, Ren G L *et al* 2010 Geological Characters and Prospecting Significace of Xiaotong Molybdenum Deposit in Western Kunlun, Xinjiang *XinJiang Geology* vol 28(1) pp 38-42.

[3] Henan institute of Geological Survey 2004 *Report Of 1:250,000 Geologic Surveys in Keketulu and Tashikuergan Tajik Autonomous County Area* (Department of Land and Resources of Henan Province, Zhengzhou) pp 17-22.

[4] He S P, Li R S, Yu P S 2010 *The Tibetan Plateau and Its Adjacent Area Precambrian Geological Map(1:1500000)* (Beijing: Geological publishing house) p 21.

[5] Yan C H, Chen C J, Cao X Z 2012. The discovery of the "Pamir-type" iron deposits in Taxkorgan area of Xinjiang and its geological significance *Geological Bulletin of China* vol 31(4) pp 549-557

[6] Ren G L, Li J Q, Wang H et al 2013 The Metallogenic Series of Iron Deposits in Bulunkou-Zankan, West Kunlun *XinJiang Geology* vol 31(4) pp 318-323.

[7] Chen L Zhang W Zhou Y Liu W Chen W 2012 Application Of High-Resolution Remote Sensing Images To Searching For Sedimentary-Metamorphic Type Iron Deposits In The Taxkorgan Area, Xinjiang *Geology and Exploration* vol 48(5) pp 1039-1048.

[8] Yang M Li J Q Gao T Zhang Z Liang N Yi H Han H H 2012 The application of WorldView-2 data in Geological Survey *Morden Mining* vol 518(6) pp 35-37.

Objected-oriented remote sensing image classification method based on geographic ontology model

Z Chu[1,2] , Z J Liu[1] and H Y Gu[1]

[1] Chinese Academy and Surveying and Mapping, Beijing, China
[2] Liaoning Technical University, Fuxin, China

E-mail: chuzhao0601@163.com

Abstract. Nowadays, with the development of high resolution remote sensing image and the wide application of laser point cloud data, proceeding objected-oriented remote sensing classification based on the characteristic knowledge of multi-source spatial data has been an important trend on the field of remote sensing image classification, which gradually replaced the traditional method through improving algorithm to optimize image classification results. For this purpose, the paper puts forward a remote sensing image classification method that uses the he characteristic knowledge of multi-source spatial data to build the geographic ontology semantic network model, and carries out the objected-oriented classification experiment to implement urban features classification, the experiment uses protégé software which is developed by Stanford University in the United States, and intelligent image analysis software—eCognition software as the experiment platform, uses hyperspectral image and Lidar data that is obtained through flight in DaFeng City of JiangSu as the main data source, first of all, the experiment uses hyperspectral image to obtain feature knowledge of remote sensing image and related special index, the second, the experiment uses Lidar data to generate nDSM(Normalized DSM, Normalized Digital Surface Model),obtaining elevation information, the last, the experiment bases image feature knowledge, special index and elevation information to build the geographic ontology semantic network model that implement urban features classification, the experiment results show that, this method is significantly higher than the traditional classification algorithm on classification accuracy, especially it performs more evidently on the respect of building classification. The method not only considers the advantage of multi-source spatial data, for example, remote sensing image, Lidar data and so on, but also realizes multi-source spatial data knowledge integration and application of the knowledge to the field of remote sensing image classification, which provides an effective way for objected-oriented remote sensing image classification in the future.

1. Introduction

Geographic ontology is the expansion and extension of ontology in the field of geographic information science, which mainly has been used to design geographic type and feature class on the geographical target domain, it provides a clear conceptual model for GIS filed so that people have better understanding of the geographical structure of the world, and avoids the great contrast existing data models and human spatial cognition, meanwhile through shared geographic concepts of a clear formal definition of geographic information sharing and interoperability, which has a wide range of applications on geographic information integration and geographic information services [1-4]. With the study of the geographic ontology, geographic ontology technology begins gradually to extend to the field of remote sensing image, which extracted the image feature information as the knowledge to build image feature base, build conceptual knowledge of geographic ontology model. For this reason some scholars use

IOP Conf. Series: Earth and Environmental Science **46** (2016) 012007 doi:10.1088/1755-1315/46/1/012007

geographic ontology model in the field of remote sensing to carry out extensive research at home and abroad. Zhang Pu [5] uses geographic ontology as the basis, with sloping topography knowledge, build on the Loess Plateau landforms of ontology modelling, and model-based ontology established in accordance with the classification rules, implement to the extraction of the Loess Plateau landforms successfully; Mariana Belgium [6] based on geographic ontology and machine learning, using image feature information, constructed the building classification rules, based on geographic ontology and random forest, and realized urban building classification; G.Forestier [7] used image features to build geographic ontology, and implemented classified feature to treat the initial mark up through geographic ontology, to achieve certification marked feature of this method through expert knowledge built by the function of knowledge (Knowledge Functions) model ,achieved type of sea surface features of interpretation. As can be seen, to achieve remote sensing image classification at the level of knowledge has become the main trend in the field of remote sensing image classification, however, how to achieve the objective modelling of feature elements, how to achieve organization and expression of the knowledge characteristics effectively, such geographic entities can conceptualization and formal expression, in addition geographical entity having cognitive unity is the key problem to be solved based on the method of knowledge. This paper is mainly based on geographic ontology theory, the ontolgy has a geographical conceptualization, explicit, formal, sharing, etc. [8-10], combined with image features knowledge base, building geographic ontology model of the study area, to achieve conceptualization and formal expression of study area feature elements ,to realize the study area of remote sensing image classification

2. The introduction of geographic ontology model method

The method mainly analyzes the research need, extracts image feature knowledge and other additional knowledge such as elevation, build image feature model, decision tree model and expert rule model based on the research feature knowledge, and build geographic ontology knowledge structure of study region, imports the OWL file form of image object to the structure, the first it uses the decision tree model to construct the initial classification to obtain initial result, and then uses the expert rule model to construct the classification to obtain the final classification result. Specific technical route shows below.

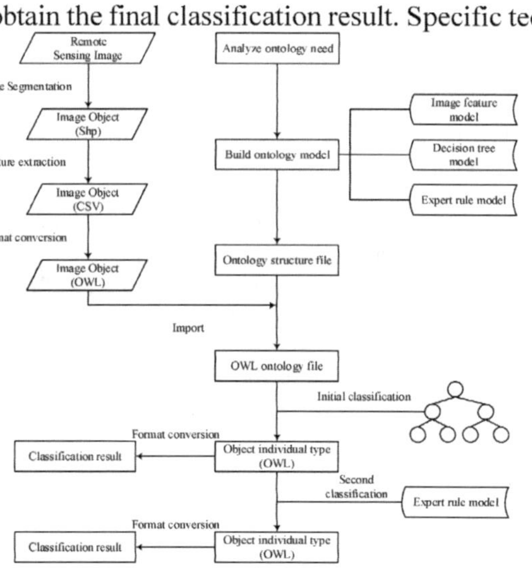

Figure.1 The classification flow chart of geographic ontology model

As can be seen from Figure 1, the main steps based on geographic ontology model classification include the following:

- The first through the image segmentation, we can obtain image object of SHP format file, then through extracting feature for the image object, we can save them as CSV format file, the last we make CSV file and OWL file conversion of image object to implement OWL formal representation of image object.

- For the study area, we could analyze ontology need, and make full use of the extracted feature knowledge to build feature knowledge base, then we should build geographic ontology structure including image feature model, decision tree model, expert rules model to get structure file and express it with OWL format.
- We will import the image object of OWL format to geographic ontology structure and use the decision tree model and expert rules model to construct XQuery queries to implement the object identification of the study are. In addition, we need make OWL format file and SHP format file conversion to get the image object of SHP format.

3. Geographic ontology model of remote sensing image

3.1. Conceptual model of image feature

The image object feature is the origin attribute of remote sensing image. How to express the object feature relationship formally, how to make use of the extracted and mined feature information knowledge to implement image classification, which is the key problem that need to be solved in the field of remote sensing classification. This section mainly bases on the conceptual knowledge of image object and the OWL language to implement to build conceptual model of image feature and express the object feature relationship formally. We would consider research need and the data characteristics of the study area to define the feature and type of image object. The feature types include spectral feature property(Spectral_Property),texture feature property(Texture_Property), shape feature property (Shape_Property) and spatial feature property(Spatialfeature_Property).

3.2. Decision tree model

The method use decision tree in OWL format to build model, which is divided into four steps.

3.2.1. The description of related concepts. It mainly include related concepts of decision tree algorithm and knowledge concepts of the study area. The decision tree concept includes Decision Tree, Root, Node and Leaf. The paper uses land cover as study object, therefore, the concept of study field mainly includes Building, Road, Water, BareLand, Vegetation.

3.2.2. Define classes and their relationships. We define the class and the class hierarchy with top-down approach, the first we determine the base class—urban land cover class (City_Land), the second which could be divided into building class (Building) and low feature class (LowFeature) with the elevation, the third the low feature class would be divided into artificial feature class (ArtificalFeature) and natural feature class (NaturalFeature), the last the artificial feature class would be divided into road class (Road) and bare land class (BareLand) and the natural feature class would be divided into water class (Water) and vegetation class (Vegetation). We can build the structure diagram of tree and establish the relationships among classes.

3.2.3. Define attributes and attributes setting. Because decision tree model mainly involves comparison of eigenvalues, the primary object properties defined include greater than (GreaterThan), greater than or equal (GreaterThanOrEqual), less than (LessThan) and less than or equal (LessThanOrEqual) four main attributes, and then we would set their domain and range of property.

3.2.4. Define individual. According to class hierarchy and class definitions, we can create individual instances based on decision tree model. The paper uses the land cover as example, which mainly creates Node1, Node2, Node3, Node4, Node5, Node6 Node7, Node8 and Node9 etc. nine individuals. We would set each individual instance, which includes class setting, judging the individual belongs to class and attributes setting including object properties setting and data properties setting.

3.3. Expert rule model

Expert rule model mainly uses the semantic knowledge and concept of the field to create flag rule, which is basically the process of the image feature representation of semantic relations, then, according to

IOP Conf. Series: Earth and Environmental Science **46** (2016) 012007 doi:10.1088/1755-1315/46/1/012007

experts' flag rules and prior knowledge, we can get expert rules, and finally, according to the expert rules, we would use XQuery / XPath query language to implement ontology reasoning. We would use SWRL language combining with prior knowledge to express flag rules and expert rules.

3.3.1. Flag rule expression. Brightness(?x, ?y), lessThan(?y, 1000) -> DarkLight(?x) Elevation(?x, ?y), lessThanOrEqual(?y, 2) -> Low(?x)RectangularFit(?x, ?y), greaterThan(?y, 0.78) -> Regular(?x) Length_Width(?x, ?y), greaterThan(?y, 2.0) -> Strip(?x) NDVI(?x, ?y), greaterThan(?y, 0.40) -> VegetationIndex(?x)Elevation(?x, ?y), greaterThan(?y, 2) -> Height(?x) Brightness(?x, ?y), greaterThan(?y, 1500) -> BrightLight(?x)RectangularFit(?x, ?y), lessThanOrEqual(?y, 0.4) -> Nonregular(?x)Length_Width(?x, ?y), lessThanOrEqual(?y, 2.0) -> Planar(?x)
The formula means that when RectangularFit is greater than 0.78, the object is Regular, when it less than or equal 0.4, the object is Nonregular; when Brightness is greater than 1500.0, the object is BrightLight, when it less than or equal 1000.0, the object is DarkLight; when Elevation is greater than 2.0, the object is Height, when it less than or equal 2.0, the object is Low; when Length_Width is greater than 2.0, the object is Strip, when it less than or equal 2.0, the object is Planar.

3.3.2. Expert rule expression. DarkLight(?x), Low(?x), Strip(?x) -> Water(?x) Height(?x), Planar(?x), Regular(?x) -> Building(?x)Low(?x), Planar(?x), VegetationIndex(?x) -> Vegetation(?x)BrightLight(?x), Low(?X), Planar(?x), Nonregular(?x) -> BareLand(?x)BrightLight(?x),Low(?x), Strip(?x) , Regular(?x)-> Road(?x)Where C (? X) represents? X is a C class individuals, P (? X,? Y) represents the attribute, x, y is a variable.We use Road as example that the object has the nature of bright, low strip which should be seen as road.

4. Geographic ontology model of remote sensing image

4.1. Experimental data
The experiment selects DaFeng City as the study area. Dafeng City is a country-level city of Yancheng City, which is located in south of Yancheng , near to Yellow Sea, land area is 3059 square kilometers. DaFeng is the coastal plain, the city ground elevation is 1.9-4.5 meters, the southern part of the terrain is wide and the north is narrow. In this experiment, the flight area of Dafeng City things strip test, in order to take full advantage of Lidar data, we could select areas with relatively more buildings as a test area. The geographical location of the experimental area is shown in Figure 2.

Figure.2 The geographical position of study region
The data preprocessing mainly includes the registration of point cloud data preprocessing, spectral image preprocessing and image and nDSM (DSM Normalized, normalized digital surface model) registration. The experimental study area includes a total of 8 flights, and there are more than 56% of the overlap between the bands. The first we need constuct LIDAR point cloud bands of geometric correction and mosaic, then the point cloud of singular points elimination, the last in terrasolid software we filter point cloud to generate DTM (digital terrain model and digital terrain model) and DSM (digital surface model), meanwhile, it is necessary to carry out geometric correction and image mosaic of the spectral image, which is mainly carried out under the ENVI software. In addition, we use generated DTM and DSM in the ENVI software to generate nDSM, and finally in the ENVI software to achieve the registration of images and nDSM.

4.2. Experimental result

In this experiment, we use the decision tree model and expert rule model to realize the classification of the features of the test area, the results is as shown in Figure3(a), Figure(b).

(a) The decision tree model results **(b)** The expert rules model results

Figure.3 The classification results

By using the confusion matrix we evaluate the accuracy of the three, the results of the evaluation are shown in table 1, table 2.

Table.1 The precision evaluation table of decision tree model

	Vegetation	Water	Road	BareLand	Building	Sum	Production accuracy
Vegetation	274	10	13	6	0	303	90.43
Water	2	34	0	0	0	36	94.44
Road	6	1	63	4	0	74	85.14
BareLand	0	1	4	40	0	45	88.89
Building	1	0	0	0	57	58	98.28
Sum	283	46	80	50	57	516	
User accuracy	96.82	73.92	78.75	80.00	100		
Overall accuracy =90.70%，Kappa =85.20%							

Table. 2 The precision evaluation table of expert rules model

	Vegetation	Water	Road	BareLand	Building	Sum	Production accuracy
Vegetation	273	6	8	5	0	292	93.49
Water	1	38	0	0	0	39	97.44
Road	3	0	69	4	0	76	90.79
BareLand	2	1	4	44	0	51	86.27
Building	1	0	0	0	57	58	98.28
Sum	280	45	81	53	57	516	
User accuracy	97.50	84.44	85.19	83.02	100		
Overall accuracy =93.22%，Kappa=89.27%							

According to the results of the classification of the experimental area, the classification results based on the geographic ontology model are significantly improved, especially in the classification of buildings, the use of Lidar data extraction of elevation information of the ontology model can be very good classification of the building, the user accuracy and production accuracy can reach more than 98%, the classification accuracy of the road, vegetation and water body is also higher than the traditional one, but there are some false classification, because the expert rule model is more complete in knowledge representation, the classification result is slightly better than the classification results of the decision tree model.

5. Conclusion

The paper focuses in the research of remote sensing image classification, for standardization and formalization expression of geographical entity unified cognition, solving the cognitive differences of geographical entity leading to inconsistencies of the classification results, we introduces the theory of geographic ontology, builds image data feature knowledge base, establishes geographic ontology model, implement feature elements conceptual and formal expression, research on the practical application of remote sensing images on the level of knowledge, carries out the experiment of remote sensing image classification based on geographic ontology. This method can objectively reflect the semantic information, and form the information of object elements, which provides an effective way for remote sensing image classification, and has a certain scientific and value.

- The paper could use the geographic ontology model to carry on the objective representation of the elements of the remote sensing image, realize the effective organization and utilization of the characteristic information of the image, express the semantic relations with the computer visual language, the objective nature of the information of object elements can be reflected more accurately.
- In this paper, OWL and SWRL are used to realize the decision tree rule modeling and expert rule modeling, to a certain extent, to achieve the classification of remote sensing image, but the relationship between the expression of object information is still flawed, in order to make up the deficiency of the existing model, more new data mining methods are introduced in the following research, at the same time, the establishment of expert knowledge of regional features, improve the expert knowledge, and gradually realize the model of universal.
- In this paper, the application of knowledge is more dependent on the image of the characteristics of knowledge, the image of the context knowledge, knowledge of the time, the geographical knowledge of the study area and the spatial relationship of knowledge related to the less, In the follow-up study, it is attempted to apply these knowledge to the framework of geographic ontology model, and to improve the construction of the model.

References

[1] Wang hong. Design and implementation of geographic information ontology based on domain knowledge[D]. WuHan University,2006.

[2] Huang yongqi. Research on spatial temporal query of agricultural geographic information based on geographic ontology and SWRL spatial temporal reasoning rules[D]. Institute of remote sensing applications, Chinese Academy of Sciences,2008.

[3] Hunag maojun, Du qingyun, Wu yunchao, Li fengdan. A preliminary study on geographic ontology and its application[J].Geography and Geographic Information Science ,2004,20(4),1-5.

[4] Jing dongsheng. Research on ontology based semantic representation and service of geospatial information[D]. Institute of remote sensing applications, Chinese Academy of Sciences,2005.

[5] Zhang pu. Study on DEM extraction method of the typical landscape of the Loess Plateau driven by Ontology[D]. Lanzhou University,2013.

[6] Mariana Belgiu,Ivan Tomljenovic,Thomas J.Lampoltsshammer,Thomas Blaschke. Ontology-Based classification of building types Detected from Airborne Laser Scanning Data[J].Remote Sensing,2014(6):1347-1366.

[7] G.Forestier,C.Wemmert,A.Pussant.Costal image interpretation using background knowledge and semantics[J].Computer&Geosciences,2013(54):88-96

[8] Zhang ying. Research on Geographic Ontology -- Research Progress and Application[J]. Standardization of Surveying and mapping, 2014,30(2):24-27.

[9] Li bin, Liu bin, Liu jiping. Review and Prospect of research on Geographic Ontology[J]. Science of Surveying and Mapping, 2015,40(4):53-57.

[10] Huang maojun. The formal representation mechanism of geographic ontology and its application in map service[D]. WuHan University,2005.

Acknowledgments

The paper was jointly supported by foundation support: Project of the National Natural Science Foundation of China(Item Number:41371406)

The Extraction of Altered Minerals by use of Hyperspectral and Aster Data and Their Comparison Research in Beketan, Altyn

H Yi*, J Q Li, G L Ren, M Yang, H H Han, T Gao and J L Yang

Xi'an Center of Geological Survey, Geological Survey of China, Xi'an, China

E-mail: joy200299@163.com

Abstract. Hyperspectral remote sensing technology has played an important role in the field of geological application in recent years. Based on the reflectance of typical altered minerals and the spectral characteristics of rocks, we tried to identify and extract the altered minerals such as calcite, dolomite, epidote, chlorite, moscovite, limonite, hematite, etc. by using of CASI/SASI data in Beketan, Altyn. Moreover, we extracted Mg-OH/CO_3^{2-}/Al-OH/Fe^{3+}/Ferric oxide from Aster data by the ratio method. Combined with known geological background and the field investigation, we found that the extracted altered mineral information was consistent with the actual distribution of the field. The altered mineral information extracted by Hyperspectral data dispalyed more mineral distribution in details. It showed that the hyperspectral remote sensing technology could provide accurate hydrothermal altered mineral information which is helpful for geological mapping work.

1. Introduction

Hyperspectral remote sensing technology has been widely used in geological, mineral resources, and related environment survey because of its high spectral resolution and high spatial resolution since 1980s [1-3]. Extracting alteration minerals and identifying lithology by remote sensing spectral information is very important in remote sensing geology field. In this paper, we extracted the altered minerals information and Fe^{3+}/Al-OH/Mg-OH/CO_3^{2-} by Hyperspectral data and Aster data. Accorded with the geological background, we evaluated the result of different data.

2. Data Parameters and their preprocessing

The hyperspectral data used in this paper was acquired by airborne hyperspectral imager CASI/SASI and Table 1 presents its Parameters [4]. Data acquisition time: July to August 2012; time of flight: 11 to 16; the weather: sunny, low cloud cover, visibility > 20 km.

High spatial resolution (1m) DEM data which was obtaining by laser radar are applied to CASI/SASI data Ortho-rectification, and then the hyperspectral data of different strips were assembled and carried out geometric correction. We superposed CASI and SASI together, with all the CASI bands reserved while the bands of SASI which had the same range of CASI's or were deeply influenced by water vapor were removed. After the atmospheric correction of overlay data, we got the high spectrum data that has been processed [5].

The Aster data has eight bands whose wavelength from 0.52- 2.43 μm. Its acquisition time: October 2016; area cloud cover pct: <1%. We reprocessed the aster data such as removing its border, inversing the reflectivity, removing the interference features and carrying out geometric correction, etc.

After that, the Aster data can be used to extract alteration information.

Table 1 Parameters of instruments used for the hyperspectral image data

	CASI	SASI
Band range	380-1050 nm	950-2450 nm
Number of pixels per line	1470	640
Continuous spectral channel	288	100
Spectral bandwidth	2.3 nm	15 nm
Frame rate	14	100
Spatial resolution	0.9 m	2.2 m

3. Geology Background

Figure 1 shows the geology environment in Beketan, Altyn. The exposed stratas are Jixianian Period Jinyanshan Group (Jxj) and Cambrian - Ordovician Hongliugou Ophiolite (∈ Ohm). The former straturm contains calcirudite(cal), dolomitic marble(dolmb), sandstone(ss) while the other consists of schist(sch), phyllite(ph), limestone(ls), siliceous rock(si). In this region, the fault structure develops well, they are mainly distributed in the nearly east- west and northeast directions. There are also intensive Magmatism, and the magmatic rocks are mainly ultrabasic rocks, diabase, and granite which were formed in the Caledonian period.

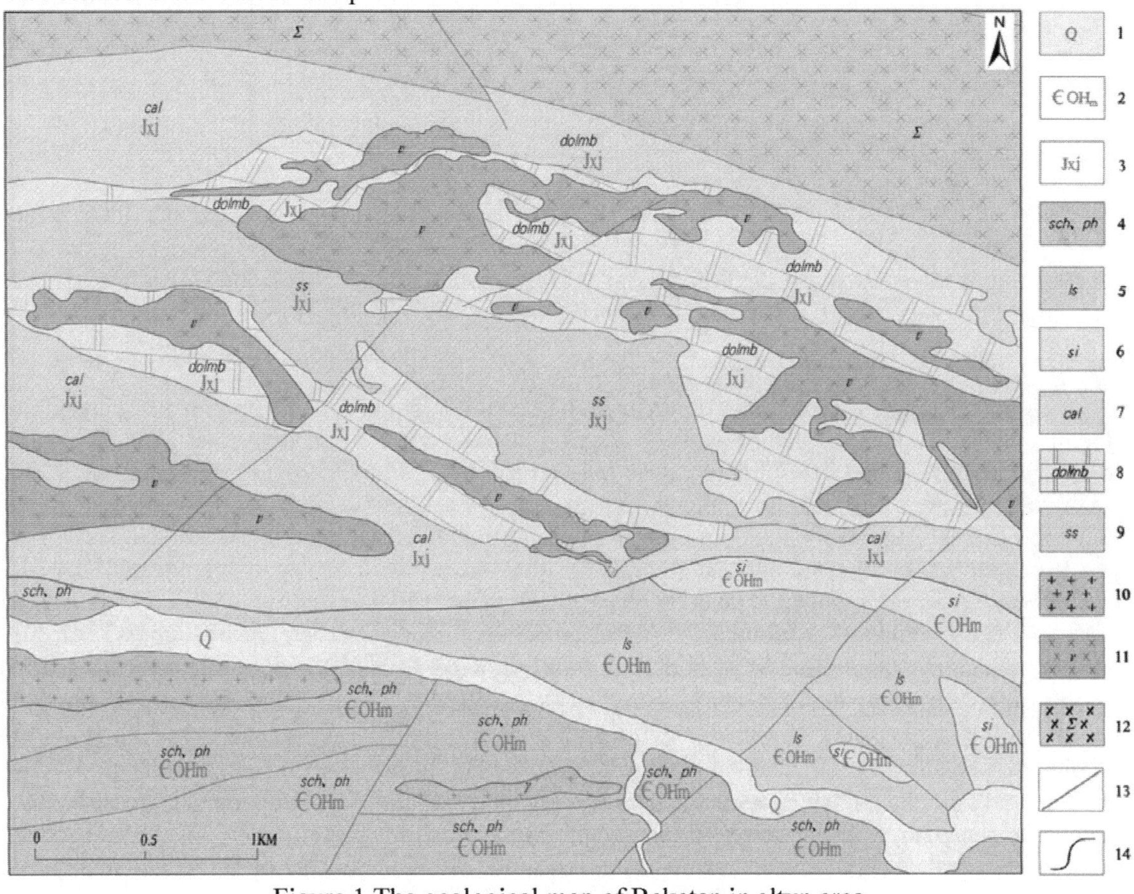

Figure 1 The geological map of Beketan in altyn area.

1.Quaternary; 2.Hongliugou Ophiolite; 3.Jinyanshan Group; 4.schist\phyllite; 5.limestone; 6.siliceous rock; 7.calcirudite; 8.dolomitic marble; 9.sandstone; 10.granite;11.diabase; 12.ultrabasic rocks; 13.fault; 14.geological boundary.

4. Alteration mineral information extraction

4.1. Spectral characteristics of typical altered minerals

Table 2 shows different spectrum curves of typical alteration minerals in USGS spectra library [6]. The mineral spectra of Mg-OH and carbonate are characterized by the maximum absorption spectrum bands of 2300 ~ 2400 nm, which is also a significant sign for the identification of Mg-OH containing minerals [7-8]. Its representative minerals are calcite, dolomite, epidote, chlorite, etc. The recognition spectrum bands of minerals containing Al-OH is located between 2165 ~ 2215nm, and the representative minerals are muscovite and illite. The identification spectrum bands of minerals containing Fe^{3+} is located near 600 ~ 900 nm, whose representative minerals are limonite, hematite, etc.

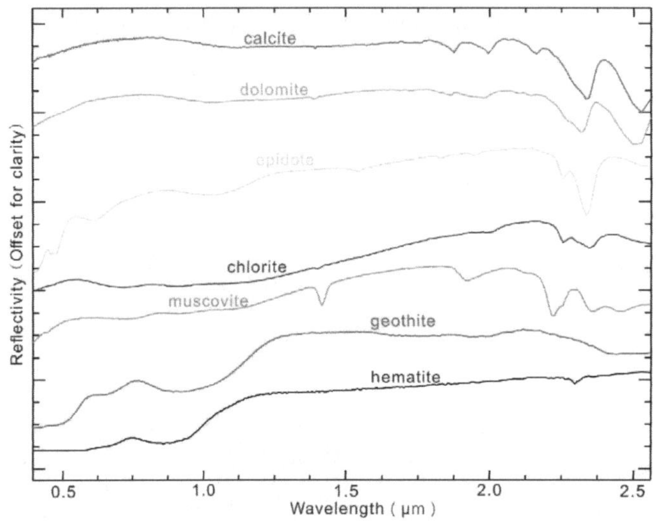

Figure 2 Typical alteration minerals spectral curves in USGS Spectral library.

4.2. Spectral characteristics of typical altered minerals

The alteration mineral information was extracted according to the absorption spectrum bands of different minerals. For Aster data, (band 7 + band 9) / band 8 was used to extract Mg-OH / CO_3^{2-} information, (band 5+band 7)/band 8 to extract Al-OH information, band 2/band 1 to extract Fe^{3+} and band 4/band 3 to extract ferric oxides information.

Figure 3 (a)(c)(e) shows the ionic groups anormaly information extracted by Aster data, and Figure 3 (b)(d)(e) presents corresponding altered mineral information extracted by hyperspectral data. Mg-OH/ CO_3^{2-} anomalies in Figure 3 (a) mainly show planar or banded distribution in dolomitic marble of Jxj strata, a few anomalies distributed in schist and phyllite of ∈Ohm strata. In Figure 3 (b), four kinds of minerals that containing Mg-OH or CO_3^{2-} are extracted. Calcite and dolomite basically exhibit the same distribution of dolomite marble of Jxj strata. Besides, chlorite and epidote are not only distributed in the schist and phyllite of ∈Ohm strata but also in the dolomite marble of Jxj strata. As shown in Figure 3 (c), Al-OH anomalies present planar distribution in the schist and phyllite of ∈ Ohm strata and granite. A few scattered points corresponding to the Al-OH anomalies are distributed in the white image area of the dolomite marble of Jxj strata. From Figure 3 (d), high-Al muscovite mainly present scattered point or banded distribution in the dolomite marble of Jxj strata. While low-Al muscovite mostly presents planar distribution in the schist and phyllite of ∈Ohm strata and granite, however, a few anomalies are strip distributed in the dolomite marble of Jxj strata, which is very agreement with that extracted by Aster. It shows that Fe^{3+} anormalies in Figure 3 (e) exhibit planar and strip distribution in the dolomite marble of Jxj strata. Ferric oxide anormalies present scatter point distribution in the schist and phyllite of ∈ Ohm strata and granite. Limonite/hematite are mainly distributed in the Jxj strata with the most widely development in the dolomitic marble, followed by in the calcirudite (see Figure 3(f)).

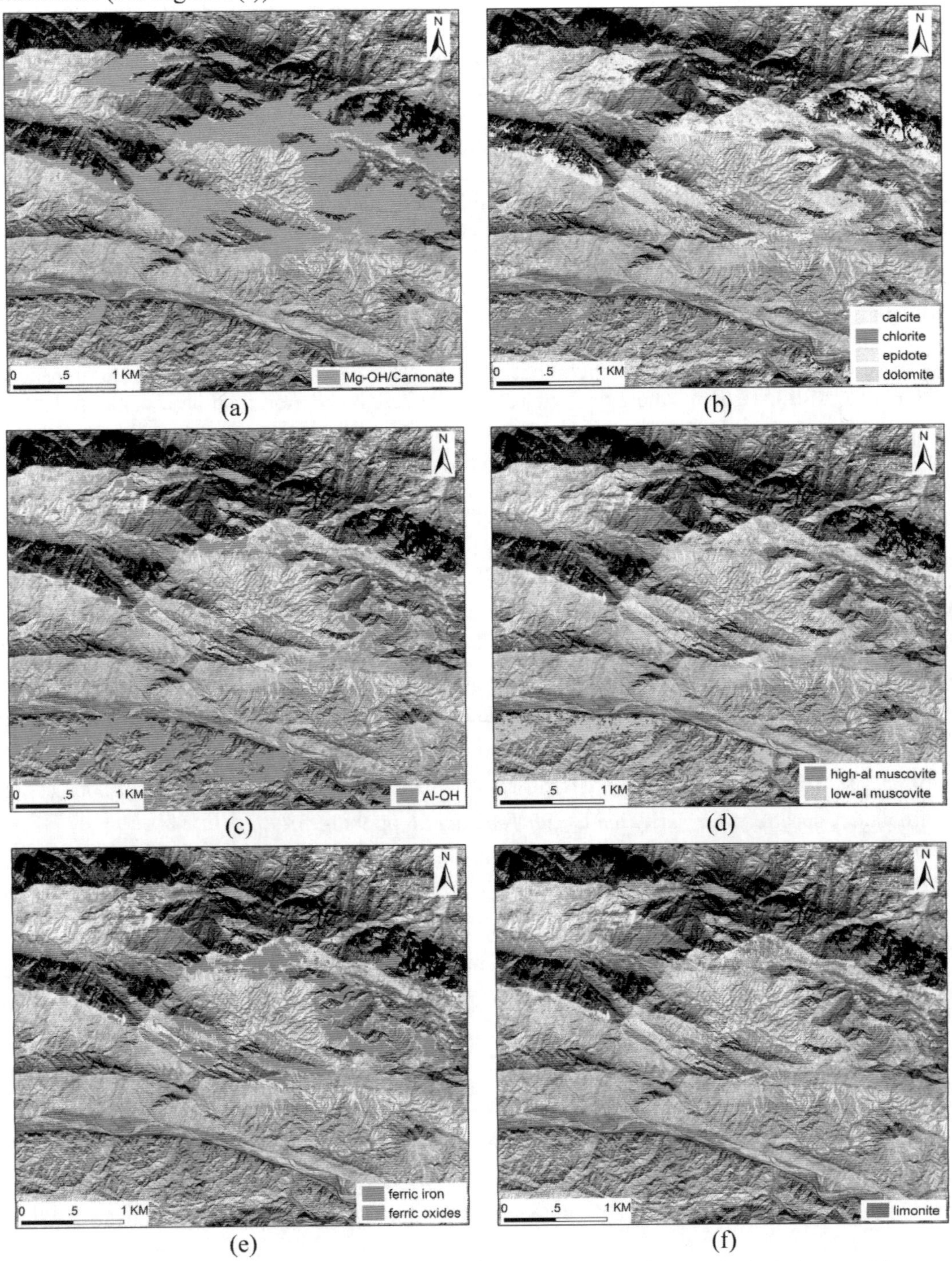

Figure 3 The distribution of Alterated mineral/ionic groups abnormaly in Beketan, Altyn.

5. Conclusions

Based on the analysis of the reflectance spectrum curve features of the typical altered minerals in USGS spectra library, we extracted the altered minerals such as calcite, dolomite, epidote, chlorite, moscovite, limonite, hematite, etc. by using of CASI/SASI data in Beketan, Altyn. Moreover, we extracted Mg-OH/ CO_3^{2-}/ Al-OH/ Fe^{3+}/ Ferric oxide from Aster data by the ratio method.

Through the analysis of known geological background, Mg-OH/ CO_3^{2-} anormalies exhibit the same distribution of carbonate stratum. In the corresponding hyperspectral extraction results, calcite, dolomite, and Mg-OH/CO_3^{2-} anormalies are similarly distributed. Epidote and chlorite are mainly distributed in the carbonate stratum and metamorphic strata. Al-OH anormalies display the same distribution as the metamorphic strata and granite body. Local Al-OH anormalies also developed in carbonate stratum. Low-Al moscovite show the similar anormaly distribution as the Al-OH anormalies. However, high-Al moscovite are mostly developed in the carbonate stratum. Fe^{3+} anormalies are mainly distributed in the carbonate stratum. Ferric oxide anormalies are mainly distributed in the metamorphic rock and granite. Limonite/hematite presents the similar distribution as Fe^{3+} anormalies.

This study indicates that based on the reflectance spectrum curve features, we extracted Aster remote sensing anormaly and the altered minerals distribution information of CASI/SASI data, which is very agreement with actual geological background. Because of the advantage of its spectrum resolution of CASI/SASI data, the extracted results are relatively more accurate compared with the traditional method, which is playing an important role in the geological mapping work.

Acknowledgement

Financial support was provided by China geological survey (No. 12120113044900). We thank CNNC Beijing Research Institute of Uranium Geology for providing CASI/SASI data.

References:

[1] Li Z Z, Yang R H, Dang F X, Zhang X F, Tan B X, Zhao H J 2009 The hyperspectral remote sensing technology and its application *Geological Bulletin of China* **28** pp 270-277

[2] Zhang Z C, Zhang X J, Hu D G 2011 Application of Hyperspectral Remote Sensing on Mineral Exploration in Dongdatan District of East Kunlun, Qinghai Province *Geoscience* **25** pp 759-767

[3] Wang K, Gu X F, Yu T 2013 Classification of hyperspectral remote sensing images using frequency spectrum similarity *Sci China Tech Sci* **56** pp 980-988

[4] Ren G L, Yang J L, Yang M 2013 Application of Hyperspectral Remote Sensing Anomaly Information on Metallogenic Prediction in the Jintanzi-Mingjingou Area of Beishan, Gansu Province *Geotectonica et Metallogenia* **37** pp 765-776

[5] Yang Y J, Zhao Y J, Qin K 2013 Application of Hyperspectral Remote Sensing Anomaly Information on Metallogenic Prediction in the Jintanzi-Mingjingou Area of Beishan, Gansu Province *Science and Technology Review* **31** pp 65-67

[6] Li M M, Xing L X, Pan J 2011 Research of Combinatory Analysis Method in Altered Information Extraction *Remote Sensing Techonology and Applycation* **26** pp 303-308

[7] Lv F J, Hao Y S, Shi J 2009 Alteration Remote Sensing Anomaly Extraction Based on Aster Remote Sensing Data *Acta Geoscientica Sinica* **30** pp 271-276

[8] Gan F P, Wang R S, Ma A N 2003 Integration for Extracting and Mineral Analysis Models for Geological Application Using Remote Sensing Data *Journal of Remote Sensing* **7** pp 207-214

An improved SIFT algorithm in the application of close-range Stereo image matching

Xuehua Zhang[1], Xiaoqing Wang[1], Xiaoxiang Yuan[1], Shumin Wang[1]

[1] Institute of Earthquake Science, China Earthquake Administration, Beijing, China 100036

E-mail: zhangxhua1990@163.com wangxiaoq517@163.com yuanxx1983@163.com wmcnu@163.com

Abstract. As unmanned aerial vehicle (UAV) remote sensing is applied in small area aerial photogrammetry surveying, disaster monitoring and emergency command, 3D urban construction and other fields, the image processing of UAV has become a hot topic in current research. The precise matching of UAV image is a key problem, which affects the subsequent processing precision directly, such as 3D reconstruction and automatic aerial triangulation, etc. At present, SIFT (Scale Invariant Feature Transform) algorithm proposed by DAVID G. LOWE as the main method is, is widely used in image matching, since its strong stability to image rotation, shift, scaling, and the change of illumination conditions. It has been successfully applied in target recognition, SFM (Structure from Motion), and many other fields. SIFT algorithm needs the colour images to be converted into grayscale images, detects extremum points under different scales and uses neighbourhood pixels to generate descriptor. As we all know that UAV images with rich colour information, the SIFT algorithm improved through combining with the image colour information in this paper, the experiments are conducted from matching efficiency and accuracy compared with the original SIFT algorithm. The results show that the method which proposed in this paper decreases on the efficiency, but is improved on the precision and provides a basis choice for matching method.

1. Introduction

Unmanned aerial vehicle (UAV) is a kind of integrated the technologies of UAV remote sensing system, flight control system, remote sensing and GPS navigation and positioning, and other new technology. Because its characteristics, such as high efficiency, less cost and low labour intensity, increasingly cause the attention of many fields. UAV remote sensing has been widely applied in precision agriculture [1, 2], fire detection [3] and characteristics of forest [4, 5], especially in the building 3D modelling of digital city [6]. In the process of images processing, the corresponding points matching is a key problem, as its result will directly affect the final subsequent precision, such as image matching, 3D reconstruction and automatic aerial triangulation[7]. Now in order to solve this problem there are two main methods: corresponding points matching based on grey correlation and corresponding point matching based on features. Based on grey correlation matching method has the advantages in calculation , but the accuracy are greatly influenced by SNR(Signal-to-Noise Ratio) of window, and there will be a low matching success rate if the grey level information is poor or SNR is small, and it is difficult to guarantee the stability of the matching accuracy[8]. By contrast, the method of based on the features have a stable matching precision, Mikolajczyk compared some representative feature matching algorithms , such as moment invariants, SIFT, etc., through the experiments and analysis of the pictures that in the cases of image geometric changes, compression and illumination change and so on, the SIFT algorithm show the best extracting result [9]. SIFT algorithm has strong stability to image scaling, rotation, shade, light and other conditions change [10], and has been successfully applied in target recognition, SFM (Structure from Motion), and many other fields. But the SIFT algorithm is based on image grey value to seek extreme value point as characteristics points and convert colour images to grayscale images, not use the rich colour information of the UAV images. Many scholars to research the SIFT algorithm in order to improve the precision, such as Jiajia Xu in combination with Harris operator, this paper proposes a fast images matching based on improved Harris - SIFT operator method [11]; Ping Xie proposed the method to improve the matching accuracy through improving the corresponding points matching[12]. Therefore, this paper will combine the images colour information to improve the SIFT algorithm. Experimental results show that the method in matching efficiency is slightly lower, but the precision is better than the original SIFT algorithm.

2. Extracting and matching feature points

2.1. Feature points extraction

The number and the precision of the control points eventually used to matching determines the accuracy of images matching [13]. For the characteristics of the UAV images such as high resolution, flight attitude instability and different light conditions, this paper uses the SIFT algorithm to detect the points. The first step is to convert colour images to grey images when the feature points are extracted, grey images are sampled fall under the different levels of images structure. In order to retain the image colour information of original image, the same degree of reducing the sampling method is adopted in this paper, so that we can get feature points corresponding to the colour information. Then using a different scale factor of the Gaussian function to each layer (Eq. 1), forming the pyramid structure under different scale space (σ). This method simulates the human eye observation from coarse to fine scenery, the outline to the details of mode of observation, to satisfy the visual invariance. In order to more effectively detect the stability of the extreme value points, we use the pixel value of adjacent two layers that after filtering to subtract to get difference of Gaussians pyramid images (DoG)(Eq. 2). Extract the extreme value points of stability of the DoG scale space, must compare the pixel value between each pixel point and all of its adjacent points, including the same scales around eight points and adjacent scale up and down in the corresponding 9 x 2 points, ensure that point for the image space and scale space under the local minimum or maximum (as shown in Fig. 1). At the same time, record the extreme value point, scale, and the coordinates of the pyramid.

$$L(x,y,\sigma)=F(x,y,\sigma)*I(x,y) \quad (1)$$

Among: (x, y) —— the coordinates of the pixel

$$F(x,y,\sigma)=\frac{1}{2\pi\sigma^2}e^{-(x^2+y^2)/2\sigma^2}$$ —— Gaussian blur function

$$D(x,y,\sigma)=L(x,y,k\sigma)-L(x,y,\sigma) \quad (2)$$

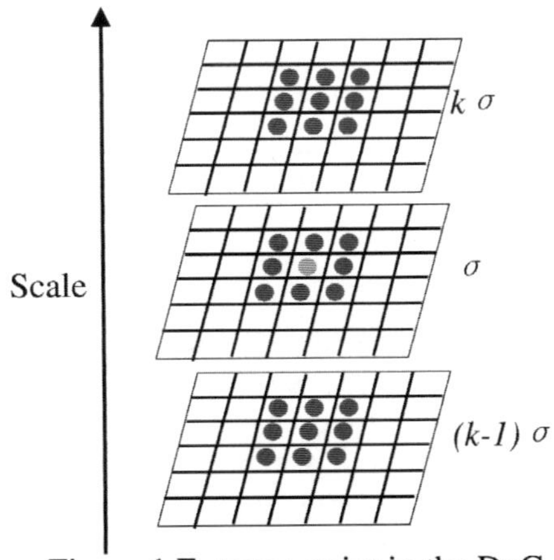

Figure 1 Extreme point in the DoG

Then get the extreme value point location is in discrete space under the rough position. Lowe David using three dimensional quadratic function to correct the point's position. In order to remove the extreme points of low contrast, we set a threshold value that the point's position correction is greater than the threshold will be deleted. In addition, DoG operator will produce strong edge effect, and need to use the eigenvalues of the Hessian matrix ratio to weed out the points which edge is not stable. The Hessian matrix of the H (Eq. 3), maximum and minimum eigenvalue is α、β respectively,**and** $\alpha = r\beta$, Tr (*H*) (trace of *H*, Eq. 4), Det (*H*) (determinant of *H*, Eq. 5):

$$H = \begin{bmatrix} D_{xx} & D_{xy} \\ D_{xy} & D_{yy} \end{bmatrix} \tag{3}$$

$$\mathrm{Tr}(H) = D_{xx} + D_{yy} = \alpha + \beta \tag{4}$$

$$\mathrm{Det}(H) = D_{xx}D_{yy} - D_{xx}^{\ 2} = \alpha\beta \tag{5}$$

$$\frac{\mathrm{Tr}(H)^2}{\mathrm{Det}(H)} = \frac{(\alpha+\beta)^2}{\alpha\beta} = \frac{(r+1)^2}{r} \tag{6}$$

Set a threshold value, if the ratio is greater than the threshold is determined that point as the edge points, delete them. After the above steps extreme value points of quality was improved, identified as the final feature points.

2.2. To determine the main direction

In order to guarantee the algorithm of rotation invariance, we need to rotate selected feature points and its local characteristics, make sure all feature points generated by the descriptor in the same direction vector. Different feature points with different local grey-level gradient distribution, so it can be determined a principal direction as the direction of the stability of local structure according to gradient distribution around the point. In the first step we have determined the feature points in the DOG pyramid, to collect all the pixels value which located the Gaussian pyramid image within 3 σ neighbourhood window of the point, press type calculation of the gradient of each pixel (direction Fig. 8 and amplitude Fig. 7).

$$m(x,y) = \sqrt{\begin{matrix}[(x+1,y) - L(x-1,y)]^2 + \\ [L(x,y+1) - L(x,y-1)]^2\end{matrix}} \tag{7}$$

$$\theta(x,y) = tan^{-1}[\frac{L(x,y+1) - L(x,y-1)}{L(x+1,y) - L(x-1,y)}] \tag{8}$$

IOP Conf. Series: Earth and Environmental Science **46** (2016) 012009 doi:10.1088/1755-1315/46/1/012009

After computing feature points and neighbourhood of gradient, using histogram method to statistics. Gradient will be divided by direction in the histogram graph, the interval is 10 degrees, a total of 36 columns. Histogram peak direction represents the main direction of the point. Lowe David thought only keep peak is greater than the principal direction of the direction of the peak of 80% as the auxiliary direction, this can set up multiple different directions at the same location and scale of feature points to increase the stability of the match.

2.3. Generate the descriptor
After the above processing each feature points have three information: position, scale and direction. SIFT algorithm is to establish a set of feature points vector as a descriptor, including this point and its surrounding point's features. Lowe David was used the feature points in the scale space of 4 * 4 window to calculate eight direction of gradient information, a total of 4 * 4 * 8 = 128 dimensional, and proved that the 4 * 4 window as the best detection window. This paper uses Lowe David's conclusion, but need to extract RGB information according to the feature points location , which makes the 128 dimensional descriptor increased to131 dimensional, makes the feature points matching is not only has the characteristics of the original SIFT algorithm, but considered the rich colour information of the images, finally get a stable matching results.

3. Experimental analysis

3.1. Experimental data
This paper used four rotorcraft UAV for data acquisition, flight radius about 5 km, the SONY A6000 16-50 mm lens, the image below for the UAV.

Figure 2 The photo of the UAV

Its wheelbase diagonal motor is 105cm, there are many types of flight mode, including artificial remote autonomous hover, independent route and go around the points of interest, the biggest task load can reach 8kg, maximum take-off weight of 14kg, battery life varies according to the load, when the load is zero it can flying 90 minutes. The cruising speed for 3-50km/h, 6 class wind resistant effect. Navigation method have three modes: GPS navigation, beidou navigation and combine GPS with beidou navigation.

The data acquisition area is Beizhen city, Liaoning province. The flight course overlap rate can reach 74%, side overlap rate was 40%, and the average speed is about 15m/s. We get a series of high quality images because the weather is good and no wind. The aircraft sorties flying height is about 120m, the images spatial resolution is 0.02m. The images like Fig. 3.

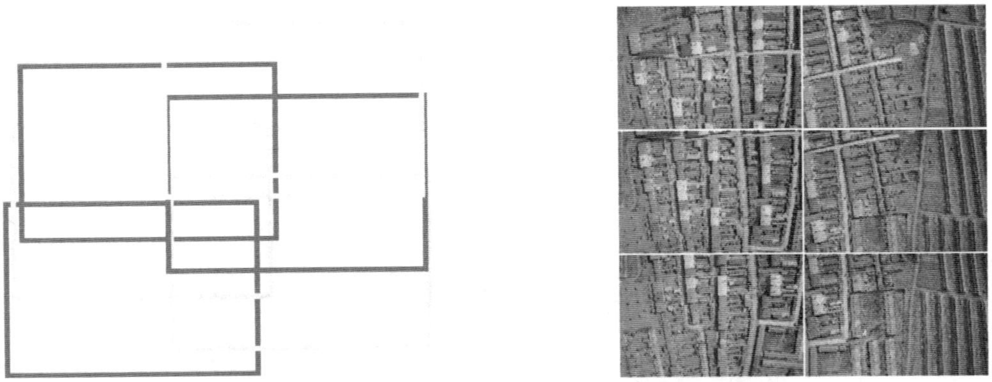

Figure 3 the sample pictures and range that we acquired

3.2. Experimental results and analysis

In this paper, we will use VS2010 compiler as a platform, with the aid of using Opencv2.4.10 function library to implement the traditional SIFT matching algorithm and SIFT matching algorithm with colour information. Purpose research discussed with colour information SIFT and effectiveness and efficiency of the traditional SIFT algorithm, therefore not use all flight images. By using two images to contrast the differences of the two methods, in the aspect of time may not be much difference, therefore, decided to use the ratio of time as a measure. Used in the experiment process parameters are considered as the best value in the Lowe David's experiments.

Through use of the two images that has the overlap area of UAV images as experiment analysis data. Below pictures are the effect pictures respectively by using traditional SIFT method and the SIFT which combined with colour information and corresponding points matching. In Fig. 4 is the images that used a traditional method of SIFT under the different scales of Gaussian pyramid and DoG. From the pictures we can see the different objects in different layers of the pyramid, conforms to the human eye to observe the features of the object. In DoG, image feature points and line is more apparent and can be good to extract the feature points in the hierarchy. The extraction of feature points set is used in the original method of matching eventually and get the matching results of the method. By the method of computing the match to get 5219 points, with the same consistency by using random sampling algorithm (RANSAC) resulting from the processing of the same number of points to 3298, this experiment shares 26.284 seconds.

Figure 4 The Gaussians and the DoG

On the basis of using OpenCv function library to improve the program, through using the function in the library's to generate colour description operator to get the corresponding points, a total of 4805(Fig. 5), after removal of gross error and two-way matching strategy received 3547 valid points, a total of 46.204 seconds.

Figure 5 Key Feature Points of the two Pictures

Through the experiment we can get the comparison of the accuracy and efficiency of two methods, experimental results are integrated in the table below for two methods.

Table 1 Comparison of two methods

method	Total corresponding points	Effective corresponding points	Elapsed time (s)
Traditional SIFT	5219	3298	26. 284
SIFT which combined with colour information	4805	3547	46. 204

From the above statistics can be seen that the improved SIFT algorithm have more effective corresponding points, and the ratio is 63.19% for total corresponding points. The effective points of the traditional SIFT algorithm is less, the ratio is 73.82%. For the same images for both effective points was similar, with the ratio is about 0.93. In terms of the efficiency of the improved SIFT algorithm is more elapsed time, it is about 1.7 times of the former, the reason is that with the colour information of the SIFT algorithm in matching dimension is more, and it is also possible that has certain influence in the program optimization.

Comprehensive the above analysis when using SIFT algorithm for UAV images matching, we can according the need to choose a different method. When to process a large amount of data, we can give priority to efficiency to choose the traditional matching SIFT algorithm. When the amount of data is small, we can consider the precision priority, SIFT algorithm with colour information is a good choice.

4. Conclusion

In this paper, we represented the traditional SIFT algorithm and the colour information SIFT algorithm, using the same UAV images to perform experiment and analyses, and found out that the colour information of SIFT algorithm is higher effective corresponding points ratio than the traditional SIFT algorithm, but lacking in efficiency, and time consuming in this experiment is about 1.7 times that of traditional SIFT algorithm, especially in a large amount of data. The reason of time consuming, on the one hand is the dimensions of the colour information of SIFT algorithm is higher when corresponding points matching, on the other hand may be the program optimization problem. This experiment will make a certain guiding role for UAV image matching using the SIFT algorithm.

Acknowledgments

The research was supported by the Fundamental Research Funds of Institute of Earthquake Science, China Earthquake Administration (Grant No. 2013IES0204).

References

[1] Valente, J. A., Sanz, D., Del Cerro, J., Barrientos, A. and de Frutos, M., 2013. Near-optimal coverage trajectories for image mosaicking using a mini quad-rotor over irregular-shaped fields [J]. *Precision Agriculture* 14, pp. 115–132.

[2] Zhang, C. and Kovacs, J., 2012. The application of small unmanned aerial systems for precision agriculture: a review [J]. *Precision Agriculture* 13, pp. 693–712.

[3] Merino, L., Caballero, F., Martnez-de Dios, J., Maza, I. and Ollero, A., 2012. An unmanned aircraft system for automatic forest fire monitoring and measurement [J]. *Journal of Intelligent & Robotic Systems* 65, pp. 533–548.

[4] Tao, W., Lei, Y. and Mooney, P., 2011. Dense point cloud extraction from UAV captured images in forest area. In: Spatial Data Mining and Geographical Knowledge Services [C], *2011 IEEE International Conference*, pp. 389 –392.

[5] Wallace, L. O., Lucieer, A. and Watson, C. S., 2012. Assessing the feasibility of UAV-based Lidar for high resolution forest change detection [C]. *ISPRS International Archives of the Photogrammetry, Remote Sensing and Spatial Information Sciences*, pp. 499–504.

[6] Tang min, Li yongshu, Lu heng. 2012. Correspondence Points Matching of UAV Images [J]. *Opto-Electronic Engineering*. 39(3): 19-24.

[7] Liu you. 2012. Research on Matching and Mosaicking Methods for UAV Image Sequence [D]. CENTRAL SOUTH UNICERSITY. pp3.

[8] Liu qingyuan, et, al. 2010. Research on key technology of digital photogrammetry based on UAV remote sensing image [J]. *Science of Surveying and Mapping*. 35(1): 28-30.

[9] Mikolajczyk K, Schmid C. Scale & Affine Invariant Interest Point Detectors [J].2004. *International Journal of Computer Vision*, 60 (1):63-86.

[10] Lowe David G. 2004. Distinctive Image Features from Scale-Invariant Keypoints [J]. *International Journal of Computer Vision*. 2(60): 91-110.

[11] Xu jiajia, Zhang ye, Zhang he. 2015. Fast image registration algorithm based on improved Harris-SIFT descriptor [J]. Journal of Electronic Measurement and Instrumentation. 1(29): 48-54.

[12] Xie ping, Zhou zhengrong, Xiao qi. 2012. A close-range image matching based on Harris corners and SIFT features [J]. Science of Surveying and Mapping, 37(4): 107-110.

[13] Schmid C , Mohr R , Bauckhage C. 2000. Evaluation of interest point detectors [J]. *International Journal of Computer Vision (S0920-5691)*, 37 (2): 151-172.

RADI's Airborne X-SAR with High Resolution: Performance, Characterization and Verification

T Shen[1], J Li[2], Z R Wang[2] and L Huang[1]

[1] Institute of Remote Sensing and Digital Earth, Chinese Academy of Sciences, Beijing, 100094, China.
[2] Beijing Institute of Radio Measurement, 100143, China

E-mail: shenting@radi.ac.cn

Abstract. X-SAR is an airborne multi-mode synthetic aperture radar (SAR) system with high-resolution, interferometer and full-polarization, developed by the Institute of Remote Sensing and Digital Earth (RADI), Chinese Academy of Sciences (CAS), funded by the CAS Large Research Infrastructures. Since 2009, the first developed stage of X-SAR system was successfully implemented to an operational SAR with high resolution (up to 0.5 meter). In May 2013, the imaging verification on flights test was carried out. The data calibration on the laboratory measurements were completed at the end of 2015. Many valuable results of imaging verification and data calibration have emphasized the quantitative microwave measurement capabilities. This paper presents the results of X-SAR system performance, characterization, optimization, and verification as carried out during the flight trials and laboratory measurement. The system performance and calibration parameters are presented such as transmitter amplitude accuracy, phase noise, system gain change with temperature variation, long-term radiometric stability. The imaging verification of the key performance parameters is discussed, including target-response function, target pairs discrimination, image noise and radiometric resolution. The example imagery of radiometric enhanced products for intensity change detection is also described.

1. Introduction

From 2009, RADI has implemented to develop an airborne multi-mode SAR system to investigate the potential for disaster areas of natural hazards such as debris flow, earthquakes and floods. At the end of 2015, the high resolution processing had been achieved to the high-quality multi-look calibrated X-SAR images, using a near real time on ground imaging processing based on GPU architecture. Now, the upgrading radar is being to realize the interferometer and full-polarization capabilities. X-SAR radar is an X-band SAR at a centre frequency of 9.0GHz with 360 MHz maximum bandwidth, housed and operated in CAS's Cessna Citation II jet. The radar was designed to the dual-channel received with the internal calibrator, which is convenient to configure the interferometer antenna pairs or the polarization switcher. This paper focuses on X-SAR performance, characterization and verification with high resolution, achieving by the flight test and laboratory measurement. Section 2 provides a general description of system overview, architecture and performances. Section 3 describes data calibration results of temperature characterization and long-term radiometric stability measured on the laboratory, while section 4 provides flight test imaging results of the geometric and radiometric measurements, meantime providing an example image of radiometric enhanced products of intensity change detection. Finally, section 5 concludes and presents an outlook on upgrade X-SAR.

2. System overview

X-SAR architecture is divided into modular units or subsystems with controlled interfaces. In this way, each unit can be updated or replaced with minimal need for modification in other units [1]. The six main units are: antenna and servo unit, high power transmitter (HPA), internal calibration unit, radio frequency (RF) subsystem, digital formatting subsystem and monitoring control subsystem. Main operating parameters at high resolution mode are listed in Table1.

Table 1. Main operating parameters at high resolution.

Parameters		Values	
Operating frequency		X (9.0 GHz)	
Antenna gain		>26 dB, HH polarization	
Bandwidth (MHz)	360	180	90
Pulse length(us)	25	41	41
Operating range(km)	19～23	25～33	31～45
Look angle	63°	70°	75°
Noise equivalent sigma zero(NESZ)		<-30dB@Max-swath	

HPA is based on a travelling wave tube amplifier (TWTA) with 4.5 KW/ 400W peak/ average transmitted power. Fig.1 shows the transmitter amplitude stability was less than 0.045 dB, the change values are stable over operational 4 hours and do not results in a measurable effect of the overall system performances.

In RF subsystem, the phase noise of local oscillator signal (7.8GHz) was below -111.65dBc/1kHz, to be minimized the lower phase noise. Meanwhile, the phase noise of transmitting signal (9GHz) was realized to below -102.9dBc/1kHz, which was exceeded -90dBc/1kHz for the design criteria.

In digital formatting subsystem, the sample of received signal data was performed by a 500MHz ADC with a 12-bit capacity, therefore the received dynamic range can be theoretical more than 72dB. Thus, the receiver gain setting is commanded based on an estimated terrain scattering values for each individual scene. No automatic gain control is implemented in X-SAR radar.

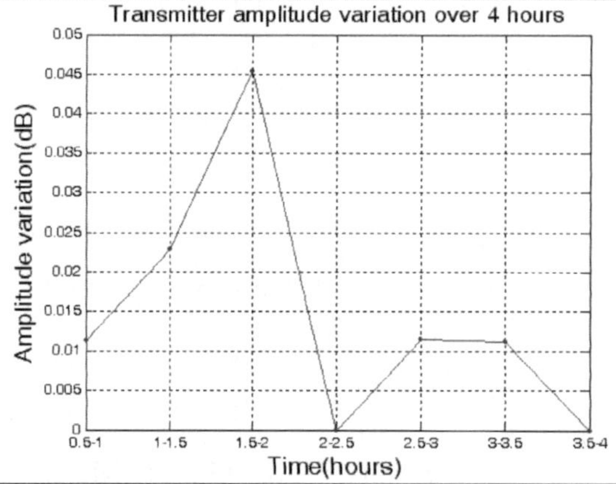

Figure 1. Transmitter amplitude variation over 4 hours.

3. Data calibration on laboratory measurements

X-SAR radar has been integrated a flexible internal calibrator which can be guaranteed the high radiometric stability of all data products, regardless of operated on board and on laboratory.

3.1. Temperature characterization

The different operational loads onboard are influenced by the cabin's temperature, which can lead to the total system characterizations dependence on a thermal. The radiometric characterizations with temperature variations can be used by the internal calibration in ground temperature test chamber.

The radiometric stability of transmitted gain (left) and system gain (right) are shown in Fig 2, the root mean square (RMS) values of transmitted gain drifts and system gain drifts are respectively below 0.26dB and 0.82dB for 0~40°C during data acquisition in high resolution up to 0.5 meter, it is no necessary to correct the amplitude change with temperature variations using on-ground characterization data, during data processing steps.

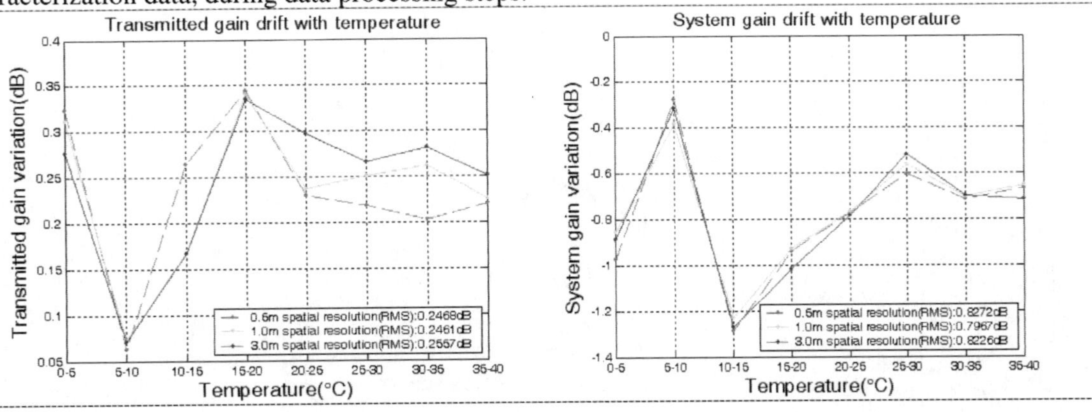

Figure 2. Transmitted gain drifts (left) and system gain drifts (right) with temperature variation.

3.2. *Radiometric stability characterization*

The long-term radiometric stability can be monitored by the evaluation of the internal calibration data [2]. Fig.3 describes the stability curve of system gain and system noise for long-term monitoring as a function of time throughout in 2014. The RMS values of system gain and system noise is respectively below 0.46 dB and 0.31 dB at 0.5 meter spatial resolution, the offsets can be convenient to corrected using on-ground internal calibration data, during data processing steps.

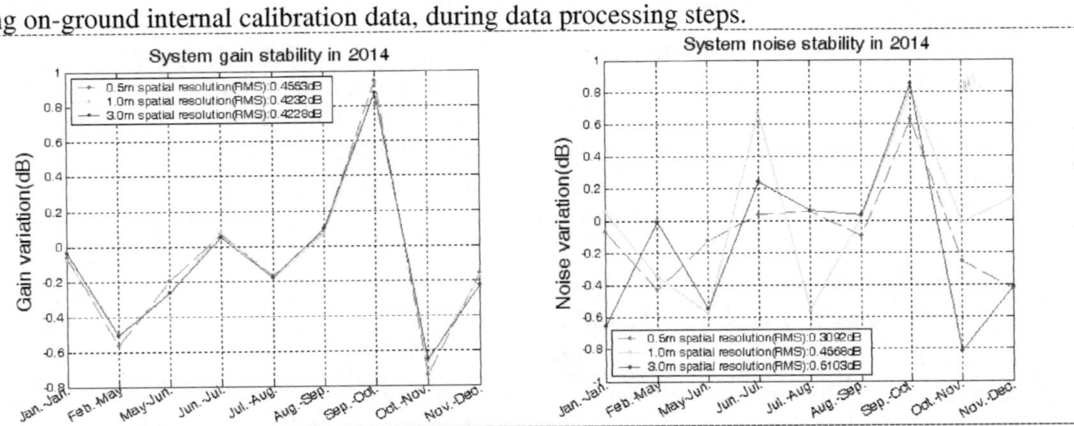

Figure 3. The long-term stability of system gain (left) and system noise (right) throughout in 2014.

4. Imaging verification on flight tests

In May 2013, X-SAR performed the first flight tests included data calibration and double side-looking strip imaging verification. We assess the geometric performance of SAR by the observed spatial resolution of corner reflectors using data collected over the calibration site [3]. Fig. 4 shows the X-SAR image (left) obtained during the test flights over the Binhai test sites, near Tianjin. The eight triangular corner reflectors marked by the red rectangular blocks, were deployed for the purposes like target resolution, radiometric and geometric calibration, located in low backscatter areas. Two pairs of rectangular corner reflectors marked by the green rectangular blocks were deployed for the target discrimination in range and azimuth directions.

Figure 4. X-SAR image (left) of the test flights over the Binhai test sites, near Tianjin. A Google Earth image inlay for comparison (right).

4.1. Spatial resolutions measurement and target pairs discrimination

Measured on eight triangular corner reflectors having the size of 0.3 meter corresponds to radar cross section (RCS) of 20.6dBm^2, the values of spatial resolutions in range and azimuth directions are listed in Table 2. The measured values of peak side lobe ratio (PLSR) and integral side lobe ratio (LSLR) are revealed the very low side lobe of the point target response function.

Table 2. Spatial resolutions measurement results.

Parameters	Value(Range × Azimuth)		
Spatial resolution (m)	0.47× 0.52	0.88×1.06	1.37×2.03
PSLR (dB)	-23.7×-22.7	-28.7×-26.2	-32.1×-27.5
LSLR (dB)	-23.1×-20.4	-24.8×-23.1	-23.8×-22.3

The discrimination for target pairs similar to optical line pairs test could be more directly evaluated of the SAR target resolution. Fig.5 describes the distinguished trend calculated curve for spacing of 1.0 meter in range and azimuth directions with target pairs, measured on the SAR image with 0.5 meter resolution in rectangular corner reflectors having the size of 0.3 meter.

4.2. NESZ measurement

NESZ remains the performance parameter to be analyzed and verified, since it does not depend on the scene dependent SNR [4]. The still waters in the red rectangular blocks of Fig. 6 depict the measured NESZ. No masking of waters was applied for a better accordance with the measured NESZ. The mean value of backscatter coefficient for still waters provides the NESZ shown in Table 3. The NESZ performance was to stay below –30 dB at the maximum swath exceeded the design criteria, which was optimized by the balance of the X-SAR radar parameters in Table 3.

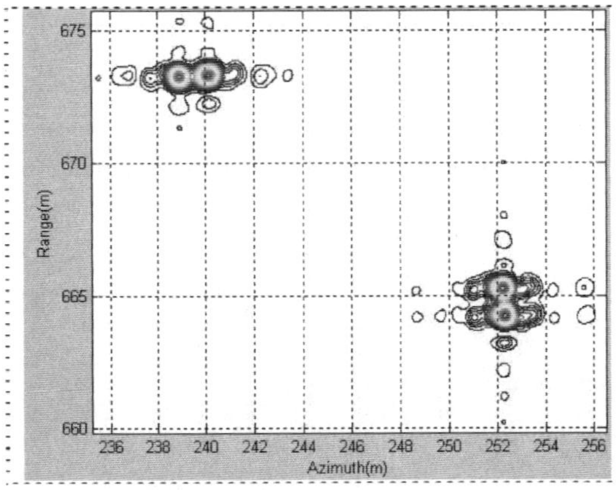

Figure 5. Distinguished trend for target pairs of the rectangular corner reflectors.

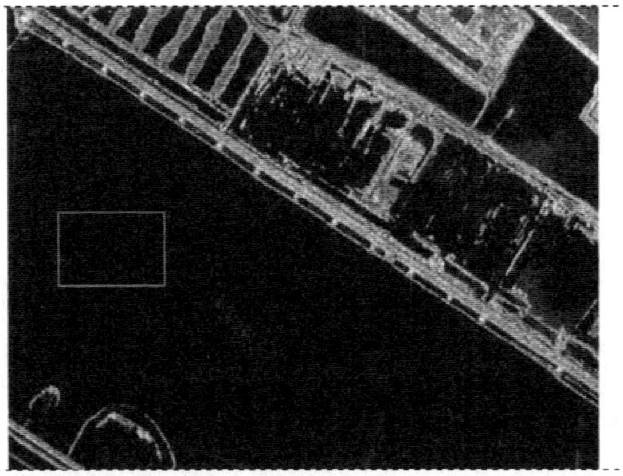

Figure 6. Measured NESZ on the still water area in the red square.

Table 3. NESZ measurement results.

Parameters	Performances		
Bandwidth (MHz)	360	180	90
Operating range(km)	19～23	25～33	31～45
NESZ(dB)	-35.36	-40.89	-34.89

4.3. Radiometric resolution measurement

The radiometric resolution is the important image-quality characteristic for radiometric enhanced images and related to the SNR and the numbers of looks [4]. The multi-look processing can be reduced the speckle noise with SAR image, defined as the equivalent numbers of looks (ENL). The radiometric resolution (RR) is a parameter to measure the terrain distinguishing capability of SAR image, which can be estimated from the measured ENL, mean value μ and standard deviation σ of image intensity, taking into account a distributed target using the following formula:

$$RR = 10\log_{10}\left(\frac{1}{\sqrt{ENL}}+1\right) = 10\log_{10}\left(\frac{\sigma}{\mu}+1\right) \quad (1)$$

As listed in Table 4, the radiometric resolution in the 360 MHz bandwidth is better for measurement increasing the numbers of looks, beyond a certain threshold will not result in significant improvement in measured radiometric resolution with 180 MHz and 90 MHz bandwidth.

Table 4. Radiometric resolution measurement results.

Bandwidth (MHz)	Mean value μ	Standard deviation σ	The numbers of looks	Radiometric resolution (dB)
360	-35.2	9.48	12	1.08
180	-25.9	4.09	40	0.64
90	-33.5	5.16	42	0.62

4.4. Radiometric enhanced products of intensity change detection

By the operational ground SAR imaging processor based on GPU architecture, the standard image data products in high resolution(up to 0.5 meter) can be already achieved less than 6 pixels in the image located precision and long-term radiometric stability better than 0.5dB. Therefore, the new radiometric enhanced products of intensity change detection can be operationally implemented. Fig. 7 shows the example imagery of vegetation cover change over Hebei, China, using two X-SAR images in 0.5 meter resolution acquired with a temporal separation of 5 days on 2013-05-21 and 2013-05-25. Approach to change detection is to be preferred by the straightforward comparison of image intensities

[5]. The difference in the reflected radar signals evidently appeared (blue) or disappeared (purple) is particularly vegetation cover, where the most changes occurred in the time between data acquisitions.

Figure 7. X-SAR example imagery of intensity change detection with the multi-looks resolution of 0.5 meter over Hebei, China. The changes with many pieces of the fields have evidently appeared in blue or disappeared in purple with a temporal separation of 5 days on 2013-05-21 and 2013-05-25.

5. Conclusions

As discussed in this paper, the imaging verification and data calibration provide the advanced high resolution quantified images of modern airborne SAR like RADI's X-SAR. The upgrading X-SAR with interferometer and polarization has been implemented and will be planned to flight test in 2016.

6. References

[1] Del Castillo M J and Larranaga Sudupe J R 2013 Design, Performance and Calibration Review of RBX. INTA New Polarimetric SAR System Germany. Proc. of the 10th European Radar Conference 235-38

[2] Brautigam B, Gonzalez J H, Schwerdt M and Bachmann M 2010 TerraSAR-X Instrument Calibration Results and Extension for TanDEM-X *IEEE Tran. on Geoscience and Remote Sensing* **48** 702-15

[3] Alexander G F, Bruce D C, Brian P H, Scott H, Cathleen E J, Thierry R M and Ronald J M 2015 UAVSAR Polarimetric Calibration *IEEE Tran. on Geoscience and Remote Sensing* **53** 3481-91

[4] Mittermayer J, Younis M, Metzig R, Wollstadt S, Marquez M J and Meta A 2010 TerraSAR-X System Performance Characterization and Verification *IEEE Tran. on Geoscience and Remote Sensing* **48** 660-76

[5] Reigber A, Scheiber R, Jager M, Prats-Iraola P, Hajnsek I, Jagdhuber T, Papathanassiou K P, Nannini M, Aguilera E, Baumgartner S, *et al* 2013 Very-High-Resolution Airborne Synthetic Aperture Radar Imaging: Signal Processing and Applications *Proc. of the IEEE* **101** 759-83

Acknowledgments

Authors wishing to acknowledge the whole X-SAR Radar development team supporting the system design, integration and laboratory measurements, special worked on flight campaign and SAR processing for their effort in the development of X-SAR systems.

An Approach of Web-based Point Cloud Visualization without Plug-in

Mengxuan Ye[1,2], Shuangfeng Wei[1,2] and Dongmei Zhang[1,2]

[1]School of Geomatics and Urban Spatial Information, Beijing University of Civil Engineering and Architecture, Beijing, China

[2]Key Laboratory for Urban Geomatics of National Administration of Surveying, Mapping and Geoinformation, Beijing, China

E-mail:weishuangfeng@bucea.edu.cn

Abstract. With the advances in three-dimensional laser scanning technology, the demand for visualization of massive point cloud is increasingly urgent, but a few years ago point cloud visualization was limited to desktop-based solutions until the introduction of WebGL, several web renderers are available. This paper addressed the current issues in web-based point cloud visualization, and proposed a method of web-based point cloud visualization without plug-in. The method combines ASP.NET and WebGL technologies, using the spatial database PostgreSQL to store data and the open web technologies HTML5 and CSS3 to implement the user interface, a visualization system online for 3D point cloud is developed by Javascript with the web interactions. Finally, the method is applied to the real case. Experiment proves that the new model is of great practical value which avoids the shortcoming of the existing WebGIS solutions.

1. Introduction

As a new means for surveying and mapping, three-dimensional laser scanning technology make it possible to gather fast a vast volume of spatial point information, which has become an important way to obtain three-dimensional spatial information due to its features such as non-contact, fast and continuously getting information of objects[1]. It is widely used in ancient architecture protection, super-large structural deformation monitoring and city surveying and mapping, etc[2]. The obtained point cloud to be processed is the basis of the subsequent data analysis, 3D mesh modeling and 3D visualization[3]. For this reason, it is necessary to seek a remote, user-friendly, point cloud visualization tool without plug-in for quick assessment of the data quality and the plan of processing[4].

A literature review indicates many web-based 3D mesh viewers are developed, but there are few web-based viewers for 3D point cloud[5], among which are mainly commercial viewers developed by the companies of laser scanner such as Leica Tru View, Faro Scene Web Share, All point Insight, etc. However, these point cloud browser online are mainly customized for the manufacturers themselves. During the development of these browsers, the most important problem lies in that, the standard developed by each company is supported by their own plug-ins. These plug-ins with poor compatibility are required to be downloaded and installed, which brings to the users a lot of inconvenience when the bandwidth is not under ideal conditions. Besides, the formats of point cloud data acquired from different scanners of the manufacturers are diverse, and these data must be converted before the visualization process, which has also become a big issue.

With the introduction of WebGL, graphic engines such as three.js[6] and Cesium[7]are applied to visualization of point cloud data. Based on these visualization tools, there are also non-commercial systems for scientific research. As an open-source web browser, Potree[8] is required to convert point cloud data locally and upload it to the server, the automation degree is low and the conversion process needs much time. Online LiDAR point cloud viewer[9], a website for visualizing point cloud, supports reading point cloud from server of the users and the data is needed to be placed on their own server, but only ASPRS LAS 1.2 and XYZ formats are allowed. Oscar Martinez-Rubi [10]et al. chose Potree[8] as prototype system to develop an open source website[11] ,which is based on WebGL and HTML5, but the data is limited to airborne LiDAR data. Zeng X[12]proposed a WebGL-based visualization method for LiDAR point cloud data, which cannot integrate with database for a variety of file formats, still didn't solve the complex problem of diverse data formats.

This paper presents a Web-based point cloud visualization method without any plug-ins. The proposed model addresses the issue of diverse point cloud data formats, which supports spatial database management and flexible extension. The visualization system is written in JavaScript and has realized online interaction of point cloud by using ASP.NET and WebGL technologies, which also intergrates the spatial database based on PostgreSQL. Experiment results indicates that the system can provide users with a good visualization experience for point cloud data in web browser and avoid the problems of plug-in installation and data conversion.

The remainder of the paper has the following structure: Section 2 presents the key techniques. Section 3 discusses the route of technology. Section 4 presents the tests and Section 5 concludes this paper and discusses the future work.

2. The key techniques

2.1. HTML5 and <Canavas> element

HTML is a standard of web development language, which had not been updated more than a decade since HTML4 released. During this decade, the web application development demands more from develop language, such as support for video, asynchronous transportation, vector data rendering, client data storage and other functions[13]. In 2014, the World Wide Web Consortium announced that HTML5 standard specification was finalized and publicly released. Browsers which supports HTML5 include Firefox (Firefox browser), IE9 and later, Chrome (Google Browser), Safari, Opera, etc.

The <canavas> tag in HTML5 can be achieved for drawing vector data[14]. This tag interface can draw points, lines, polygons, text, raster images without any third-party plug-in by using JavaScript. At the same time, attributes and many events of <canavas> tag allow JavaScript to respond to some events to realize graphic and analysis operations.

2.2. WebGL and Three.js

As part of the family of HTML5, WebGL[15] is a web 3D graphics standard developed by Khronos Association , it is originated from the Mozilla Foundation and currently developed by developers of WebGL Working Group of the Khronos and other corporations such as Google, Opera, Mozilla and Apple. WebGL is an OpenGL-based JavaScript programming interface which provides rendering function in the browser. WebGL runs on the Canvas element in HTML5.

Three.js[6] is an open source 3D graphic engine based on WebGL, which allows javaScript to operate GPU in the browser to realize true 3D scenes. It provides basic framework for WebGL-based visualization of point cloud data, which greatly simplifies the developing process. Therefore, in this paper, Three.js is applied as graphic engine for visualization.

3. The route of technology

The route of technology are as follows: Firstly, building database for the raw point cloud data; Secondly, exchanging data between database and web browser; Finally, creating and visualizing three-dimensional scenes. The process is described in details below.

3.1. Building the database

PostgreSQL is an open-source object-relational database server[16]. It is fast, lightweight and support many file types with rich client interfaces. Accordingly, PostgreSQL is chosen as the spatial database in our work. The steps of building the database are summarized as follows:

- Establishing the project in PostgreSQL and creating tables to store data in the database.
- In order to visualize different formats of point cloud data files, coordinates XYZ and RGB color information extracted from raw data will be saved.
- C ++ Libpq interface is chosen as the connection between application and database, the maximum transmission speed of which in campus network can reach to 40MB / s.

The building process is shown in **Figure 1**.

3.2. Data exchange

Since the database has been built, the point cloud data information stored there need to transferred into the browser for rendering. Web Service is used in this period to complete the mission. Web Service can be converted to a web application that can be used by other applications. Basic platform of Web Services is XML + HTTP. By using the ASP.NET AJAX, Web Service can be called with above-mentioned Libpq embedded in the script in order to obtain data.

The process can be reformulated with more details in **Figure 2.**

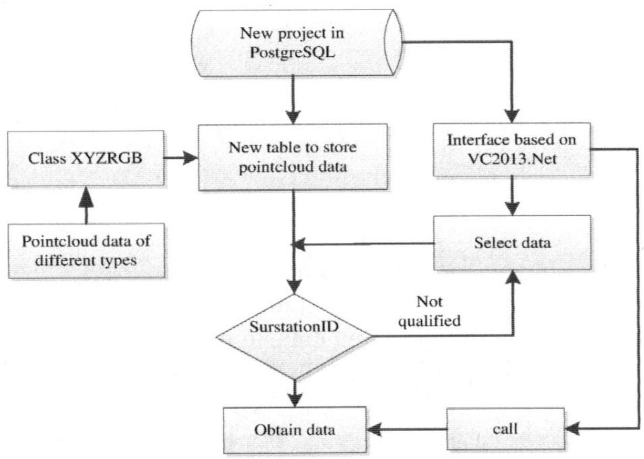

Figure 1. Database construction

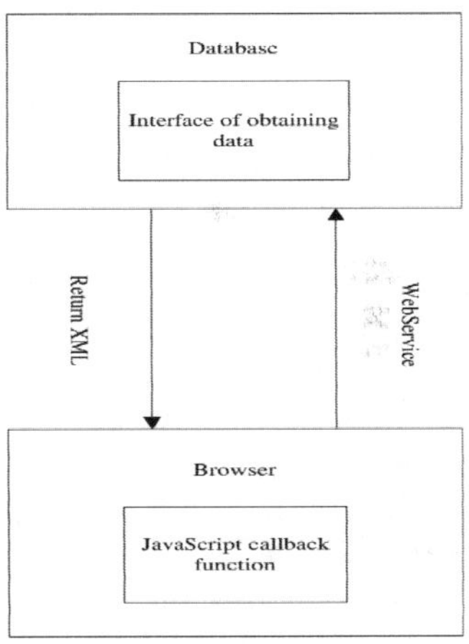

Figure 2. Data exchange

3.3. 3D scenes

To show complete 3D scenes, four essential components are required including Scene, Renderer, Objects, Camera.

In order to build WebGL scene by using three.js, a HTML page is needed to be created firstly. Three.js library is required to be introduced in the <head> tag. Also, a WebGL renderer is created for loading the scene in the <body> tag. The scene is a container, which is used to store and keep tracking of the objects to be rendered. The Renderer completes conversion process from two-dimensional plane to three-dimensional space of the computer window and then the scene is displayed to the users. After the rendering of three-dimension elements, it is ready to write the script of the relevant code in Javascript. At the beginning, the coordinates and color information of point cloud is obtained from database provided by the callback function.

This research is based on Particle System for point cloud rendering. Considering point cloud as a whole, the system allows manipulating point cloud efficiently such as translation, rotation and scale changing. These interactions need to be added to the camera of the scene with its parameters adjusted, so that the observer can browse the object in the scene from different angles and level of details.

The 3D objects is presented in the Scene by using 'scene.add ( )' function provided by Three.js to load point cloud data, then various components of the above object are initialized and the scene began rendering.

4. Experiment and Analysis

In this paper, the point cloud data of Beijing University of Civil Engineering and Architecture (hereafter referred as BUCEA) is adopted as the experimental data. For the development platform, we choose Visual Studio 2013 to build the visualization system with PostgreSQL database integrated. ASP.NET and WebGL technologies are also applied to realize the system's function.

The system consists of three modules: database, web server and the client. PostgreSQL is applied as spatial database. At the same time, the solution uses C# as the programming language to implement server-side response to a variety of operations of the client. The client, using FireFox as a graphic platform, avoids developing another visual interface. The system is running as Figure 3.

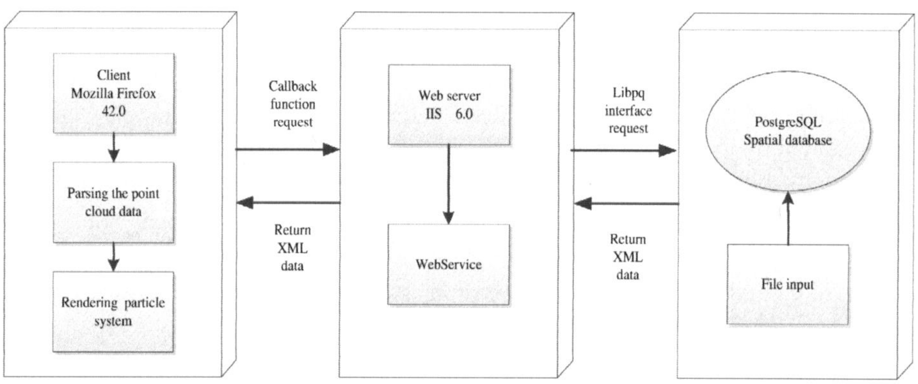

Figure 3. Point cloud management system

Database: It provides interface as the connection between the systems and spatial database. It's responsible for unified organization and management of point cloud data. To a certain extent, it plays the role of a bridge and link in the relationship between database and browser. The database, PostgreSQL, is just a container for spatial data and Libpq is the conversion channel for data processing.

Server: It is supported by the Web server. Web server provides a runtime environment for the development, deployment, operation, and management of web application services. Web server can also support managing spatial data. In this process, web server receives a request from client, which is processed in WebService at the same time, then returns it to the client after obtaining the results.

Client: Browser acquires data by initiating Ajax Web Service interface. Web Service is responsible for obtaining background data, then the processed data is sent to the browser. As the interface of the system, the Client is the forefront of this system through which users can get provided WebGIS. The system has transferred part of the analysis function to the client, which can increase WebGIS interactive experience and reduce the pressure on the Web server.

The actual operation effect of the built system in FireFox browser window is shown in **Figure 4**. In order to easily observe and verify the visual effects, panoramic image is also given in **Figure 5**. The number of points rendered in the scene is 4,102,504 and interactive operation runs smoothly.

Figure 4. 3D point cloud of BUCEA.

Figure 5. Panoramic photo of BUCEA.

In the interface, the upper left is the monitor of FPS, while the right is the control menu, in which the size, color and other attributes of the point cloud can be adjusted, effects are compared in **Figure 6** and **Figure 7**.

Figure 6. Color: green, particle size: 0.01.

Figure 7. Color: red, particle size: 2.81.

5. Conclusions

In this paper, a solution of point cloud data visualization based on WebGL has been introduced and the WebGIS system has been built and tested. Combined with the actual requirement, the paper presents the system development platform, choice of development tools, system structure and functions, and key techniques. Finally, point cloud data visualization based on web is achieved. Test demonstrates that, for online users, it is capable of providing fluent experience when browsing point cloud data. Also, Aiming at the data formats and plug-in problems often occurs in development of publishing system, the spatial database is integrated, which can be used in the actual operating period.

However, due to time and technique reasons, there are many problems to be continue studied in the future. Follow-up studies will focus on the following aspects:

- Large amount of data to read and display. When the point cloud data is massive, the page loading become difficult and improving the performance of massive data reading and visualizing will be an emphasis of future research;
- Comparison with similar research. Due to time limit and other factors, no comparison was conducted to other point cloud publishing system, efficient comparisons is also the object of the next step.

References

[1] ZHANG, R J.,WANG, Y M.,LI, D R.,2006. Technologies of rapid coping with large amount of 3D laser scanning data. Science of Surveying and Mapping, China, 31 (5), pp.93-94
[2] LIU, J N.,ZHANG, X H., 2003. Progress of Airborne Laser Scanning Altimetry. Geomatics and Information Science of Wuhan University, China, 28(2),pp. 132-137
[3] E. Maravelakis, A. Konstantaras, A. Kritsotaki, D. Angelakis, M. Xinogalos, 2013. Analysing User Needs for a Unified 3D Metadata Recording and Exploitation of Cultural Heritage Monuments System, Advances in Visual Computing, Lecture Notes in Computer Science, Volume 8034, pp. 138-147
[4] J. Fernandez, A. Singhania, J. Caceres, K. Slatton, and R. K. M Starek, 2007. An overview of LiDAR processing software Technical report, Geosensing Engineering and Mapping, Civil and Coastal Engineering Department, University of Florida
[5] M. De La Calle, D. Gómez-Deck, O. Koehler and F. Pulido, 2011. Pointcloud visualization in an open source 3d glob3. International Archives of the Photogrammetry, Remote Sensing and Spatial Information Sciences, Volume XXXVIII-5/W16, 2011ISPRS Trento 2011 Workshop, Trento, Italy
[6] http://threejs.org/
[7] http://cesiumjs.org/
[8] http://potree.org/wp/
[9] http://lidarview.com/
[10] Oscar Martinez-Rubi, Stefan Verhoeven, Maarten van Meersberge et al, 2015.Taming the beast: Free and open-source massive point cloud web visualization
[11] http://ahn2.pointclouds.nl/
[12] Zeng , X., 2012.WebGL based LiDAR Point clouds visualization .Journal of Hunan University of Science & Technology, China, 27(4) , pp. 60-64
[13] Maravelakis E, Konstantaras A, Kabassi K, et al, 2014. 3DSYSTEK Web-based Point Cloud Viewer[C]// International Conference on Information, Intelligence, Systems & Applications, Iisa. IEEE, pp.262-266
[14] Xu Z., Zhu J., 2011.Research of WebGIS based on HTML5 and JSON[C]// Computer Science and Network Technology (ICCSNT), 2011 International Conference on. IEEE, pp.1714-1717
[15] WebGL Specification. https://www.khronos.org/registry/webgl/specs/1.0/
[16] Li ,X F.,2014.Research on Storage and Query for the Fine 3D Spatial Data. Master's Thesis, Beijing University of Civil Engineering and Architecture, China

Ackonwledgments

This work is supported by The Importation and Development of High-Caliber Talents Project of Beijing Municipal Institutions (CIT&TCD201404070), Beijing Natural Science Foundation(No.8142014), State Key Development Program of Basic Research of China (No. 2012CB725301), the National Natural Science Foundation of China (No. 40801159) and Key Laboratory for Urban Geomatics of National Administration of Surveying, Mapping and Geoinformation (No. 20111215N) .

IOP Conf. Series: Earth and Environmental Science **46** (2016) 012012 doi:10.1088/1755-1315/46/1/012012

A hybrid change detection analysis using high-resolution remote sensing image

Q Q Xu[1,2], Z J Liu[1], M Z Yang[2], H C Ren[1,2], C Song[1,3] and F F Li[1,4]

1 Institute of Photogrammetry and Remote Sensing, Chinese Academy of Surveying and Mapping, Beijing 100830, China

2 Lanzhou Jiaotong University, Gansu Provincial Engineering Laboratory for National Geographic State Monitoring, Lanzhou 730070, China

3 China University of Geosciences, Wuhan 430074, China

4 School of Geomatics Liaoning Project Technology University, Fuxin 12300, China

E-mail: xuqiangqiang2014@foxmail.com

Abstract. In order to reduce noise and improve the accuracy of the final change results, in this paper, we presented a hybrid change detection method based on combining pixel- and object-schemes, Firstly, the method obtained the orthogonal difference images using the pixel-based iteratively reweighted multivariate alteration detection (IR-MAD) algorithm, additionally in the process of iterative weighting, we applied the regularized scheme to stable the generalized characteristic equation for the multispectral data. Consequently, image segmentation algorithm was used to extract the image objects where the changes occurred. Finally, object-based classification method was applied to determinate the types of changes. In order to validate the effectiveness and feasibility of the proposed approach, a simple case was done by using the Horgos Port local multi-temporal and multispectral high-resolution image data in Xinjiang. Compared to the pixel-level IR-MAD, the experimental results showed that the overall accuracy has been improved, moreover successfully reduced noise and pseudo small changes in the final result.

1. Introduction

With the rapid development of remote sensing technology, the high-spatial resolution imagery has been available from commercial operators, providing unique opportunities for change detection. Compared with the traditional low-resolution images, high resolution images with the richer spatial information and the detail information of the feature are widely used[1]. Land-cover and land-use change information is important because it is widely used in various directions, including deforestation damage assessment, disasters monitoring, urban outspread, urban planning, and land cover change. Singh[2] defined change detection(CD) as "the process of identifying differences in the state of an object or phenomenon by observing it at different times". The general objectives of the remote sensing image change detection is consist of identifying the geographical location and type of changes, quantifying the changes and assessing the accuracy of change detection results[3-5].

At present, high-resolution remote sensing image change detection mainly use the object-oriented analysis techniques. Chen et al. summarized the state-of-the-art object-oriented change detection methods, including image-object change detection, class-object change detection, multi-temporal-object change detection, and hybrid change detection[6]. The hybrid change detection involved the use of both object and pixel paradigms, which was proposed by Carvalho et al. and the authors concluded that this approach was insensitive to geometric misregistration and atmospheric discrepancies between the multitemporal images[7]. Al-Khudhairy et al. applied pixel-based PCA and image differencing for the high-spatial resolution imagery. The change images were then analysed by an object-based classification method, which improved upon the pixel-based change detection[8]. McDermid et al. and Linke et al. transformed multispectral images into wetness bands, which were effective for detecting forest disturbance. This transformation was followed by a pixel-based image differencing using wetness information with an object-based classification applied to the changed areas[9-10]. Niemeyer et al. reported a similar procedure[11]. However, there is a problem that it is difficult to focus change information by means of the common pixel-based change detection.

In this paper, in order to reduce small and spurious changes which are introduced by the inconsistent delineation of objects, and better focus change information, we combine pixel- and object-based schemes to conduct the change detection for the high resolution remote sensing. The pixel-based iteratively reweighted multivariate alteration detection(IR-MAD) could better focus change information[12], and the orthogonal MAD variables are considered to be very suitable for detecting change of the generalized difference image. In addition, Geographic Object-Based Image Analysis(GEOBIA) is new and evolving paradigm[13]. Therefore, we combine pixel-base IR-MAD and GEOBIA to acquire better change detection result.

2. Method

Based on image preprocessing— the image registration and the relative radiometric correction, we combined pixel-based and object-based schemes—a hybrid change detection for high-spatial resolution remote sensing change detection. Firstly, the IR-MAD variables that represented orthogonal different images for the subsequent processing were produced by the pixel-based IR-MAD algorithm. Consequently, the GEOBIA was applied to extract change object. In the end, the object-based post-classification method was used to obtain the change matrices indicating the "from – to" changes. The overall flow chart was shown in the **Figure1**.

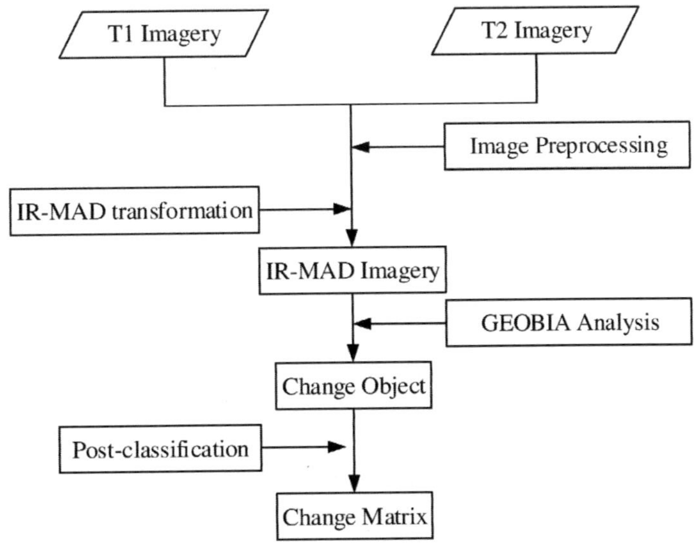

Figure 1. The overall flow chart.

2.1 The pixel-based IR-MAD for obtaining orthogonal different image

The multivariate alteration detection (MAD) algorithm was proposed by Nielsen[14]. The MAD method was based on the canonical correlation analysis (CCA) technique. With the CCA, we could calculate the canonical variates and subtract them from each other. Simultaneously, the canonical variates could be obtained by solving the generalized eigenvalue problem for $p \times 1$ image data X from one point in time and $q \times 1$ image data Y from another point in time ($p \geq q$).

$$\mathbf{a}^T X = a_1 X_1 + \cdots + a_p X_p \tag{1}$$

$$\mathbf{b}^T Y = b_1 Y_1 + \cdots + b_q Y_q \tag{2}$$

or

$$\begin{bmatrix} 0 & \Sigma_{12} \\ \Sigma_{21} & 0 \end{bmatrix} \begin{bmatrix} a \\ b \end{bmatrix} = \rho \begin{bmatrix} \Sigma_{11} & 0 \\ 0 & \Sigma_{22} \end{bmatrix} \begin{bmatrix} a \\ b \end{bmatrix} \tag{3}$$

Among the equation, Σ_{11} is the $p \times p$ variance-covariance matrix of X, Σ_{22} is the $q \times q$ variance-covariance matrix of Y and Σ_{12} is the $p \times q$ covariance matrix between X and Y, other $\Sigma_{21} = \Sigma_{12}^T$. $\mathbf{a}$ is the eigenvector involving the weights with which to multiply X from the one time and $\mathbf{b}$ is also the eigenvector including the weights with which to multiply Y from another time.

In the MAD transformation, by means of maximizing variance, the change information could be focused. Thus, we must minimize the canonical correlation $\rho = Corr\{\mathbf{a}^T X, \mathbf{b}^T Y\}$. However, we need to get the best coefficients combinations $\mathbf{a}$ and $\mathbf{b}$. A more well-known formulation of the CCA problem is given by the coupled eigenvalue problems.

$$\Sigma_{12}\Sigma_{22}^{-1}\Sigma_{21}\mathbf{a} = \rho^2\Sigma_{11}\mathbf{a} \quad (4)$$

$$\Sigma_{21}\Sigma_{11}^{-1}\Sigma_{12}\mathbf{b} = \rho^2\Sigma_{22}\mathbf{b} \quad (5)$$

In order to conduct the hybrid change detection, we calculate the canonical variates $U_i = a_i^T X$ and $V_i = b_i^T Y$, and then the MAD change detector as the orthogonal difference image $Z_i = U_i - V_i$. The MAD variates have variances $Var\{MAD_i\}=Var\{U_i-V_i\}=2(1-\rho_i)$. Hence, the revers ordering which maximizes the MAD variance $Var\{MAD_i\}$ in the low order MAD variates which are the main differences among the high order canonical variants.

To build a better background of no-change against which to detect change and make the change information better focus, the iteratively reweighted (IR) scheme had been applied the MAD transformation to put high weighs on observations that exhibited little change over time. The sum of squared for the standardized MAD variates would approximately follow a χ^2 distribution with p degrees of freedom, i.e. approximately.

$$T_i = \sum_{i=1}^{p}\left(\frac{Z_i}{\sigma_{Z_i}}\right) \in \chi^2(p) \quad (6)$$

$\sigma_{Z_i} = \sqrt{2(1-\rho_{p-i+1})}$ ideally is the standard deviation of the no-change observations. In the iteratively reweighted scheme, first start, we assign the same weight (=1) to all pixels. Based on statistics calculations, we choose to weight pixel j in the next iteration by w_j which is a measure of no change, namely the probability of finding a greater value of the χ^2 value in Equation 6

$$w_j = P\left\{>\sum_{i=1}^{p}\left(\frac{Z_i}{\sigma_{Z_i}}\right)^2\right\}_j \simeq P\{>\chi^2(p)\} \quad (7)$$

Since the IR-MAD variates are invariant to affine transformation to the original data including linear and affine radiometric normalization or calibration. This makes them well generalized multivariate orthogonal differences among all variables at the two time points of acquisition. Additionally, to solve the ill-conditioned variance-covariance matrices problem and obtain a stable generalized eigenvectors, we apply regularization in CCA.

$$\begin{bmatrix} 0 & \Sigma_{12} \\ \Sigma_{21} & 0 \end{bmatrix}\begin{bmatrix} a \\ b \end{bmatrix} = \rho\begin{bmatrix} \Sigma_{11}+k_1\Omega & 0 \\ 0 & \Sigma_{22}+k_2\Omega \end{bmatrix}\begin{bmatrix} a \\ b \end{bmatrix} \quad (8)$$

where k_1 and k_2 determine the amount of regularization and Ω is designed to minimize as the function of wavelength.

2.2 Geographic object-based image analysis of image objects

GEOBIA has become a paradigm for remote sensing image analysis. In this paper, we apply the GEOBIA on the IR-MAD orthogonal difference images. Firstly, the IR-MAD images are segmented by the multi-scale segmentation algorithm in the e-Cognition software. Consequently, these features are extracted from the image objects for the segment images, including the geometrical features (width, area and compactness et.al), the spectral features (mean and variance et.al), the texture features (GLCM) and the context information; the Object-oriented change detection is used to the segmented IR-MAD images.

2.3 object-based post-classification for change matrices

In order to obtain the "from-to" change information, namely, the change matrices, we independently classify the IR-MAD image objects, and the classified objects are obtained using the method of supervised classification, moreover to prepare for the further analysis and utilization.

3. Experimental results and analysis

In order to verify the feasibility of the proposed method, The two time ZY3 high resolution satellite image was used for the test in Horgos Port of Xinjiang, the two time original image shown in figure 2.

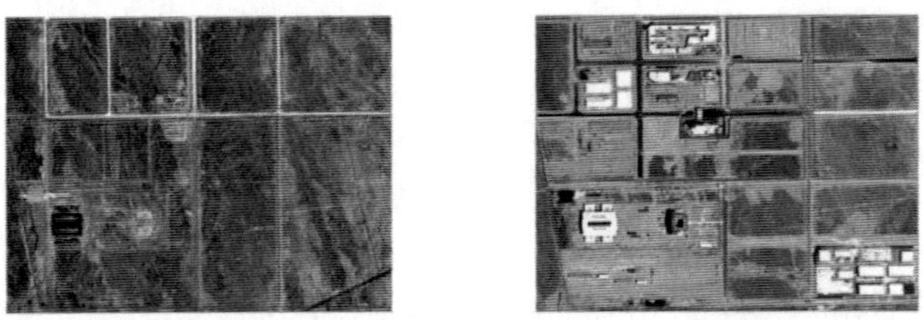

(a) (b)

Figure 2. 2013 and 2015 ZY3 high-resolution images in the Horgos Port Xinjiang.

Firstly, the IR-MAD orthogonal difference images were obtained by the pixel-level IR-MAD algorithm, the results showed in figure 3 (a). Secondly, these IR-MAD images are segmented and conducted the feature extraction which can be a very good description of the image, the scale level is 60, the results showed in figure 3 (b). Finally, the object-based post-classification change detection results showed in figure 3 (c), which could be seen from the figure. Different colors represented different changes from the bare to other land use class.

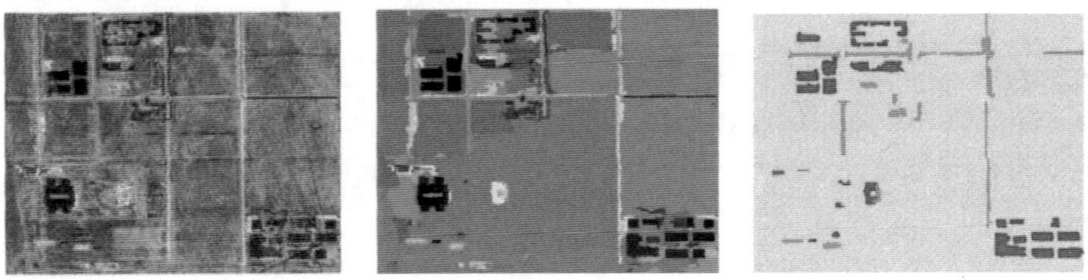

(a) (b) (c)

Figure 3. the hybrid change detection results for IR-MAD images(a), segmented images(b) and from-to changes(c).

4. Conclusion and outlook

By combining the pixel-level IR-MAD algorithm and the object-based change detection method, we take full advantage of the pixel-level IR-MAD algorithm to obtain a more focused change information, in addition, by means of the GEOBIA advantage in the analysis of the high-resolution remote sensing images we obtain a better change detection results for the multi-temporal high-resolution remote sensing images. However, there are some problems here, such as the low degree of automation and the error transmission as a result of many steps are involved in the hybrid change detection, it remains unclear how the final change results are influenced by the different combinations of pixel-based and object-based schemes.

5. References

[1] Dou W 2003 Exploration Study and Comparative Implementation Method for Object-Oriented Remote Sensing Image Analysis (Nanjing: Nanjing University) pp 1-41
[2] Singh A 1989 Int. J. Remote Sens 10 898–1003
[3] Coppin P, Jonckheere I, Nackaerts K, Muys B and Lambin E 2004 Int. J. Remote Sens. 25 1565–96.
[4] Im J and Jensen J R 2005. Remote Sens. Environ. 99 326–40
[5] Macleod R D and Congalon R G 1998 Photogramm. Eng. Remote Sens. 64, 207–16
[6] Gang C, Geoffrey J H, Luis M T 2012 int. j. Remote Sens. 33 4434-57
[7] Carvalho L M T, Acerbi F W, Scolforo J R, Mello J M and Oliveira A D 2007 Proc. Inter. Workshop Analysis Multitemporal Remote Sens. Image vol 5 (Leuven: Belgium) pp 18–20
[8] Al-khudhairy D H A, Caravaggi I and Glada S 2005 Photogramm. Eng. Remote Sens. 71 825–37
[9] Mcdermid G J, Linke J, Pape A, Laskin D N, Mclane A J and Franklin S E 2008 Can.J. Remote Sens. 34 462–6
[10] Linke J, Mcdermid G J, Laskin D N, Mclane A J, Pape A, Cranston J, Hallbeyer M and Franklin S E 2009 Photogramm. Eng. Remote Sens.75 981–95
[11] Niemeyer I, Nussbaum S and Canty M J 2005 Proc. IEEE Inter. Geoscience Remote Sens. Symp. (Seoul:South Korea/IGARSS) p 4
[12] Nielsen A A 2007 IEEE trans. Image process. 16 463-78
[13] Blaschke T, Geoffrey J H, Maggi K 2014 ISPRS J. Photogramm. Remote Sens. 87 180–191.
[14] Nielsen A A, Conradsen K and Simpson J J 1998 Remote Sens. Environ. 64 1–19

Acknowledgments

This work was jointly supported by the Research and special funds projects for mapping geographic information nonprofit industry (approved No. 201512027) and project supported by the National Natural Science Foundation of China (Grant No. 41371406).

Airborne LIDAR point cloud tower inclination judgment

Chen liang[1,2] and Liu zhengjun[1] and Qian jianguo[2]

[1] Chinese Academy and Surveying and Mapping, Beijing, China
[2] Liaoning Technical University, Fuxin, China

E-mail: 18201249765@163.com

Abstract. Inclined transmission line towers for the safe operation of the line caused a great threat, how to effectively, quickly and accurately perform inclined judgment tower of power supply company safety and security of supply has played a key role. In recent years, with the development of unmanned aerial vehicles, unmanned aerial vehicles equipped with a laser scanner, GPS, inertial navigation is one of the high-precision 3D Remote Sensing System in the electricity sector more and more. By airborne radar scan point cloud to visually show the whole picture of the three-dimensional spatial information of the power line corridors, such as the line facilities and equipment, terrain and trees. Currently, LIDAR point cloud research in the field has not yet formed an algorithm to determine tower inclination, the paper through the existing power line corridor on the tower base extraction, through their own tower shape characteristic analysis, a vertical stratification the method of combining convex hull algorithm for point cloud tower scarce two cases using two different methods for the tower was Inclined to judge, and the results with high reliability.

1. Introduction.

Towers entire power system is a vital part of the accord is possible in a variety of complex weather conditions frequency between the wire and overhead transmission lines and towers with magneto telluric and electrical insulation safety requirements[1]. In the electrical inspection operations, inclined tower, tower base, and other parts of the state judgment is an important content.

Among them, the tower tilted factors are mainly the following aspects: when in the role of external damage or special meteorological environment, on both sides of the tower wire discontinuity, Both sides of the tower to create tension differential, transmission line pole Tower base will force imbalance due to its inclined slip and other phenomena, in this case the event will affect the normal operation of the transmission line. Therefore, to ensure the normal operation of the company's power grid, power lines should be checked regularly in order to detect the presence of hidden dangers and defects tower to prevent accidents. As China's power equipment intelligent upgrading and deepening of industrialization mature, intelligent power grid and gradually shift to the business model, transmission line towers inclined more and more serious harm, how do the Inclined tower measurement, high-voltage transmission line tower inclined state to take preventive measures we need to explore and become the subject of in-depth study [2].

At present, the power line tower inclination measurement methods mainly vertical corridor method[3-7], a method theodolite, plane mirror method, a terrestrial laser measurement and online monitoring inclined tower. Development of UAV technology for the development of the grid inspection technology provides a new tool for development. UAV power line inspection apparatus having a small investment, relatively low cost of inspection, automation, intelligent features, with significant technical and economic advantages and electrical inspection characteristics, and can be avoided manned patrol and complicated geographical conditions under manual inspection of security issues. From airborne radar scan point cloud power line corridor, based on the existing tower automated extraction algorithm, based on the tower elevation data and projection plane fitting algorithm for point cloud tower inclination judgment[8-9].

2. Analyzing inclined tower principle

Tower formula inclination

$$q = SH \times 1000 ‰$$

The formula q-- inclination; S-- tilt value, mm; H-- tower top or measurement points to the ground height, mm.

Normal tower Inclined the maximum allowed value (see table1) [10]

Table.1 The property of image feature and the corresponding number

category	Reinforced concrete pole	Iron tower
Inclination tower (Including the deflection)	1.5%	0.5%（It applies to more than 50m high Iron tower and）
	1.0%	1.0%（The following applies to 50m height Iron tower）
Cross arm skewness		1%
Iron tower adjacent contacts between the curvature of the main material >0.2%		

3. Characteristics of laser point cloud tower

There are many drawbacks to explore the use of spatial information technology to solve artificial transmission line power line inspection method, Chinese Academy of Surveying and Mapping, the PLA General Staff 60th Institute, Beijing University of Guangdong Power Grid Corporation Electric Power Research Institute, aerospace joint Wuhan University conducted a large-scale application of the use of the UAV power line inspection. Data source of this article a number of areas from Guangdong 110KV-500KV high-voltage transmission line inspection sweep the surface of the laser point cloud data. As shown in figure 1) tower cloud point coordinates.

Figure 1. Schematic point cloud tower

Figure 2 by the analysis of various types of towers, towers are generally divided into tower head, tower body, the tower legs of three parts, tower head from the tower legs up tower sectional dramatic changes (broken line appears) over part of the tower head, If there are no dramatic changes in cross-section, the lower cross arm above the lower chord part of tower head, the tower head contains insulators, ornaments, etc., the whole point cloud tower head irregular distribution. Basis of the above in the first paragraph of the tower called the tower legs. Section of the tower and the tower between the legs called tower body, tower body extracted by setting a threshold value, extraction tower cloud point above the lowest point between 2m to 15m tower body is the point cloud point cloud, (according tower Depth freely set) tower body is generally shown as a rectangular table rules are as follows.

Figure 2. Tower tower point cloud

4. Overall inclined tower monitoring processes and methods

4.1. Experimental data

In the lidar data, first, from a large number of the original point cloud point cloud isolated tower. Point cloud tower separation method is to use Liu Zhengjun like patent "an airborne LIDAR data extraction and power lines fast reconstruction method" (patent number: ZL201210002838.5) used in the method. Secondly, the point clouds extracted tower tower body has a regular shape, head to the adoption of the rules of the point cloud tower body extracted tower was straight lateral edges, tower body will be projected onto a two-dimensional point cloud in all directions each perpendicular to the horizontal plane plane, the final calculation of the point cloud projecting tower body inclined to the side edge of each of two-dimensional direction of the plane, to determine the effect of tilting tower. Concrete tower tilt determination flow chart shown in figure 3.

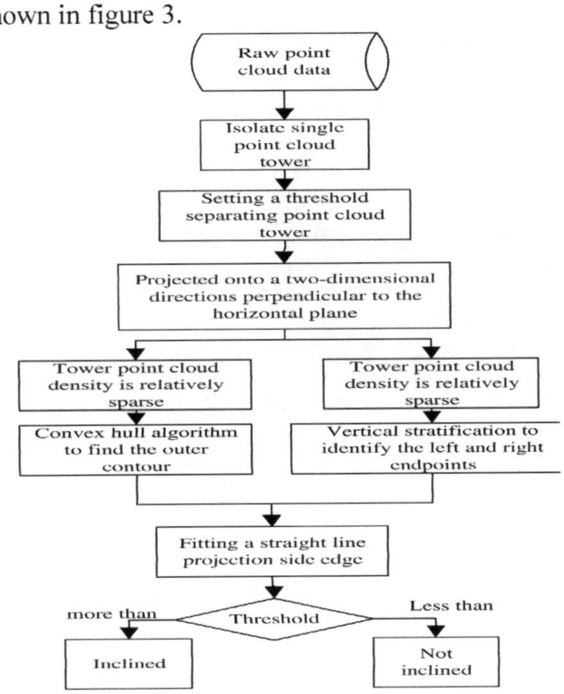

Figure 3. Tower of the flow chart

4.2. Projection

Laser point cloud data itself is three-dimensional data, since the tower body has a rectangular point cloud data table structure is very regular, and the side edge in a straight line, if tower body is extracted four side edge in the three-dimensional situation is very difficult, with respect to the three-dimensional space dimensional planar side edge extraction is relatively simple and accurate, so tower body will be a three-dimensional projection [44] in all directions perpendicular to the geoid (in all directions: in this paper between 0 ° to 180 ° every 10 ° as a unit 18 projection angle can be selected in this direction, if the guarantee greater precision, you can select a smaller angle) as shown in figure 4.

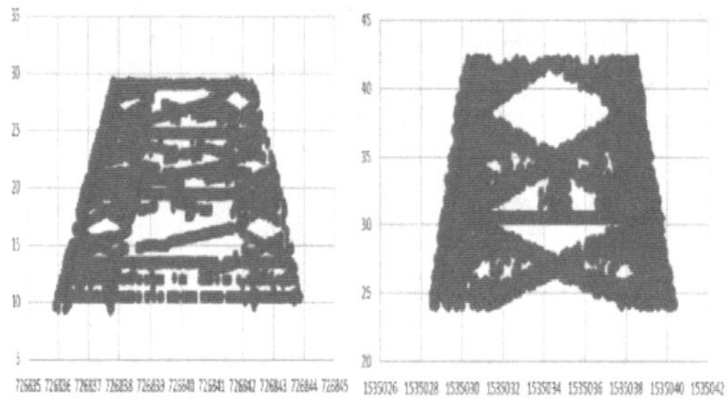

Figure 4. Two dimensional plane point cloud tower tower

4.3. Extraction side ribs

UAV mounted laser radar scanning power line corridor, due to machine vibration, weather, sweep the laser surface scanner hardware and other reasons can lead to a power line tower scanning incomplete, incomplete. Aiming point cloud density in both cases using the two methods is projected onto a two-dimensional plane left and right side of the ribs on both sides of tower body linearly cloud point extraction.

1) Point cloud density is relatively normal

Shown in figure 5, when the original tower normal point cloud data-intensive, projected onto a two-dimensional plane of the presentation rules similar isosceles trapezoid side edge and dense point cloud rendering linear fitting point of the line, Fitting line points the better, So the use of vertical stratification.

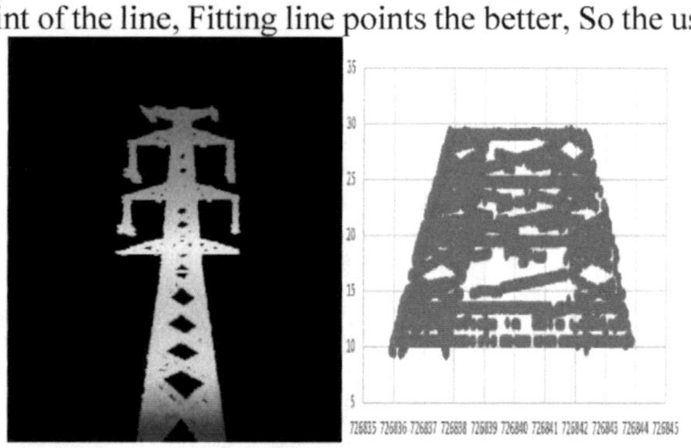

Figure 5. Tower body good point cloud diagram

Vertical stratification: the approximate isosceles trapezoid rule than the two-dimensional plane to the horizontal center line about hierarchical, left and right point cloud rendering right angle trapezoidal, hierarchical set threshold, according to a threshold to identify the left and right sides of the leftmost and rightmost coordinate point cloud abscissa corresponding to the minimum and maximum values, to the right and left side edge extraction.

As shown in figure 6, will be projected onto a two-dimensional plane data-intensive vertical stratification reached renderings, to achieve the effect of the extraction side of the ribs, lateral ribs for straight after fitting.

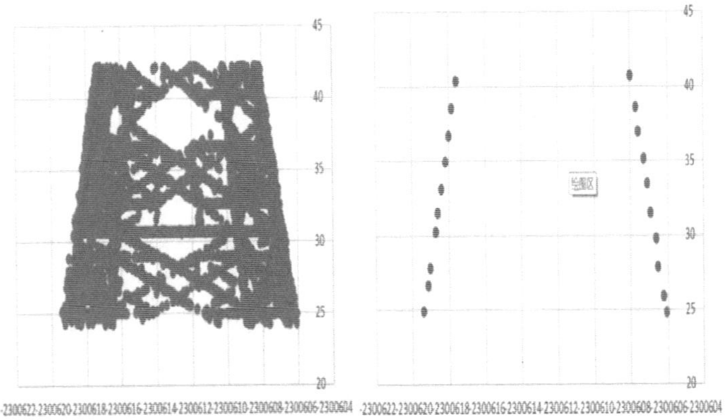

Figure 6. Schematic diagram of the structure of vertical stratification

2) Point cloud density is relatively sparse case

Shown in figure 7, the right of the original point cloud tower missing chunks of data, projected onto a two-dimensional plane is still the case of missing data, since part of the point cloud can not extract the missing tower body side edge all point clouds, and it is difficult fitted to complete the tower side of the trunk edge, so the use of convex hull algorithms.

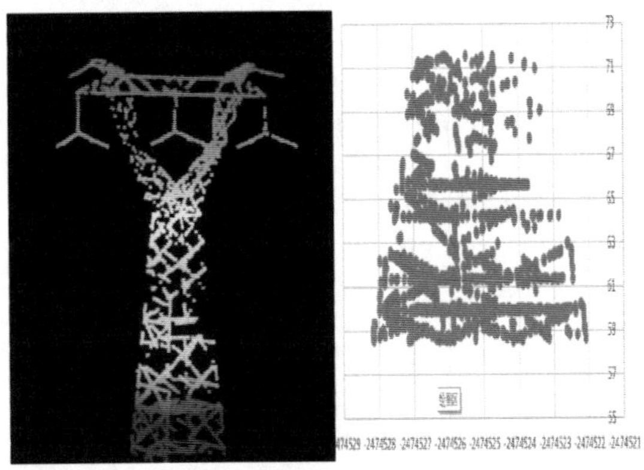

Figure 7. Schematic diagram of tower body loss

Convex hull Definition: Let S be the intersection of a plane, closed s all minimum convex polygon vertices, called s convex hull, expressed as CH (S). Vertex CH (S) on, sometimes also called the S pole. Define the convex hull of planar point set point set that contains the smallest convex set that point as a point of focus part of convex polygon vertices, any one of the convex polygon edges, not all of the points are concentrated in the edge of the point side on the same side [45].

Shown in figure 8,9, will be projected onto a two-dimensional convex hull of the scarce data presented renderings, achieved little cloud of scattered peripheral contour extraction.

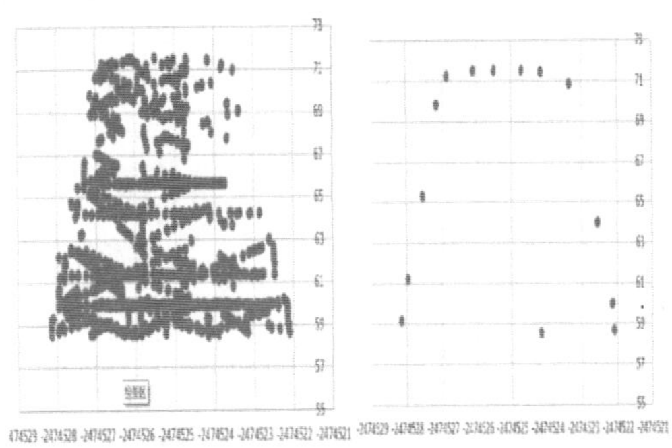

Figure 8. Schematic diagram of constructing convex hull

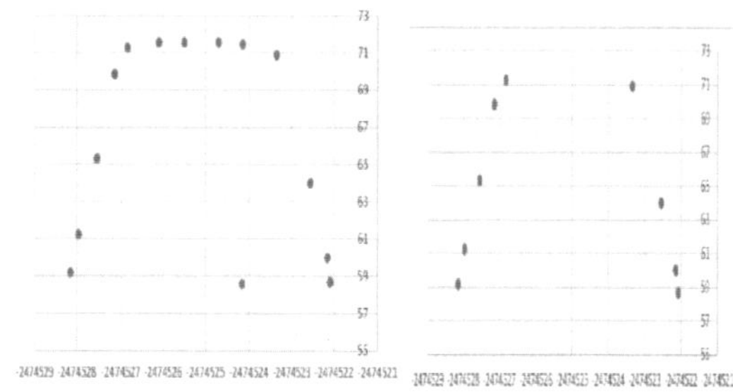

Figure 9. Schematic diagram of the structure of vertical stratification

4.4. Extraction side ribs

We usually ask when the general slope of the line obtained by plotting the manner determined linear equation $y = ax + b$ of a slope and intercept b, but more in the measurement data and scattered when the same data to different people handle the results are not the same, so that the calculated a and b is relatively large errors. Using least squares fitting line solution to the large amount of data, and obtain a slope and intercept b is unique. Least squares fit to the scarcity and dense point cloud above two cases have reached the edge extraction side effects, by bilateral side edge least squares linear fit to determine its lateral edge inclined slope.

4.5. Analyzing inclined tower

Tower tilted inclination value divided by the tower top or measurement points to ground level, as shown below for the calculation principle tower tilt, ca tower in a vertical plane normal position, cb segment for the tower tilted position, ab to tower angled horizontal length, the inclination of the tower $\tan\angle c = ab / ac$, $\angle c$ is the desired angle of inclination.

For peripheral side edge contour extraction point cloud is still using the vertical stratification method, illustrated as below figure 10.

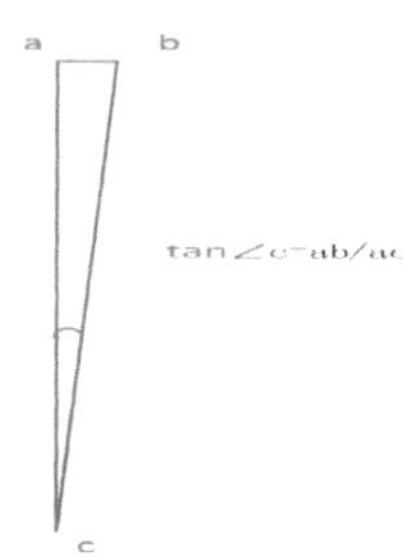

Figure 10. Sketch map of tangent angle theory

By defining judgment can be converted to point cloud tower inclination, that tower to find the inclination angle of the point cloud can be. As shown below(see figure 11) there are two flat, horizontal surface normal level 1, level 2 is a tower which was inclined plane, can draw a horizontal plane through the 1,2 angle of inclination angle of the tower, according to the geometric relationship between the availability of $a = ((b + a) - (b-a)) / 2$.

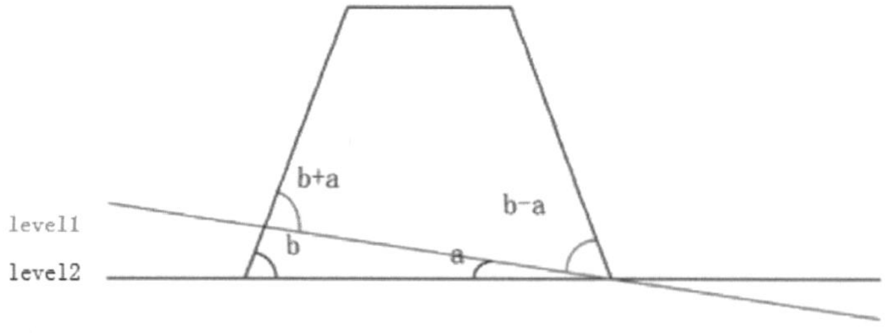

Figure 11. Schematic diagram of tower inclination theory

As shown below is projected onto a two-dimensional plane towers tower schematic point cloud using the least squares method can be determined slope tower tower AB and DC two side edges, AB side edge slope tan∠B, CD side edge slope tan∠C turn can be obtained ∠B and ∠C angle size. Through the above analysis, the difference between the figure 12 and ∠C ∠B half of the absolute value of the tangent of the inclination the tower is, it is judged that the tower tilted formula tan (| ∠B-∠C | / 2).

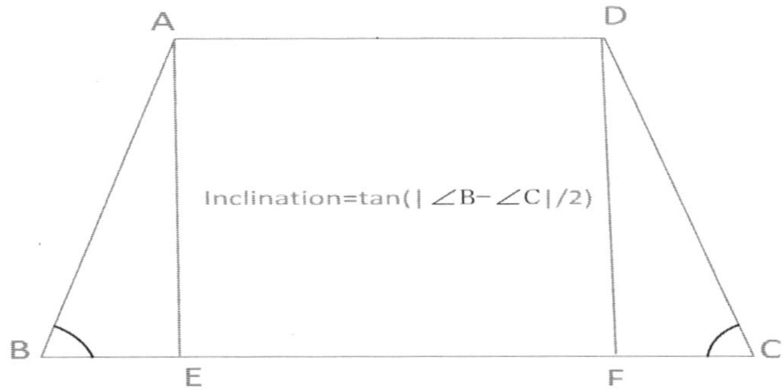

Figure 12. Tower inclination transformation diagram

4.6. Achievements Exhibition

Table.2 Results show schematic diagram

No.	XOY vertical projection plane angle (degrees)	The inclination angle (degrees)
1	0	0.113521
2	10	0.146535
3	20	0.091209
4	30	0.092416
5	40	0.054037
6	50	0.016442
7	60	0.167541
8	70	0.08295
9	80	0.005290
10	90	0.065680
11	100	0.012954
12	110	0.049968
13	120	0.199206
14	130	0.056920
15	140	0.035524
16	150	0.085106
17	160	0.031889
18	170	0.168842
18 projection plane of maximum tilt angle (degrees)		0.199206

50m or less tilt the tower is less than 0.01	0.003477
Whether Inclined tower	No

As shown above (see table2) is a point in an area of the test tower in Guangzhou cloud of output achievement, achievement in the form of txt text output, text output is included in each of the inclination angle with the horizontal plane of the projection angle and the corresponding projection, the projection plane 18 is inclined angle the maximum inclination of the tower and the tower tilted whether information. Judging by the principle known tower tilt tower tilt less than 50m less than 0.01, known tower tilt angle 0.199206, can be obtained inclination tan (0.199206) = 0.003477 <0.01, therefore not inclined tower.
More than a single tower tilt measurement, for practicality, optimized and improved for more than one tower tilt judgment, the raw point cloud data into a plurality of towers and tower center position coordinates, the results of its output, as shown below(see table 3), one for each tower tower inclination information, location coordinates tower Center, 18 projection plane and the inclination angle of maximum inclination and judgment basis (examples include five towers, the tower may be more inclined to judge).

Table.3 Results show schematic diagram

Tower Center position (X, Y, Z)	Maximum 18 projection plane inclination angle	0.01 tower tilt angle, the tilt tower	Whether Inclined tower
10795.080000 7702.760000 40.23	3.907276	0.068301	Yes
10501.690000 7641.690000 45.60	0.118568	0.002068	No
10154.970000 7566.350000 41.10	0.375367	0.006551	No
9881.050000 7505.010000 39.50	0.287791	0.005023	No
9591.460000 7445.540000 47.00	0.331600	0.005788	No

References

[1] Yang jingbo,Li maohua,Yang fengli. The study of transmission line tower structure Progress [J]. Power System Technology, 2008,32 (22): 77-83.
[2] Wang ruigang. On optimized tower tilt measurement and preventive measures [J]. Management & Technology, 2013 (12).
[3] Li yunni, Du yong, Shen xiaojun. Comparison of several transmission line towers inclination measurement methods [J]. Hubei Electric Power, 2014,38 (10): 55-58.
[4] Song jiming, Song huasong, Wang yiwen. Methods of measurement at the Special Terrain tower tilted rate [J]. Power System Technology, 2010,34 (12): 119-222.
[5] Song huaSong. Measurement plane mirror inclined tower rate [J]. CONSTRUCTION, 2008, (6): 196-197.
[6] Wang Dengfeng Power tilt tower design real-time monitoring and early warning system [J] coal, 2012,05 (15): 35-37.
[7] Zhou kehong, Zhang ye. Transmission tower inclination-line monitoring system [J]. Guangdong Electric Power, 2013,26 (7): 57-61.
[8] Liu zhengjun, Peng xiangyang, Guo xiaolong. Power line inspection data acquisition and processing technology [M]. China Electric Power Press, 2015.
[9] Liu zhengjun, Liang jing, Zhang jixian. Rapid Extraction of spatial domain segmentation airborne LiDAR data transmission line [J]. Journal of Remote Sensing, 2014,18 (1): 61-76.

[10] Sun liang, He hongmao.Tower inclinometer Application to Prevent the Collapse of the line [J] Inner Mongolia Electric Power Technology, 2008,26 (2): 43-45.

Temporal-spatial distribution of archaeological sites in the Nihewan-Huliu Basin during the Paleolithic-Neolithic and Iron Age in northern China

J X Zhang[1], X Y Wang[2], R X Yang[2], X Z Li[1] and W Zhang[1]

[1] Key Laboratory of Space Utilization, Technology and Engineering Center for Space Utilization, Chinese Academy of Sciences, Beijing, 100094, China
[2] Key Laboratory of Digital Earth Science, Institute of Remote Sensing and Digital Earth, Chinese Academy of Sciences, Beijing, 100094, China
E-mail: jxzhang@csu.ac.cn

Abstract. The Nihewan-Huliu Basin is one of the great regions of human evolution, located in geographical transition zone. This study reveals the temporal-spatial distribution of archaeological sites based on the DEM (Digital Elevation Model) data, and analyses the changes for four periods (Paleolithic-Neolithic transition, Neolithic Age, Bronze Age and Iron Age). These sites expanded from the Sanggan River to the Huliu River and from the central and lower Huliu River to its upper reaches. The population and cultural development in the Nihewan Basin gradually lagged behind the Huliu River Baisin after the Paleolithic-Neolithic transition Age, because of the environmental impacts. The results may aid the understanding and study of the cultural heritage and civilization evolution in northern China.

1. Introduction

It is well known that there are a large number of Paleolithic sites in the famous Nihewan Basin [1][2][3], but many important Neolithic cultural sites are not well known, especially the sites in the adjacent Huliu River Basin. The Huliu River is a major tributary of the Sanggan River; it flows through Guangling County of Datong City in the northeast of Shanxi Province, Yu County and Yangyuan County (core area of the Nihewan Basin) of Zhangjiakou City in the northwest of Hebei Province, and into the Sanggan River in Yangyuan County. The study area will be called the Nihewan-Huliu Basin in this paper.

From the 1920s[4], archaeologists worldwide have investigated the northern parts of Hebei Province many times, especially the Paleolithic archaeology in the Nihewan Basin, resulting in great achievements and world-renowned discoveries[1][2][3][5][6][7]. From the 1970s, detailed Neolithic archaeological surveys have been carried out in Yu County[8][9][10], resulting in important achievements and proof of prehistory cultural exchanges. However, the temporal-spatial evolution study is not enough, and is not consistent with the significant archaeological status of the Nihewan Basin and the Huliu River Basin. The Nihewan-Huliu Basin holds almost successive human evolution relics from the Paleolithic Age to the Han Dynasty (approximately 21 cal ka BP ago to AD 220 in this paper), extending even to the present, so they should be studied as a whole to reveal the significance of man-land relationships.

With the gradual increased investigation, it is time to reveal the temporal and spatial distribution of archaeological sites based on the DEM (Digital Elevation Model) data, and their relationship with the environments. The results may provide valuable results for the study of ancient human activities, cultural heritage, prehistoric civilization in northern China [10] and so on. This topic is attracting more and more attention.

In this paper, firstly, the regional setting and archaeological sites will be introduced; then, the spatial distribution of published sites will be mapped for four periods and will be analyzed in detail; finally the relationship between the temporal-spatial distribution and the environments will be simply explained.

2. Regional setting and data

2.1. Regional setting

The Nihewan-Huliu Basin is located in the southeast of the Inner Mongolian Plateau, the northeastern part of the Loess Plateau, and the northwest of the North China Plain. The basin is surrounded by hills and mountains, whose altitude varies from approximately 0.9 to 2.8 km a.s.l. The geographic scope is approximately 39.56-40.38N, 113.85-115.06E, an area of approximately 6,300 km^2. The position of the basin in northern China and the satellite imagery of the intermontane basin are shown as Figure 1.

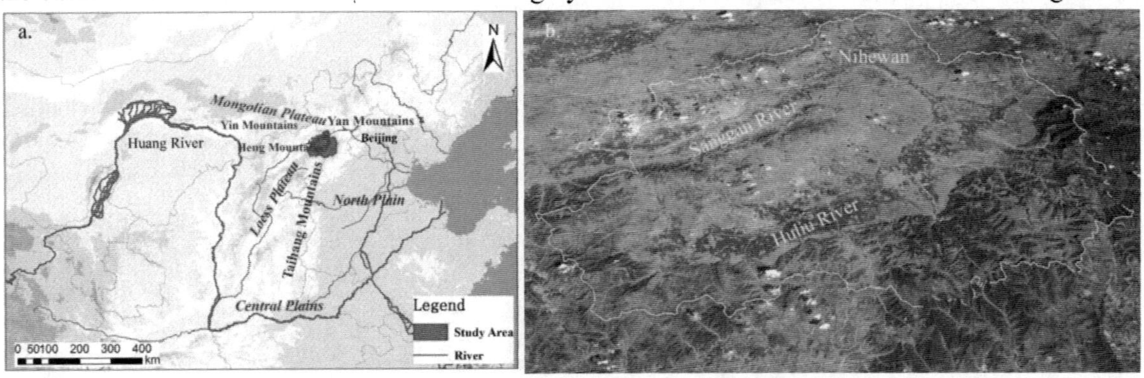

Figure 1. The position mapping of the Nihewan-Huliu Basin. (a) The position in northern China with the background DEM dataset from the CHGIS (China Historical GIS, http://www.fas.harvard.edu/~chgis); (b) the satellite imagery of the Nihewan-Huliu Basin from RSCM (Remote Sensing Cloud Mart, http://www.rscloudmart.com) with elevation supported by the ASTER GDEM 30 m

2.2. Archaeological data

In this paper, 175 archaeological sites are collected from the "Atlas of Chinese Cultural Relics" edited by the State Administration of Cultural Heritage in China (Chinese: Guojia Wenwuju, abbreviated: GWJ) [11][12] and the published data [3][2][7][13][14], including 61 sites (including the Small Stone and Microlithic sites) in Yangyuan County, 101 sites in Yu County, and 13 sites in Guangling County, from the late phase of the Late Paleolithic Age to the Han Dynasty. These sites could be classified into four periods and groups according to archaeological cultures, such as the Paleolithic-Neolithic transition Age, the Neolithic Age, the Bronze Age and the Iron Age. The dates and corresponding typical sites are listed in Table 1.

Table 1. Archaeological cultures and corresponding typical sites

Periods	Dates	Typical Sites
Paleolithic-Neolithic transition	approximately 21-8 cal ka BP in this paper	Xibaimaying[2], Yiduquan [2], Meigou, and Weidipo [15], Yujiagou [7][2], Hutouliang Group [6], Youfang [2], Jijitan [2][3], Erdaoliang [2], and so on [2][15].
Neolithic Age	approximately 10-4 cal ka BP	Jiangjialiang[2][10], Heitupo[2], Yujiagou[2], Sishilipo[8], Sanguan[11][8], Pipazui[11]. Qianbao, Fanpo, Zhuangke, Shaizilingluo, Dongshuiquan, and so on [11]
Bronze	approximately 2-0.5	Sanguan[9][11], Shaizilingluo[11], Fanpo[11], Zhuangke[11],

Age	ka BC	"Qianbao, Dashuimentou, Sishilipo, Sanguan, Longfengpo"[9], and so on [11].
Iron Age	770 BC–220 AD	Sanguan [11], Sishimutan[11], Xiakangzhuang [11], and so on.

3. Spatial-temporal distribution of archaeological sites

The number of studied sites is perhaps different from ancient times, but the spatial distribution before the Han Dynasty in the Nihewan-Huliu Basin could be mapped if enough sites are collected. The temporal and spatial distribution is shown in Figure 2 and Figure 4 based on the ASTER GDEM 30 m. There are approximately 22 sites in the Paleolithic-Neolithic transition Age (among them, approximately 16 Mirothic sites), 57 sites in the Neolithic Age, 50 sites in the Bronze Age, 101 sites in the Iron Age; some sites cover multiple periods. The number of sites in different periods is shown in Figure 3 (P-N: Paleolithic-Neolithic, N: Neolithic, B: Bronze, I: Iron).

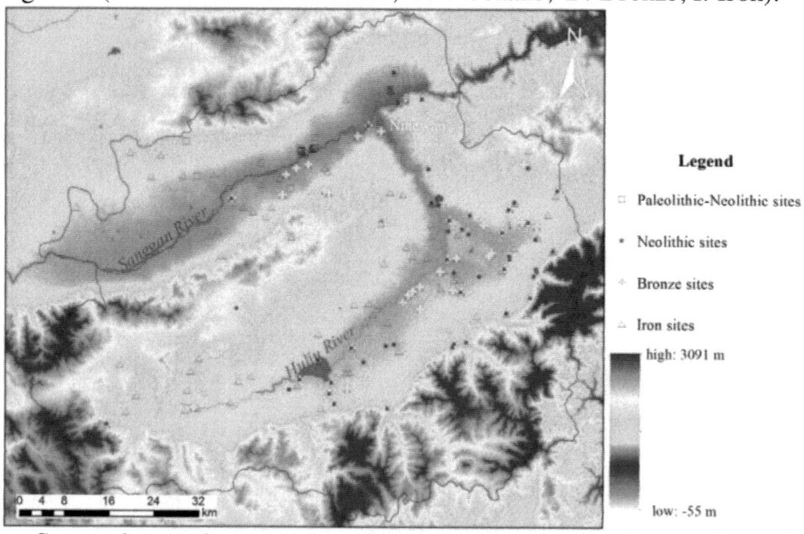

Figure 2. The distribution of archaeological sites in different stages.

Black squares: sites in the Paleolithic-Neolithic transition, Red points: sites in the Neolithic Age, White crosses: sites in the Bronze Age, Triangles: sites in the Iron Age.

Some observations can be made based on the distribution from the Microlithic Age to the Han Dynasty, as shown in Figure 2: (1) archaeological sites are mainly located in the valleys and tributaries of the Sanggan River and the Huliu River; (2) the distribution gradually expands from the Sanggan River basin to the Huliu River basin; (3) the sites in the Huliu River basin gradually expand from the region of the river's bend to the western headwater streams in the Heng Mountains; (4) before the Neolithic Age, the sites concentrated in the Nihewan Basin, however, in the Neolithic Age, only a few were located there, and since then, the Nihewan Basin became gradually depressed.

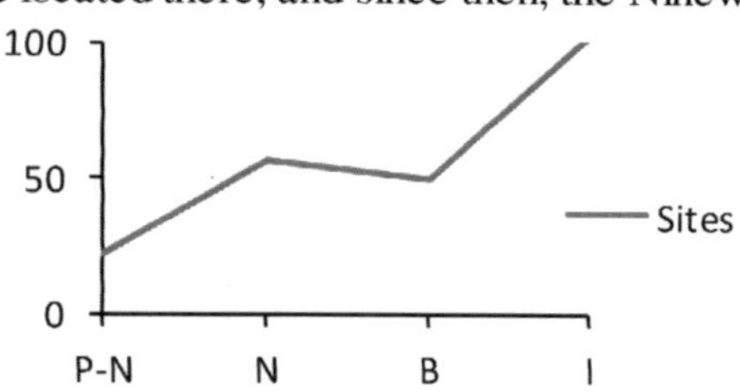

Figure 3. The number of archaeological sites in different stages

In the Paleolithic-Neolithic transition Age, the Paleolithic technology was predominant for a long time. The Small Stone and Microlithic culture appeared successively and coexisted in the Nihewan Basin, and the Microlithic culture became prominent gradually before the early Neolithic Age.

In the Neolithic Age, the archaeological sites were mainly concentrated in the east of the Huliu River's bend region and south of the upper river. At least 15 sites were located in the piedmont during this period, of which 11 sites were located at the foot of the Xiaowutai Mountains, 1 site at the front of the Heng Mountains, 1 site in the south of the mountains between the Sanggan River and Huliu River, and 2 sites at the front of the eastern mountains of the Huliu River. Fourteen of fifteen sites were located over 1 km above sea level.

Compared with the Neolithic Age, the number of archaeological sites did not change very much in the Bronze Age, while the scope became a little larger, extending to the south bank of the Sanggan River and Huliu River. Note that some piedmont sites disappeared during this period.

In the Iron Age, the total number of archaeological sites increased substantially. There were 76 sites in the Spring and Autumn and Warring States Periods, which were reduced to 55 sites in the Han Dynasty, as shown in Figure 4. In the Han Dynasty, more than 10 sites in the Sanggan River region disappeared and 13 sites appeared in the headwater of the Huliu River and the western mountains, indicating expansion and development to the west of the Huliu River Basin. The Huliu River Basin flourished because the plain surrounded by mountains was suitable for agriculture and cultivation. Comparatively, the Nihewan Basin lagged behind the Huliu River Basin after the Paleolithic-Neolithic transition Age, especialy it was more obvious after the Han Dynasty.

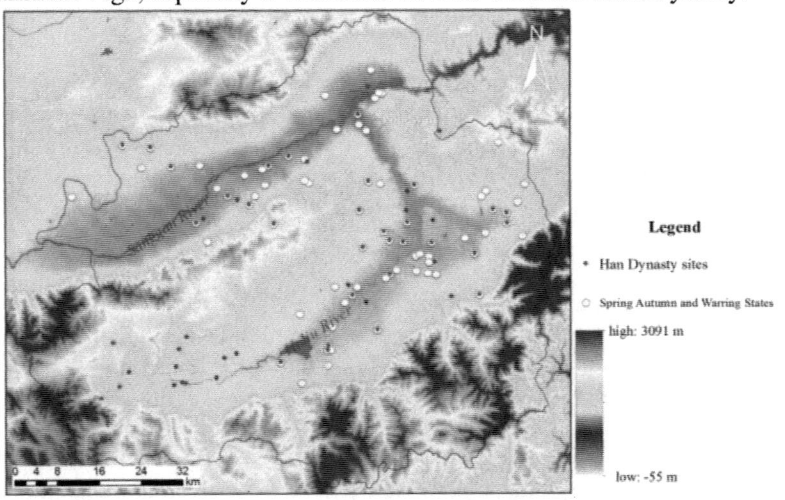

Figure 4. The distribution of archaeological sites in the Iron Age.

Circle points: sites in the Spring and Autumn and Warring States Periods, Red points: sites in the Han Dynasty

4. Relationship between the temporal-spatial distribution and the environments

4.1. Relationship between the archaeological sites and the environments

There are significant correlations between the spatial distribution of cultural relics and the Holocene environmental changes. As observed from the distribution mapping of archaeological sites, ancient relic densities had been high in the Nihewan-Huliu Basin since the late phase of the Late Paleolithic Age, with an especially unprecedented boom after the early Neolithic Age.

In the late phase of the Late Paleolithic Age, the climate was very cold and dry in the LGM (Last Glacial Maximum, approximately 26.5-19 ka BP) [16].The Holocene optimum reached a maximum during approximately 10-7 cal ka BP in north-central and northern east-central China [17].There were noticeable climatic changes during 4.6-4.3 cal ka BP [18] or the abrupt climatic shift at 4.5 cal ka BP [19]. After the changes and events, the general climate became cold and dry. Neolithic cultural transformation occurred in approximately 4.2-4 cal ka BP ago [19].

Abrupt unsuitable climate changes and cold events might have led to the migration of population sites, and even cultural turnover. Spatial foothill sites moving to higher elevations might be associated with wet weather or flooding in the Neolithic Age, because a relatively large amount of precipitation was likely to cause larger water areas and river flooding in the typical intermontane basin. When the water levels dropped and stabilized, the population moved again to the river banks or the wetlands in the Bronze Age and the Iron Age. There was a survival suitable platform with a relatively higher altitude at the corner of the Huliu River, where the general water expansion could be avoided, leading to a relatively high population density for several periods.

The relics from different periods, different types and different cultural backgrounds show close relationships between the archaeological cultures and the environmental changes. Comprehensive comparisons of the site number, climate and vegetation in the different periods are shown in Figure 5.

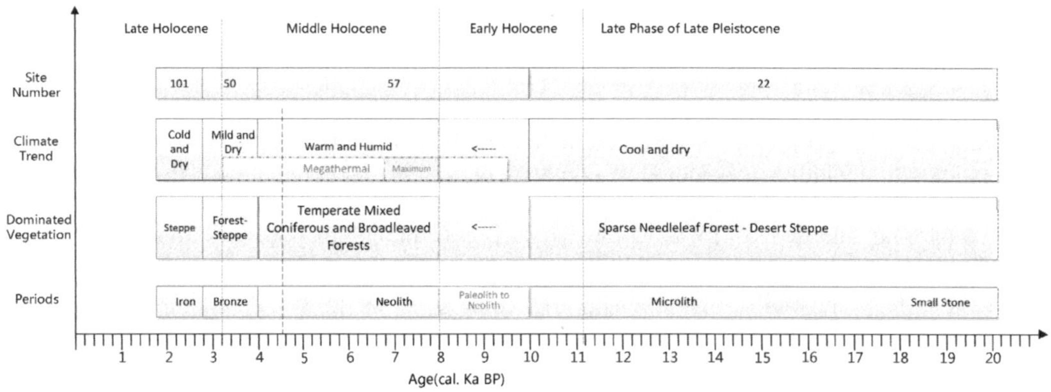

Figure 5. Comprehensive comparisons in different periods

The temporal and spatial distributions of archaeological sites were in good agreement with the main trend of the environmental changes. The archaeological sites were substantially increased in middle Neolithic Age. The agricultural and animal husbandry were more developed and the living technologies were improved although the cold trend of the climate, so the archaeological sites were almost increased and expanded in the Iron age.

4.2. The temporal-spatial distribution under the regional differences

There were differences in the regional environments of the study area. Compared with the Huliu River Basin, the vegetation and plants in the Nihewan Basin (Yangyuan Basin) were more sensitive to climate changes and human activities. In the late Paleolithic and early Neolithic Age, dominated by gathering and hunting, some forests grew in low mountainous areas [20], animals and plants were more prosperous and suitable for human living and survival in the Nihewan Basin.

After the development of new tools in the Neolithic Age, the relatively narrow geographical areas on the two sides of the Sanggan River were not conducive to large-scale original cultivation. Because of cold and drought events in the late Neolithic Age, some low mountainous forests quickly disappeared with human activities and never recovered [20]. The temporal and spatial distributions of archaeological sites show that cultural activities were significantly reduced, especially from the late Neolithic Age in the Nihewan Basin, when dense populations moved to the relatively flat and wide Huliu River Basin. Agricultural technologies had some ability to reclaim land from the Iron Age; although populations slightly increased on the basis of the sites in the Nihewan Basin, large-scale development still did not occur because of the limitations of the geographical environments. The Nihewan Basin was more uncultured than the Huliu River Basin after the Han Dynasty. In contrast, the Huliu Rriver Basin was suitable for various types of productions and livelihoods, so population and multicultural activities developed well.

5. Conclusions

The Nihewan-Huliu Basin witnessed almost successive human cultural evolution and important ancient activities. The temporal and spatial distributions of archaeological sites are consistent with the main trend of the environmental changes, the sites and reflected cultural activities gradually moved from the Nihewan Basin to Huliu River Baisn. The research results will provide valuable clues for much future researches regarding cultural responses to geographical and climatic environments, prehistoric cultural exchanges, and even the origin and evolution of the Chinese civilization.

References

[1] Zhu R X, An Z S, Potts R and Hoffman K A 2003 Magnetostratigraphic dating of early humans in China *Earth-Science, Reviews* **61** pp 341-359.

[2] Xie F, Li J and Liu L Q 2006 *Paleolithic Archeology in the Nihewan Basin* (Shijiazhuang,

Hebei: Huashan Art and Literature Press) (in chinese).
[3] Xie F 2006 Nihewan Basin (Beijing: Cultural Relics Press) (Chinese with English abstract).
[4] Barbour G B 1925 The deposits of the Sankanho Valley *Bulletin of Geological Society of China* **4** pp 53-55.
[5] Zhu R X *et al.* 2004 New evidence on the earliest human presence at high northern latitudes in northeast Asia *Nature* **431** pp 559-562.
[6] Gai P and Wei Q 1977 Discovery of the Late Palaeolithic Site at Hutouliang, Hebei *Vertebrata Palasiatica* **15** pp 287-300 (in chinese).
[7] Xia Z K, Chen F Y, Chen G, Zheng G W, Xie F and Mei H J 2001 Environmental background of evolution from the paleolithic to neolithic culture in Nihewan Basin, North China *Science in China(Series D)* **44** pp 779-788.
[8] Zhang Z P and Qiao L, 1992. A Study of the HouGang 1 culture *Acta Archaeologica Sinica* **3** pp 261-280 (Chinese with English abstract).
[9] Jiang G 2008 The Chronology, Genealogy and Patterns of archaeological cultures in the Southern Yanshan Area during the early second millennium BCE. *The Shanxi and Shaanxi Plateau and Yan Mountain During the 2nd Millennium BCE* ed J H Yang and G Jiang (Beijing: Science Press) pp 111-147 (Chinese with English abstract).
[10] Guo D S (translated by Shi G Q) 2010 *In Search of the Five Emperors* (Shenyang in China: Liaoning People's Publishing House) (Chinese edition), (United Kingdom: Design Book Limited) (English translated edition) 2012.
[11] Guojia Wenwuju (State Administration of Cultural Heritage) 2013 *Atlas of Chinese Cultural Relics: Hebei Volume* (Beijing: Sinomaps Press) (Chinese with English summary).
[12] Guojia Wenwuju (State Administration of Cultural Heritage) 2006 *Atlas of Chinese Cultural Relics: Shanxi Volume* (Beijing: Sinomaps Press) (Chinese with English summary).
[13] Hebeisheng Wenwuju (Hebei Provincial Administration of Cultural Heritage) 2015 *Shi qian yi zhi (Prehistory Sites)* (Beijing: Science Press) pp 173-188 (in Chinese).
[14] Hebeisheng Wenwuju (Hebei Provincial Administration of Cultural Heritage) 2015 *Gu dai yi zhi ji mu zang (Ancient Sites and Tombs)* (Beijing: Science Press) pp 86-92 and pp 157-160. (in Chinese).
[15] Mei H J 2006 The Late Paleolithic Localities at Meigou and Weidipo in the Nihewan Basin *Acta Anthropologica Sinica* **25** pp 299-307 (Chinese with English abstract).
[16] Clark P U, Dyke A S, Shakun J D, Carlson A E, Clark J, Wohlfarth B, Mitrovica J X, Hostetler S W and McCabe A M 2009 The Last Glacial Maximum *Science* **325** pp 710-714
[17] An Z S, Porter S C, Kutzbach J E, Wu X H, Wang S M, Liu X D, Li X Q and Zhou W J, 2000. Asynchronous Holocene optimum of the East Asian monsoon *Quaternary Science Reviews* **19** pp 743-762.
[18] Jin G Y and Liu D S (Liu Tungsheng) 2002 Mid-Holocene climate change in North China, and the effect on cultural development *Chinese Science Bulletin* **47** pp 408-413.
[19] Wu W X and Liu T S 2004 Possible role of the "Holocene Event 3" on the collapse of Neolithic Cultures around the Central Plain of China *Quaternary International* **117** pp 153-166.
[20] Hu Y N, Xu Q H, Zhang S R, Mu H S, Li Y and Li M Y 2015 Environmental changes and human impact since 6400 cal. a B.P. from Jijitan profile in Nihewan Basin *Quaternary Sciences* **35** pp 946-959 (Chinese with English summary).

Acknowledgments
This research was supported by the National Science Foundation of China (NSFC grant no. 41271427) and the International Scientific and Technological Cooperation Program of China (ISTCP grant no. S2013GR0477). We thank ASTER GDEM, RSCM, and CHGIS for the background imagery. The ASTER GDEM dataset was provided by International Scientific & Technical Data Mirror Site, Computer Network Information Center, Chinese Academy of Sciences, (http://www.gscloud.cn). Thanks to Dr. Luo Lei for meaningful proposals.

EOS situational data shared service mechanism

L Lv[1], Q Xu[1], C Z Lan[1], Q S Shi[1], W J Lu[1] and W Q Wu[2]

[1]Institute of Surveying and Mapping, Information Engineering University, 66 Longhai Road, Zhengzhou, Henan 450001, China
[2]61243 Troops, Urumqi, Xinjiang 830006, China

E-mail: lvliangvip@163.com

Abstract. With the rapid development of aerospace and remote sensing technology, various high-resolution Earth Observation Systems (EOS) are widely used in economic, social, military and other fields and playing an increasingly prominent role in the construction of Digital Earth and national strategic planning. The normal operation of the system is the premise of high quality data acquisition. Compared with the ground observation mode, EOS itself and the surrounding environment are more complex, and its operation control mainly depends on all kinds of Space Situational Awareness (SSA) data acquisition and analysis. SSA data has more extensive sources, larger volume, stronger time-effectiveness and more complicated structure than traditional geographical spatial data. For effective data sharing and utilization, combined with the analysis of data types and structures, a SSA data sharing identity language SSDSML is designed based on the extensible mark-up language XML, which realizes a comprehensive description of satellites and their attributes, space environment, ground stations, etc. Then EOS situational data shared service mechanism is established and provides a powerful data support for the normal operation of the system.

1. Introduction

The normal operation of the system is the prerequisite to obtain data with high quality. Compared with other systems, EOS itself and the surrounding environment are more complex. So awareness data has more sources, greater amount, stronger timeliness, and more complex structure. Due to different programming languages and architectures, the integration becomes very difficult, and this contradiction is particularly obvious in the sharing of spatial situation data. Traditional data sharing mechanism is often limited in conversion between spatial data with different structure, which mainly includes intermediate data format conversion, the standard spatial data exchange format conversion and direct access method[1][2].

The fundamental way to solve the multi-source and heterogeneous spatial situation data sharing lies in the open expression of perceptual data in Web environment. Based on XML, GML and SVG, Zhou Wensheng, etc.[3] created the expression system of open geospatial data. Distributed spatial data modeling and visual expression description language(DSDML) is proposed by Wang Peichao, etc.[4] based on XML, KML and GML, which has the function of geographical annotation, vector, raster, DEM data loading and can provide roaming control and distributed data integration capabilities. Lunar spatial data description and integration method is established by Tan Li, etc. [5] based on Digital lunar mark-up language(DLML), which can complete spatial data (including three-dimensional model) description, scientific results data description, geographic information tagging, scene roaming control and distributed data integration description. Based on the detailed analysis of the data types and structures, this paper designs a kind of space situational data sharing mark-up language Space situational data shared mark-up Language (SSDSML) based on the syntax and format of extensible mark-up language XML. And then EOS situational data shared service mechanism is established through the network service, which provides a powerful guarantee for the normal operation of the system.

2. Situational awareness data analysis and modeling

2.1. Situational awareness data composition

Data is the main body of sharing and a comprehensive understanding of the composition and characteristics is the prerequisite for sharing. The complexity of the environment around EOS, the diversity of space objects and particularity of space activities make the situational awareness data elements more complex. Generally speaking, the data is divided into two parts including environmental elements and spatial entities[6]. Figure 1 shows the composition of EOS situational awareness data elements.

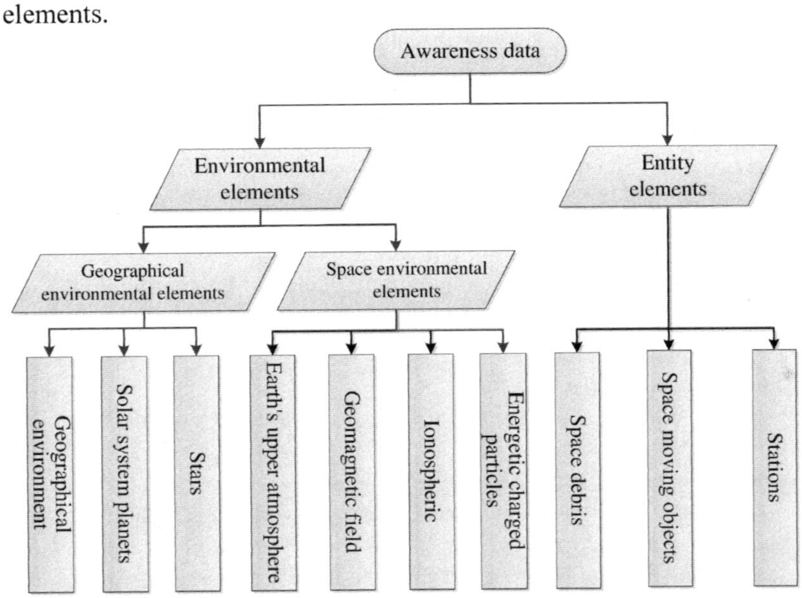

Figure 1. Situational awareness data composition

The basic geographical space environment is the foundation of any space activities, including not only the remote sensing images, vectors, DEM and other geographical environmental data, but also the background data of the solar system and other stars. Space environmental factors refer to the natural environment around the spacecrafts, including the Earth's upper atmosphere, Earth's magnetic field, ionosphere and energetic charged particles. Spatial entities mainly involve space debris, satellites and many ground stations for EOS.

2.2. Situational awareness data modeling

There are kinds of situational awareness data whose type are complex. If the model is to be customized for each element, it is not only a huge workload, but also causes the redundancy of the system structure. Through the analysis of all kinds of data, what can be found is that all of them have certain spatial and temporal characteristics no matter Earth's atmosphere, magnetic field and satellites, while difference embodies in the different geometries. As Figure 2 shows, in this paper, all elements are abstracted to an object of the base class, which is composed of three elements, property, chronology and geometry.

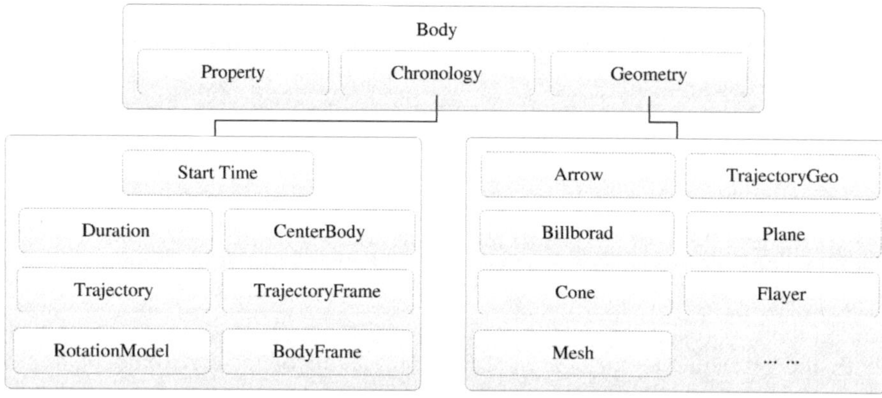

Figure 2. Situational awareness data modeling

Property is relatively simple, including the name of object, the color of logo and other information. Chronology information is responsible for the management of spatial-temporal data, which determines the position and attitude. Taking into account the multi-segments of ephemeris, ephemeris of track is syncopated according to the running time. Each section of the track also has its own characteristics, including the start time and the end time of track, the center of the object, the trajectory and its coordinate system, the rotation and its coordinate, etc. Geometric shapes can be divided into the arrow type, the bulletin board type, column type and so on according to the types of targets. Each one corresponds to a specific type of SSA data. The unified data organization method is simple in operation, convenient in management, great in expansibility and high in execution efficiency, and can ensure the reusability of the program.

3. SSDSML description function

3.1. SSDSML grammatical structure and rules

The syntax of SSDSML inherits from the XML standard and is defined using XML schema. The functional design refers to the DSDML[4]. SSDSML element design takes into account not only the composition of data elements, but also the operation of viewpoint and data integration. As shown in Figure 3, object is the base class element. Body element is the description for the target ontology model, which contains property, chronology and geometry. Viewpoint is used to control the position, attitude and posture of the field of view. Data integration element mainly considers three forms of access including database, network connections, and other services.

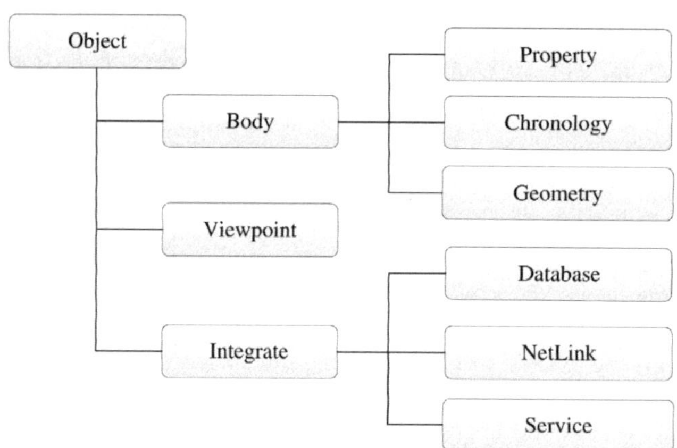

Figure 3. SSDSML element definition

3.2. Spatial data description

In the last chapter, we analyse the structure of EOS situational awareness data and carry out the unified data modeling. The description of spatial data is fully included in SSDSML, including attributes, ephemeris and geometry. Property elements complete the description of the ownership, type, attribution and summary information and so on, which is a record of the fixed information.

The description of ephemeris elements is rather special, because it also contains certain characteristics of the motion of the target track in addition to general stationary or moving objects. The description of the spatial data is more complex, especially in the track which can be divided into fixed trajectory, model trajectory and point track. They correspond to the ground station stationary targets, the observation satellites and debris, ground and near-earth targets respectively. In addition, ephemeris elements need to record the object coordinates, including local and orbit coordinate systems.

SSDSML contains various geometric shape description. Due to the complexity of data, in addition to the basic shape such as point, line, surface, and polygon and so on, it also supports the description of the shape of arrow, bulletin board, ionosphere and geomagnetic field, particle system and so on. In 3D scene, MDL format is used to describe the model data, which contains the description of the components, the attitude, the size and the proportion coefficient of model.

3.3. Viewpoint control description

Taking into account the needs of the 3D scene browsing, SSDSML design fully considers roaming. In order to achieve the precise control of roaming process, viewpoint elements are defined to describe the position, attitude and view, which can supply solution to solve the user's multi angle, full range of observation.

3.4. Distributed integrated description

EOS data sharing mechanism should include support for other data sources. Through the network link, SSDSML integrates the local data, the network database, TCP, UDP, OGC services and other SSDSML data sources. Data are integrated to form a new data source, thereby serve the visual application.

4. Prototype system architecture and application

4.1. Application architecture

EOS situational awareness data sharing is a new direction, and the sharing level is still relatively low now. Most systems mainly use local data and their ability to integrate data is insufficient. Based on GIS data sharing mechanism, SSDSML is used for scene construction, viewpoint roaming and data integration and so on. It can greatly enrich the content of EOS guarantee and supply a powerful support for the system operational control. Figure 4 shows the application architecture of SSDSML, which is composed of application layer, platform layer, analysis layer and data layer. Data sharing mainly depends on parsing layer and has nothing to do with the specific platform. It is responsible for parsing SSDSML files or other operations and obtaining situational awareness data elements and viewpoint roaming requirements. Then the data is sent to specific situational awareness platform to complete the scene construction and viewpoint control, thus the typical application scenarios are completed.

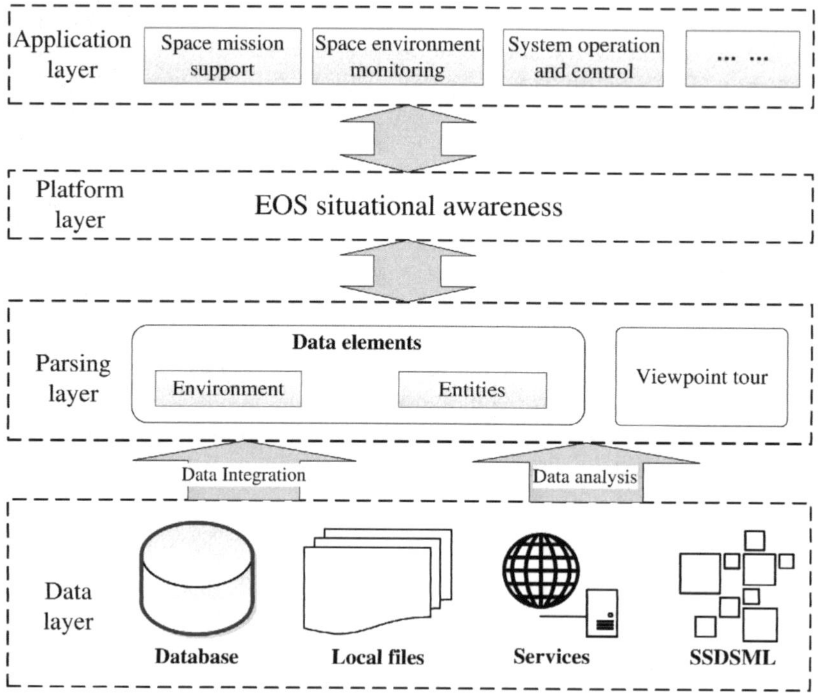

Figure 4. SSDSML Application Architecture

4.2. Prototype system

In this paper, the original space situation visualization system InSpace is upgraded and improved based on the SSDSML. The original InSpace system has complete functions and can realize environmental factors (including planets and space environment, imagery, topographic, vector) and spatial entities (space debris, spacecraft and ground station) loading, display and fundamental analysis. However, the ability of data sharing is insufficient because it is a stand-alone version. After the update InSpace is constructed by C/S architecture under the support of SSDSML and can complete the integration of multiple distributed data sources, greatly improving the timeliness of the shared. Then it provides a strong support for the operation and control of EOS. Figure 5 shows the InSpace system main interface, Figure 6, Figure 7 shows the data visualization results.

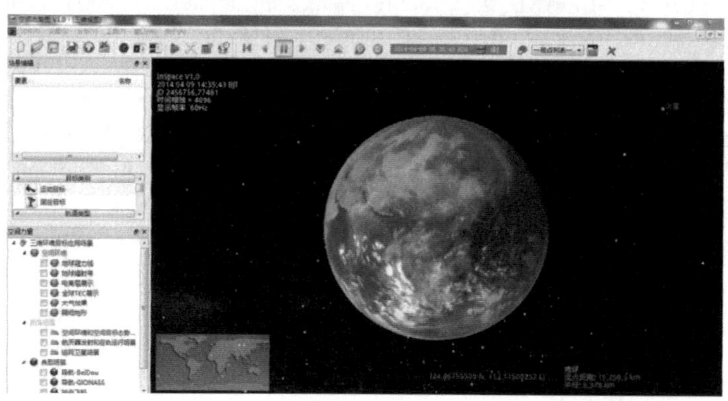

Figure 5. InSpace system main interface

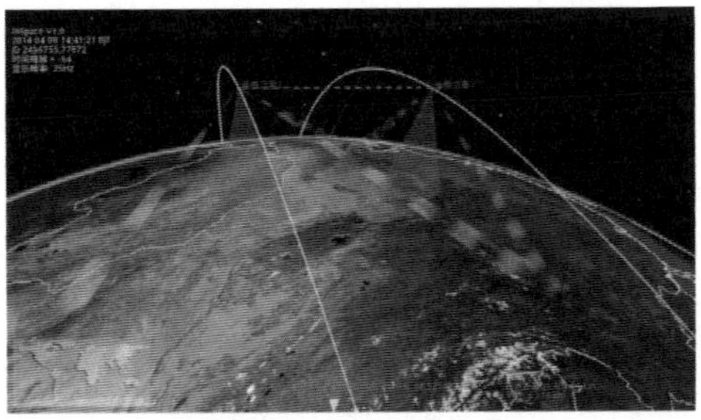

Figure 6. Cooperative observation of Earth Observation Satellites

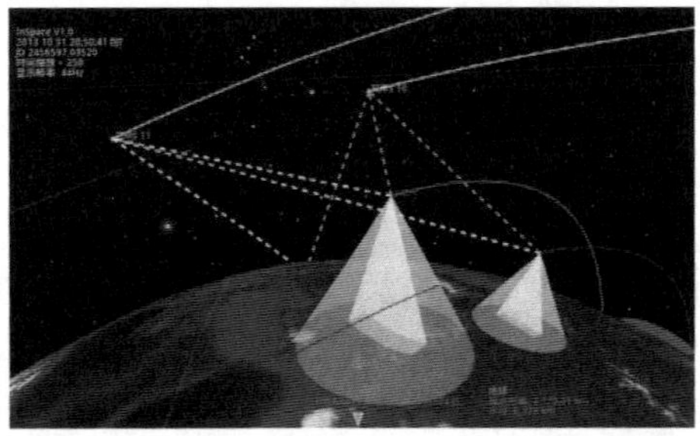

Figure 7. Ground coverage and satellite communication link

5. Conclusion

The control of EOS cannot be separated from the support of situational awareness data and the effective sharing of data can greatly improve the efficiency and accuracy of guarantee. The SSDSML provides the unified description of multi-source, multi-type situational awareness data. Experimental results show it can provide powerful support for all kinds of space applications and lay a solid foundation for the sharing mechanism. The kinds of elements of SSDSML will be extended based on development of the platform function constantly, and the description and analysis of the massive data will be researched next.

References

[1] Yu X 2011 *Research on Spatial Data Sharing Based on GML3.1* (Wuhan: Wuhan University Press) pp 12-15

[2] Wang Y D, Gong J Y, Huang J T and Deng Y J 2000 *Acta Geodaetica et Cartographica Sinica* **29** 143-144

[3] Zhou W S, Mao F, Hu P 2004 *Geomatics and Information Science of Wuhan University* **29** 43-47

[4] Wang P C, Zhu X Y, Su K H, Chen J 2009 *Geomatics and Information Science of Wuhan University* **34** 659-662

[5] Tan L 2011 *Research on key technologies for digital Moon platform based on G/S model* (Chengdu: Chengdu University of Technology Press) pp 51-68

[6] Lv Liang 2014 *Research on space situational picture construction and visualization* (Zhengzhou: Engineering in Information Engineering University Press) p 18

A Hierarchical Object-oriented Urban Land Cover Classification Using WorldView-2 Imagery and Airborne LiDAR data

M F Wu [1, 2], Z C Sun [2, 3], B Yang [1], S S Yu [2, 4]

[1] Hunan Normal University, Changsha Hunan 410006, China
[2] Key Laboratory of Digital Earth Science, Institute of Remote Sensing and Digital Earth, Chinese Academy of Sciences, Beijing 100094, China
[3] Hainan Key Laboratory Earth Observation, Sanya Hainan 572029, China
[4] Shandong University of Science and Technology, Qingdao Shandong 266510, China

Email: sunzc@radi.ac.cn

Abstract. In order to reduce the "salt and pepper" in pixel-based urban land cover classification and expand the application of fusion of multi-source data in the field of urban remote sensing, WorldView-2 imagery and airborne Light Detection and Ranging (LiDAR) data were used to improve the classification of urban land cover. An approach of object-oriented hierarchical classification was proposed in our study. The processing of proposed method consisted of two hierarchies. (1) In the first hierarchy, LiDAR Normalized Digital Surface Model (nDSM) image was segmented to objects. The NDVI, Costal Blue and nDSM thresholds were set for extracting building objects. (2) In the second hierarchy, after removing building objects, WorldView-2 fused imagery was obtained by Haze-ratio-based (HR) fusion, and was segmented. A SVM classifier was applied to generate road/parking lot, vegetation and bare soil objects. (3) Trees and grasslands were split based on an nDSM threshold (2.4 meter). The results showed that compared with pixel-based and non-hierarchical object-oriented approach, proposed method provided a better performance of urban land cover classification, the overall accuracy (OA) and overall kappa (OK) improved up to 92.75% and 0.90. Furthermore, proposed method reduced "salt and pepper" in pixel-based classification, improved the extraction accuracy of buildings based on LiDAR nDSM image segmentation, and reduced the confusion between trees and grasslands through setting nDSM threshold.
Keywords: object-oriented; classification; land cover; WorldView-2; LiDAR

1. Introduction

Very high resolution (VHR) remote sensing images are often used for urban land cover classification at a local scale, because detailed information in urban complex scene is presented in VHR images. But there are lots of challenges for deriving urban land cover types from single source VHR images [1,2], such as different objects having spectral similarity, same objects having spectral difference, shadows of tall buildings and large tree crowns, etc. Single source images can only provide a partial information, e.g. spectral information, of urban objects, however, more comprehensive information of urban area is not presented. Therefore, many researches are attempting to integrate multi-source data for urban land cover mapping. Authors reported that the fusion of moderate resolution images and VHR images was helpful for reducing mixed pixels and improving classification accuracy in large area [3,4]. The combined use of radar data and optical high resolution image were valuable for distinguishing high albedo roads and bare soils, reducing mixed pixels, and identifying objects in the shadows correctly [2,5,6]. Furthermore, the combination of Light Detection and Ranging (LiDAR) data and VHR optical imagery has been applied to urban land cover mapping in the emergence [7-9]. Because of adding derived height information from LiDAR data, the integration of LiDAR data and VHR images improved the separation of ground objects and non-ground objects, and LiDAR data was effective for extracting buildings and trees more completely and accurately [10,11].

Except for the synergistic use of multi-source data, some advanced approaches were proposed and applied to improve the classification accuracy of urban area. In order to reduce the "salt and pepper" noise in pixel-based classification, object-oriented classification method was used, based on a suitable image segmentation scheme. Object-oriented approach can make use of more comprehensive image information, such as spectral, spatial geometry and texture information and can effectively present the characters of urban objects. Many researches were attempting to use the parametric and non-parametric classifier, hierarchical rule-based technique in object-oriented approach to improve classification accuracy [12,13]. Object-based classification reduced the misclassification from spectral variation of individual objects, spectral similarity from different objects and the "salt and pepper" noise [14].

In our study, a hierarchical object-oriented classification approach based on the combine used of WorldView-2 imagery and airborne LiDAR data was proposed and applied to urban area. A comparison analysis of three approaches, including pixel-based from only WorldView-2 image, non-hierarchical object-oriental from fused data, and proposed hierarchical object-oriented from fused data, was carried out. The objective of this study is to expand the application of object-oriented approach using VHR optical image and airborne LiDAR data in the field of urban land cover classification.

2. Datasets and Method

In this study, an urban area was selected as the study area, located in Pingdingshan City, Henan Province, China. Collected WorldView-2 imagery was acquired on August 21, 2014, consisting of one panchromatic (PAN) band with the 0.5 meter resolution and eight multispectral (MS) bands (including Costal Blue, Blue, Green, Yellow, Red, Red Edge, NIR1, NIR2) with 2 meter resolution. Collected airborne LiDAR data was acquired on August 19, 2013 and was provided in ASCII format, including the X, Y, and Z coordinates and delivered first return intensities. The ground sample distances of LiDAR point clouds ranged from 0.12 to 0.4 meter, and the mean point density was 23.0 points/m^2. LiDAR intensity point clouds provided the height information of image objects, including derived digital surface model (DSM) and digital elevation model (DEM). Normalized DSM (nDSM) image, with 0.5 meter spatial resolution, which was generated by subtracting DSM with DEM, represented the above-ground feature and reduced the impact of the topography (Figure 1).

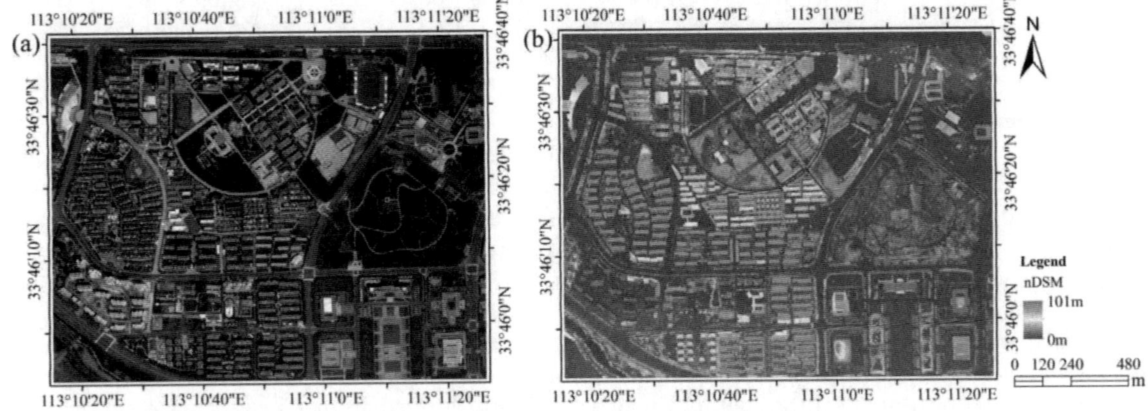

Figure 1. The WorldView-2 and LiDAR nDSM image of the study area.
(a): WorldView-2 fused imagery (band5,band3,band2 compound); (b): LiDAR nDSM data

Figure 2 showed the data procession for classifying urban land cover in this study, including data pre-processing, hierarchical object-oriented classification, and accuracy assessment and classification results analysis.

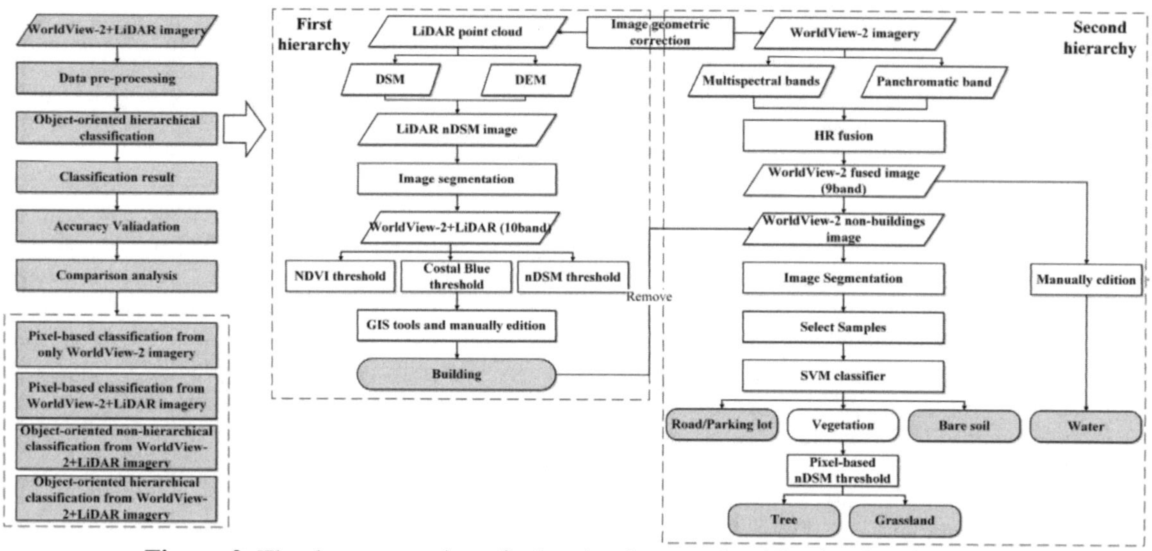

Figure 2. The data processing of urban land cover classification in this study.

2.1 Data pre-processing

During the data pre-processing, WorldView-2 MS and PAN ortho-rectified images were fused using a Haze-and-ratio-based (HR) fused scheme. The HR fused method, proposed by Jing [15], belonged to a PAN modulation (PM) based fusion method and took into account haze. HR fused scheme was presented as Equation (1).

$$\frac{M_{i,f} - h_i}{M_i - h_i} = \frac{PAN - h_p}{P_L - h_p} \tag{1}$$

Where $M_{i,f}$ $(i = 1..n)$ denoted the fused i th MS band, M_i was the corresponding i th MS band, PAN represented the high-resolution PAN image. h_p and h_i were related to atmosphere effects in PAN band and in the i th MS band, respectively. P_L was a spatially degraded PAN image by averaging the high-resolution PAN pixels. A fused eight WorldView-2 MS image, with 0.5 meter resolution, was generated through HR fusion.

WorldView-2 MS fused bands and PAN band were projected to a Universal Transverse Mercator (UTM) coordinate system with the corresponding rational polynomial coefficients (RPC) sensor camera model, and were registered to the LiDAR nDSM image (UTM projection), with less than 1 pixel of the Root Mean Square (RMS).

2.2 Hierarchical object-oriented classification

In this study, according to visual survey and the experience of predecessors [8,9], the land cover types in our study area were classified into six classes, including building, tree, road/parking lot, grassland, bare soil and water. Building and tree belong to non-ground object, and road/parking lot, grassland, bare soil and water belong to ground object. The water was masked before classification due to the little area proportion of the study area. The processing of classification included two hierarchies.

2.2.1 First hierarchy: LiDAR nDSM image segmentation and building extraction. Suitable image segmentation was a key factor of object-oriental classification and information extraction. In this study, the multi-resolution segmentation method, which was wide-applied in the image segmentation, was applied to objects classification. Different scale parameter, color/elevation criterion and shape criterion (including compactness and smoothness) were set for different land cover types. The building extraction was implemented, because of significant height and shape difference between buildings and other objects in LiDAR nDSM image. The multi-resolution segmentation was used in LiDAR nDSM image, and selected parameters were set through lots of experiments (scale: 30, elevation: 0.9, shape:

0.1, compactness: 0.5, smoothness: 0.5). Segmented objects were overlaid on 10band images (including eight fused WorldView-2 MS bands, one WorldView-2 PAN band and LiDAR nDSM data) for extracting the building's characters. Then a threshold-based approach was applied to building extraction. We recognized that the nDSM of buildings was higher than other ground objects, but the height of some low-rise buildings was similar to trees. The Normalized Difference Vegetation Index (NDVI) value of most buildings was lower than the vegetation (including trees and grasslands), but some buildings with blue roofs have a similar NDVI value to vegetation. However, the Costal Blue value of buildings with blue roofs was higher than vegetation significantly. Therefore, the intersection of nDSM, NDVI and Costal Blue thresholds were acquired and a little manually edition was carried out to generate the result of building extraction.

2.2.2 Second hierarchy: WorldView-2 image segmentation and objects classification. In the second hierarchy, building objects were masked. Although the height of ground objects (including road/parking lot, grassland, bare soil and water) were similar, it was considerable to discriminate those ground objects based on spectral and spatial properties. Therefore non-building image was classified based on the WorldView-2 image segmentation. The multi-resolution segmentation was used to WorldView-2 9band fused image (including eight fused MS bands and one PAN band) and selected parameters were defined (scale: 50, color: 0.9, shape: 0.1, compactness: 0.6, smoothness: 0.4). The Support Vector Machine (SVM) classifier was carried out in the segmented WorldView-2 9band image, with the kernel of Radial Basis Function (RBF). Based on Sun's conclusions, it was suitable that the Gamma parameter G was set to 0.1, and the penalty parameter C was set to 100 [16]. Selected samples include roads/parking lots, bare soils and vegetation. Trees and grasslands were merged into vegetation due to the complex shape and unsuitable scale of the segmentation. Selected features of objects include mean value of eight MS bands and one PAN band, as well as the length/width (the ratio of length and width of objects) of segmented objects, because the length/width of roads was larger than other objects. After SVM classification, an nDSM threshold was set to separate trees and grasslands from the vegetation based on pixel scale. Finally, the extracted buildings in the first hierarchy and the classified objects in the second hierarchy were merged and recoded to generate final classification result. The data processing was carried out on the platform of eCognition software (version 8.7).

2.3 Accuracy assessment and classification results analysis
As for accuracy validation, 510 random points were selected as the reference data in the WorldView-2 fused image. The number of created random points was 72 of buildings, 117 of trees, 160 of roads/parking lots, 141 of grasslands, and 20 of bare soils. The confusion matrix was constructed and the producer's accuracy (PA), the user's accuracy (UA), the overall accuracy (OA) and the overall kappa coefficient (Kappa) were calculated to analyse the classification accuracy. A visual comparison and accuracy analysis were carried out to four classification results: pixel-based classification from only WorldView-2 imagery (9band), pixel-based classification from WorldView-2 imagery and LiDAR nDSM data (10band), object-oriented non-hierarchical classification from 10band image, and proposed object-oriented hierarchical classification from 10band image. SVM classification was applied to all classification, and the scale parameters of object-oriented non-hierarchical classification was the same as segmented WorldView-2 imagery in the second hierarchy.

3. Results and analysis
According to the former methods, in the first hierarchy, the NDVI threshold was set as 0.2, Costal Blue threshold was set as 500, and nDSM threshold was set as 2.4 meter based on a lots of manual experiments. The intersection of NDVI greater than 0.2 and Costal Blue less than 500 were removed. The part of nDSM greater than 2.4 meter were extracted from removed intersection, and a little manual edition was carried out to generate building objects. In the second hierarchy, nDSM threshold was set as 2.4 meter to separate trees ($\geq$2.4 meter) and grasslands ($<$2.4 meter) from classified vegetation. Table 1 listed the accuracy validation of classification results.

Table 1. The accuracy validation of urban land cover classification in this study

Classification Method	Accuracy	Non-ground		Ground		
		Building	Tree	Road/Parking lot	Grassland	Bare soil
Pixel-based classification from only WV-2 imagery	PA(%)	44.44	70.94	68.75	66.67	70.00
	UA(%)	56.14	65.35	73.83	60.65	63.64
	OA(%)	65.29				
	OK	0.54				
Pixel-based classification from WV-2 and LiDAR image	PA(%)	90.28	88.89	76.88	88.65	65.00
	UA(%)	80.25	83.87	92.48	80.65	76.47
	OA(%)	84.31				
	OK	0.79				
Object-oriented non-hierarchical classification from WV-2 and LiDAR image	PA(%)	84.72	68.38	89.38	68.09	50.00
	UA(%)	91.04	67.23	83.63	67.13	100.00
	OA(%)	76.47				
	OK	0.68				
Proposed object-oriented hierarchical classification from WV-2 and LiDAR image	PA(%)	94.44	94.02	88.75	95.74	90.00
	UA(%)	95.77	95.65	94.04	91.22	72.00
	OA(%)	92.75				
	OK	0.90				

PA: the producer's accuracy; UA: the user's accuracy; OA: the overall accuracy; OK: the overall kappa coefficient; WV-2: WorldView-2

Table 1 showed that the lowest overall accuracy was yield in the Pixel-based classification from WorldView-2 imagery alone (OA: 65.29%, OK: 0.54). As for the classification from WorldView-2 imagery alone, because the confusion between building and road/parking lot was serious, PA and UA of the building were only 44.44% and 56.14%; PA and UA of road/parking lot were only 68.75% and 73.83%, respectively. In addition, the mixture between tree and grassland was also obvious due to the similar reflectance of near-infrared band.

Adding LiDAR nDSM data, the PA and UA of the building (PA: 90.28%, UA: 80.25%), road/parking lot (PA: 76.88%, UA: 92.48%), tree (PA: 88.89%, UA: 83.87%) and grassland (PA: 88.65%, UA: 80.65%) improved significantly in the pixel-based classification, because of height difference from building and road/parking, tree and grassland. However, the accuracy of bare soil increased inconspicuous and the PA of the road/parking lot was lower than the building. The phenomenon indicated that the problem of confusion between bare soil and road/parking lot was not solved efficiently.

In terms of object-based non-hierarchical classification from WorldView-2 and LiDAR image, the overall accuracy (OA: 76.47%, OK: 0.68) was lower than pixel-based classification from two data. Non-hierarchical classification based on object scale reduced the noise of classification results, however, a scale parameter of the segmentation were unsuitable to all classes. Much tiny area including different classes were missed, which lead to the decreasing of classification accuracy, especially the tree (PA: 68.38%, UA: 67.23%) and grassland (PA: 68.09%, UA: 67.13%).

The proposed object-oriented hierarchical classification from WorldView-2 and LiDAR data was the best accuracy, compared to other three approaches (OA: 92.75%, OK: 0.90). The PA and UA of the tree (PA: 94.02%, UA: 95.65%) and grassland (PA: 95.74%, UA: 91.22%) improved significantly because of setting nDSM threshold. Hierarchical segmentation for different land cover types was helpful for improving detailed land cover classification.

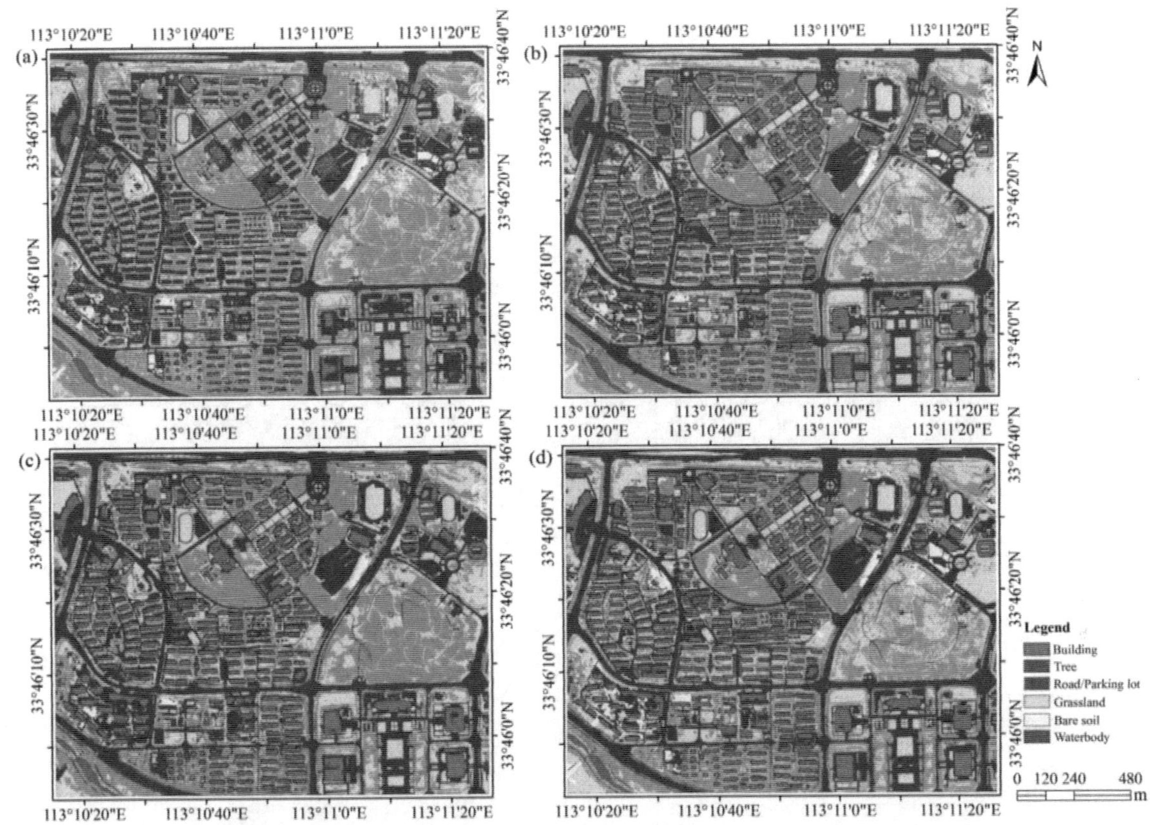

Figure 3. The classification results of urban land cover in this study.
(a): pixel-based from only WV-2 imagery; (b): pixel-based from WV-2 and LiDAR image; (c): object-oriented non-hierarchical from WV-2 and LiDAR image; (d): proposed object-oriented hierarchical from WV-2 and LiDAR image

Figure 3 showed the four classification results of our study area. Non-LiDAR classification cannot separate ground and non-ground objects with similar spectral characters effectively, such as building and road/parking lot, tree and grassland (Figure 3(a)). After adding LiDAR height information, the misclassification between building and road/parking lot was decreased obviously, and buildings were extracted completely (Figure 3(b)). Object-oriented classification reduced much "salt and pepper" noise but some detailed information was ignored. The non-hierarchical classification caused the extracted buildings incomplete and irregular, then many trees and grassland were merged and indistinguishable (Figure 3(c)). Proposed object-based hierarchical classification had the best visual presentation (Figure 3(d)).

Furthermore, the detailed comparison of four classification results was performed in Figure 4. As for extracting buildings (Figure 4a), the confusion between building and road was the most serious in the pixel-based classification from WorldView-2 image alone (Figure 4 a(2)). After adding LiDAR nDSM data, the building's extraction had a pleasant performance, however the some noise made extracted buildings incomplete (Figure 4 a(3)). Object-based non-hierarchical classification from two data decreased the noise, many irregular buildings were occurred in classification result. Because different scale parameters of image segmentation was suitable for different land cover types, single scale of segmentation cannot fit closely the boundary of objects (Figure 4 a(4)). Proposed method solved following problems effectively and extracted more regular and complete buildings (Figure 4 a(5)). Similarly, for roads extraction (Figure 4 b(1)-b(5)), Proposed method reduced the uncertainties and errors, including the misclassification from spectral similarity to road and building, speckle noises, and unsuitable scale parameters of segmentation. In terms of classified vegetation (Figure 4c), due to non-height information, classified trees and grasslands had an inferior performance in the pixel-based classification from WorldView-2 image alone (Figure 4 c(2)). Because of the overlarge scale of segmentation, many detailed presentation of trees and grasslands was incorporated (Figure 4 c(4)). Trees and grasslands were identified accurately using nDSM feature based on pixel scale (Figure 4 c(3), c(5)). Noticeably, for shaded objects, proposed method detected accurately the vegetation in the shadows. Other three methods had the mixture between shaded vegetation and low albedo roads (Figure 5 d(1)-d(5)). In summary, proposed method in the study was valuable for improving the accuracy of urban land cover classification.

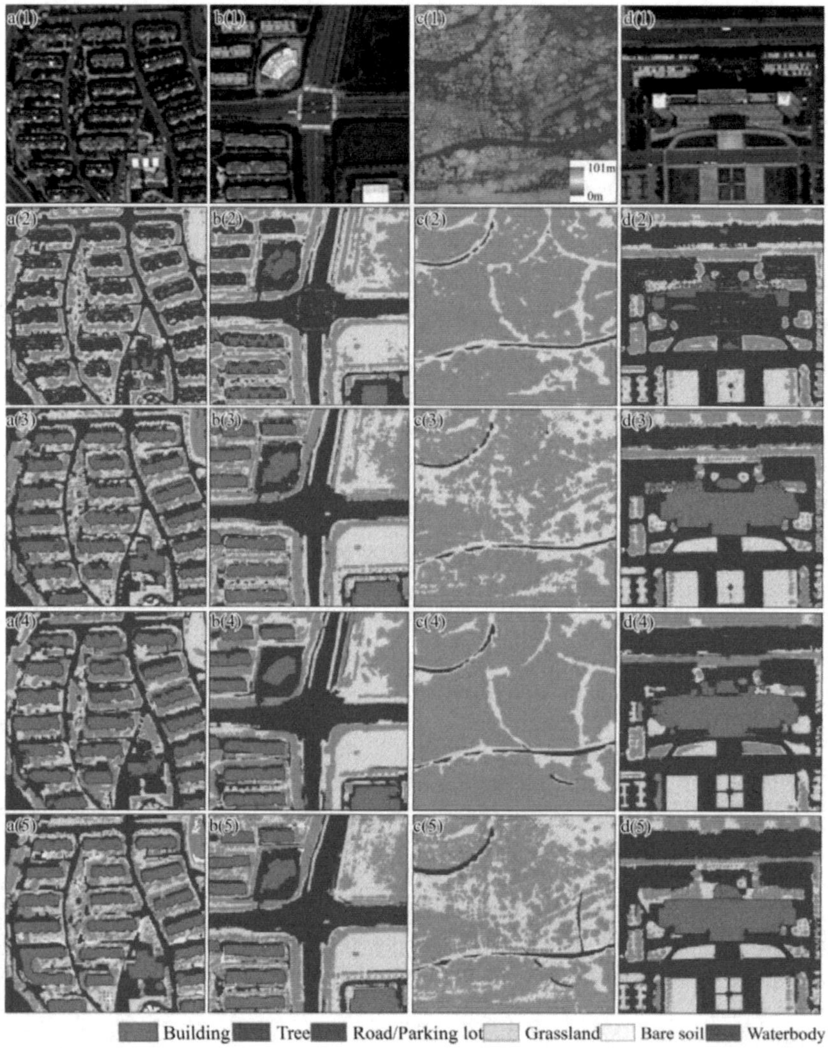

Figure 4. The detailed performance of the four classification methods.
WorldView-2 image: a(1), b(1), d(1); LiDAR nDSM data: c(1); Pixel-based classification from WorldView-2 imagery alone: a(2), b(2), c(2), d(2); Pixel-based classification from WorldView-2 and LiDAR image: a(3), b(3), c(3), d(3); Object-oriented non-hierarchical classification from WorldView-2 and LiDAR image: a(4), b(4), c(4), d(4); Proposed object-oriented hierarchical classification from WV-2 and LiDAR image: a(5), b(5), c(5), d(5)

4. Conclusions

In this study, based on the combined use of WorldView-2 imagery and airborne LiDAR data, an object-oriented hierarchical method was proposed and applied to urban land cover classification. Primary finds were obtained: (1) Proposed method yield the best accuracy and visual performance compared to other three approaches (OA: 92.75%, OK: 0.90). (2) Buildings extraction was complete and regular based on LiDAR nDSM image segmentation due to height and shape difference from buildings and other objects. Adding LiDAR height information reduced the confusion between buildings and roads/parking lots. (3) Compared with pixel-based method, object-oriented classification decreased the noise but ignored some detailed information. Unsuitable scale parameters of the segmentation decreased the accuracy of classified trees and grasslands. Correct and detailed discrimination between trees and grasslands was presented based on nDSM threshold in the vegetation. (4) Proposed method reduced the misclassification between shaded vegetation and low-albedo roads, which was helpful for identifying the objects in the shadows in the high resolution image. In conclusion, proposed method explored the potential of LiDAR data and hierarchical object-based classification in the urban land cover classification. However, proposed method increased the accuracy of bare soils non-clearly. How to improve the performance of classified bare soils will be further studied.

Acknowledgments

The paper is supported by the National Natural Science Foundation of China (No.41201357). The authors wish to acknowledge the experts providing the WorldView-2 imagery and airborne LiDAR data used for the study, and the anonymous reviewers for providing helpful suggestions that greatly improved the manuscript.

References

[1] Dengsheng L, Scott H and Emilio M 2011 Impervious surface mapping with Quickbird imagery. *Int. J. Remote Sens.* **32** 2519-2533
[2] Huadong G, Huaining Y, Zhongchang S, Xinwu L and Cuizhen W 2014 Synergistic use of optical and PolSAR Imagery for urban impervious surface estimation *Photogramm. Eng. Remote Sens.* **80** 91-102
[3] Mohapatra R P and Changshan W 2010 High resolution impervious surface estimation: an integration of IKONOS and Landsat-7 ETM+ imagery *Photogramm. Eng. Remote Sens.* **76** 1329-1341
[4] Tammy E P and James B C 2013 Comparing urban impervious surface identification using Landsat and high resolution aerial photography. *Remote Sens.* **5** 4942-4960
[5] Gaia Vaglio L, Veraldo L, Qi C, Leila G, Fabio Del F, Antonio B, David C, Beccy W, Jeremy L and Riccardo V 2013 Optical and SAR sensor synergies for forest and land cover mapping in a tropical site in West Africa. *Int. J. Appl. Earth Obs.* **21** 7-16
[6] Yuanzhi Z, Hongsheng Z and Hui L 2014 Improving the impervious surface estimation with combined use of optical and SAR remote sensing images. *Remote Sens. Environ.* **141** 155-167
[7] Yunhao C, Wei S, Jing L and Zhongping S 2009 Hierarchical object oriented classification using very high resolution imagery and LiDAR data over urban areas. *Adv. Space Res.* **43** 1101-1110
[8] Jungho I, Zhenyu L, Jinyoung R and Lindi J Q 2012 Impervious surface quantification using a synthesis of artificial immune networks and decision/regression trees from multi-sensor data. *Remote Sens. Environ.* **117** 102-113
[9] Shezhou L, Cheng W, Xiaohuan X, Hongcheng Z, Dong L, Shaobo X and Pinghua W 2016 Fusion of airborne discrete-return LiDAR and hyperspectral data for land cover classification. *Remote Sens.* **8** 1-19
[10] Wai Yeung Y, Ahmed S and Nagwa E 2015 Urban land cover classification using airborne LiDAR data: A review. *Remote Sens. Environ.* **158** 295-310
[11] Lei C, Shuhe Z, Wenquan H and Yun L 2015 Building detection in an urban area using LiDAR data and QuickBird imagery *Int. J. Remote Sens.* **33** 5135-5148
[12] Xiaoxiao L, Soe W M, Yujia Z Chritopher G, Xiaoxiang Z and Billie L Turner I 2014 Object-based land-cover classification for metropolitan Phoenix, Arizona, using aerial photography *Int. J. Appl. Earth Obs.* **33** 321-330
[13] Yuanrong H, Xinxin Z and Lizhong H 2016 Object-based distinction between building shadow and water in high-resolution imagery using fuzzy-rule classification and artificial Bee Colony optimization. *J. Sens.* **2016** 1-10
[14] Soe W M, Patricia G, Anthony B, Susanne G and Qihao W 2011 Per-pixel vs.object-based classification of urban land cover extraction using high spatial resolution imagery. *Remote Sens. Environ.* **115** 1145-1161
[15] Linhai J and Qiuming C 2009 Two improvement schemes of PAN modulation fusion methods for spectral distortion minimization. *Int. J. Remote Sens.* **30** 2119-2131
[16] Zhongchang S, Huadong G, Xinwu L and Linlin L 2011 Estimating urban impervious surfaces from Landsat-5 TM imagery using multilayer perceptron neural network and support vector machine. *J. Appl. Remote Sens.* **5** 1-17

Extracting potential bus lines of Customized City Bus Service based on public transport big data

Yibin Ren[1], Ge Chen[1, 2], Yong Han[*1, 2], Huangcheng Zheng[1]

[1]Qingdao Collaborative Innovation Center of Marine Science and Technology, College of Information Science and Engineering, Ocean University of China, China
[2]Laboratory for Regional Oceanography and Numerical Modeling, Qingdao National Laboratory for Marine Science and Technology, China

E-mail: yonghan@ouc.edu.cn

Abstract. Customized City Bus Service (CCBS) can reduce the traffic congestion and environmental pollution that caused by the increasing in private cars, effectively. This study aims to extract the potential bus lines and each line's passenger density of CCBS by mining the public transport big data. The datasets used in this study are mainly Smart Card Data (SCD) and bus GPS data of Qingdao, China, from October 11th and November 7th 2015. Firstly, we compute the temporal-origin-destination (TOD) of passengers by mining SCD and bus GPS data. Compared with the traditional OD, TOD not only has the spatial location, but also contains the trip's boarding time. Secondly, based on the traditional DBSCAN algorithm, we put forwards an algorithm, named TOD-DBSCAN, combined with the spatial-temporal features of TOD.TOD-DBSCAN is used to cluster the TOD trajectories in peak hours of all working days. Then, we define two variables P and N to describe the possibility and passenger destiny of a potential CCBS line. P is the probability of the CCBS line. And N represents the potential passenger destiny of the line. Lastly, we visualize the potential CCBS lines extracted by our procedure on the map and analyse relationship between potential CCBS lines and the urban spatial structure.

1. Introduction

Public transport is one of the most common ways of daily travel [1]. In recent years, Customized City Bus Service (CCBS) becomes a new popular mode of public transportation service. It not only improves the comfort of public bus, but also improves the efficiency of commuting. It is conducive to easing traffic congestion and environmental pollution caused by increasing of private cars. So, the promotion of using CCBS is very significant for urban management. In China, CCBS was first introduced and implemented in Qingdao in August 2013 and it spread to other cities quickly. By 2015, there have been more than 30 cities across the country running CCBS systems [2]. Currently, the way of setting a CCBS line is "from bottom to top": passengers send the origins and destinations of their trips to public transportation management department through APP or Web page; bus department collects all requests and decides which lines to be set up. This way is limited by network and mobile devices. Thus, if the bus management department takes the initiative to analyze the potential CCBS lines, there will be an effective complementary way to improve the use of CCBS.

The way to discover potential CCBS lines is finding the similarities of transits between different passengers. So, we need to grasp the daily travel characteristics of passengers. Smart Card Data (SCD) records the travel process of the cardholder, widely covered and quickly accessed [3]. We can mine the passengers' origin and destination (OD) from consumption records of SCD and further extract the potential CCBS lines by analyzing spatio-temporal characteristics of OD trajectories. Mostly, SCD doesn't record the boarding station when cardholder brushing the card. We need to combine SCD with bus GPS data to get passenger's boarding station [4]. Then, we can deduct the alighting station based on origins by trip chain model [5, 6]. A series of study work have been done to extract OD information and analyze the spatio-temporal characteristics of passengers. Nishiuchi H et al [7] analyze variations in trip patterns to understand how passengers' daily travel patterns vary temporally and spatially among one month. Ma X et al [8] use the K-Means++ clustering algorithm and the rough-set theory to cluster and classify travel pattern regularities. And they develop a data-driven platform for online transit performance monitoring [9]. Tao S et al [10] analyze the travel differences between BRT and non-BRT passengers. Zhao J et al [11] study the spatio-temporal travel patterns of individual passengers in metro system of Shenzhen, China. Recently, some researchers focus on the travel characteristics of specific groups. Long et al [12] analyzes the travel characteristics including travel frequency, commute time, as well as jobs-housing location and their change of underprivileged residents in Beijing by SCD. And they also study the travel characteristics of 4 kinds extreme public transit riders (early birds, night owls, and tireless/recurring itinerants) [13].

Existing studies mentioned above are focusing on OD deduction, commuting analysis and travel regularity discovering etc. The studies about discovering similarities between passengers public travel behaviors which is aiming to extract potential CCBS lines are a little fewer. Therefore, this study aims to extract the potential CCBS lines and each line's passenger density by mining the public transport big data. We propose temporal-origin-destination (TOD) data structure and extent the traditional DBSCAN (Density-Based Spatial Clustering of Applications with Noise) algorithm in both spatial and temporal dimensions, forming TOD-DBSCAN algorithm. Then, we use the TOD-DBSCAN to find the similarly travel patterns and extract the potential CCBS lines. At last, we visualize the potential CCBS lines on the map by a Web GIS system and make some discussion.

The remainder of the paper is structured as follow: Section 2 introduces the public transport big data used in this study. Section 3 expounds the procedure to extract potential CCBS lines, including construction of TOD, TOD-DBSCAN algorithm, and extraction and visualize of potential CCBS lines. In Section 4, we come to the conclusion of this study.

2. Public transport big data

The dataset used in this study is provided by the Qingdao Public Transportation Group, which has around 5000 buses on the core roads of Qingdao. The dataset includes SCD data, bus GPS data and scheduling table of drivers. The SCD data contains more than 50 million transactions, collected from 1.2 million smart cards between October 11th and November 7th 2015. The format of SCD record is shown in table 1. The bus GPS data is an important auxiliary data to extract boarding station of cardholders, because it records the time when the bus arriving and leaving the station, table 2. And the scheduling table of drivers records the relationship between bus and driver, table 3. It is the bridge to link SCD and bus GPS Data.

Table 1. Data structure of SCD

Field Name	Description
CardID	Unique number for each card.
RouteName	Line name of this transaction.
DriverID	Driver's ID number.
ActTime	Detial time of the transaction.

Table 2. Data structure of bus GPS data

Field Name	Description
BusID	Unique number of bus.
RouteName	Line name of this bus.
StationNum	Station number of this line.
ActTime	Detial time of the action.
Flag	Arriving or leaving.

Table 3. Data structure of schedule table

Field Name	Description
BusID	Unique number of bus.
RouteName	Line name of this bus.
DriverID	Number of driver on this bus.

3. Extracting potential CCBS lines by TOD-DBSCAN algorithm

3.1. Construction of TOD

In order to construct the TOD, we should extract OD trajectory firstly. The Automatic Fare Collection System of Qingdao is flat fares: passengers pay a fixed price and they brush the card only when they boarding on the bus. So, we need to extract the passenger's origin (boarding station) firstly and then deduce the destination (alighting station) based on the origin.

As shown in table 1, the SCD record doesn't contain boarding station. The bus GPS data records arriving time and leaving time of each station. So, we can calculate the boarding station by matching card time with arriving and leaving time of the bus. The scheduling table contains the corresponding relationship between the driver and the bus. Thus, we can match SCD record with bus GPS data through schedule table (Step 1 in figure1.) to get the corresponding relationship between SCD and bus GPS. Actually, some passengers brush the card after the bus leaving station. So, we use the arriving times of two adjacent stations as time threshold, figure1. Further, for the deduction of alighting station, we use the trip chain model which is based on two assumptions[14]: ① The previous trip destination is nearby the bus stop of next trip origin; ② Passengers return to the first boarding station of the day at the end of the day. They are widely adopted by most studies.

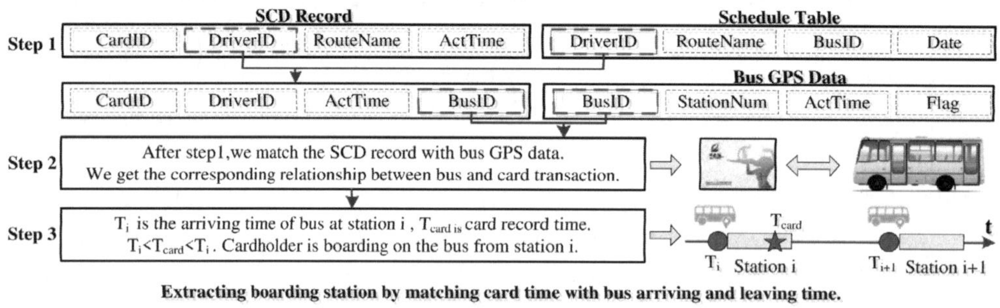

Figure 1. Steps to extract boarding station

After above processes, we got the passengers' OD trajectories. But a complete OD maybe divided into several ODs by transfer stations. We need to find the transfer behavior and merge them to a complete one. We compute the time interval between current boarding behavior and last alighting behavior. If the time interval is less than 30 minutes, we merge two ODs to one, (figure 2). Since we have obtained the corresponding relationships between transactions and bus, and the alighting station is known, so the alighting time is the time when the bus leaving from the alighting station.

Based on three datasets mentioned in Section2, we extract OD trajectories which including boarding and alighting station, boarding and alighting time. In order to analyze the spatial and temporal similarity of OD trajectories, we propose TOD (temporal-origin-destination) data structure :{ ID, T_B, (X_O, Y_O), (X_D, Y_D)}. ID is the card ID, T_B is the boarding time at origin station, (X_O, Y_O) is the coordinate of origin station, and (X_D, Y_D) is the coordinate of destination station. Taking card "2***26" (figure 2) for example, its TOD model is {2***26, 2015-10-12 07:13:15, (120.34783, 36.0822), (120.42315, 36.07415)}. Because the alighting time is impacted by various factors, we don't take it as a factor to analyze the similarity between OD trajectories. And it's not included in TOD. Then, we will detail the algorithm to cluster TOD trajectories.

Raw OD

CardID	Date	Boarding time	Alighting time	OD
2***26	2015.10.12	07:13:15	08:17:17	Line 125:Taidong ->Qingdao University
2***26	2015.10.12	08:24:32	08:37:56	Line 227:Qingdao University->Qingda Yilu

(08:24:32 - 08:17:17) = 7.25 minutes < 30 minutes, merging two OD trajectories.

Final OD

CardID	Date	Boarding time	Alighting time	OD
2***26	2015.10.12	07:13:15	08:37:56	Line 125:Taidong -> Line 227: Qingda Yilu

Figure 2. Modifying raw OD trajectories

3.1 TOD-DBSCAN algorithm

We analyze the spatial and temporal similarities of OD trajectories by clustering operation. DBSCAN algorithm is not required to define the number of clusters or identify arbitrarily shaped clusters because higher-density records are more likely to be grouped into a cluster [15]. It needs two parameters: distance threshold (ε) and the minimum number of points (*MinPts*). With two parameters, it calculates the connected relation between points iteratively, forming result clusters. In 2014, Kieu L M [16] used DBSCAN algorithm to cluster OD trajectories, including 3 steps: clustering alighting points, clustering boarding points and clustering boarding time. He extended the traditional DBSCAN algorithm from one dimension (point) to three dimensions (line segment with time information). But, this method clusters identical OD trajectories repeatedly, increasing the clustering time, especially for long time series public transport big data. To solve this problem, we analyze the characteristics of public transport OD trajectories, designing TOD-DBSCAN algorithm. The new algorithm contains two parts: spatial clustering and temporal clustering. The ε is no longer just a space distance. It becomes a combination of spatial distance (l) and temporal distance (t).

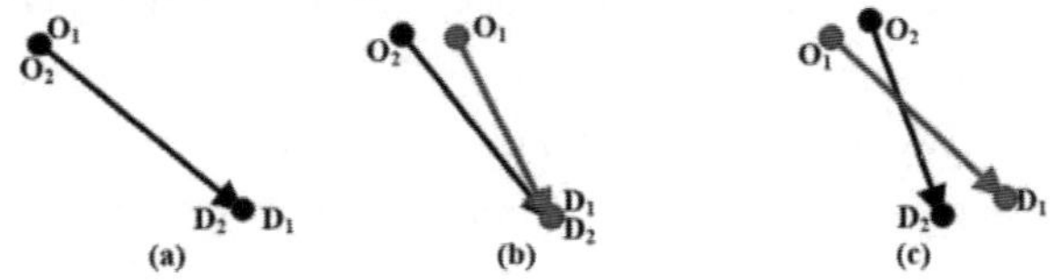

Figure 3. OD trajectories which are similar in spatial

We illustrate the spatial clustering process firstly. OD trajectories which are similar in spatial are show in figure 4: complete overlap (figure 3(a)), only one point overlap (figure 3(b)) and complete under lap but both terminals are within limited distance (figure 3(c)). So, in the process of OD clustering, we should judge whether origins and destinations between different OD trajectories are within the distance threshold respectively. The bus stations in the city form a fixed points set. And both the origin and the destination are all belong to this points set. So, we take the unique origins and unique destination of all OD trajectories instead of all origins and destinations as the cluster objects to find the spatial similarities. Take the situation in figure5 for example. There are 10 OD trajectories in figure 4(a). Using the method of Kieu L M, we have to clustering 20 points, even if there are some duplicate points. But in our procedure, firstly, we extract the unique points collection of origins and destinations: $\{O_1,O_2,O_3,O_4,O_5,O_6\}$ and $\{D_1,D_2,D_3,D_4,D_5\}$ (figure 4(b)). Then, we cluster two point collections respectively by DBSCAN, froming 2 origin clusters (OC_1, OC2) and 3 destination clusters (DC_1, DC2, DC3) (figure 4(c)). And the number of clustering objects is reduced from 20 to 11. Lastly, for each group in origin clustered results, we select OD trajectories whose origins belong to the group and cluster the OD trajectories based on result clusters in step 2, forming the final spatial clustering results (figure 4(d)). Taking OC_1 for example, there are 5 OD trajectories (OD-ID:1,2,3,7,10) whose origins are included in OC_1. And the 5 destinations (D_1,D_2,D_3,D_3,D_4) are clustered into 3 groups(DC_1,DC_2,DC_3) in step 2. So, combining the origin clusters with destination clusters, these 5 OD trajectories are clustered into 3 groups (Cluster ID: 1, 2, and 3). In this way, all the 10 OD trajectories are clustered into 5 groups by spatial clustering process.

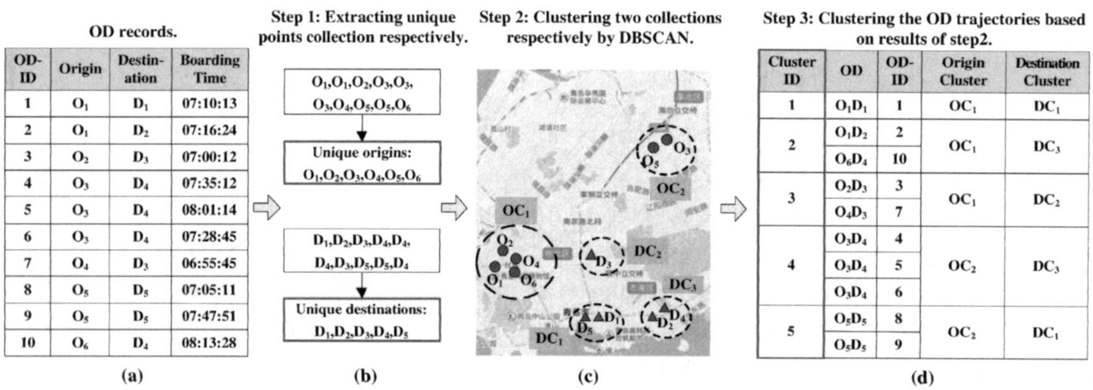

OD-ID	Origin	Destin-ation	Boarding Time
1	O_1	D_1	07:10:13
2	O_1	D_2	07:16:24
3	O_2	D_3	07:00:12
4	O_3	D_4	07:35:12
5	O_3	D_4	08:01:14
6	O_3	D_4	07:28:45
7	O_4	D_3	06:55:45
8	O_5	D_5	07:05:11
9	O_5	D_5	07:47:51
10	O_6	D_4	08:13:28

Cluster ID	OD	OD-ID	Origin Cluster	Destination Cluster
1	O_1D_1	1	OC_1	DC_1
2	O_1D_2	2	OC_1	DC_3
	O_6D_4	10		
3	O_2D_3	3	OC_1	DC_2
	O_4D_3	7		
4	O_3D_4	4	OC_2	DC_3
	O_3D_4	5		
	O_3D_4	6		
5	O_5D_5	8	OC_2	DC_1
	O_5D_5	9		

Figure 4. Spatial clustering process of TOD-DBSCAN algorithm

Temporal clustering is relatively simple. After above processing, OD trajectories which are similar in spatial are clustered into one spatial group. Temporal clustering is clustering all trajectories from one group in time dimension. As mentioned above, we use the boarding time to represent the temporal information of OD trajectory. And the boarding time is recorded in TOD, so we just need to set the boarding time as cluster variable, clustering TOD trajectories belong to one spatial group using DBSCAN algorithm with the time distance. For example, OD trajectories whose ID numbers are 4, 5 and 6 are clustered into the 4th group (cluster ID is 4 in figure 4(d)). After temporal clustering, they are clustered into 2 groups ($cluster_1$: OD_4 and OD_6; $cluster_2$: OD_5) with the time distance 15mins finally (figure 5).

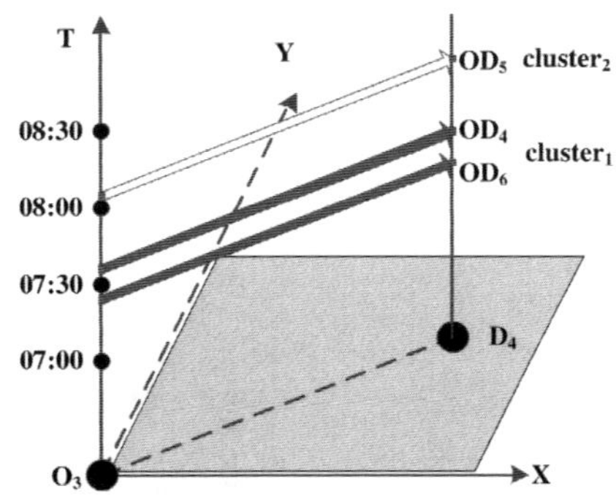

Figure 5. Temporal clustering process

3.2 Extracting potential CCBS lines using TOD-DBSCAN

The main idea to extract potential CCBS lines is to discover the similarities between OD trajectories of working day. We use TOD-DBSCAN algorithm to cluster TOD trajectories. Our analysis objects are TODs whose T_B values are in morning peak period (06:30am ~ 08:30am). In order to determine the spatial distance threshold, we change the threshold value to cluster the unique boarding stations collection (a total of 2673 stations), setting l as 100m, 250m and 500m respectively. The *MinPts* in this stage is 0, because a single station should also become one group. The results are shown in Figure 6. The number of clusters are 1759 (l=100m), 1089 (l=250m) and 352 (l=500m). The spatial clustering achieves a better result with 250m. When l is 100m, the clustering results are too scattered. And when it increases to 500m, the results are too concentrated. Both two values can't be very good to characterize the spatial proximity between stations. So, we use the 250m as spatial distance threshold. In temporal clustering stage, we set the value of t as 15 minutes. And the weekends of four weeks aren't included. The *MinPts* is 300, because if the passenger destiny less than 15 per day on average, there is no need to set the line.

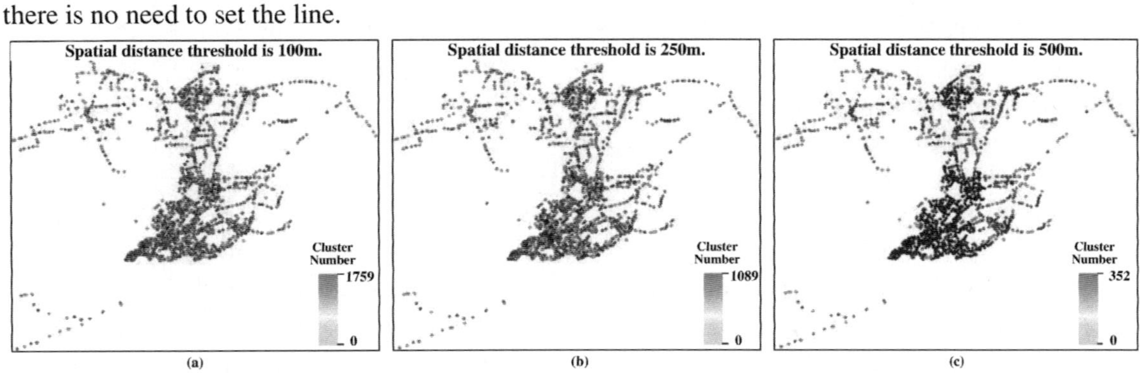

Figure 6. Clustering results of unique boarding stations with different values of l

After spatial and temporal clustering, we define two variables P and N. P is the probability of the CCBS line, and it can calculate by equation 1: P_i is the probability of line i, M_i is the number of days in group i, M_a is the number of all working days during the dataset's time span (it's 20 in this study). If P_i is less than 0.6(3 days a week), there will be no need to set the potential CCBS line. N is the number of potential passengers of the line. It is calculated by equation 2, S_i is all the passengers in group i, and M_i is the number of days in group i. Generally, most groups contain multi origins and destinations, and we select the one with the highest frequency as the origin or destination of CCBS line. The departure time of the line is also determined in the same way and we set 5mins as time unit to half-adjust it.

$$P_i = \frac{M_i}{M_a} \tag{1}$$

$$N_i = \frac{S_i}{M_i} \tag{2}$$

Table 4. CB lines extracted by our procedure

ID	Origin	Destination	P	N	Time
1	Shiji Park	International Arts and Crafts City	1	56	07:15:00
2	Xinglong Yi Road	News Center	1	54	07:20:00
3	Changcheng Road	Jiangxi Road	1	54	06:40:00
4	North Station of Yan'an Road	Lily Garden	1	53	07:30:00
5	East Bus Station	Ferry	1	52	07:00:00
...	......	......	...	...	......
49	Jingkou Road	Hongkong Zhong Road	1	20	07:05:00

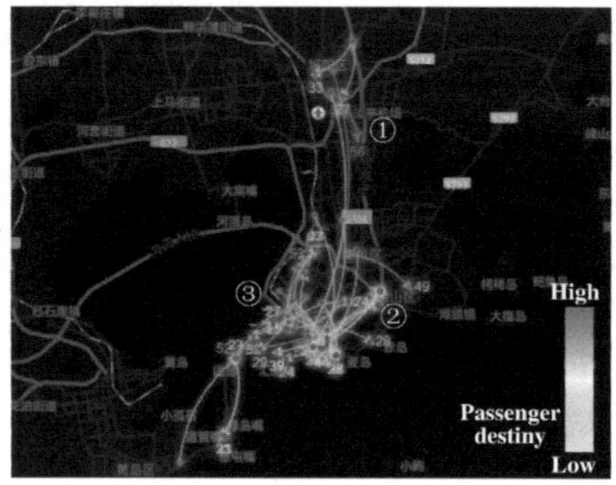

(a)

Figure 7. CCBS line results extracted by our procedure

The details of the results are shown in table 4. There are 49 potential CCBS lines extracting by our procedure. And they mainly contain 3 directions: from north to south (① in figure 7 (a)), from east to south and west (② in figure 7 (a)) and from Taidong to Qingdao North Station (③ in figure 7 (a)). This phenomenon is related to the urban structure of Qingdao. The southern coastal line is the most prosperous region in Qingdao, containing a series of companies, shopping malls and tourist areas. And the northeast part of the city is a new residential area with a relatively low housing price, attracting a lot of young people. So, they have to across the city for work. And the Taidong is a representative residential of the old city. Traditional heavy industry is distribution near the north railway station and it is the main work place of the older generation. The potential CCBS lines are corresponding to the urban structure of Qingdao.

4. Conclusion

This study aims to extract potential CCBS lines by mining public transport big data. For this goal, we propose the TOD data structure and extend the traditional DBSCAN algorithm to TOD-DBSCAN. We cluster all the TOD trajectories from October 11th to November 7th 2015 by TOD-DBSCAN algorithm. A total of 49 potential CCBS lines are extracted from the clustering results. They are corresponding to the urban structure and visualized in a Web GIS system. Next step, we will cooperate with the public transportation department to validate the key lines.

Acknowledgement

This work is supported by Science and Technology Development Plan of Qingdao; National High Technology Research and Development Program of China (863 Program): System of Ocean Observation Based on Shipboard UAV (Grant NO. 2014AA09A511).

References

[1] Carmien S, Dawe M, Fischer G, Gorman A and Kintsch A 2005 Socio-technical environments supporting people with cognitive disabilities using public transportation *ACM.TOCHI* **12** 233-262

[2] Liu T and Ceder A 2015 Analysis of a new public-transport-service concept: Customized City Bus Service in China *Transport. Policy* **39** 63-76

[3] Pelletier M, Trépanier M and Morency C 2011 Smart card data use in public transit: A literature review *Transport. Res. C.* **19** 557-568

[4] Ma X, Wang Y, Chen F and Liu J 2012 Transit smart card data mining for passenger origin information extraction *J. Zhejiang Univ. - Sci. C* **13** 750-760

[5] Barry J, Newhouser R and Rahbee A 2002 Origin and destination estimation in New York City with automated fare system data *Transport.Res.Rec.* **1817** 183-187

[6] Trépanier M, Tranchant N and Chapleau R 2007 Individual trip destination estimation in a transit smart card automated fare collection system *J INTELL TRANSPORT S* **11** 1-14

[7] Nishiuchi H, King J and Todoroki T 2013 Spatial-temporal daily frequent trip pattern of public transport passengers using smart card data *International Journal of Intelligent Transportation Systems Research* **11** 1-10

[8] Ma X, Wu Y J, Wang Y and Liu J 2013 Mining smart card data for transit riders' travel patterns *Transport.Res.C* **36** 1-12

[9] Ma X, Wang Y 2014 Development of a data-driven platform for transit performance measures using smart card and GPS data *J.Transp.Eng*, **140**(04014063)

[10] Tao S, Corcoran J, Mateo-Babiano I and Rohde D 2014 Exploring Bus Rapid Transit passenger travel behaviour using big data *Appl Geogr* **53** 90-104

[11] Zhao J, Tian C, Zhang F and Xu C 2014 Understanding temporal and spatial travel patterns of individual passengers by mining smart card data *IEEE 17th International Conference on Intelligent Transportation Systems (ITSC)* p 2991-2997

[12] Long Y, Shen Z 2015 Profiling underprivileged residents with mid-term public transit smartcard data of Beijing *Computer Science* 2015 169-192

[13] Long Y, Liu X, Zhou J and Chai Y 2015 Early Birds, Night Owls, and Tireless/Recurring Itinerants: An Exploratory Analysis of Extreme Transit Behaviors in Beijing, China *Computer Science* 2015(1502.02056)

[14] Li D, Lin Y, Zhao X and Zou N 2011 Estimating a transit passenger trip origin-destination matrix using automatic fare collection system *International Conference on Database Systems for Advanced Applications* (Springer Berlin Heidelberg) p 502-513

[15] Ester M, Kriegel H P, Sander J and Xu X 1996 A density-based algorithm for discovering clusters in large spatial databases with noise *Kdd.* **96** 226-231

[16] Kieu L M, Bhaskar A, and Chung E 2014 Transit passenger segmentation using travel regularity mined from Smart Card transactions data *Engineering* 2014

A classifying method analysis on the number of returns for given pulse of post-earthquake airborne LiDAR data

Jinxia Wang Aixia Dou Xiaoqing Wang Shusong Huang Xiaoxiang Yuan

Institute of Earthquake Science，China Earthquake Administration. Beijing 100036，China

E-mail: wangjxwake@163.com axdothy@163.com wangxiaoq517@163.com hssong1989@126.com yuanxx1983@163.com

Abstract. Compared to remote sensing image, post-earthquake airborne Light Detection And Ranging (LiDAR) point cloud data contains a high-precision three-dimensional information on earthquake disaster which can improve the accuracy of the identification of destroy buildings. However after the earthquake, the damaged buildings showed so many different characteristics that we can't distinguish currently between trees and damaged buildings points by the most commonly used method of pre-processing. In this study, we analyse the number of returns for given pulse of trees and damaged buildings point cloud and explore methods to distinguish currently between trees and damaged buildings points. We propose a new method by searching for a certain number of neighbourhood space and calculate the ratio(R) of points whose number of returns for given pulse greater than 1 of the neighbourhood points to separate trees from buildings. In this study, we select some point clouds of typical undamaged building, collapsed building and tree as samples from airborne LiDAR point cloud data which got after 2010 earthquake in Haiti MW7.0 by the way of human-computer interaction. Testing to get the R-value to distinguish between trees and buildings and apply the R-value to test testing areas. The experiment results show that the proposed method in this study can distinguish between building (undamaged and damaged building) points and tree points effectively but be limited in area where buildings various, damaged complex and trees dense, so this method will be improved necessarily.

Key words: Light Detection And Ranging (LiDAR); point cloud; number of returns for given pulse; trees; damaged buildings; ratio

1. Introduction

The earthquake occurred in Haiti caused houses collapsed and damaged, roads and communication lines interrupted, that make difficult to rescue. The recognition of damaged buildings is an important step and a judgement index of earthquake intensity [1]. Obtaining information of damaged buildings quickly and accurately can support emergency rescue after earthquake [2]. Now the surface information of the post-earthquake area are most obtained from optical image which contains plane information and is only useful in sunny days.

LiDAR (Light Detection And Ranging) is a new measuring technique which developed in recent years. The ranging accuracy of Airborne radar measurement technology can be 11cm (500m height) [3] and now be widely used in the detection of surface information, restoration and reconstruction of the model. Airborne LiDAR can acquire point cloud for the detection of seismic secondary disaster like surface deformation, rupture, landslide and barrier lakes [4]. The discrimination and extraction for damaged buildings based on airborne LiDAR point cloud is an important method to the disaster assessment [5]. Experts distinct between the undamaged buildings, collapsed buildings and trees mostly used elevation, slope, and vector of point cloud. However, it is difficult to distinguish damaged buildings from trees because elevation distribution of points on damaged buildings is similar to that of trees which is uneven and coarse. So this study analyzed the number of returns of different surface features and propose a classification method based on the ratio of point cloud echo times.

2. Method

2.1. The ratio of point cloud echo times （the ratio of points whose number of returns for given pulse greater than 1 of the nearby points, R）

Point cloud data generally contain three-dimensional coordinate information, echo signal intensity value and multiple echo information. Among them multiple echoes information includes two aspects: one is the total echoes number of pulse that point belongs to, and the other is that the order of point in the total number of echoes. Different feature goals have different echo times, when the laser pulse irradiated to the flat building roof or bare ground surface echo is generated once, and when the pulse irradiation to the vegetation, forms multiple echoes since pulses can penetrate vegetation. Multiple echoes is an important feature of vegetation in the LiDAR data [7].

According to the echo times information of buildings and trees, this paper analysed the proportion of points whose echo time is greater than 1 in trees, undamaged buildings and collapsed buildings point cloud. Presented one method of feature point cloud classification by calculate the ratio (R) of points (M) whose echo time is greater than 1 in a certain number (N) of nearby points (Formula 1).

$$R = M/N \tag{1}$$

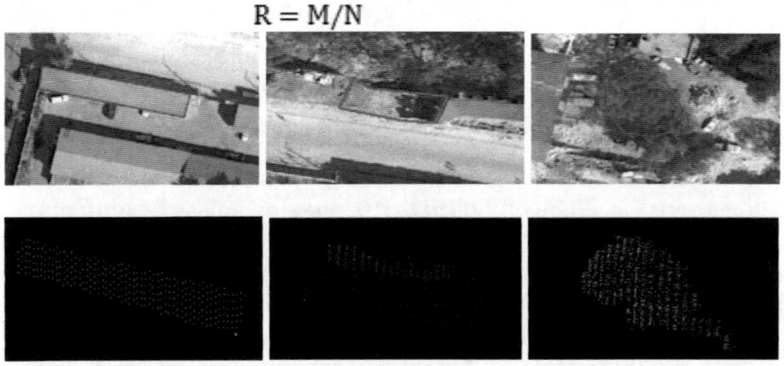

Figure 1.The echo times of undamaged building, damaged building and tree

2.2. Verification and analysis on threshold of R

Combining hyper-spectral image in the study area (a) and (b) (Fig 3), manually select a sample of 50 undamaged buildings 50 collapsed buildings and 50 trees area. Count respectively the weight of points whose echo time is greater than 1, the results shown in Figure 2a. We can see that weight in undamaged building and damaged building samples mostly in the range of 0-0.1, but the weight in the vast majority trees area is about greater than 0.1. Normal distribution curves of R (Figure 2b)for three types of terrain are built based on 150 samples to define the threshold, we get the intersection point A(0.0038,12.001) of curves of undamaged and damaged building and point B(0.083,0.986) of curves of damaged building and tree. Therefore the possibility of misclassification for building and tree is least when R_d= 0.0038 and R_t=0.083 and this paper makes R_d= 0.0038, R_t=0.083 as segmentation threshold of undamaged and damaged buildings, buildings and trees separately.

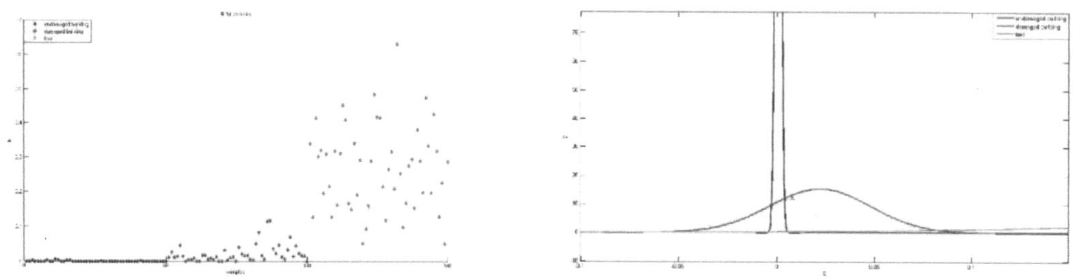

a b

Figure 2. Sample distribution by R

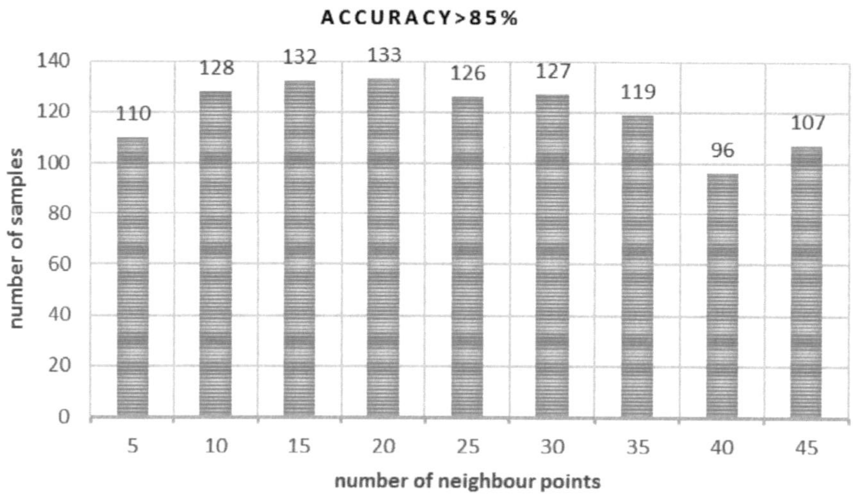

Table 1. Number of samples whose accuracy >85%

To avoid the influence of different quantity points of samples on the ratio factor R, improve echo number information on the accuracy of the LiDAR point cloud classification of damage by earthquake and realize ground point automatic classification based on the ratio of point cloud echo times . The method used in this paper is to search around each point a certain number of nearby points Calculation the weight (R) of points whose echo time is greater than 1 in this area. By selecting different number (N) of nearby points, carry on comparison test in the case of split factor threshold R_d= 0.0038 and R_t =0.1. Considering the density of point cloud and planar area of buildings and trees in the study area, we select 5-50 valid points in the neighbourhood. Use this method to select a different neighbour points, for each sample point cloud classification, and calculate the accuracy (A) of sample classification. Finally this paper statistics the number of samples whose correct rate is more than 85% in 150 samples and the result is shown in Table 1. When the neighbourhood points n= 20, the number of samples whose accuracy above 85% is 133 which is the highest proportion. Therefore, this paper selected n_0 = 20, R_d= 0.0038, R_t= 0.083 as point cloud classification threshold.

3. Data and Analysis

3.1. Data and study area

After 2010 Haiti 7.0 earthquake, Digital Globe, GeoEye, NOAA, USGS, the World Facility, obtain a satellite remote sensing images after the earthquake rapidly, meanwhile they collected pre-quake image, DEM, basic geographic information data., to provide the Government of Haiti disaster response decision supplementary information. This study took about 0.8km^2 region near the Haitian capital, airborne LiDAR point cloud data financed by Globe Facility for Disaster Deduction and

Recovery (GFDDR), January 21, 2010 aerial point cloud data, point cloud density 2-5points/ m^2. Auxiliary optical remote sensing images using the GeoEye-1 satellite imagery from Digital Globe Company obtained at January 13, 2010, resolution was 0.5m. Housing construction in this area have diverse types, such as large multi-story buildings, apartments, hotels, churches and others, they have large construction area, distributed sparsely ,among them there are several strains of trees and small woods; In this area there are also simple low, single house smaller, they intensive distributed on 0.5m satellite imagery, there between less trees. The study area covered almost all urban housing construction types, are representative.

(a) (b)

Figure 3.Figure with test area (undamaged building: blue, damaged building: red, tree: green)

3.2. Pre-processing of point cloud

In the process of point cloud data acquisition, point cloud contains some noise points because of defects of the scanner and environmental disturbance. The process of eliminating noise points is called point cloud de-noising. In this study we use point cloud denoising algorithm of local spatial distribution statistics to remark and eliminate points whose local points density has large difference from the overall points density based on the software of Points Cloud Magic [8]. Then we should filter points to split the ground point and the non-ground point, in this study we used three filtering algorithms: mathematical morphology filtering method, the slope filtering method and CSF（Cloth Simulation Filter）filtering method. Mathematical morphology filtering method is using the erosion and expansion of mathematical morphology operation to remove the cloud high point cloud and keep low point cloud to achieve the purpose of extracting ground[9]; Slope filtering method is based on adjacent ground point and feature point elevation mutation characteristics to extract ground point[10]; The CSF algorithm first make the original point cloud inversion, then has a certain hardness of cloth covered on it and get the terrain of the approximate estimate, finally estimate the terrain and separate ground points and un-ground points[11]. We filter the point cloud of test area 1 shown as Fig.2a by the above methods and get ground points, from result comparing: CSF filtering can mostly separate ground points shown as Fig.2d; Morphological filtering has good recognition effect for edge of different features shown as Fig.2b, but the effect of the recognition of whole feature is worse comparing with CSF filtering; Slope filtering can only extract a bit of ground points shown as Fig.2c. So this paper used the CSF filtering algorithm to get ground points. Then we extract un-ground points by eliminating ground points from point cloud.

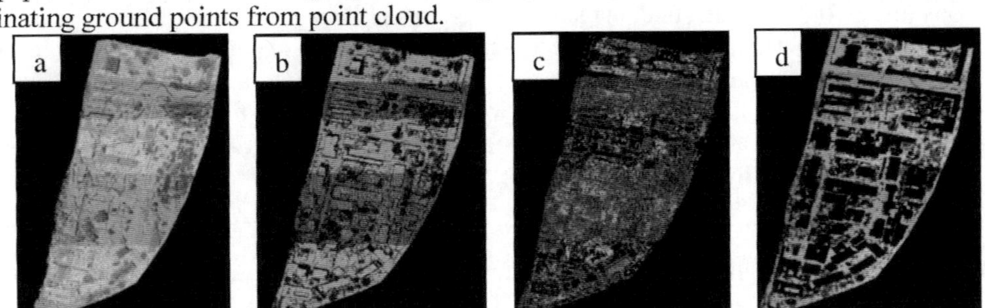

Figure 4. Original point cloud of test area1 and filtering results by different filter

IOP Conf. Series: Earth and Environmental Science **46** (2016) 012018 doi:10.1088/1755-1315/46/1/012018

4. Classification of test area

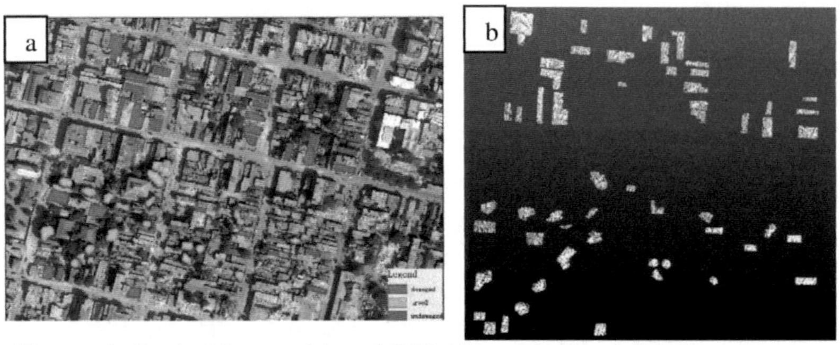

Figure 5. Optical image (a) and LiDAR point cloud (b) of test area

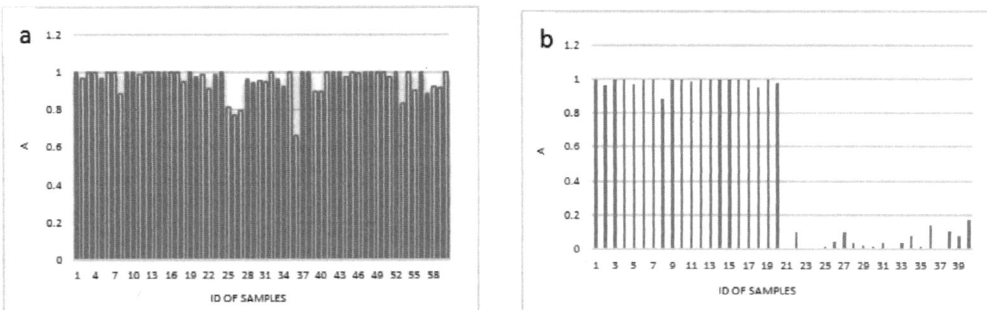

Table 2. Accuracy distribution of test samples

To verify the efficacy of the ratio of point cloud echo times and the threshold factor, This paper selects approximately 6200m^2 area where at northwest of the presidential palace as a test area(Figure 5.a). This area is with strong representation which contains the undamaged buildings, collapsed buildings and trees. This paper extracts un-ground points from points in test area by filtering and get sixty samples outline which contains buildings and trees based on high resolution remote sensing images of pre- and post- earthquake, finally point clouds of test samples (Figure 5.b)are obtained by outlines based on the Point Cloud Magic. To classify test point clouds, the ratio of point cloud echo times (R) and accuracy (A) of classification are calculated by method described above, if $R>0.0083$, the point is judged as tree, conversely is building, finally accuracy distribution of all samples is obtained shown as Table 2.a, the number of $A<0.8$ is 3 and $A>=0.9$ is 53; Considering the laser beam irradiation formed a certain area spot on the surface, part of the laser irradiate the house part irradiate on the ground or surface features near on the edge of house, that generates multiple echoes, so in this study samples whose accuracy more than 0.9 are classified correctly. Kappa coefficient calculated based on classification results and actual attributes of test samples from high resolution images in this study is 0.83. Therefore the classification result is almost the same as the reality, the classification accuracy of the method is high. To distinguish undamaged buildings from damaged buildings, this paper takes $R_d=$ 0.0038 as the threshold. The classification accuracy distribution of 40 building samples is shown as Table 2.b, there are 20 samples whose accuracy is less than 0.2. Therefore the classification method can't distinguish buildings efficiently. In conclusion, the classification method based on the ratio of point cloud echo times has a weak ability to extract damaged buildings but efficient to distinguish buildings from trees, especially damaged buildings.

5. Discussion

The above results show that in a certain area the ratio (R) of points whose echo times is greater than 1 can reflect land features, effectively distinguish between buildings and trees, especially have better recognition between damaged buildings and trees, and can realize the house monomers' preliminary division. This method only use after the earthquake Airborne LiDAR point cloud data for damage extraction compared with remote sensing auxiliary or before and after the earthquake change detection method, can avoid the lack of before earthquake data and multivariate data matching problems, easier and more convenient. And the use of ratio of points echo times (R) can better distinguish between trees and collapsed buildings, improve the accuracy of the earthquake damage extraction. But we found in the study that, due to the collapsed buildings have different degree and various forms, the factor is difficult to accurately determine the collapsed building range boundaries, result classification error. In addition, in leafy trees areas, multiple echoes have lower chance to produce, so there will be some misclassification in that area. Therefore, it can improve classification accuracy to combine with normal vector and intensity. Besides the method takes monomer samples as study object which are obtained by manual cutting, so how to get feature monomer automatically still needs to explore.

6. Conclusion

Based on the 7.0 earthquake 2010 Haiti Airborne LiDAR point cloud data, this paper counted the proportion that points whose echo times greater than 1 in buildings, collapsed buildings and trees point cloud, introduced the ratio of echo times factor R, and determined classification threshold, used the method that for each point search a certain number of nearby points and calculate the proportion of point echo times greater than 1to identify surface features. The results showed that the ratio factor R for echo times can reflect the characteristic of buildings and trees, compared to other earthquake damage classification factor, this factor can improve distinction between building especially the collapsed buildings and trees. And it can initial realize the division of a single building, and laid a foundation for earthquake damage classification and extent judgment based on after the earthquake Airborne LiDAR data.

Acknowledgments

The research was supported by National Natural Science Foundation of China for funding. (41404046).

References

[1] Guo H D, Lu L L, Ma J W, et al. An improved automatic detection method for earthquake-collapsed buildings from ADS40 image. Chinese Sci Bull, 2009, 54: 3303－3307, doi: 10.1007/s11434-009-0461-3

[2] Wang Caifeng. Detection of collapsed building by classifying airborne laser scanner data[D]. Southwest Jiaotong University, 2013.

[3] Lai Xudong 2010 The basic principle and application of airborne laser radar (Beijing: Publishing House of Electronics Industry) p45

[4] DOU Ai-xia and MA Zong-jin 2013 Automatic Identification Approach of Building Damages Caused by Earthquake Based on Airborne LiDAR and Multispectral Imagery Remote Sensing Information (Vol.28,No.4) p104

[5] Richard C. Labiak and Jan A.N. van Aardt Automated method for detection and quantification of building damage and debris using post-disaster LiDAR data

[6] Zhang Xiaohong 2007 Airborne laser radar measurement technology theory and method (Wuhan: Wuhan University) p94

[7] Gong Liang 2011 Research on Classification of Airborne LiDAR Data (PLA Information Engineering University)

[8] www.lidar.radi.ac.cn

[9] Sui Lichun 2010 Filtering of Airborn LiDAR Point Cloud Data Based on the Adaptive Mathematical Morphology Acta Geodaetica et Cartographica Sinica (Vol.39,No.4) p391

[10] Zhang Hao 2009 Filtering of Airborne LIDAR Data Based on Pseudo-Grid Concept and Modified Slope Filtering Algorithm Journal of Geomatics Science and Technology (Vol.26,No.3)p225

[11] Wang Cheng 2015 Handbook of Points Cloud Magic

Non-destructive Analysis Chlorophyll Content of Different Genotypes of Poplars Based on Hyperspectral Reflectance Data

S Jin[1,2,*], Y Dian[2],R Wang[1,2], L Peng[2], X Liu[2],Z Zhou[2],S Zhong[2],Y Wang[2]

[1] State Key Laboratory of Vegetation and Environmental Change, Chinese Academy of Sciences, Beijing 100093, China
[2] College of Horticulture and Forestry Sciences, Huazhong Agricultural University, Wuhan 430070, China

E-mail: jschaon@163.com

Abstract. Leaf Chlorophyll content (Ct) indicates plant physiological status and can be detected by hyperspectral measurements. However, it is difficult to conclude that different genotypes of same species have the same relationship with the hyperspectral data. The aim of this paper was to test that whether the different genotypes of same species have the similar relationship with hyperspectral reflectance. First of all, spectral reflectance of populus simonii (Populus simonii Carr) and I-72 poplar (Popul*us euramericana* cv. 'San Martino I-72/58') were collected by spectrometric meter, and then extract chlorophyll index (CI) and other 11 types of vegetation indices from the hyperspectral reflectance data. At last, relationships between different vegetation indices and Ct of the two genotypes of poplar were compared. Results show that (1) the relationships between SPAD value and Ct are different in the low and high Ct level, we can choose proper vegetation index, REPIG, mSR705 and SDr/SDb et al to predict the Ct value. (2) Meanwhile, we can use PSSRb and PRI to distinguish fine difference between different genotypes.

1. Introduction

In the process of light energy absorption of photosynthesis, chlorophyll absorbs light energy and has a direct relationship with the energy use of vegetation. Chlorophyll content has a good correlation with photosynthetic capacity of vegetation, developmental stages and nitrogen status, they can usually indicate nitrogen stress, photosynthetic capacity and vegetation growth phase (especially the aging stage) [1]. Using spectrophotometer in lab to determinate chlorophyll content is a traditional method. It's a time consuming method and also damage causing to vegetation. In addition, it is likely to loss chlorophyll in the process of transportation from the field to the laboratory and sample preparation [2]. Thus, quick and accurate measurement of plant leaf chlorophyll content has important application value in many areas such as ecology, agriculture and forestry [3].

For the purpose of fast and non-destructive determination of content of vegetation pigment and other biochemical components, there are some devices such as SPAD 502 and method for rapid determination of chlorophyll relative content value of vegetation leaf [4]. In the recent years, hyperspectral technology develops rapidly. Due to its more and narrower band, it provides a more convenient, fast, non-destructive method for collecting data and estimating leaf or canopy absolute chlorophyll content [5]. Through the determination of the green leaf reflectance, transmittance and absorption, Portable hyper-spectrometer is a non-invasive method of determination of chlorophyll, which makes the hyperspectral technology has a unique role in vegetation chlorophyll content evaluation. Portable spectrometer uses its delicate high spectral resolution, establishing regression equation between chlorophyll content and spectral reflectance in some characteristic bands, as the basis of prediction. Vegetation spectral index is based on the assumption that vegetation spectral behavior, unique biophysical function and biological physical properties are related, it is likely to estimate vegetation parameter more accurately, and reduce the influence of external factors. Therefore, vegetation spectral information characteristics can be used to estimate chlorophyll content [6-7]. Gitelson et al. find there is a significant correlation between chlorophyll content and CIrededge based on three bands of vegetation index, R^2 is more than 0.9[8]. Horler et al. show that reflection spectrum in the red edge position may indicate the change of vegetation chlorophyll concentration [9]. Because the first order differential data of the red edge is discrete, the red edge position extracted from traditional algorithm causes larger deviation on chlorophyll content. Miller et al. [10] and Cho and Skidmore [11] respectively use Gaussian optical model and linear extrapolated method to successfully correct a red edge location algorithm, which has achieved good effect on the estimation of chlorophyll. Blackburn [12-13] find that Canopy and leaf scales pigment concentration have extremely significant correlation with the first and second order derivative of the spectral reflectance, and construct the suitable chlorophyll spectral index for estimating chlorophyll by using the characteristic absorption bands of Chlorophyll in 470nm, 635nm and 680nm. However, Sims and Gamon [14] suggest that the reflectance of 550nm, 700nm and 680nm are more sensitive to chlorophyll from his further study. There are other vegetation indices such as NDVI, RVI and PRI are often used for remote sensing of chlorophyll density monitoring, and all have good results [15]. Current research on chlorophyll high spectral estimation has made great progress, the spectral parameters method considers its physical mechanism in constructing index , at the same time, requires less band ,leading to its widespread use[16].But even so, the spectral index is influenced by vegetation type, developmental stage, background environment and other factors[17]. Its sensitivity to the different vegetation, different growth period and internal theory remain to be further in-depth study.

There are many studies focus on the single crop, such as rice, corn, and soybean and so on, but lack of the analysis of different forest tree species, especially plants of different genotypes. We need to know whether they have similar patterns between two different genotypes, although they are both poplars. On the foundation of these, this paper chooses two different genotypes of poplars and tries to study whether hyperspectral reflectance data can apply to non-destructive test of chlorophyll content, and answer the pattern is similar or not.

2. Materials and Methods

2.1. The study area and Materials collection

Research was conducted at Huazhong agricultural university campus, Wuchang district of Wuhan city, Hubei province (30° 28′ 15.79 ″ N, 114° 21′ 37.23″ E). Populus simonii (Populus simonii Carr) and I-72 poplar (Populus euramericana cv. 'San Martino I-72/58') are selected as the experimental materials. Collecting different ages leaves in Windless cloudy weather or sunny in the morning or the evening. We choose 3 trees of each poplar, and use long reach chain saw to get 4 branches of every tree in 4 different directions in the intermedial height. Selecting leaves from the top branch to the bottom, then measure the SPAD value of every leaves. Ensuring leaves include different age stages and are all normal growth, free from disease, insect. Then choose proper leaves and put them into the ice box with the valve bag.

2.2. Spectrum data collection

Spectrum data was collected through PSR-3500 Spectral Evolution. The spectrometer wavelength range is 350 ~ 2500nm and spectral resolution is 3.5nm (350-1000nm), 10nm @ 1500nm, 7nm @ 2100nm. Before using leaf clip to determinate spectrum, instrument need to be optimized by using the white plate side. While measuring, put the leaf between the leaf clips, whose underlying surface is black. Keeping the leaf fill the whole viewing angle, avoid leaf veins, test 6 times per leaf and 10 spectrum curves per times. The average 10 sets of data as the spectral reflectance of each time, the average data of 6 times as the sample final spectral reflectance. The instrument needs to be optimized at intervals of about 10 minutes.

2.3. Chlorophyll content data collection

The absolute chlorophyll content of per leaf was measured by traditional alcohol extraction method. Punching every measured leaf to get as many disks (n) with a puncher whose diameter is 0.87 cm, calculating those disks total area and grind them with 95% alcohol, SiO2, CaCO3, filtering and keeping constant volume to 25 ml, measuring absorbance of chlorophyll a and b at the wavelength of 665 nm and 649 nm respectively by using spectrophotometer of UNICO-UV 2102C, finally converting to their absolute content (mg/cm2) according to the equation 1.

$$Ca = 13.95A_{665} - 6.88A_{649}$$
$$Cb = 24.96A_{649} - 7.32A_{665} \qquad (1)$$
$$Ct = (Ca + Cb) \times \frac{25 \times (10)^{-3}}{n \times \Pi \times (\frac{0.87}{2})^2}$$

Where n is the number of disk, Ct is the total content of chlorophyll (mg/cm2).

2.4. Vegetation index and Model verification

Numerous vegetation indices have been used to study the relation between spectral index and chlorophyll. This research chooses the following spectral indices, Chlorophyll index (CI) [8], Pigment specific simple ratio (PSSRb) [12], Simple ratio index modified on SR705 (mSR705) [14], Normalized difference index(ND705)[15], Simple ratio index(SR)[15], Photosynthetic reflectance index(PRI)[15], Ratio vegetation index(RVI)[18], Ratio of reflectance at 750 nm to 550 nm(GMI) [19], Red edge reflectance ratio index(VOG3) [20], Ratio of reflectance at 750 nm to 710 nm(ZM)[21], Radio of the red-edge integral areas and the blue-edge integral areas(SDr/SDb) [22], Red edge position obtained from inverted gaussian fitting technique(REPIG) [23].

Randomly select 2/3 the total sample to build regression equations, and the other as the verification sample. We use the verification sample to calculate y′ according to the regression equation, which is the estimated value of actual y, than make correlation and significance test with the y ' and the y.

3. Results

3.1. The relationships between total chlorophyll content (Ct) and SPAD values of two different genotypes of poplars.

From the fig 1, we found (1) When the Ct value is the same, Populus simonii shows bigger SPAD than I-72, it tells we cannot treat the two different genotypes poplar as one type. (2) The linear relationship between Ct and SPAD is perfect when the SPAD is lower than 30 or more than 30, but not good within the scope of the entire. What is more, the SPAD value indicates relative Ct value. All in all, we cannot use SPAD value to predict the Ct value accurately.

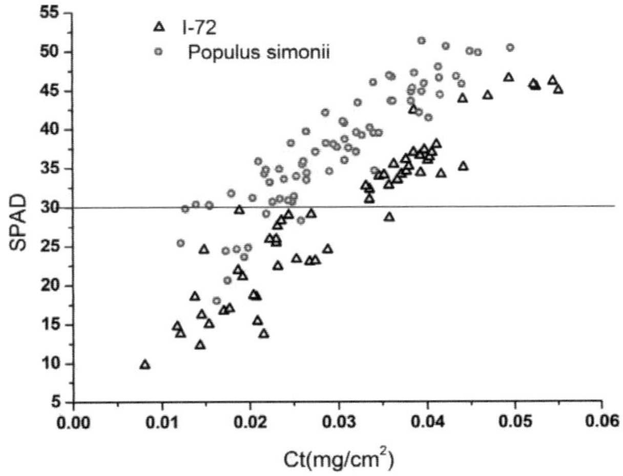

Fig 1. The Ct and SPAD distribution map of two different Genotypes

3.2. The relationships between total chlorophyll content (Ct) and vegetation index of two different genotypes of poplars.

Using Ct of populus simonii, I-72 and the combined to fit linear, logarithmic (log) and exponential (exp) curve with the above vegetation indices, the results are shown in table 1. It is observed that(1) the exponential model of Ct and SDr/SDb index have best fitting effect of populus simonii, R^2= 0.85; the exponential model of Ct and mSR705 index have best fitting effect of I-72, R^2 = 0.90; the logarithmic model of REPIG index have best fitting effect of the combined two kinds poplars, R^2=0.77. (2) We find the three biggest R^2 in all the models, and they are REPIG, mSR705 and SDr/SDb. Meanwhile, we find the two smallest R^2 in all the models, and they are PSSRb and PRI.

Table 1. The fitting results R2 of the total chlorophyll and vegetation indices

Ct	populus simonii			I-72			populus simonii and I-72		
R^2	linear	log	exp	linear	log	exp	linear	log	exp
CI	0.84^{**}	0.78^{**}	0.80^{**}	0.87^{**}	0.78^{**}	0.82^{**}	0.72^{**}	0.69^{*}	0.69^{**}
PSSRb	0.76^{**}	0.73^{**}	0.72^{**}	0.71^{**}	0.68^{**}	0.68^{**}	0.36^{**}	0.38^{**}	0.40^{**}
mSR705	0.83^{**}	0.76^{**}	0.83^{**}	0.87^{*}	0.76^{**}	**0.90^{**}**	0.75^{**}	0.70^{**}	0.76^{**}
ND705	0.82^{**}	0.78^{**}	0.76^{**}	0.88^{**}	0.84^{**}	0.82^{**}	0.68^{**}	0.68^{**}	0.65^{**}
SR	0.81^{**}	0.78^{**}	0.83^{**}	0.87^{**}	0.86^{**}	0.88^{**}	0.70^{**}	0.72^{**}	0.71^{**}
PRI	0.49^{**}	0.49^{**}	-	0.39^{**}	0.52^{**}	-	0.36^{**}	0.45^{**}	-
RVI	0.84^{**}	0.76^{**}	0.84^{**}	0.83^{**}	0.71^{**}	0.86^{**}	0.59^{**}	0.55^{**}	0.63^{**}
GMI	0.84^{**}	0.76^{**}	0.84^{**}	0.83^{**}	0.71^{**}	0.85^{**}	0.60^{**}	0.56^{**}	0.64^{**}
VOG3	0.84^{**}	0.77^{**}	-	0.85^{**}	0.74^{**}	-	0.72^{**}	0.67^{**}	-
ZM	0.83^{**}	0.77^{**}	0.83^{**}	0.88^{**}	0.79^{**}	0.89^{**}	0.72^{**}	0.68^{**}	0.73^{**}
SDr/SDb	0.84^{**}	0.75^{**}	**0.85^{**}**	0.81^{**}	0.67^{**}	0.88^{**}	0.71^{**}	0.63^{**}	0.75^{**}
REPIG	0.79^{**}	0.76^{**}	0.79^{**}	0.88^{**}	0.89^{**}	0.88^{**}	0.75^{**}	**0.77^{**}**	0.75^{**}

* * mean (double side) significantly related at the level of 0.01,* mean (double side) significantly related at the level of 0.05.

It is observed from fig 2 that (1) almost all the indices of the I-72 and populus simonii change with the Ct obviously. (2)Generally, the SR, PRI, VOG3 and SDr/SDb index of populus simonii is less than I-72 poplar, while the other 9 indices value of populus simonii are all bigger than I-72.

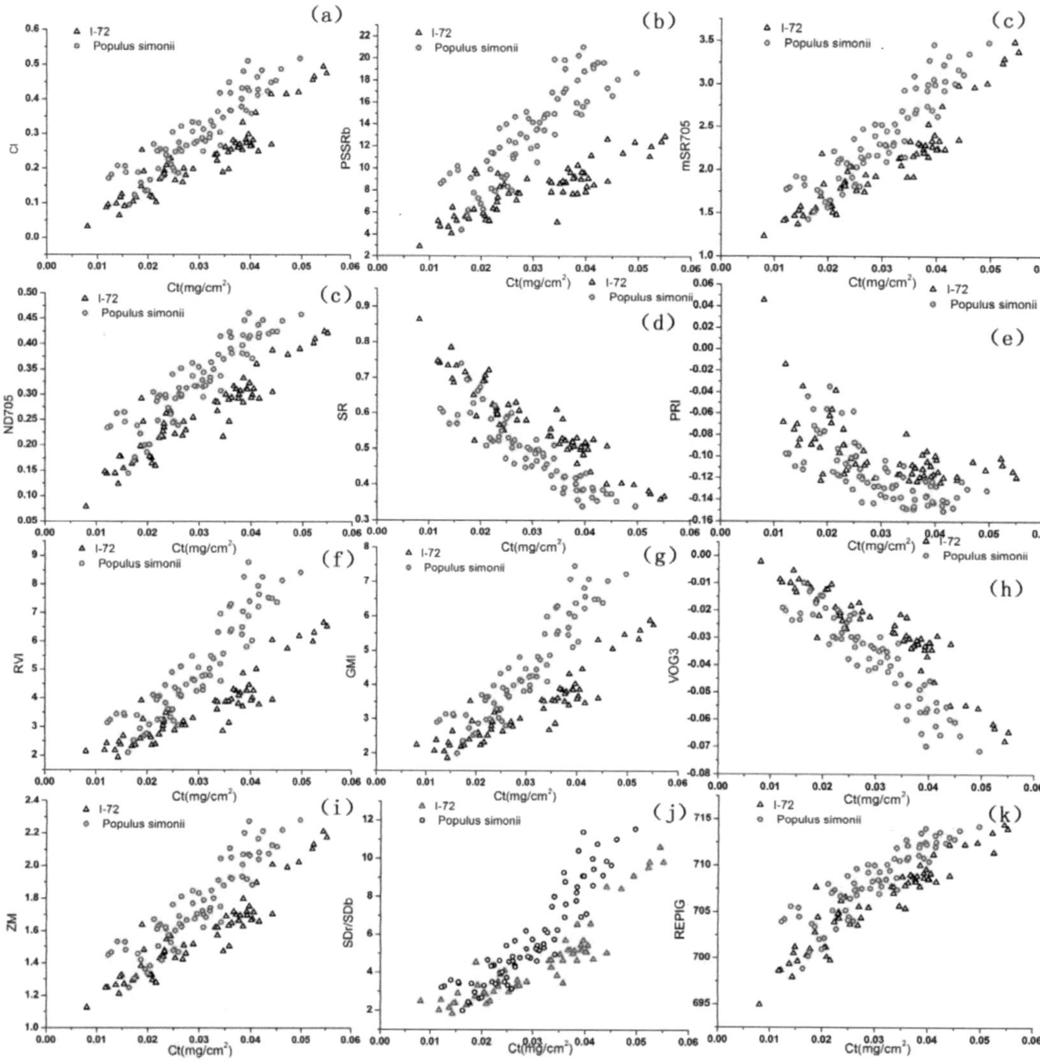

Fig 2. The Ct and multiple vegetation indices distribution map of two different genotypes of poplars

4. Discussion

We cannot use SPAD value to predict the Ct value accurately at all level, because it shows saturation causing by leaf structure and other circumstance as well as it only express relative content.The correlation of the two poplar combined Ct and 12 vegetation indices are all significant, we can choose proper vegetation indices to predict the Ct value, such as REPIG, mSR705 and SDr/SDb et al, their R^2 is relative higher. Meanwhile, two genotypes of poplars showed a certain difference on the spectral index, for example PSSRb and PRI et al., of which PRI was proved to be able to reflect differences between different genotypes of rice species [24]. Thus, we can find fine difference using PSSRb and PRI hyperspectral index if necessary.

5. Conclusion

This paper studied the relationships between Ct and SPAD and the 12 vegetation indices of two different genotypes of poplars. The results are as follows:(1)the exponential model of Ct and SDr/SDb index have best fitting effect of populus simonii, R^2 =0.85;the exponential model of Ct and mSR705 index have best fitting effect of I-72, $R^2 = 0.90$; the exponential of Ct and mSR705 index have best fitting effect of the combined two kinds of poplars, $R^2 = 0.77$. (2) Almost all the indices of the I-72 and populus simonii change with the Ct obviously, and showed a consistency rule, which suggests that these spectral indices of different genotypes poplars changes similarly. (3) The SR, PRI and VOG3 index of populus simonii are little less than I-72 poplar, and the other 9 indices value of populus simonii are all little bigger than I-72,but the correlation of the two poplar combined Ct and 12 vegetation indices are all significant. That is to say, we can ignore genotypes to a certain extent, we also can distinguish genotypes using hyperspectral measurements when applying in Precision agriculture and other similar research filed.

References

[1] Uddling J, Gelang-Alfredsson J, Piikki K and Pleijel H 2007 *Photosynth. Res.* **91** 37-46
[2] Chappelle E W, Kim M S and McMurtrey J E 1992 *Remote. Sens. Environ.* **39** 239-47
[3] Li Z, Liu H and Zhang F 2003 *Plant. Nutr. Fert. Sci.* **9** 401-5
[4] Richardson A D, Duigan S P and Berlyn G P 2002 *New. Phytol.* **153** 185-94
[5] Wu J 2011 *Spectrosc. Spect. Anal.* **31** 2305-12
[6] Curran P J, Dungan J L, Macler B A and Plummer S E 1991 *Remote. Sens. Enviro.* **35** 69-76
[7] Feng W, Yao X, Tian Y, Cao W and Zhu Y 2008 *Crop. Pasture. Sci.* **59** 748-60
[8] Gitelson A A, Keydan G P and Merzlyak M N 2006 *Geophys. Res. Lett.* **33**
[9] Horler D, Dockray M, Barber J and Barringer 1983 *Adv. Space. Res.* **3** 273-7
[10] Miller J, Hare E and Wu J 1990 *Remote. Sens-basel.* **11** 1755-73
[11] Cho M A and Skidmore A K 2006 *Remote. Sens. Enviro.* **101** 181-93
[12] Blackburn G A 1998 Remote. Sens. Enviro. 66 273-85
[13] Blackburn G A 1998 Int. J. Remote. Sens. 19 657-75
[14] Sims D A and Gamon J A 2002 *Remote. sens. Environ.* **81** 337-54
[15] Wu C and Wang X 2015 *Sci. Silvae. Sinicae* **3** 49-56
[16] Chun Y Y, Zheng N, Ji H W, Liang Y L and Wen J H 2005 *J. Remote. Sens.* **9** 742-51
[17] Meng Q Y, Dong H, Qi M, Wang J L and Zhao J 2012 *Spectrosc. Spect. Anal.* **32** 2218-22
[18] Rouse J, Haas R, Schell J, Deering D, Harlan J. Monitoring the vernal advancements and retrogradation of natural vegetation. NASA/GSFC, Final Report, Greenbelt, MD, USA. 1974:1-137.
[19] Gitelson A A and Merzlyak M N Remote estimation of chlorophyll content in higher plant leavs 1997 *Int. J. Remote. Sens.* **18** 2691-7
[20] Zarco-Tejada P J, Miller J R, Noland T L, Mohammed GH and Sampson P H 2001 *IEEE. T. Geosci. Remote.* **39** 1491-507
[21] Gupta R, Vijayan D and Prasad T 2003 *Adv. Space Res.* **32** 2217-22
[22] Feng W, Yao X, Tian Y, Cao W and Zhu Y 2008 *Crop. Pasture. Sci.* **59** 748-60
[23] Miller J, Hare E and Wu J 1990 *Remote. Sens-basel.* **11** 1755-73
[24] Wang H, Chen J, Lin H and Yuan D 2010 *Remote. Sens. Lett.* **1** 223-229.

IOP Conf. Series: Earth and Environmental Science **46** (2016) 012020 doi:10.1088/1755-1315/46/1/012020

An Effective Algorithm Research of Scenario Voxelization Organization and Occlusion Culling

Guangling Lai [1], Lu Ding [1], Zhiyuan Qin [1], Xiaochong Tong [1]

(1.Information Engineering University, Zhengzhou 450001, China)

E-mail: 923809012@qq.com

Abstract. Compared with the traditional triangulation approaches, the voxelized point cloud data can reduce the sensitivity of scenario and complexity of calculation. While on the base of the point cloud data, implementation scenario organization could be accomplishment by subtle voxel, but it will add more memory consumption. Therefore, an effective voxel representation method is very necessary. At present, the specific study of voxel visualization algorithm is less. This paper improved the ray tracing algorithm by the characteristics of voxel configuration. Firstly, according to the scope of point cloud data, determined the scope of the pixels on the screen. Then, calculated the light vector came from each pixel. Lastly, used the rules of voxel configuration to calculate all the voxel penetrated through by light. The voxels closest to viewpoint were named visible ones, the rest were all obscured ones. This experimental showed that the method could realize voxelization organization and voxel occlusion culling of implementation scenario efficiently, and increased the render efficiency.

1. Introduction

Voxel or three-dimensional pixels is short for volume pixels. Conceptually, voxel is to three-dimensional space what pixel is to two-dimensional space. It is the smallest unit of digital data in three-dimensional space, and mainly used in 3 D imaging, scientific data and medical imaging, etc. Compared to the traditional triangulation method, voxelized point cloud data is a kind of spatial-oriented data organization mode, the voxelized space is fixed which have constant space number, position, structure and relationships. This method could overcome some exist shortages, such as unable to express the object's internal structure; sensitivity to the complexity of the scene; when there are too many objects in the scene, triangulation efficiency is low and unable to implement dynamic scene update efficiently. This method owns a very important practical significance [1].

In order to improve the render efficiency without affecting the expression, visible cutting technology was usually used to express the voxel scenario. Depth buffer algorithm (Z-Buff) is mainly used to compare all the surfaces' depths of each pixels on the projective plane[2][3]. So that it could distinguish the visible surface, process each object surface of a polygon point by point and it is appropriate for the only polygon contained scene. However, two caches are needed in this method, and some unnecessary calculations were often executed; a pixel can only find one visible surface, and unable to deal with the transparent surfaces. A-Buff algorithm based on the Z-Buff, it establish a surface linked list for each pixel, not only can put each pixel's intensity values of multiple surfaces into consideration, and also can eliminate the aliasing of object's boundary[4]. BSP algorithm uses plane to divide the space, with the viewpoint position and the plane's visibility, divides the special objects into two divided spaces, then process them with a recursion, and establish a visibility binary tree eventually. In the scene display, draw them in accordance from front to back. Key of this approach lies in the establishment of the tree structure. When scene or view point is changed, the tree structure needs to be rebuilt. So it was suitable for the static scene establishment[5]. Octree algorithm is a rather common spatial data organization structure[6]. When the observed object is described, the accordance from front to back was commonly used to map the octree nodes to the observed surface, and to eliminate the hidden surfaces. However, this approach is hard to manage dynamic scenes, when the scene is updated, the corresponding parts needed to be rebuilt either . Ray-tracing algorithm was originated from the ray-casting algorithm [7]; shade relations are mainly determined by means of the intersection of light and surface [8]. This method can realize real-time rendering of scene, but occlusion relations needs to be rejudged in each changing scene. And if there are more surfaces, it contained too many intersection times and larger calculated amounts by using triangular mesh to reconstruct the surface with large-scale scene point cloud data.

In order to implement the real-time updated display of dynamic scene, on the basis of advantages and disadvantages of the mentioned methods, this paper fully embodied the characteristics of scene regularization (each surface of voxel is paralleled to the coordinate plane, coordinate value could be quickly obtained, and it is easy for intersection calculation), and proposed a new ray-tracing algorithm.

2. Voxelization of point cloud data

The original point cloud data are some structurelessness discrete points. In order to be applied better, point cloud data needs to be organized. Method of this paper is to voxelized them, the main steps were as follows:

Step one: used comparative law to determine the whole point cloud data's minimum and maximum values in the three coordinate directions x, y, z:x_{min}, x_{max}, y_{min}, y_{max}, z_{min}, z_{max}.

Step two: according to the distribution range and resolution requirements of the point cloud data, set the voxel size l;

Step three: use type (1) to judge each point position of point cloud data(x, y, z) on voxels. After calculated through all the points, only display the voxels which contained point cloud data. Then, voxelization of point cloud data was done.

$$\begin{cases} i = floor(\frac{x-x_{min}}{l}) \\ j = floor(\frac{y-y_{min}}{l}) \\ k = floor(\frac{z-z_{min}}{l}) \end{cases} \tag{1}$$

It can be seen by the whole process of voxelization, the level of similarity between voxelized scene and original scene mainly depends on the size of voxel, the smaller scale, the higher similarity. But a greater amount of data may lead a lower efficiency about the display of voxelized scene. In order to improve the efficiency of display, a common way is using occlusion cutting technology to insure occlusion relationship, reduce the draw number of voxel.

3. Occlusion cutting technology of voxel

In order to implement the real-time rendering of scene, this paper made full use of the regularity characteristics of voxelized scene, and then adopted the ray-tracing theory to determine the occlusion relationship. First, calculate light vector corresponded to each image point in the world coordinate system. Then, get the intersections of light and voxels, determine all the voxels which penetrated by light. The voxels which closest to viewpoint and contained point cloud data were named visible ones, the rest were all obscured ones. In the end, drew the visible voxels only. From the above principles, it could be seen there were three factors that influenced the efficiency of ray-tracing algorithm[9]: (1) the number of light; (2) intersection speed; (3) the number of intersection. Therefore, the efficiency could be improved by exacting the scope of pixels, reducing the amount of light; and took advantage of the regular distribution of voxels to reduce the number of intersection and improve the intersection speed. Specific methods were as follows.

3.1. Calculation of light vector

Use ray-tracing algorithm to determine the occlusion relationship. First, we need to calculate and get the light vector which refers to a direction vector in the world coordinate system, which started from the coordinate of screen pixel and ended in the coordinate of camera's position.

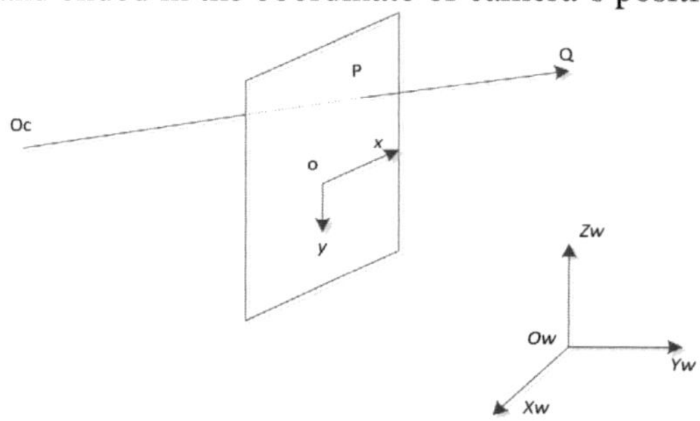

Figure 1 light vector diagram

In figure 1 O_c (x_c, y_c, z_c) stands for the coordinate of camera in the world coordinate system $O_w - X_w Y_w Z_w$, $o - xy$ coordinates refers to the image plane coordinate system, $p(x_p, y_p)$ represents a pixel, $Q(x_q, y_q, z_q)$ is the coordinate of point p on the world coordinate system. So the light vector $\vec{n}(a, b, c)$crossed pixel point p was calculated by the following formulas:

$$sum = \sqrt[2]{(x_c - x_q) * (x_c - x_q) + (y_c - y_q) * (y_c - y_q) + (z_c - z_q) * (z_c - z_q)} \quad (2)$$

$$\vec{n} = (x_c - x_q, y_c - y_q, z_c - z_q)/sum \quad (3)$$

3.2. Confirmation of the pixel range

In the visual expression of the scene, not each pixel on the screen contains scene information. Thus, determine the effective pixel range on screen can reduce the amount of light, so as to improve the efficiency.

Using comparative method to determine the minimum and maximum of all the point cloud data in x, y, z coordinate directions: $x_{min}, x_{max}, y_{min}, y_{max}, z_{min}, z_{max}$, then convert $(x_{min}, y_{min}, z_{min})$ and $(x_{max}, y_{max}, z_{max})$ in world coordinate system to screen coordinate system. The range determined on screen is the range of pixels which used for ray-tracing.

3.3. Visibility judgment

Key link of ray-tracing algorithm was to determine the whole voxels went through by light, as shown in figure 2.

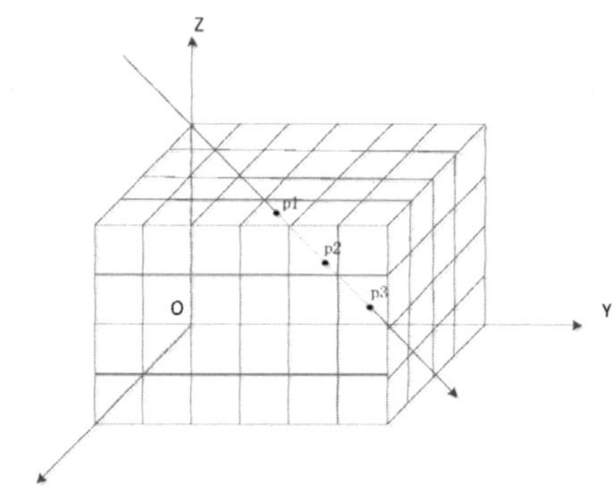

Figure 2 occlusion cutting diagram

The way how to confirm the voxels which went through by light was as follows.

Set the direction vector of light vector $\vec{n} = (a, b, c)$, $Q(x_q, y_q, z_q)$ was set to any pixels' coordinate that converted to the world coordinate system, $O_c(x_c, y_c, z_c)$ was camera's position. ON the basis of regularity characteristic of voxel, the cross point intersected by light and a voxel's upper surface or lower surface (as the points $p1$ and $p2$ shown in figure 2) could all be calculated. The formula is shown in (4)-(7), (x_i, y_i, z_i) stands for intersection coordinates. By computing the intersection in formula (1), voxels went through by a vertical direction light could be determined.

$$z_i = z_{max} - i * l \quad (i = 0,1,2, \dots.) \tag{4}$$

$$k = (z_i - z_q)/c \tag{5}$$

$$x_i = x_q + k * a \tag{6}$$

$$y_i = y_q + k * b \tag{7}$$

l stands for length, width and height of a voxel in formula (4).

After obtaining voxels in z direction, we still need to judge the voxels passed by light in each layer. And this problem could be treated as a two-dimensional plane problem, as shown in figure 3.

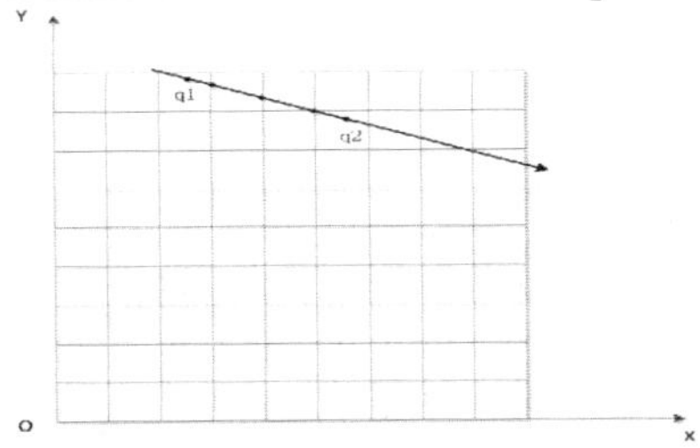

Figure 3 Occlusion cutting plane sketch

Based on z, according to the equally distributed characteristics of voxel, calculate the cross points intersected by light and two adjacent planes, mark them (x_1, y_1, z_1) and (x_2, y_2, z_2) (as $p1$ and $p2$ shown in figure 2). Projected these two points on the plane xy, took advantages of the projection relationship to obtain the coordinates $q1(x_1, y_1)$ and $q2(x_2, y_2)$. (a, b) is the direction vector of light on the xy plane. Thus, cross point intersected by light and each grid boundary could be obtained.

For example, net values in y direction were an arithmetic progression, so the following could be obtained.

$$y_k = k * l + y_{min} \tag{8}$$

$$x_k = a * (y_k - y_1)/b + x_1 \tag{9}$$

In formula (8), $k = 0,1,2 \dots$ until x_k, y_k is smaller than x_2, y_2.

After x_k, y_k was got, according to the fixed z value and formula (1), voxel corresponded to this point could be calculated, and that is also the voxel passed by light.

Change the value of z until the point cloud contained voxel had been found, then went ahead to determine another occlusion relationship of voxel by next light.

4. The experimental results and analysis

This experiment used a city point cloud data, with its range was a cuboid with 1740.25 meters long, 1723.75 meters wide, and 183.14 meters high. As shown below in figure 4.

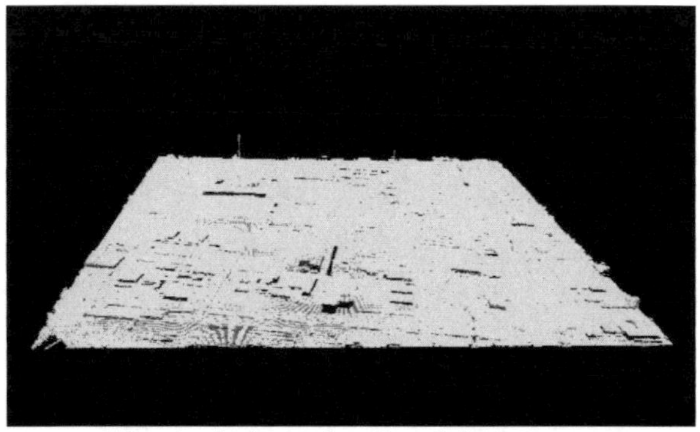

Figure 4 city point cloud data

After voxelizing the point cloud data with the method mentioned before, the result was shown in figure 5. To make sure the accuracy of ray-tracing algorithm, projection of a single voxel shouldn't be less than a pixel. In the experiment, size of each voxel was 8 meters, the resolution is less. Smaller voxel could make the voxelized scene as similar as the actual scene; their occlusion relationship remains the same.

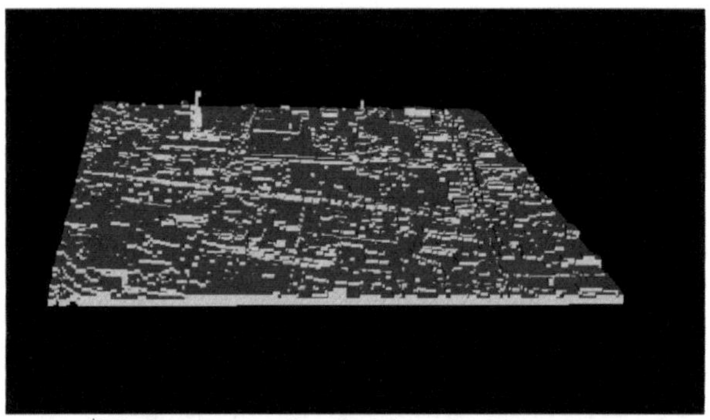

Figure 5 voxelized point cloud scenes

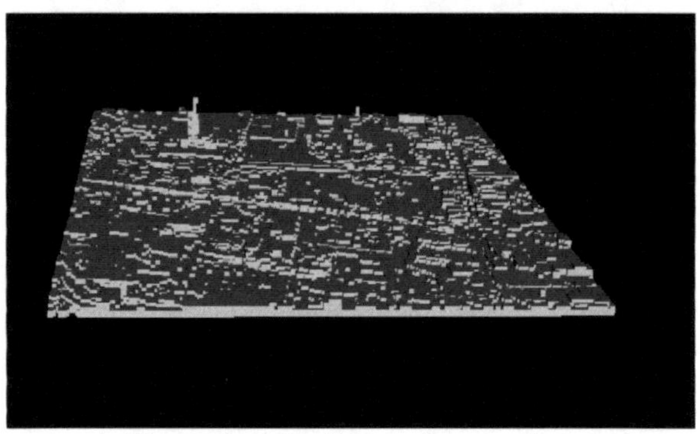

Figure 6 occlusion cutting results

Figure 5 represents the voxelized result; and figure 6 showed the visible voxels after confirming occlusion and cutting relationships. Specific quantities of occlusion and cutting were shown in Table 1.

Table 1 statistics of the occluded voxel

The total number of voxel	The visible number of voxel	Occlusion rate
39713	31722	20.12%

It could be told that, without influencing the expressive effect, method in this paper reduced the draw number of voxel efficiently, and improved the efficiency.

5. Conclusion

In this paper, voxelization about city point cloud data of a big scene was made. And on this fundament, made a full use of the regular distribution character of voxel; and adopted a modified ray-tracing method in occlusion and cutting study. The experimental results indicated that the proposed method could implement the voxelization of scene quickly; on the basis of voxelization, it could be very efficient to determine the occlusion relationship between voxels, reduced the draw number, and improved the draw efficiency.

References

[1] LI J 2005 Voxel imaging technology and application *Harbin Engineering University Press* pp 6-20

[2] Greene N, Kass M and Miller G 1993 Hierarchical Z-buffer visibility *Conference on Computer Graphics and Interactive Techniques* 27 pp 231-238

[3] Greene, Ned, Kass and Michael 1994 Error-bounded antialiased rendering of complex environments *Conference on Computer Graphics and Interactive Techniques* pp 59-66

[4] Schilling A, Stra&# and Er W 1993 EXACT: algorithm and hardware architecture for an improved A-buffer *Conference on Computer Graphics and Interactive Techniques* pp 85-91

[5] Fuchs H, Kedem Z M and Naylor B F 1980 On Visible Surface Generation by a Priori Tree Structures *Acm Siggraph Computer Graphics* 14 pp 124-133

[6] Egenhofer and Max J 1991 Book Review: The design and analysis of spatial data structures. Hanan Samet: Addison-Wesley Publishing Company, Reading, MA, 1989, 493 pp. ISBN 0-201-50255-0 *Isprs Journal of Photogrammetry & Remote Sensing* 46 pp 49-51

[7] Wald, Ingo, Kollig, Thomas, Benthin and Carsten 2002 Interactive global illumination using fast ray tracing *EG Workshop on Rendering Eurographics Association* pp 15-24

[8] ZHOU P 2012 Reach onray tracing based photorealistic global illumination issues *Shandong University* pp 3-10

[9] ZHAO S and LI X 2006 Algotithm of efficient ray tracing of complex scenes *Computer Engineering* 32 pp 224-225

A Global "Natural" Grid Model Based on the Morse Complex

Hongbin Wang[1], Xuesheng Zhao[2], Xinying Zhu[3] and Jiebiao Li[1]

[1]Beijing Research Institute of Uranium Geology, Beijing, China
[2]Geoscience and Surveying Engineering College, China University of Mining and Technology (Beijing), Beijing, China
[3]Esri China Information Technology Co. Ltd., Beijing, China

E-mail: whbcumtb@163.com

Abstract. In the exploration and interpretation of the extensive or global natural phenomena such as environmental monitoring, climatic analysis, hydrological analysis, meteorological service, simulation of sea level rise, etc., knowledge about the shape properties of the earth surface and terrain features is urgently needed. However, traditional discrete global grids (DGG) can not directly provide it and are confronted with the challenge of the rapid data volume growth as the modern earth surveying technology develops. In this paper, a global "natural"grid (GNG) model based on the Morse complex is proposed and a relatively comprehensive and theoretical comparison with the traditional DGG models is analyzed in details as well as some issues to be resolved in the future. Finally, the experimental and analysis results indicate that this distinct GNG model built from DGG is more significant to the advance of the geospatial data acquisition technology and to the interpretation of those extensive or global natural phenomena.

1. Introduction

Recently, with the advancement in geospatial data acquisition technology, spatial data at high spatiotemporal resolution about the earth surface are being collected. However, the large data volume posts a great challenge to the traditional methods of terrain modelling and analysis, especially at the global scale. On the other hand, knowledge about the shape properties of the earth surface and terrain features is urgently needed in the exploration and interpretation of some extensive or global natural phenomena such as environmental monitoring, climatic analysis, hydrological analysis, meteorological service, simulation of sea level rise and so on. For instance, the distribution and orientation of mountains are the critical factors in the field of global climatic analysis. Additionally, river basins over the world play an important role on the global water pollution monitoring and control [1].

Over the years, many discrete grid models for the global surface have been proposed, called discrete global grid (DGG), such as Quadrangle Grid [2], Quaternary Triangular Mesh (QTM) [3], or Hexagon Grid [4] and Voronoi/TIN grid [5], as shown in Figure 1. As a result, they are qualified to simulate the global surface [6], but the DEM constructed by them can neither directly reveal the terrain natural features nor provide the knowledge about the shape properties of the earth surface. Actually, they are mainly applied to the organization or index of the global spatial data, thematic map production and 3D visualization [7]. On the contrary, the Global "Natural" Grid (GNG) model partitions the earth surface into "natural" cells by using the natural terrain feature elements, such as feature points (pits, peaks and saddles) and feature lines (such as ridges and ravines), which possibly provide a practicable data model and analysis method to explore and understand those natural phenomena mentioned above.

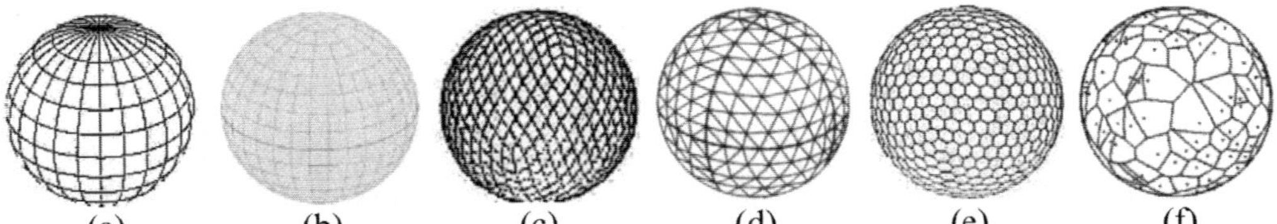

(a) (b) (c) (d) (e) (f)

Figure 1. Traditional discrete global grids (DGG): (a) latitude and longitude grid, (b) DQG [8], (c) diamond grid [9], (d) QTM; (e)hexagon grid, (f) Voronoi/TIN grid

The topological method based on Morse theory [10,11] has been recognized as an important tool for shape analysis and understanding in many applications, including geography, physics, chemistry and medicine [12], because of the ability of feature extraction, multi-scale analysis and forming an abstract or compact description of a surface shape. Over the years, based on the Morse complex [13,14], lots of algorithms for terrain morphology have been presented (see [12,15] for more details). However, most of them and the related applications are limited to local terrain datasets homeomorphic to a 2-disk. In this paper, a GNG model based on the Morse complex is proposed for the whole earth surface which is topological equivalent to an open 2-ball.

The remainder of this paper is organized as follows. In Section 2, some basic notions and the development of Morse Theory and Morse complex are introduced briefly. In Section 3, a modeling approach for the GNG is proposed. After that, an experiment is designed to validate this method and a relatively comprehensive and theoretical comparison with the traditional DGG is further analyzed in details. Following, some issues or challenges for the GNG are discussed. Finally, some concluding remarks are drawn.

2. Morse theory and Morse complex

Morse theory is a powerful tool to capture the morphology of manifold on which a scalar function f is defined. Let f be a C^2-differentiable real-valued function defined over a domain D in the 2D space. A point p of D is a critical point of f if and only if the gradient of f vanishes on p. Function f is said to be a Morse function when all its critical points are non-degenerate (i.e., if and only if the Hessian matrix Hesspf of the second derivatives of f at p is non-singular: the determinant of Hesspf is not zero). This implies that the critical points of f are isolated. The number of negative eigenvalues of Hesspf is called the index of a critical point p. There are three types of non-degenerate critical point p: a minimum (pit), a saddle (pass), or a maximum (peak) if and only if p has index 0, 1 or 2, respectively. Points which are not critical are said to be regular.

An integral line of f, going through a point p, is a maximal path which is everywhere tangent to the gradient of f. Integral lines that converge to (originate from) a critical point p of index i form an i-cell ((2-i)-cell) called the descending (ascending) cell of p. The collection of all descending cells form a Euclidean cell complex, called descending Morse complex (Figure 2a), and the collection of all ascending cells form also a Euclidean cell complex, called ascending Morse complex (Figure 2b), which is dual with respect to the descending Morse complex. A Morse function f satisfying the Morse-Smale condition, that is, descending and ascending Morse complexes intersect only transversally, is called a Morse-Smale function. This means that the intersection of the edges of the descending and ascending cells is a saddle point. In this case, we can define a complex, called the Morse-Smale complex, whose cells are the connected components of the intersections of the descending and ascending cells (Figure 2c). Integral lines connecting saddle points to other critical points are called separatrices. All separatrices forms a network, called the critical net. It is the 1-skeleton of the Morse-Smale complex.

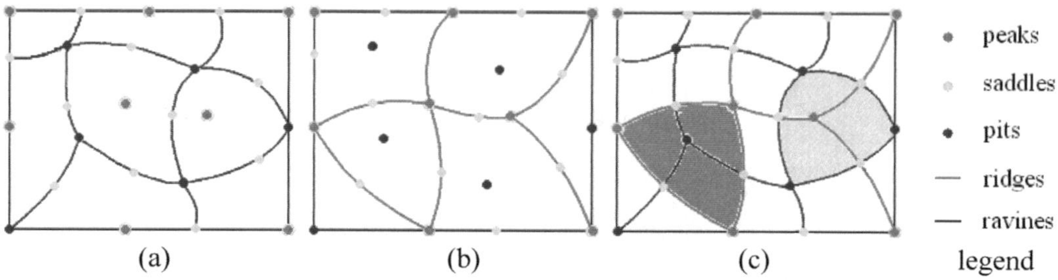

Figure 2. Morse and Morse-Smale complex: (a) ascending Morse complex, (b) descending Morse complex, (c) Morse-Smale complex

As mentioned above, the fundamental restrictive condition with the Morse theory is the assumption that surface is twice differentiable and that the Hessian matrix of partial derivatives is everywhere non-singular. However, real surfaces are seldom sufficiently smooth enough, especially in the discrete case (eg. the DEM). As a result, to harmonize the contradiction between the theory and application, main two extensions of the classic Morse theory are proposed, that is, piece-wise linear Morse theory [16] and discrete Morse theory [17]. Figure 3 indicates the development of Morse theory and Morse complex.

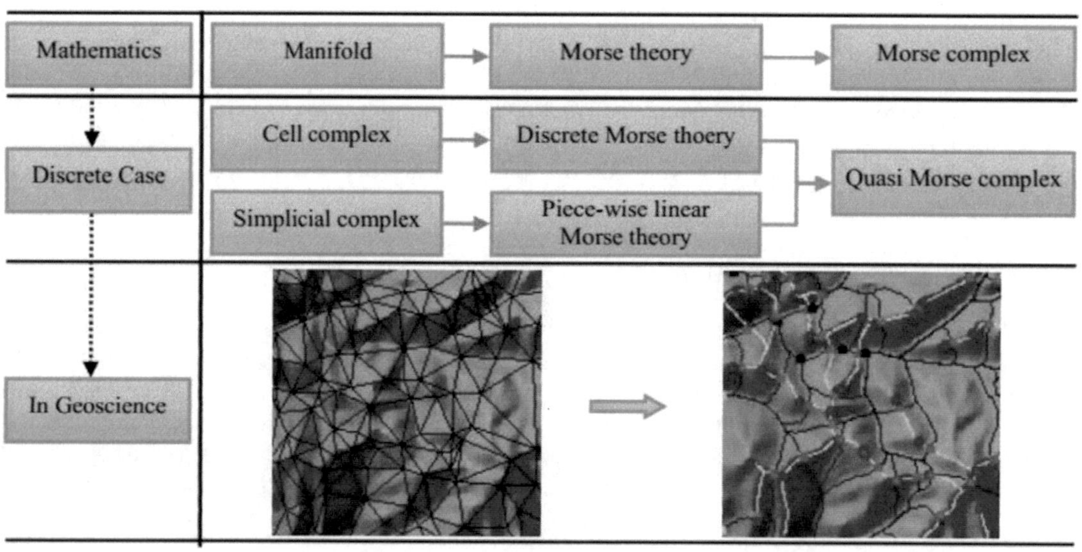

Figure 3. The development of Morse theory and Morse complex

In terrain analysis, the separatrix starting from saddle to peak is called the ridge line (or ridge), while the one starting from saddle to pit is called the ravine line (or ravine). Theoretically, it is impossible that valley and ridge line intersect at regular points and that the isolated critical points occur. Meanwhile, the earth surface is just a 2-manifold, where Morse theory can be used directly. However, as shown in Figure 4, when applying the existing algorithms for local terrain to the global DEM, some abnormal situations will occur, such as thevarious local intersections between a ridge and a ravine and some ridges passing across the ocean without the marine identification.

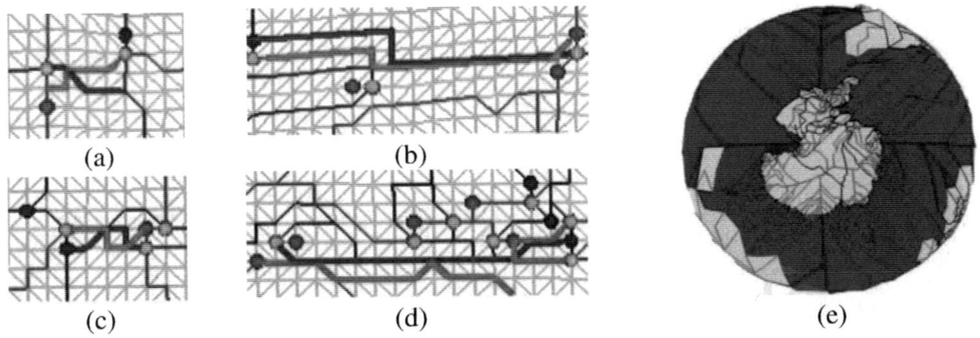

Figure 4. The abnormal when applying the existing construction algorithms of Morse complex to the global DEM: (a) single-point intersection, (b) intersections forming a line, (c) multiple intersections, (d) multi-intersections between a ravine and some ridges, (e) some ridges passing across the ocean

3. A Modeling Approach of GNG

To effectively avoid those defaults discussed above (in Figure 4), a GNG modelling method is proposed in this section. Generally speaking, this method mainly consists of three steps: Firstly, based on the restricted quadtree [18], a variable-resolution geometric representation is constructed to simulate the global surface; Secondly, the feature elements are extracted and an original GNG model is constructed; Thirdly, the topological simplification or refinement operators [13,14] are directly applied to the original GNG model recursively to establish a multi-resolution representation. Here, some details about the geometrical simulation and the original GNG construction are our focus.

3.1. A Variable-resolution Geometrical Simulation of the Earth Surface

Since the data volume of the global DEMs is huge, a variable-resolution representation of the earth surface is more appropriate to simulate the earth surface with fine expression in land area while rough expression in marine region (about 71% of the whole earth surface). However, some cracks or T-junctions may occur, which is adverse to the extraction of the feature elements. As a result, in view of these two aspects, a triangulated global diamond grid based on the restricted quadtree [18] is adopted to construct a seamless expression of the global DEMs.

Specifically, according to the principle of the global diamond subdivision [19], the earth surface is uniformly partitioned to a certain level at first. After that, the maximum elevation difference of each diamond vertexes is calculated and compared with the given threshold, and then taking advantage of the algorithm of restricted quadtree tiling [18], the diamond node is determined to be recursively refined, until the maximum elevation difference does not exceed the threshold value any more. As a result, the representation level of the flat ocean region is progressively shallower than that of the fluctuant land area, where the level between two adjacent diamonds exceeds one at most. Finally, leaf node is triangulated according to the adjacencies of the quadtree nodes.

3.2. A Modified Algorithm of the original GNG

To the various local intersections, the duality of the Morse Complex is considered here. The main principle is as follows:

From the dual structure of the ascending and descending Morse complex (in Figure 2), it can be observed that the boundary of an ascending cell is consisted of ravines and saddles, and that the ridges starting from the boundary saddles to the central peak route within the cell. Dually, the boundary of a descending cell is consisted of ridges and saddles, the ravines starting from the boundary saddles to the central pit route within the cell.

By this principle, a modified algorithm for the original GNG is designed, which mainly consists of three steps. Firstly, one of Morse complex, that is to say, descending or ascending Morse complex is constructed by the existing algorithm (for example, the simple STD algorithm [20] without floating-point computations), after the topology relationship of each two adjacent vertexes and two adjacent triangles as well as vertex and triangle is established in the triangulated global diamond grid. Here, all the marine areas are identified first of all and each is regarded as a pit region according to the common sense. Secondly, ridges (ravines), that is, the separatrix of the dual Morse complex is extracted in every constructed cell to prevent those intersections. The saddles belonging to a cell are classified into two groups: the boundary ones or the internal ones. So, a cell saddle whether located on the boundary or not need to be determined. What's more, there may be some macro-saddles, that is, one feature line passes two saddles and the path between the two saddles (called macro-saddle line) plays the dual role of both ravine line and ridge line. Therefore, all the macro-saddles are identified and any of the macro-saddle lines is extracted according to the corresponding path established in the first stage. Furthermore, it should be noticed that the ravines starting from the saddles located on the boundary of the marine regions are not extracted and the end point of each ravine is a pit or a macro-saddle or a point of some marine area. Finally, regarding the feature lines extracted in the second stage as the boundary, the dual complex of the former is constructed and the topological representation of the earth surface, i.e. the original GNG is constructed. Figure 5 shows the pseudo codes of this algorithm.

Input: global triangular terrain grid

Step1 Extract every marine region as a "pit" according to the height;
Step2 Extract all the peaks and pits in the land area and construct descending Morse complex by the existing algorithm (eg. the STD algorithm [20]);
Step3 Extract saddles within every descending cell. **IF** the current saddle is not on the bondary of the marine region, **THEN** extract the ravines according to the location of the saddle:
case1 **IF** the current saddle is within the current cell, **THEN** calculate the paths in the current cell until meet a pit or a saddle on the boundary of the marine region;
case2 **IF** the current saddle is on the boundary of the current cell, **THEN** test it whether one of the macro-saddle;
case2.1 **IF** yes, **THEN** calculate the paths by the cell boundary in step 2 until meet the other saddle;
case2.2 **IF** no, **THEN** calculate the paths in the current desceding cell until meet a pit or a saddle on the boundary of the ocean;
Step4 Construct the ascending Morse complex cells regarding the ravines and the boundary of marine areas extracted in step 3 as the cell boundary.

Output: the GNG model overlaping both the ascending and descending Morse complexes

Figure 5. The pseudo codes of this modified algorithm of the original GNG

4. Experiments and Comparison Analysis

In this section, an experiment system with C++ and OSG (Open Scene Graph) developed by using GTOPO30 data is designed to validate the feasibility and correctness of our approach for the GNG model.

Figure 6a shows a triangulated variable-resolution global diamond grid is constructed based on the restricted quadtree, where the resolution of the land areas is higher than that of the marine regions and there are no cracks and T-junctions as well as thin and long triangles. Here, the earth surface is uniformly partition into the diamond grid at level 3 first of all, and then each diamond node is refined recursively to the max level 7, if the maximum elevation difference of the diamond vertexes exceeds the given threshold 10.0 m.

The saperatrix of original GNG by our modified algorithm from that variable-resolution global triangulated terrain grid is shown in the Figure 6g. From the view of structure, it can be regarded as the overlap of the global descending Morse complex (in Figure 6d) and the global ascending Morse complex (in Figure 6e), while Figure 6b and 6c are the GNG cells. In hydrology, each descending cell is a catchment area and all the rivers converge into the sea. What's more, Figure 6f indicates that some islands occur in the marine areas which represent as isolated peaks, that is, no saddles and ravines connect them. Meanwhile, there are also some saddles to the marine minimum but no paths, locating at the boundary of some marine region. Besides, from the Figure 6g, 6h and 6i, we can see that a multi-resolution GNG is achieved by applying 2000-step and 6000-step topological simplification operators respectively to the original model, while the inverse of this simplification is the refinement.

Further, according to this modelling method and the experimental results, we can see some relationship between the GNG and traditional DGG models (eg. the Quaternary Triangular Mesh [3]), that is, traditional DGG provides a fundamental geometrical model by the simulation of the earth surface because the separatrices of the GNG are extracted from the global DEM by the DGG. As a result, the veracity and accuracy of the separatrix extraction depend on the simulated global DEM. Meanwhile, the GNG is the resegmentation and classification of the cells of traditional DGG models, according to the hydrological characteristics of the global terrain surface.

On the other hand, there are also some distinguishing characterastics of the GNG comparing with the traditional DGG, which is discussed and analyzed as follows:

1) From the perspective of the theoretical basis, Morse theory provides a strict mathematical foundation for the GNG model and supports a knowledge-based approach to analyze, understand and visualize the behaviour in space and time of the scalar field [21]. In fact, GNG can capture the topological structure information related to the level sets of the height function defined on the earth surface and provides a knowledge-based approach to analyze, visualize and understand the earth surface behaviour through a global perspective.

2) To the way of surface description, GNG can describe the shape of the earth surface by those fundamental feature elements (pits, peaks, ravines and ridges), which effectively act as landmarks and can be used to get a representative coverage of the entire earth surface. Therefore, it is an abstract and compact topological description rather than the geometrical formulation by the traditional DGG, which usually contains data redundancy and cannot directly reveal the nature feature elements. Moreover, nature features at the different levels can be analyzed and the main terrain morphological structure can be visualized through the multi-resolution representation of GNG by the topological simplification or refinement, which we can see from the Figure 6g, 6h and 6i.

3) In terms of the complex cell, as shown in Figure 6b and 6c, GNG decomposes the earth surface into irregular "natural" cells rather than regular "data" cells by the traditional DGG models. Since the cells of GNG or Morse complex are the segmentation of the terrain into regions of influence of minima and maxima or into regions of uniform gradient field, they relates closely to the hydrological characteristics of the earth surface. For example, the descending Morse complex of a minimum corresponds to a basin in hydrology [22]. However, the adjacency relationship among the cells of GNG, coded by Morse Incidence Graph [23], seems to be more complex than the one among the cells of the DGG models.

4) As for the aspect of the practical application, the background and purpose of the traditional DGG models is to overcome some shortcomings such as data broken, geometric distortion, when the plain grid models are applied to the organization and visualization of the global multi-scale spatial data [24,25]. However, since the separatrices of the GNG model are consisted of terrain feature lines and points, it provides the terrain morphological information or knowledge about the shape properties of the earth surface which is necessary to the exploration and interpretation of those natural phenomena mentioned in Section 1. In addition, as any descending Morse complex corresponds to a basin, a topological simplification operation [26] of a descending Morse complex can be considered as the simulation of the phenomenon that water overflows along the lowest saddle when the corresponding basin is flooded. As a result, when applied in the field of the rise-or-fall simulation and analysis of global sea levels, GNG is a better option than the traditional DGG models (eg. the QTM [27]), because of the grid cell, granularity and calculation method.

A brief comparison of GNG with the traditional DGG models is summarized in Table 1.

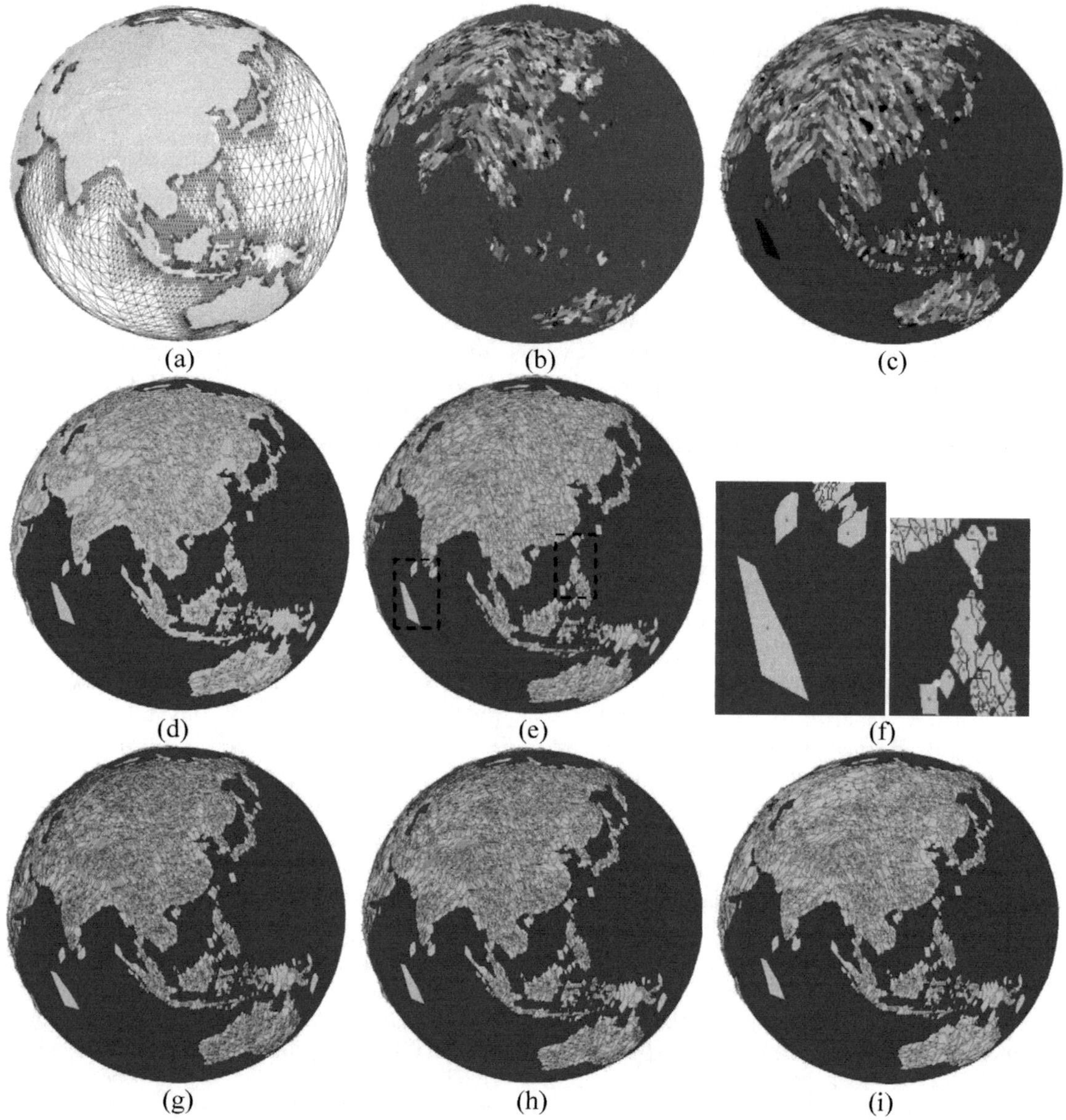

Figure 6. The original construction and multi-scale representation of GNG: (a) variable-resolution triangular mesh, (b) cells of descending Morse complex, (c) cells of ascending Morse complex, (d) sepratrix of the global descending Morse complex, (e) sepratrix of the ascending Morse complex, (f) local enlarged details of degeneration in the dotted region in figure (e), (g) the original global natural grid without simplification, (h) 2000-step simplification, (i) 6000-step simplification

5. Challenges for the GNG model

Although a distinctive GNG model is proposed in this paper, there are still some issues or challenges to be considered in the future works:

5.1. Simplification operations for local degeneration of GNG

As mentioned above, there may be some local degeneration in the GNG model. However, the connection of the feature points coded by the Morse incidence graph is the precondition and foundation of the topological simplification and construction of the multi-scale Morse complexes [23]. As a result, the generalization and refinement operators proposed by [13,14] are no longer suitable for the global terrain morphology. So, some further extension and improvement of those operators needs to be done, in order to construct and achieve an overall and correct multi-resolution representation of the GNG model more than that of the mainland of earth surface. Here, the marine areas may be used to resolve this issue. After that, a combined geometrical and topological multi-scale representation of the GNG just can be built based on the MMT [28] or HFT [29].

5.2. Construction based on the discrete Morse theory

Here, the GNG model was built from the global triangular terrain mesh, just based on the piece-wise linear Morse theory. However, based on the discrete Morse theory, a large number of algorithms (see [12,15] for more details) have been designed and proposed from the local DEM, including both the triangular irregular network (TIN) and the regular square grid (RSG). Actually, the discrete Morse theory overcomes the inherent default of the piece-wise linear Morse theory [12] and may be more suitable for the parallel computation of the Morse complex in the case of a large data sets [30,31]. Therefore, based on the discrete Morse theory the construction and parallel computation of the GNG model from other cell types of the DGG may be one of our further work directions.

5.3. Local refinement operation for GNG

In the field of the global GIS, the whole and expansive earth surface is usually regarded as the study object and the size of data volume is very large, especially under the background of the rapid development of the modern earth observing technology. However, because of the difference of major perspective, some small objects such as house, rabbit, human body and molecule are usually considered at the most existing literature on the Morse complex in computer graphics, biology, physics, chemistry and so on. As a result, the local update operation can be achieved by the reconstruction of the Morse complex in these fields, which is impossible to the GNG model. Consequently, a scientific and effective approach of local update and refinement for the GNG model is particularly needed in global GIS, in order to meet the practical application requirements at the different scales and the development trend of the modern earth observing technology.

Additionally, there are also some other issues to be resolved such as the real-time LOD (level of details) visualization and specific application schema in case of the analysis and understanding of those related nature phenomena.

6. Concluding remarks

In this paper, a global "natural" grid (GNG) model base on the Morse complex is proposed. As a compact structure, GNG segments the earth surface by those fundamental feature elements (pits, peaks, ravines and ridges) and is consisted of the natural cells, which can directly reveal the natural features and provide the knowledge about the shape of the whole earth surface. As a result, it is of great significance to the development trend of the modern earth surveying technologies and has a closer link to the analysis and understanding of some extensive or global nature phenomena. The main work of this paper is as follows:

1) A global natural grid (GNG) based on the Morse complex is proposed, of which the modeling approach includes the simulation of the earth surface, original construction and multi-resolution representation.

2) A relatively comprehensive and theoretical comparison between the GNG and traditional DGG is analyzed in details. In brief, traditional DGG provides a geometrical foundation of the feature extraction and cell grouping for the GNG. However, there are also some individual distinguishing characters of the GNG, such as the theoretical foundation, description way, model cell, and background of application.

3) Some issues or challenges for the GNG to be considered in the future works are also discussed and given out, mainly including the multi-resolution representation, the construction and parallel computation based on the discrete Morse theory, the local update and refinement operation.

Table 1. The comparison of GNG with the traditional DGG models

Model	Relationship	Difference					
		theoretical foundation	description	cell	adjacency coding	feature revealing	applications
GNG	traditional DGG provides a foundation of cell grouping for the GNG	rigorous	compact and topological	irregular natural cells	based on Morse incidence graph (MIG)	able	environmental monitoring, climatic and hydrological analysis, meteorological service, simulation of sea level rise, etc.
Traditional DGG		relatively loose	accurate and geometrical	regular data cells	mostly based on a tree	disable	organization or index of global spatial data, thematic map production and visualization

References

[1] Deng H, Wang Q, Cai Q, 1998. Watershed ecology - new discipline, new idea and new approach. *Chinese Journal of Applied Ecology*, **9**(4) : 443-449.

[2] Seong J C, 2005. Implementation of an equal-area gridding method for global-scale image archiving. *Photogrammetric Engineering & Remote Sensing*, **71**(5) : 623-627.

[3] Chen J, Zhao X, and Li Z, 2003. An algorithm for the generation of Voronoi diagrams on the sphere based on QTM. *Photogrammetric Engineering & Remote Sensing*, **69**(1) : 79-89.

[4] Sahr K, 2008. Location coding on icosahedral aperture 3 hexagon discrete global grids. *Computers, Environment and Urban Systems*, **32**(3) : 174-187.

[5] Mostafavi M A, Gold C, and Dakowicz M, 2003. Delete and insert operations in Voronoi/Delaunay methods and applications. *Computers & Geosciences*, **29**(4) : 531-537.

[6] Zhou Q, Li D, 2001. Reference model of the digital Earth. *From Digital Image to Digital Earth*, Pess: Wuhan Technical University of Surveying and Mapping, Wuhan, pp.88~95.

[7] Zhou C, Ou Y, Ma T, 2009. Progresses of geographical grid systems researches. *Progress in Geography*, **28**(5) : 657-662.

[8] Zhao X, Cui M, Li A, Zhang M, 2009. An adjacent searching algorithm of degenerate quadtree grid on spherical facet. *Geomatics and Information Science of Wuhan University*, **34**(4) : 479-482.

[9] Zhou L, Sheng Y, Lin B, Lü G, Zhao Z, 2014. Diamond discrete grid subdivision method for spherical surface with icosahedron. *Acta Geodaetica et Cartographica Sinica*, **43**(12) : 1293-1299.

[10] Milnor J, 1963. *Morse theory*. Princeton Univ, Press.

[11] Smale S, 1960. Morse inequalities for a dynamical system. *Bulletin of American Mathematical Society*, **66**: 43-49.

[12] Floriani D L, Fugacci U, Iuricich F, Magillo P, 2015. Morse complexes for shape segmentation and homological analysis: discrete models and algorithms. *Compute Graph Forum 2015*, http://dx.doi.org/10.1111/cgf.12596.

[13] Floriani L D, Magillo P, Vitali M, 2010. Modeling and generalization of discrete Morse terrain decompositions. In *20th International Conference on Pattern Recognition*, Istanbul, Turkey, August 23-26, pp.999-1002.
[14] Danovaro E, Floriani D L, Vitali M, et al., 2007. Multi-scale dual Morse complexes for representing terrain morphology. *Proceedings of the 15th ACM International Symposium on Advances in Geographic Information Systems*, 220-227.
[15] Čomić L, Floriani D L, Magilo P, Iuricich F, 2014. Morphological modeling of terrains and volume data. *Springer Briefs in Computer Science*. Springer.
[16] Banchoff T, 1970. Critical points and curvature for embedded polyhedral surfaces. *American Mathematical Monthly*, **77**(5) : 475-485.
[17] Forman R, 1998. Morse theory for cell complexes. *Advance in Math*, **134** : 90-145.
[18] David H, 2002. An efficient, hardware-accelerated, level-of detail rendering technique for large terrains. Master Thesis, University of Toronto.
[19] Zhao X, Bai J, and Wang Z, 2007. An adaptive visualized model of the global terrain based on QTM. *Acta Geodaetica et Cartographica Sinica*, **36**(3) : 316-320.
[20] Magillo P, Danovaro E, Floriani L, Papaleo L, Vitali M, 2009. A discrete approach to compute terrain morphology. *Computer Vision and Computer Graphics - Theory and Applications, Communications in Computer and Information Science*, Volume **21**, Part 1, I, 13-26.
[21] Biasotti S, De Floriani L, Falcidieno B, Frosini P, Giorgi D, Landi C, Papaleo L, Spagnuolo M, 2008. Describing shapes by geometrical-topological properties of real functions. *ACM Computing Surveys*, **40**(4), 12 : 1-87.
[22] Magillo P, Floriani D L, Iuricich F, 2013. Morphologically-aware elimination of flat edges from a TIN. *Proceedings 21st ACM SIGSPATIAL International Conference on Advances in Geographic Information Systems (ACM SIGSPATIAL GIS 2013)*, Orlando, Florida, USA, November 5-8.
[23] Čomić L, Floriani L, and Iuricich F, 2012. Dimension-independent multi-resolution Morse complexes. *Computers & Graphics*, **36** : 541-547.
[24] Kolar J, Ilsøe P, 2005. Flexible terrain representation using runtime surface reconstruction. WSCG'2005, Jan. 31-Feb. 4, Plzen, Czech Republic.
[25] Hu P, Liu P, Hu H, Yang C, 2005. Metric space of earth information and global GIS. *Geomatics and Information Science of Wuhan University*, **30**(4) : 317-321.
[26] Edelsbrunner H, Harer J, and Zomorodian A, 2001. Hierarchical Morse complexes for piecewise linear 2-manifolds. In *Proceedings 17th ACM Symposium on Computational Geometry*, ACM Press, 70–79.
[27] Hou M, Jiang X, Zhao X, Xing H, 2012. Analysis and simulation of sea level rise base on QTM, *Geography and Geo-Information Science*, **28**(1) : 35-38.
[28] Danovaro E, De Floriani L, and Vitali M, et al., 2010. Multiresolution Morse triangulations, *Proceedings of the 14th ACM Symposium on Solid and Physical Modeling*, Haifa, Israel, *SESSION: Geometric modeling and processing*, 183-188.
[29] Iuricich F, 2014. Multi-resolution shape analysis based on discrete Morse decompositions. PhD thesis, University of Genova -DIBRIS, Italy.
[30] Shivashankar N, Senthilnathan M, Natarajan V, 2012. Parallel computation of 2d Morse-Smale complexes. *IEEE Transactions on Visualization and Computer Graphics*, **18**(10) : 1757-1770.
[31] Gyulassy A, Bremer P, Pascucci V, 2012. Computing Morse-Smale complex with accurate geometry. *IEEE transactions on visualization and computer graphics*, **18**(12) : 2014-2022.

The Discrete Fourier Transform on hexagonal remote sensing image

Yalu Li, Jin Ben, Rui Wang, Lingyu Du

Information Engineering University

E-mail:1505282011@qq.com

Abstract: Global discrete grid system will subdivide the earth recursively to form a multi-resolution grid hierarchy with no Overlap and seamless which help build global uniform spatial reference datum and multi-source data processing mode which takes the position as the object and in the aspect of data structure supports the organization, process and analysis of the remote sensing big data. This paper adopts the base transform to realize the mutual transformation of square pixel and hexagonal pixel. This paper designs the corresponding discrete Fourier transform algorithm for any lattice. Finally, the paper show the result of the DFT of the remote sensing image of the hexagonal pixel.

1. Introduction

With the rapid development of earth observation technology, Human obtain remote sensing information of big data characteristics, so remote data mining and application becomes the focus of attention. Global discrete grid system will subdivide the earth recursively to form a multi-resolution grid hierarchy with no overlap and seamless which help build global uniform spatial reference datum and multi-source data processing mode which takes the position as the object and in the aspect of data structure supports the organization, process and analysis of the remote sensing big data. Only three basic polygons-triangles, rectangles and hexagons can seamlessly tile plane[2]. Since the adjacent triangles differ in directions, the triangle is not the optimal scheme for plane partition. So far the rectangular grid is the most commonly used . Because square pixel image is symmetry, a small amount of calculation ,easy to store and implement and many existing image processing algorithm. However, the image of square pixels also have many disadvantages, such as aliasing effects, quantization errors, inconsistency in adjacent cells and less relevance with visual process and so on. In the image processing , compared with square pixel ,hexagonal pixel have a higher degree of circular symmetry ,consistent contiguity, greater angular resolution, and require less storage space and computing[1,5]. Above all, this article will research remote-sensing image of hexagonal pixels. Since the Fourier transform has been widely used in feature extraction, spatial frequency and filtering, image restoration and texture analysis. This article firstly describes a clear and simple DFT suitable for n-dimensional lattice L .The space domain of the DFT is a system of coset representatives of the quotient group generated by the lattice L and the sublattice of L. Then we apply this result on the image of hexagonal pixel.

2. The DFT of any lattice

This paper will introduce DFT which is suitable for any lattice. Mathematical set which can undertake DFT is a N order finite Abelian group D. Abelian group is D additive group.

$\widehat{D}$ denote the character group $\mathrm{Hom}(D, I_N)$ of group D. D denote space domain and $\widehat{D}$ denote frequency domain. Fourier transform is a powerful tool for spatial domain to frequency domain.

A lattice in n dimensional European space is a linear combination of linearly independent vector. Take these n vectors as base vector of the lattice. The row vectors of base matrix are base vectors of the lattice L.denote that L_0 is a n dimensional sublattice of the lattice L. Because the lattice L and the lattice L_0 are both Abelian group, The quotient group L/L_0 is also a finite Abelian group[3].

Given $L_0 < L$ and $u \in L$, $\overline{u}$ denote the coset of u in L/L_0.Similarly given $v \in {L_0}^*$, $\overline{v}$denote the coset of v in ${L_0}^*/L^*$. Spatial domain is $D = L/L_0$ and frequency domain is $D^* = {L_0}^*/L^*$,and order $N = |G| = |G^*|$. the DFT of spatial domain into frequency domain is a linear transformation[3,4,6] , as shown in the following type:

$$(\mathcal{F}a)(\overline{v}) = \frac{1}{\sqrt{N}} \sum_{\overline{u} \in D} a(\overline{u}) e^{-2\pi i \langle u,v \rangle} \tag{1}$$

The inverse Fourier transformation:

$$(\mathcal{F}a^*)(\overline{u}) = \frac{1}{\sqrt{N}} \sum_{\overline{v} \in D^*} a^*(\overline{v}) e^{2\pi i \langle u,v \rangle} \tag{2}$$

Theorem If the quotient group L/L_0 have elementary divisor $N_1, N_2, \cdots, N_d$, DFT and inverse DFT can be reduced to computing multiple one dimensional DFT. The running time needed for this result is:

$$O\left(N \sum_{i=1}^{n} \frac{\log Ni}{N_1 N_2 \cdots N_{i-1}}\right)$$

Proof: Proof the result that DFT can be reduced to computing multiple one dimensional DFT, the proof of inverse DFT is similar. Given a n dimensional lattice L and a sublattice L_0 ,the base vector of L with respect to L_0 compose the set $\{\vec{v}_1, \vec{v}_2, \cdots, \vec{v}_n\}$ .Let $M = \{i_1\vec{v}_1 + i_2\vec{v}_2 + \cdots + i_n\vec{v}_n : 0 \le i_j < N_j, 0 \le j \le n\}$,the$N_1, N_2, \cdots, N_n$are elementary divisors of L/L_0.Hence M is a system of coset representatives of the quotient group L/L_0. The base vectors of L with respect to L_0 compose the base matrix W_L.The base vectors of L compose the base matrix W. The diagonal elements of the diagonal matrix B are composed of the corresponding L/L_0 elementary divisors. As the base vector of L_0 is integer linear combination of the base vector of L, there is an integer matrix T to make TW a base matrix of L_0. $P_1, P_2, \cdots, P_a$and$Q_1, Q_2, \cdots, Q_b$ is the elementary matrix used to diagonalize matrix T. So you can get the diagonal matrix B:

$$B = P_a \cdots P_2 P_1 T Q_1 Q_2 \cdots Q_b$$

With the relationship between matrix W_L and the matrix W:

$$W_L = (Q_1 Q_2 \cdots Q_a)^{-1} W$$

Let $Q = Q_1 Q_2 \cdots Q_b$ and $P = P_a \cdots P_2 P_1$. As the row vector of TW is the base vector of L_0 and V is a unit matrix, the row vector of the matrix $W_{L_0} = P(TW)$ is also the base vector of L_0,similarly the row vector of the matrix $W_L = Q^{-1}W$ is also the base vector L.Therefore

$$BW_L = (PAQ)(Q^{-1}W) = P(TW) = W_{L_0} \tag{3}$$

The row vector of the matrix of W_L is the base vector of L with respect to L_0,and the corresponding divisors are on the diagonal of the matrix B.

The dual lattice of the lattice L is:

$$L^* = \{\mathrm{y} \in R^n : \langle \mathrm{x}, \mathrm{y} \rangle \text{is an integer for all} \mathrm{x} \in L\}$$

$\langle \mathrm{x}, \mathrm{y} \rangle$ is the inner product of a standard European space. The following three properties is:

- $L^* < L_0{}^*$
- $\widehat{L/L_0} \cong L_0{}^*/L^*$
- If W is the base matrix of L,$W^{-T}$is the base matrix of L^*.

As the quotient group of L/L_0 is the space domain of DFT, according to the above second properties , the quotient group $L_0{}^*/L^*$ is the frequency domain.

The base vector $\{\vec{v}_1, \vec{v}_2, \cdots, \vec{v}_n\}$ of L with respect to L_0 is the row vector of the base matrix W. According to the above third properties ,we get The base vector of L_0^* with respect to L^* is the row vector $\{\vec{v}_1^*, \vec{v}_2^*, \cdots, \vec{v}_n^*\}$ of the base matrix $(BW)^{-T}$.According to

$$(\langle \vec{v}_j, \vec{v}_k^* \rangle) = (BW)^{-T} W^T = B^{-1} \tag{4}$$

Therefore

$$\langle \vec{v}_j, \vec{v}_k^* \rangle = \begin{cases} \frac{1}{N_j} & \text{if } j = k \\ 0 & \text{otherwise} \end{cases}$$

So

$$\left\langle \sum_{i=1}^{n} j_i v_i, \sum_{i=1}^{n} k_i v_i^* \right\rangle = \sum_{i=1}^{n} \frac{1}{N_i} j_i k_i \tag{5}$$

$J = (j_1, j_2, \cdots, j_n)$ denote $j_1\vec{v}_1 + j_2\vec{v}_2 + \cdots + j_n\vec{v}_n$ and $K = (k_1, k_2, \cdots, k_n)$ denote $k_1\vec{v}_1^* + k_2\vec{v}_2^* + \cdots + k_n\vec{v}_n^*$. The formula (5) in the above formula (1):

$$\begin{aligned}(\mathcal{F}a)(K) &= \frac{1}{\sqrt{N}} \sum a(J) e^{-2\pi i \frac{j_1 k_1}{N_1}} e^{-2\pi i \frac{j_2 k_2}{N_2}} \cdots e^{-2\pi i \frac{j_n k_n}{N_n}} \\ &= \frac{1}{\sqrt{N}} \sum_{j_n=0}^{N_n-1} e^{-2\pi i \frac{j_n k_n}{N_n}} \\ &\times \left(\cdots \left(\sum_{j_2=0}^{N_2-1} e^{-2\pi i \frac{j_2 k_2}{N_2}} \left(\sum_{j_1=0}^{N_1-1} a(J) e^{-2\pi i \frac{j_1 k_1}{N_1}} \right) \right) \right)\end{aligned} \tag{6}$$

$0 \leq j_1 < N_1, \cdots, 0 \leq j_n < N_n$.

Let

$$b_1(k_1, j_2, \cdots, j_n) = \sum_{j_1=0}^{N_1-1} a(j_1, j_2, \cdots, j_n) e^{-2\pi i \frac{j_1 k_1}{N_1}},$$

$$b_2(k_1,k_2,j_3,\cdots,j_n)=\sum_{j_2=0}^{N_2-1} b_1(k_1,j_2,\cdots,j_n)e^{-2\pi i\frac{j_2k_2}{N_2}},$$

$$\cdots$$

$$b_n(k_1,k_2,\cdots,k_n)=\sum_{j_n=0}^{N_n-1} b_{n-1}(k_1,k_2,\cdots,k_{n-1},j_n)e^{-2\pi i\frac{j_nk_n}{N_n}}$$

So

$$\begin{aligned}(Fa)(k)&=b_n(k_1,k_2,\cdots,k_n)\\&=\sum_{j_n=0}^{N_n-1} b_{n-1}(k_1,k_2,\cdots,k_{n-1},j_n)e^{-2\pi i\frac{j_nk_n}{N_n}}\end{aligned} \tag{7}$$

Because the run time of one dimension DFT of size N_i is $O(N_i\log N_i)$.So the complexity of the above formular is:

$$O\left(N\sum_{i=1}^{n}\frac{\log Ni}{N_1N_2\cdots N_{i-1}}\right)$$

3. pixel conversion

Current image acquisition and display devices are all based on the square grid structure. In order to do discrete Fourier transform on the hexagonal pixel remote sensing image, firstly it needs to transform the square pixel image into a hexagonal pixel image.

The base vectors of traditional square pixel image are $\vec{a}_1=(1,0)$ and $\vec{a}_2=(0,1)$.The base vectors of hexagonal are $\vec{v}_1=(1,0)$ and$\vec{v}_2=\left(-\frac{1}{2},\frac{\sqrt{3}}{2}\right)$.The matrix C is the transition matrix which can transform square pixel into hexagonal pixel. The formular of the base transformation is as shown as below :

$$(\vec{v}_1,\vec{v}_2)=(\vec{a}_1,\vec{a}_2)C \tag{8}$$

Hence the transition matrix C is:

$$C=\begin{pmatrix}1 & \frac{-1}{2}\\ 0 & \frac{\sqrt{3}}{2}\end{pmatrix}$$

Let the vector $\boldsymbol{x}=m_1a_1+m_2a_2=n_1v_1+n_2v_2$,Therefore

$$\begin{aligned}\boldsymbol{x}&=(\vec{a}_1,\vec{a}_2)\begin{pmatrix}m_1\\ m_2\end{pmatrix}=(\vec{v}_1,\vec{v}_2)\begin{pmatrix}n_1\\ n_2\end{pmatrix}\\&=(\vec{a}_1,\vec{a}_2)C\begin{pmatrix}n_1\\ n_2\end{pmatrix}\end{aligned}$$

So

$$\begin{pmatrix}m_1\\ m_2\end{pmatrix}=C\begin{pmatrix}n_1\\ n_2\end{pmatrix} \tag{9}$$

According to the above formulas of the base transformation and coordinated transformation ,it can resample square pixel image to the desired hexagonal pixel remote sensing images. This paper uses bilinear interpolation to resample . The hexagonal images resampled are shown as below:

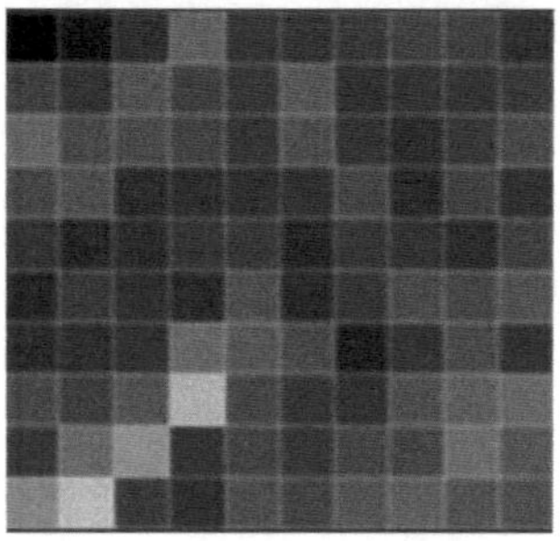

Figure 1.square pixel image

Figure 2 hexagonal pixel image

4. The DFT of the Hexagonal lattice

Remote sensing image is a function which is defined on a finite set A of the two dimensional lattice H.The set A must be nature Abelian group.H_0 is a sublattice of H.the set A is a system of coset representatives of the quotient group H/H_0. the quotient group H/H_0 is the space domain of DFT. The vector $\vec{v_1}=(1,0)$ and $\vec{v_2}=\left(-\frac{1}{2},\frac{\sqrt{3}}{2}\right)$ are base vectors of hexagonal lattice H with respect to H_0 and the corresponding divisors are N_1 and N_2.So the corresponding formular of DFT of hexagonal lattice.

$$\begin{aligned}(\mathcal{F}a)(K) &= \frac{1}{\sqrt{N}}\sum a(J)e^{-2\pi i\frac{j_1k_1}{N_1}}e^{-2\pi i\frac{j_2k_2}{N_2}} \\ &= \frac{1}{\sqrt{N}}\sum_{j_2=0}^{N_2-1}e^{-2\pi i\frac{j_2k_2}{N_2}}\left(\sum_{j_1=0}^{N_1-1}a(J)e^{-2\pi i\frac{j_1k_1}{N_1}}\right)\end{aligned}$$

$0\le j_1<N_1, 0\le j_2<N_2$.

Let

$$b_1(k_1,j_2)=\sum_{j_1=0}^{N_1-1}a(j_1,j_2)e^{-2\pi i\frac{j_1k_1}{N_1}},$$

$$b_2(k_1,k_2)=\sum_{j_2=0}^{N_2-1}b_1(k_1,j_2)e^{-2\pi i\frac{j_2k_2}{N_2}}$$

So

$$\begin{aligned}(Fa)(k) &= b_2(k_1,k_2) \\ &= \sum_{j_2=0}^{N_2-1}b_1(k_1,j_2)e^{-2\pi i\frac{j_2k_2}{N_2}}\end{aligned}$$

The experimental result is shown as below:

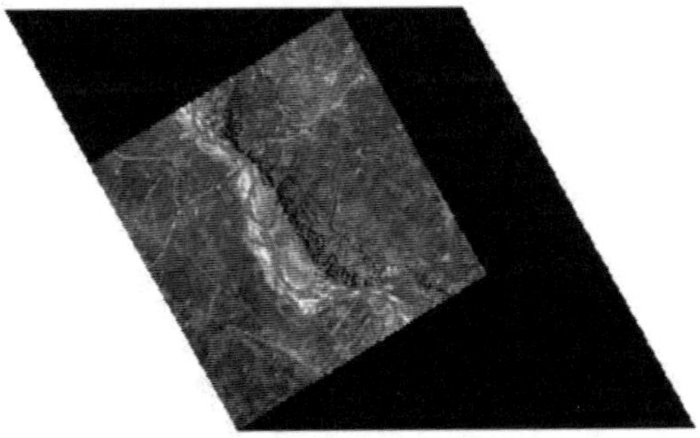

Figure 3.The source hexagonal pixel image

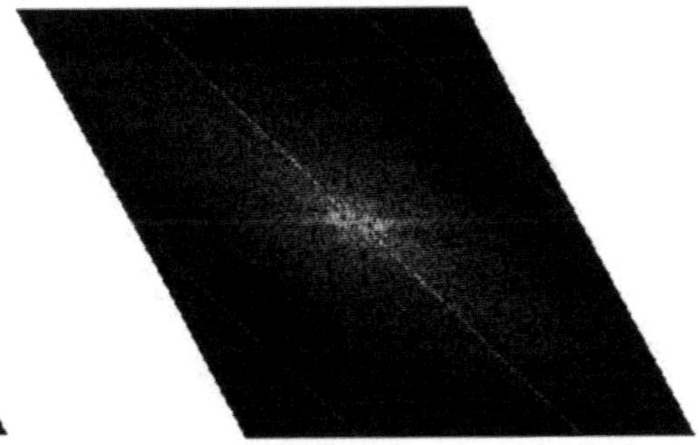

Figure 4.The result of DFT

5. Conclusion

This paper firstly describes DFT suitable for n-dimensional lattice and then introduces base the DFT on a particular subdivision structure of hexagonal pixel remote sensing image .

6. References

[1] Asharindavida, F., Hundewale, N., & Aljahdali, S. (2012). Study on hexagonal grid in image processing. *Proc. ICIKM*.

[2] SAHR K, WHITE D, KIMERLING J. Geodesic Discrete Global Grid Systems[J]. *Cartography and Geographic Information Science*, 2003,30(2):121-134.

[3] Vince A, Zheng X. Computing the discrete Fourier transform on a hexagonal lattice[J]. *Journal of Mathematical Imaging and Vision*, 2007, 28(2): 125-133.

[4] Zapata J L, Ritter G X. Fast Fourier transform for hexagonal aggregates[J]. *Journal of Mathematical Imaging and Vision*, 2000, 12(3): 183-197.

[5] Staunton R C, Storey N. A comparison between square and hexagonal sampling methods for pipeline image processing[C]//1989 *Advances in Intelligent Robotics Systems Conference. International Society for Optics and Photonics*, 1990: 142-151.

[6] Guessoum A, Mersereau R M. Fast algorithms for the multidimensional discrete Fourier transform[J]. *Acoustics, Speech and Signal Processing,* IEEE Transactions on, 1986, 34(4): 937-943.

Acknowledgments

Portions of this research were supported by National Natural Science Foundation of China (Grant NO. 41271391) and by Special Financial Grant from the China Postdoctoral Science Foundation (Grant NO. 2013T60161). The research has not been subjected to the Committee of National Natural Science Foundation or the Committee of Postdoctoral Science Foundation's peer and administrative review and, hence, does not necessarily reflect the views of them.

A Integrated Service Platform for Remote Sensing Image 3D Interpretation and Draughting based on HTML5

Yiping LIU[1], Qing XU[1], Heng ZhANG[1], Liang LV[1], Wanjie LU[1] and Dandi WANG[1]

1.Information Engineering University. Zhengzhou, Henan, China

Abstract. The purpose of this paper is to solve the problems of the traditional single system for interpretation and draughting such as inconsistent standards, single function, dependence on plug-ins, closed system and low integration level. On the basis of the comprehensive analysis of the target elements composition, map representation and similar system features, a 3D interpretation and draughting integrated service platform for multi-source, multi-scale and multi-resolution geospatial objects is established based on HTML5 and WebGL, which not only integrates object recognition, access, retrieval, three-dimensional display and test evaluation but also achieves collection, transfer, storage, refreshing and maintenance of data about Geospatial Objects and shows value in certain prospects and potential for growth.
Keywords: Integrated Service Platform, 3D Interpretation and Draughting, HTML5

1. Introduction

As the foundation of Remote Sensing Image Processing, Remote Sensing interpretation and draughting can interpret and extract ground objects from images accurately and completely, it can dig out the value of the data and pave the way for the follow-up work. Thus, even in the case of developed and prevalent digital processing technology of various remote sensing data, interpretation and draughting technology for Remote Sensing images is still one of the most fundamental, common, effective and important Remote Sensing information processing technologies.

At present, most of the remote sensing training systems are based on single machine, platform or plug-in, such as the Remote Sensing Image Group Interpretation System designed by Liu Yalan, Yan Shouyong et al[1], Remote Sensing Image Inter pretation Training System designed by Song Yingjin et al[2] and Remote Sensing Image Interpretation System Based on "Platform / Plug-in mechanism" designed by Tan Xiangrui et al[3], having the problem of different interpretation criteria , single function, high plug-in dependency , closed system and bad integration. The emergence and development of HTML5 that has features of compatible equipment, offline storage, free plug-in three-dimensional network display etc, provide a new way in to solve these problems. Then a 3D interpretation and draughting integrated service platform for multi-source, multi-scale and multi resolution geospatial objects is established based on HTML5 and WebGL, achieving the functions of collection, transmission, storage, updating and maintenance of target information data. It all-directionally and integrally making operators carry out geographical space target interpretation, access, retrieval, 3D display and the testing and evaluation of training via the wide area network and improves the operational level of the operator and work efficiency quickly.

2. The overall design of platform

2.1. The objectives and functions

The platform is designed based on wide area network with "Two Points and One Line" as pivot (the administrator and the operator for two points, the geographical space target interpretation learning as the line).A integrated human-computer interaction network system is completed to collect, transfer, store, update and maintain the geo spatial object data by Web, which will greatly improve the efficiency of the collection, management and sharing about the target information data. At the same time, the administrator can maintain, manage and update the system. It also can effectively monitor, count and analyze the training results in the background.

The platform mainly consists of 4 parts: composition and characteristics of spatial objects, design of geo spatial object database, training evaluation of typical target interpretation and draughting based on Network, 3D visualization of geo spatial object network. The "Administrator" and "Operator" two user login mode is adopted by the platform, having respective different functions and permissions. The overall system function structure is shown in figure 1.

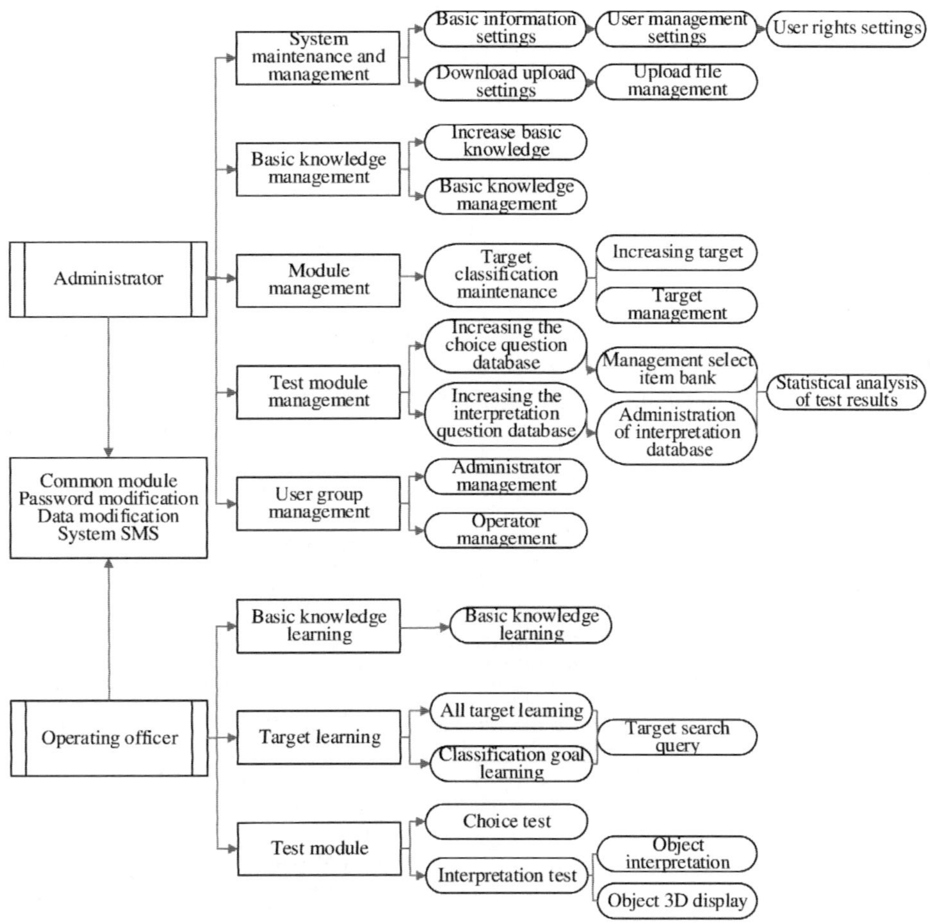

Figure 1. Structure of the Platform Function

2.2. The overall structure of the platform

This platform mainly consists of two parts, the server and the browser. The browser mainly includes the login window, the basic knowledge learning window, the geographical space target interpretation and annotation window, the 3D display window and other parts, providing a visual operating interface. The server side mainly refers to the data layer and web service layer. The data layer mainly includes image data, 3D model data, object map data, basic knowledge information data, etc., all stored in a certain format. The whole platform is built based on JavaScript, WebGL and other languages, developed by JetBrains WebStorm 10.0.3 System, whose Web Service layer is constructed by the open source web application server Apache Tomcat. The overall architecture is shown in figure 2.

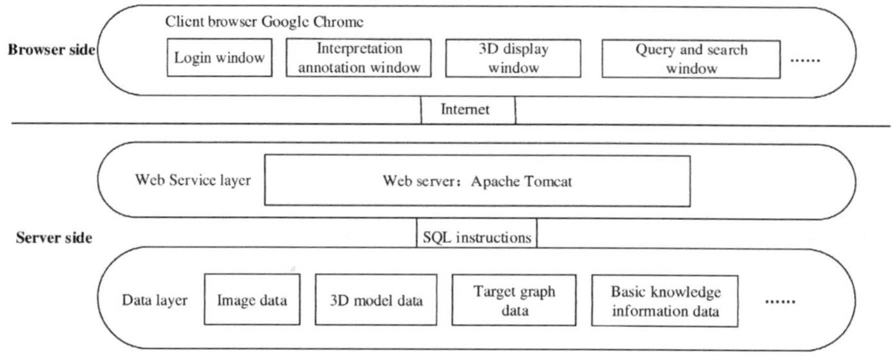

Figure 2. Overall Structure of the Platform

3. The implementation of the platform

3.1. The analysis of typical geo spatial objects and establishment of database

It forms the knowledge information database by classification and arrangement the basic knowledge about the target, the content of which mainly includes the definition, the basic knowledge, the main points and the rules of the target acquisition, etc and it is supplemented by the way of the administrator to add and maintain the basic information knowledge database.

The platform database consists of two parts: the business database and the non-business databases, as shown in Figure 3. The business database is the establishment of typical spatial object database, including the target image database, the target map database, the target ground photo database, the 3D model database, the basic knowledge database and so on; non-business database includes user database, authority distribution database, etc.

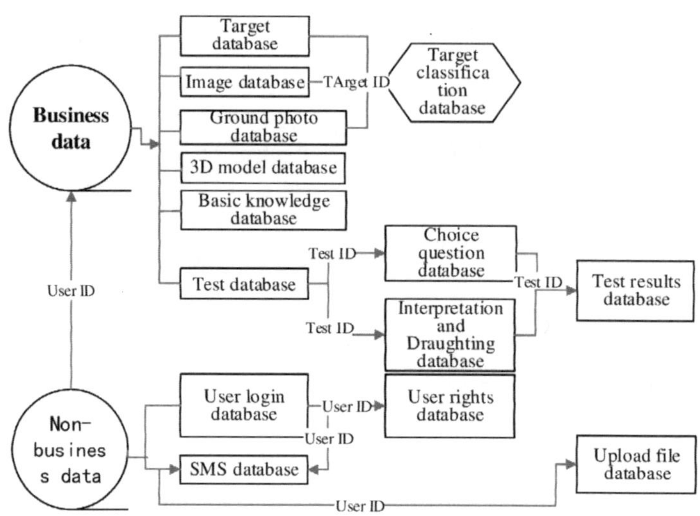

Figure 3. Structure of the Database

3.2. Interpretation and draughting of remote sensing images based on flash

After entering the system training, the operator chooses the target to be trained, and then the image and the object graph are displayed by the image database and the corresponding target database, which is shown in Figure 4. Fast Fuzzy Query and Classification-based Retrieval of arbitrary objects are designed, which is more convenient for users to learn and browse.

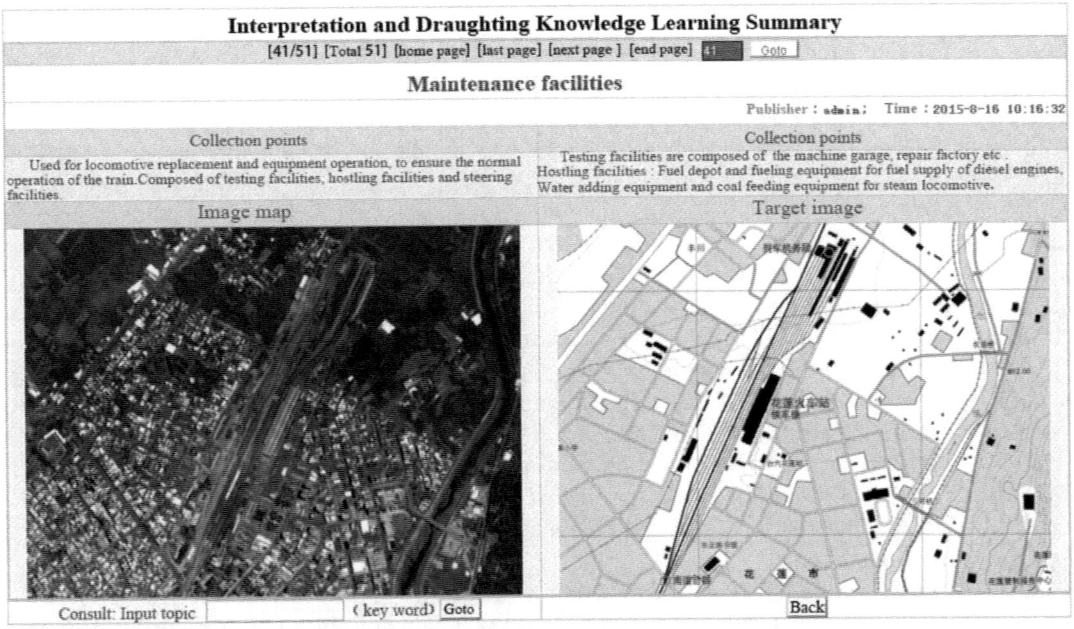

Figure 4. the Module of Interpretation Knowledge

Flash is an interactive animation design tool, as the main programmable vector image processing technology in the current Web browser. There is no distortion in the operation of the vector maps that are made by Flash zooming in and zooming out. Its characteristics of the vector and the nature of the network applications, can be very good to meet the requirements of spatial data graphics display very well. This module mainly uses the Flash+HTML5 technology. The main principles and processes are shown in figure 5.

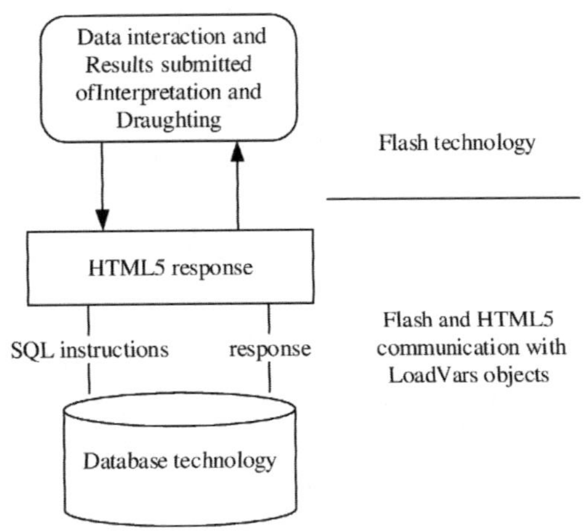

Figure 5. the Principle of Interpretation Training

3.3. 3D visual interpretation of geo spatial objects based on HTML5

The function of this module is mainly realized by the HTML5+WebGL+ third party library ht.js.

HTML5 is another revolutionary technology after HTML4 introduced by W3C organization[4]. It is the open source that makes HTML5 different from HTML4. Therefore, it supports most of the

IOP Conf. Series: Earth and Environmental Science **46** (2016) 012023 doi:10.1088/1755-1315/46/1/012023

browser and has significant cross platform characteristics. In addition, HTML5 also adds a lot of new features, such as equipment compatibility, offline storage, free plug-in 3D network display, etc. The Canvas elements in HTML5 makes it possible to display high performance animation in the browser. It designs an API supporting the browser calling graphics card to draw, which can make the JavaScript complete the draughting of 2D and 3D elements in the web by calling the WebGL[5]. The direct rendering of 3D graphics in the web by Canvas elements and WebGL not only has the feature of 3D hardware acceleration but also possesses the characteristic of free plug-in[6].

WebGL is a 3D drawing standard, which can provide hardware 3D accelerated rendering for Canvas HTML5. So you can create more complex 3D Web scenes without the support of the rendering plug-in[7].

HT is a one-stop solution for the development and application of graphical interface based on HTML5, of which the core library is ht.js, containing rich graphical interface development package such as generic component, topology component, 3D rendering engine and so on. It provides sets of visual design tools based on HTML5 such as vector editor, topology editor and 3D scene editor, and provides a powerful 3D graphics engine based on WebGL technology. The component showing 3D view in HT is ht.graph3d.Graph3dView. The structure of Graph3dView interface DOM is composed of the bottom of the DIV elements and rendering layer Canvas elements. The bottom of the DIV element can derive from getView () function. The rendering layer Canvas elements can derive from GetCanvas () function. Default interaction events of HT are added to the underlying DIV element and can also directly use the returned elements of getView () to add an event listener to achieve user custom interactive extensions.

Three dimensional mode with .Obj format contains .Obj files and.Mtl files. OBJ file describes

the model vertex, surface and map coordinates and other related information of geometric models. The material information of model such as texture images, colors and so on is described by the additional MTL material file, generally using .Mtl suffix. It's more convenient to realize visualization on Web page by importing the three-dimensional model with the texture using the ht.Default.loadObj () function. Part of the code of aircraft model visualization is as follows:

```
    ht.Default.loadObj('obj/plane.obj', 'obj/plane.mtl', {
    cube: true,
    center: true,
    shape3d: 'plane',
    finishFunc: function (modelMap, array, rawS3) {
       var node = new ht.Node();
       node.s3(70,70,70);
       node.p3(-300, 0, 0);
       node.s({
              'shape3d': 'door',
                });
       dataModel.add(node);
    }
});
```

For example, in the interpretation and draughting of the radar target, the 3D model and 2D image of radar can be observed at the same time in the visual window. Its 3D model can be controlled by the mouse wheel to complete the movement, scale operation, which will form a full range of 3D visual perception of the target. The display window of the radar target is shown in figure 6.

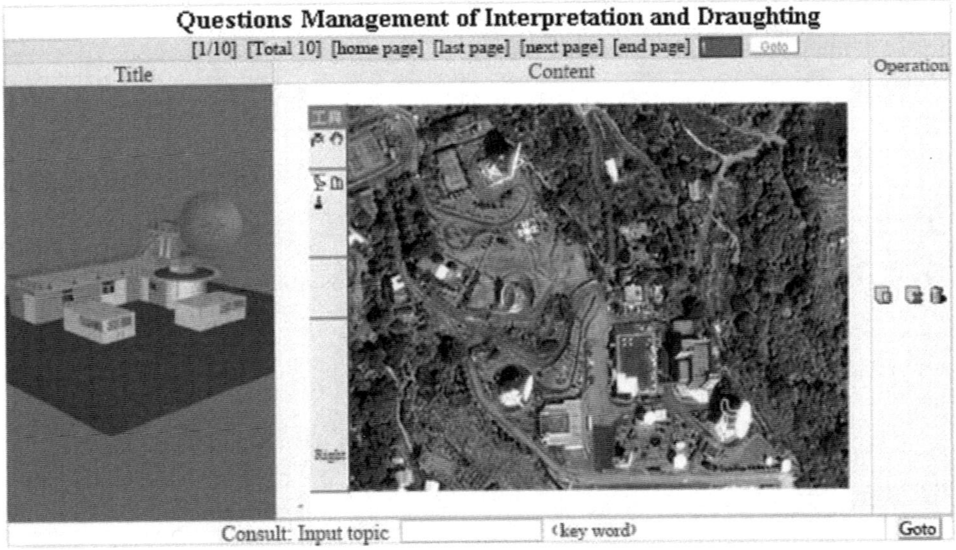

Figure 6. the Show to 2D and 3D Interpretation of Radar

4. The conclusion

This paper introduces an integrated platform of geographic space target interpretation and draughting based on HTML5, achieving the integrated service of autonomous learning, access, retrieval, geographical space target interpretation and 3D display, test and result evaluation in the network environment. It effectively solves the existing problems in the traditional platform such as inconsistent standards, single function, dependence on plug-ins, closed system and low integration level. The platform with the advantages of standardization, free plug-in, cross platform, openness and so on has a strong development potential.

References

[1] Yan ShouYong, Liu Yalan, EI Chengjie, et al. The theory and method of grouping interpretation technology for remote sensing images [M]. China Ocean Press, 2007:20-21.

[2] Zhang Wenshi, Song Yingjin, KANG Xin. The Design and Realization of Training and Application Systems of the Remote Sensing Interpretation[J]. Journal of Zhengzhou Institute of Surveying and Mapping, 2006, 23(1):19-22.

[3] Tan Xiangrui, Li Haitao, HAN Yanshun, et al. Design and implementation of remote sensing image interpretation system using platform/ plug-in mechanism. [J]. Remote Sensing Information, 2013, 28(5):73-76.

[4] Wu Ting, Zhu eizheng. Research and implementation of image intelligence processing system[J].Computer Engineering and Design, 2011, 32(11):3925-3928.

[5] Liu Haina. The research on panorama technology based on HTML5[D]. Zhengzhou: Zhengzhou University, 2013： 10-12.

[6] Wang Desheng. 3D WebGIS construction and application based on HTML5 and WebGL [D]Fujian: Fujian Normal University,2014:7-8

[7] Liu Xinxing. Space environment three-dimensional imaging research and application based on HTML5[J]. Chengdu： University of Electronic Science and Technloogy of China, 2013:18-20.

The Coverage Analysis for Moon-based Platform at Three-Polar Regions on Earth

Hanlin YE[1], Huadong GUO[1] and Guang LIU[1]

[1] Institute of Remote Sensing and Digital Earth, Chinese Academy of Sciences, Beijing, China

E-mail: liuguang@radi.ac.cn

Abstract. More and more attention has been paid to taking the Earth as a whole for researching. Though space-borne and airborne platform have acquired various data from the Earth, the existing Earth observation system lack the ability of long-term continuous observation at a global scale. We propose a new platform, Moon-based platform, which is used for observing Earth from the Moon and discuss the coverage performance for observing Three-polar regions. Three-polar regions is characterized by its large scale and need long-term observation. Moon-based platform is the ideal platform. The coverage performance of the Moon-based platform depends on Moon orbit parameters, the attitude of the Moon and the attitude of the Earth. The position of the Moon are calculated from Jet Propulsion Laboratory ephemerides. The attitude of the Moon calculated from the libration Euler angles and the attitude of the Earth derived from the Earth orientation parameters. After introducing the coordinate system transformation, a preliminary coverage geometry are conducted. With the help of coverage geometry model, the simulation about Three-polar regions is presented. The result shows that the Moon-based platform has the advantages of large observing areas, long observation time windows and rich observing angles combination.

1. Introduction

In order to get the comprehensive data of the Earth, scientists have been studying and developing Earth observation science and technology. Though space-borne and airborne platform have the ability of acquiring observing data of the Earth, when taking the Earth as a whole for researching, we will have to face the problems of temporal coherence and spatial continuity in the global scale. With the tide of lunar exploration and exploitation, Moon is the only natural satellite of Earth, with the advantages of large observing areas, long period time of observation and rich observing angles combination, turning out to be an ideal platform for observing large regions on Earth. However, the spatial coverage performance, temporal coverage performance and angular coverage performance from Moon-based platform on these large regions remains largely unknown.

The analysis of coverage performance for Moon-based platform including spatial, temporal and angular coverage performance serve to realize the important characteristics of observing from the Moon. The spatial coverage for Moon-based platform is defined as a region on Earth where the Moon-based platform is seen at a minimum predefined elevation angle. The temporal coverage refers to the start time, the end time and the duration of observing a region on Earth. The angular coverage offers great significant parameters to the multi-angle remote sensing and the observation system to detect the Earth's energy balance.

Observing the Earth from the Moon is not a new conceptual observation method. In 1972, Apollo 16 astronauts used a far-ultraviolet camera to obtain the Earth's atmosphere and corona images and spectra[1]. At the early 21st century, Johnson[2] and Hamill[3] proposed that deploying visible or near infrared instruments on the Moon to observe the Earth. They discussed some advantages and limiting factors, such as spatial and temporal resolution. Guo discussed that deploying different kinds of sensors on the Moon have the capacity to acquire data for researching global change[4].

At present, some scholars are researching the coverage performance of the Moon-based platform. In 2012, Zhang proposed the Moon-based Earth observation simulation technology based on the Jet Propulsion Laboratory(JPL) ephemerides, preliminarily analysing the regularity of observed regions[5]. The Moon-based observation simplified geometry model is shown and simulation of cloud, cyclones and water vapor dynamic is presented from his research. But errors in Earth orientation tend to be larger than those based on Earth orientation parameters(EOP) that describe irregularities in the rotation of the Earth. In 2014, Ding studied the coverage performance analysis of Earth observation from Moon-based for global change detection[6]. The point-by-point method is used to calculate the coverage indicators on the basis of orbit simulation and spherical ground triangulation. The approach can evaluate more precise results but he neglected the libration of the Moon. This could cause a great error of evaluate the angular coverage performance.

This paper discusses the coverage performance for Moon-based platform at Three-polar regions on Earth. We used one-year position of Sun, Earth and Moon in the Earth centered inertial coordinates and one-year libration Euler angles of the Moon based on JPL ephemerides. We correct the Earth attitude by using the EOP and the Moon attitude is corrected by libration Euler angles. Then, we establish the coverage simplified geometric model and analysis the coverage performance of Three-polar regions. Finally, we attempt to better understand the coverage performance from Moon-based platform on Three-polar regions.

2. Three-polar Observation and Coverage

The Arctic, Antarctic and Tibet Plateau are known as the Three-polar regions with the characteristic of large area. They are all the world's largest storage of cold and carbon which are the sensitive regions of global environment change. The Three-polar regions is a dynamic system including many interacting components that are constantly changing. The interaction and feedbacks of Three-polar regions among these components are complex and register high variability in time and space. To understand the global sensitive factors of Three-polar regions on temporal and spatial variation, it has been recognized the need to obtain detailed data at a large scale. For this purpose, many agencies launched the Earth observing system satellites that are currently monitoring many of the characteristics of Three-polar regions like temperature or snow cover[7]. It is well known that the structure and the functioning of Three-polar regions are largely determined by the regularities of global sensitive factors. But this sensitive factors regularity suffers frequent nonlinear changes that gives more complexity to the system and introduces uncertainty. For this reason, the assessment of how global sensitive factors respond to global environment change depend strongly on the time scale[8]. The analysis of these observing data set allows for modelling and predictions that provide valuable information for researching.

Though many studies on polar regions or Tibet Plateau exists, there is still no researches on taking the Three-polar regions as a whole. It is feasible to deploy sensors on the Moon to observe Three-polar regions on Earth. The coverage performance from Moon-based observation at Three-polar regions on Earth is important. For the large regions, the special geographical position and the necessity of the study purposes, Three-polar regions are suitable for our study.

3. Moon-based Platform Coverage Geometry

The Moon-based platform devoted to Earth observation coverage geometry presents peculiar features which are different from airborne platforms or spaceborne platforms. First of all, the line-of-sight vector representation and the coverage observation geometry are quite different; moreover the platform, on which the sensors will be built up, is not only a spaceborne platform, but a whole, quite stable celestial body. So we need to develop a new coverage geometric model combining existing methods of spaceborne observation geometry.

3.1. Dynamic Models for Moon-based Platform

In case of a Moon-based platform, the coverage analysis must begin with the definition of Earth-Moon relative motion model. For the sake of simplicity both Earth and Moon were assumed spherical. On this basis very high precision lunar ephemeris DE series and EPM series exploitable for Lunar exploration[9], and the lunar and planetary theory JPL DE421 were more elaborated. Therefore, DE421 was selected by us because DE421 is a general purpose ephemeris, including the Moon libration and Earth nutation[10].

It is worth emphasizing that the Earth and Moon cannot be considered as point-like bodies and their orbital motion must be disentangled from their intrinsic rotations. The DE421 ephemeris provides the Moon three positions coordinate components of the geocentric coordinate system as a function of the barycentric dynamical time (TDB) expressed in Julian centuries starting from J2000. To determine the attitude of the Moon, DE421 were applied to calculate the three libration Euler angle. International Astronautical Union (IAU) include the rotation matrix of Earth orientation and to derive the observed target position in an inertial reference frame, the Earth rotation was account for.

From the point of view of the orbital motion of Moon theory should include construction of five local coordinate systems attached to: (1) the inertial geocentric coordinate system; (2) the Earth-Centered Earth-Fixed coordinate system; (3) the selenographic coordinate system; (4) the inertial selenocentric coordinate system; (5) the Moon-Centered Moon-Fixed coordinate system. To derive this position in an inertial geocentric coordinate system, it is necessary to introduce the relevant transformations of these coordinate systems.

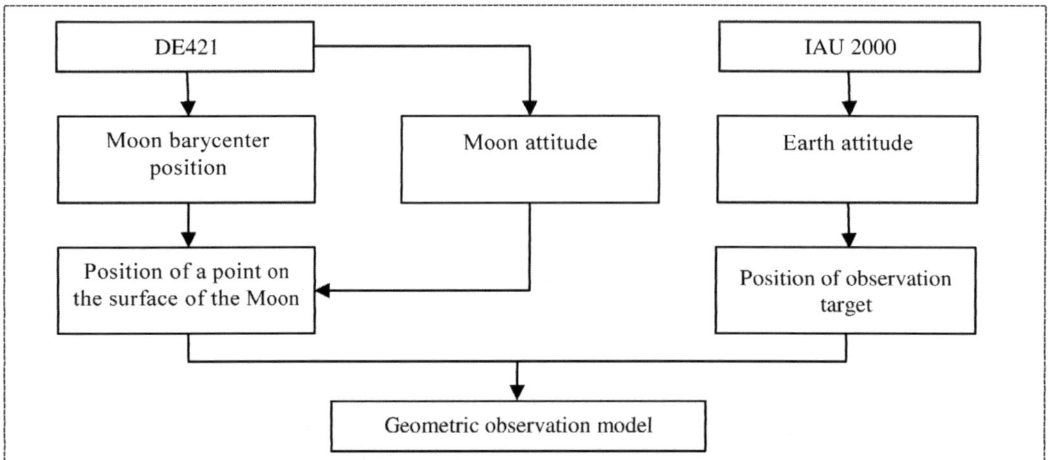

Figure 1. Simplified geometry model of Moon-based observation

3.2. Coverage Geometry

It is worth noting that, it is more difficult to calculate the coverage of Moon-based platform. Because optical sensors on the Moon-based platform can't be observed at night. The sunlit area on Earth is a hemisphere and the Moon-based observation coverage is the overlapping area. Therefore, we define a spatial coverage as the observed proportion of the observed region.

The basic geometry between the Moon-based platform sensor and observed target is depicted in Fig 2. The points indicate the Moon-based platform sensor (S), observed target (T) and then the third is the Earth's center (O). The nadir point is indicated by S'. These two lines SP_1 and SP_2 passing at point P_1 and P_2 represent horizon planes. R represents the radius of Earth and h represents the distance from Moon-based platform to nadir point. There are another four variables in Fig. 2: ε is elevation angle of observed target, α is nadir angle, λ is geocentric angle of observed target T and η is the angle between the line ST and the line SO. As soon as R, h and ε_{min} are known, the λ_{max} can be calculate by these following equations:

$$sin\alpha = \frac{R}{R+h}$$
$$sin\eta_{max} = sin\alpha cos\varepsilon_{min}$$
$$\lambda_{max} = 90^{\circ} - \varepsilon_{min} - \eta_{max}$$

The most needed parameter is the λ. This parameter will be used to judge whether can be observed by Moon-based observation.

$$arccos\frac{R_t \cdot R_s}{|R_t||R_s|} < \frac{\pi}{2}$$
$$\lambda = arccos\frac{R_t \cdot R_m}{|R_t||R_m|} < \lambda_{max}$$

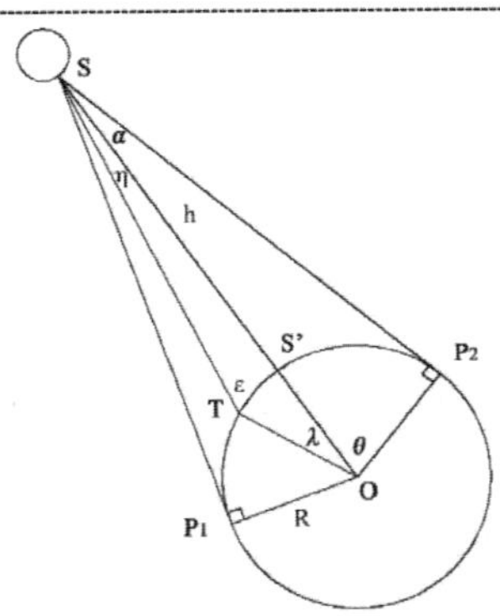

Figure 2. This is the spatial coverage geometry for Moon-based observation (not in scale for sake of clarity).

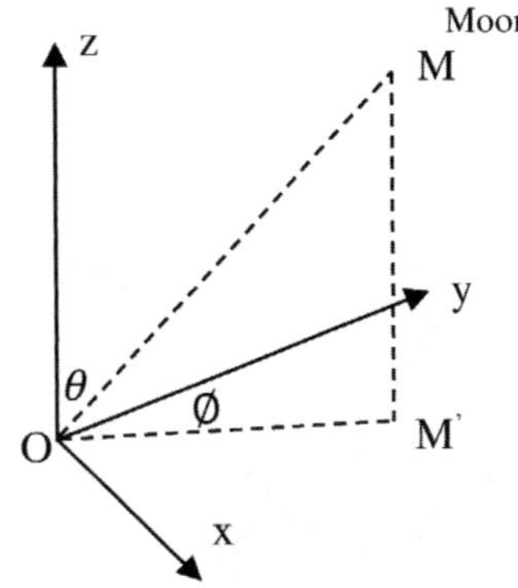

Figure 3. This is the angular distribution of Moon-based platform on Earth.

We construct a technique diagram of Moon-based observation for optical sensors spatial and temporal coverage analysis. The technology diagram of Moon-based coverage simulation includes three parts: the calculation of real-time locations of Earth and Moon; rendering of basic scene; the calculation of coverage time matrix. In the next step, we import the coordinates of test area and find out whether the regional boundary points are observed. We define the observed point with number '1' and unobserved point with '0', forming a Boolean Matrix. After all matrixes of boundary points are obtained, we calculate the intersection of them.

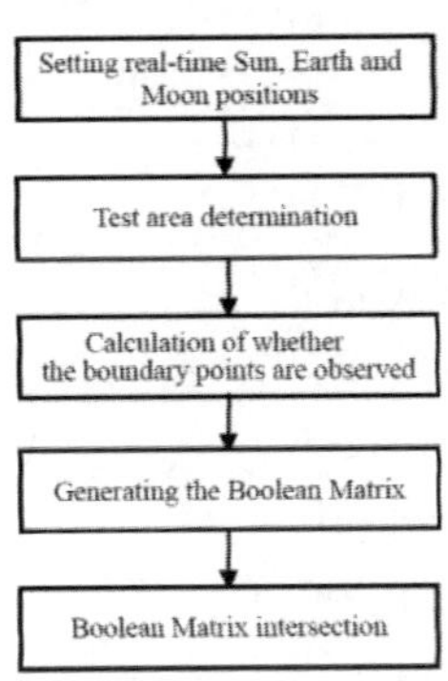

Figure 4. This is the technique diagram of Moon-based Earth observation for optical sensors coverage analysis.

The observation angle of Moon-based platform refers to the relative azimuth angle and observer viewing zenith angle. The observer viewing zenith θ is the angle of axis Z and vector OM and $\emptyset$ is the relative azimuth angle defining the azimuth angle position of the observer relative to the solar plane(Fig. 3).

4. Moon-based Platform Coverage Simulation

We use the actual Moon orbit data to simulate the time windows, observation ranges and observation angles when observing Three-polar regions from Moon-based platform and calculate coverage through a mask.

In one calendar year (366 d), 8 observations are completed on the Arctic region and 9 observations are completed on the Antarctic region (Table 1, 2). Mean duration of Antarctic region are 13 Days and mean duration of Arctic region are 10 Days. These shows that Moon-based platform serves the sustained observing needs of the polar regions. It can be seen from Table 1 that all observation are completed from March to September in 2016. From Table 1 and Table 2, there are one more observation time when observing Arctic regions. Because nadir point in the northern hemisphere is longer than the duration it in the southern hemisphere. Explained by the Kepler's second law, the asymmetric shape of the orbit lead to the observing characteristic.

Figure 5. This is the simulated image of Tibetan Plateau.

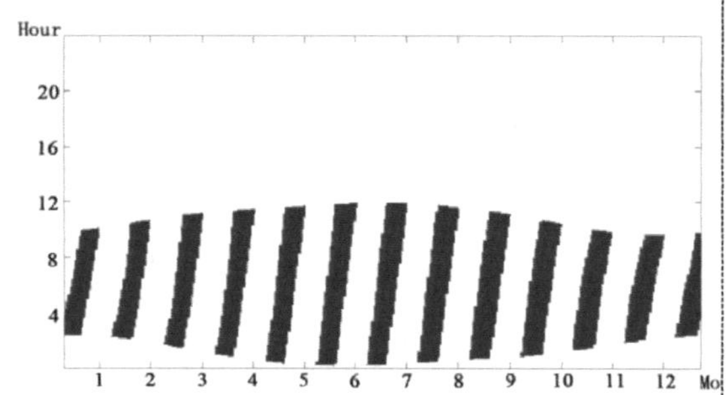

Figure 6. This is the simulated image of Tibetan Plateau.

Figure 5 shows the coverage simulation of Moon-based observation on Tibetan Plateau in 2016. Tibetan Plateau is located in the mid-latitude regions of the northern hemisphere, so it can be completely covered by one observation of Moon-based platform. Tibetan Plateau is observed 310 times in 2016. About 1471 hours will be observed in 2016 and the time coverage reached 16.7%. The longest observation time window will be on June 5. It can be seen from Figure 6 that we can observe the whole Tibetan Plateau every month. Because the direction of the Moon revolves in the same direction with the Earth's rotation. In August, there will be the more time observed. Besides, it also can be observed continuously. This is an excellent opportunity to observe Tibetan Plateau in order to obtain more information.

Table 1. The time windows when observes the Arctic regions.

	Start Time(UTCG)	Stop Time(UTCG)	Duration(Sec)
1	20 Mar 2016 20:54:38	23 Mar 2016 04:01:25	198406
2	6 Apr 2016 15:31:12	19 Apr 2016 10:10:28	1103956
3	4 May 2016 02:20:30	16 May 2016 17:07:32	1090021
4	31 May 2016 11:34:31	13 Jun 2016 01:28:50	1086858
5	27 Jun 2016 18:55:02	10 Jul 2016 10:48:06	1093983
6	25 Jul 2016 01:16:12	6 Aug 2016 20:01:21	1104309
7	21 Aug 2016 08:13:06	3 Sep 2016 04:11:33	1108707
8	17 Sep 2016 17:01:24	21 Sep 2016 21:49:38	362894

Table 2. The time windows when observes the Antarctic regions.

	Start Time(UTCG)	Stop Time(UTCG)	Duration(Sec)
1	1 Jan 2016 15:09:56	15 Jan 2016 00:21:46	1156309
2	29 Jan 2016 00:47:09	11 Feb 2016 08:10:34	1149804
3	25 Feb 2016 09:00:08	9 Mar 2016 18:03:02.	1155774
4	30 Sep 2016 22:27:31	14 Oct 2016 17:30:28	1191777
5	28 Oct 2016 04:43:38	11 Nov 2016 04:51:08	1210049
6	24 Nov 2016 11:33:04	8 Dec 2016 14:46:17	1221192
7	21 Dec 2016 19:48:15	1 Jan 2017 00:00:00	879104

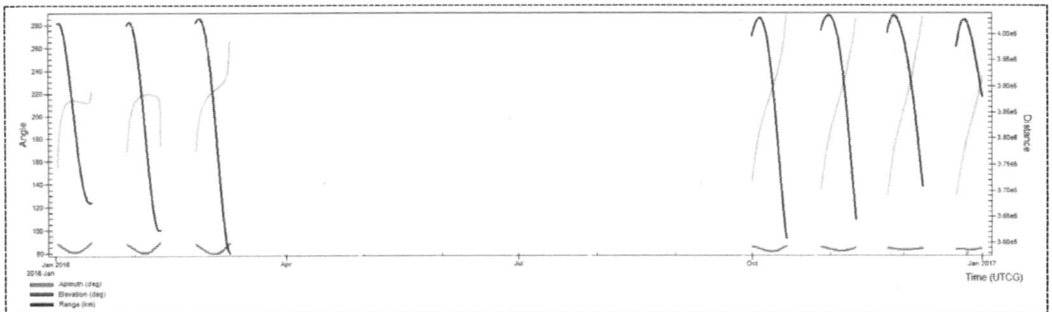

Figure 7. The angular coverage of Antarctic regions.

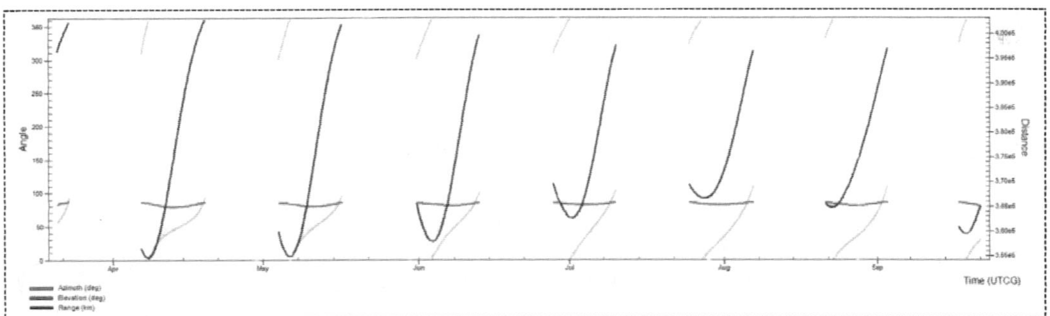

Figure 8. The angular coverage of Arctic regions.

Figure 7 and Figure 8 shows the angular coverage of polar regions. It is obvious that observing by Moon-based platform offers a much angular coverage, which is of great significance to the multi-angle remote sensing and observation system to detect the Earth's energy balance.

5. Conclusion

This paper presents a new concept of Moon-based observation. In monitoring Three-polar regions, Moon-based platform has a good performance in coverage, including temporal, spatial and angular. The result shows that Moon-based observation has large spatial coverage advantages in the research of Three-polar regions. The time windows changes with respect to the latitude of regions so the Tibet Plateau region has more observing time than Arctic and Antarctic region. Much angular coverage of Three-polar regions is shown. The multi-angles observation have contribution to the observation system to detect the Earth's energy balance. Besides, the Moon-based platform also provides long time-series data, which is greatly exceeds other platforms. In general, the Moon-based platform have a good performance in coverage of Three-polar regions on Earth. The simulation method will be essential part of foreseen future work.

6. References

[1] Carruthers G and Page T 1972 Apollo 16 far-ultraviolet camera/spectrograph: Earth observations *J. Science* vol **177** **788**–791
[2] J R Johnson, P G Lucey, T C Stone and M I Staid 2007 *Visible/near-infrared remote sensing of Earth from the moon*, NASA Advisory Council Workshop on Science Associated with the Lunar Exploration Architecture White Papers
[3] Patrick Hamill 2007 *"A lunar Earth observatory"*, http://www.lpi.usra.edu/meetings/LEA/whitepapers/HamillWhitePaper2.pdf
[4] Guo H D, Ding Y X and Liu G 2013 Conceptual Study of Lunar-based SAR for Global Change Monitoring *J. Science China: Earth Sciences* vol **57** 1771-1779
[5] Zhang D W 2012 Study on Methodology of Lunar-based Earth Observation for Global Change *D.* Shanghai: East China Normal University 13-40.
[6] Ding Y X, Guo H D and Liu G 2014 Coverage Performance Analysis of Earth Observation from Lunar Base for Global Change Detection *J. Journal of Hunan University (Natural Sciences)* vol **41** 96-102
[7] Høyer J L, Borgne P L, Eastwood S 2013 A bias correction method for Arctic satellite sea surface temperature observations. *J. Remote Sensing of Environment* vol **146** 201-213
[8] Kimball J S 2008 Earth Observation of Global Change: The Role of Satellite Remote Sensing in Monitoring the Global Environment *J. Eos Transactions American Geophysical Union* vol **89** 294-294
[9] Aleshkina, E Y 2002 Lunar numerical theory and determination of parameters k_2, d_m from analysis of LLR data *J. Astronomy & Astrophysics* vol **394** 717-721
[10] Folkner W M, Williams J G, Boggs D H 2009 The Planetary and Lunar Ephemeris DE 421 *J. Interplanetary Network Progress Report* vol **42** 1-34

Acknowledgments
This research was supported by the National Natural Science Foundation of China (41590852, 41590853), National Key Technology R&D Program of China (2012BAC16B01), Strategic Pioneer Program on Space Science, Chinese Academy of Sciences (XDA04077200).

Text Content Pushing Technology Research Based on Location and Topic

Dongqi Wei [1,2], Jianxin Wei [3,4], Naheman Wumuti[4] and Baode Jiang[2]

[1] Xi'an Center of Geological Survey, CGS, Xi'an,China
[2] National Engineering Research Center for geographic Information System, China University of Geosciences(Wuhan), Wuhan,China
[3] College of Resources and Environment Sciences, Xinjiang University, Urumqi , China
[4] Xinjiang Land and Resources Information Center, Urumqi,China

Abstract. In the field, geological workers usually want to obtain related geological background information in the working area quickly and accurately. This information exists in the massive geological data, text data is described in natural language accounted for a large proportion. This paper studied location information extracting method in the mass text data; proposed a geographic location—geological content—geological content related algorithm based on Spark and Mapreduce2, finally classified content by using KNN, and built the content pushing system based on location and topic. It is running in the geological survey cloud, and we have gained a good effect in testing by using real geological data.

Keyword. text data; content pushing; big data; location; topic;

1. Introduction

The arrival of the big data era has brought a good opportunity to give full play to the value of geological data. Big data is a kind of high-level information productivity; it promotes the change of the information production mode and also promotes the development of geological data application mode, meets the practical needs of geological workers is the key to large data technology to landing in the field of geological survey. Field work is the characteristic of geological survey, how to perceive the location of the field geological workers and the related topics (such as formation, structure, geography situation, etc.) in the vicinity, to form an active and fast information content pushing capability, is a new direction of geological data applications, is also the embodiment of intelligent geology survey [1] and it has high research value and application prospects.

Long before the arrival of the big data era, industry using traditional data (here refers to the digital data) production methods and has accumulated abundant outcome data. Some of the information contained in these data has a high value for field workers, who work in the field, but this information is often collected in advance before out of the field, and it is hard to obtain real-time in the field.

Therefore, it is the key point of this paper that how to sort out the information point of mass geological data reasonably and design the fast content pushing system for the field workers.

Finally, it should be noted that this paper uses the concept of "content" to represent a specific data and information, which is the minimum information processing unit of the system. Geological information is the expression of geological information in a particular location space, it is clear that the geological content will be associated with some geological location. In addition, the geological content and geological location of the original geological data are recorded in the text, and the geological position is a relative position concept, which is difficult to be quantified (coordinate), therefore, how to establish the geographic location—geological location—geological content related method is the key problem in this paper.

2. Architecture

First of all, the data source of the system is a huge amount of unstructured geological data accumulated in the early stage, which is mainly presented in the form of geological survey report. To make the system has the ability of intelligent information pushing service for field workers, this requires the system consists of data service center, the wireless network (or communication satellites), satellite positioning system, mobile terminal. Its complete system structure and service mode is shown in figure 1.

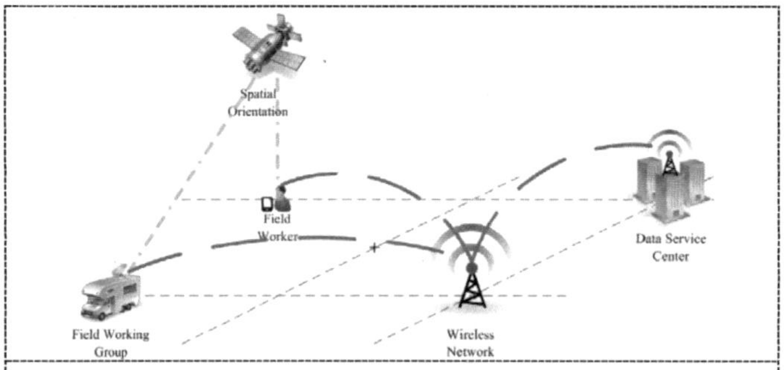

Figure 1. System Architecture and Service Model.

System obtains the position information of the field geological workers, according to the theme of its subscription, through wireless communication network to push content information in the vicinity of its position. Data service center is the core part of the system with respect to other facilities. The data service center mainly provides content pushing services, location-based services and GIS services. These services are through a series of basic components to support top application. The system structure is shown in figure 2. Its important composition includes:

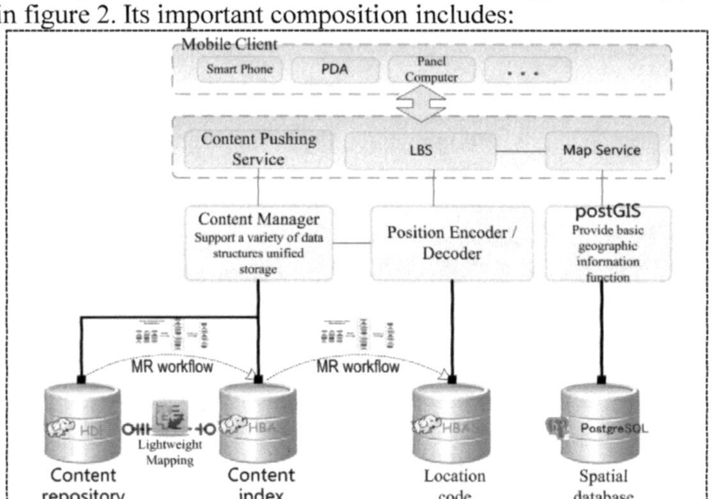

Figure 2. System Architecture.

Location-based service (LBS): LBS obtains user's location coordinates, uses position encoder to encoder it, then submits it to location code database which is realized by the HBase, through a scan operation in the database to find geological location points which are close to its location. The position code database can quickly filter out irrelevant information, reduce the candidate set, and improve the accuracy of the information acquisition.

Content pushing service: First, the content manager receives a candidate set of geological positions provided by the LBS, at this point, each geological location contains a number of geological information; content manager's role is based on the user's interest points, using collaborative filtering and other recommendation techniques to select the information that users may be concerned about in the geological location points, so as to improve the efficiency of content pushing service.

Map Service: Data visualization background component, be responsible for presenting the pushing results to the user in a map, at the same time, completing the necessary human-computer interactive function.

The basic components of the system are composed of the content repository, the content index database, the position code database and the spatial database. The former three are built on the Hadoop ecosystem, through a number of MapReduce workflow to complete the data location information acquisition, the association mapping of geographic location - geological location- geological content, the reconstruction of the position information coding and other important data processing process. The architecture of the whole system takes the content repository [2] as the core, realized the unified description model of geological data, and it can be regarded as the data warehouse of data analysis. Content index database is the lightweight mapping of the content repository, which is easy to find and locate the data quickly.

3. Research on Key Technology

In view of the particularity of industry demand and data structure, In this paper, based on the existing location-based content pushing technology [4,5], it is necessary to seek to solve the problem of unstructured geological data automatic acquisition, organization and management, analysis and service links. In the automatic acquisition and organization of data management, this paper adopts the content tree model of document [2]. The key technologies of the analysis processing and the service links are the key focus of this study section.

3.1. Data Processing Main Flow Based on Hadoop

In this paper, data objects are the geological survey reports of rich text (text and pictures). According to the method of literature [2]to model the content of this kind of data, in this way, a geological report is expressed as a document structured based on the paragraph content item subtree, all data form a content tree with the same root node. To simplify the problem, assume that each report is initially independent of each other, that is, any two subtrees can only be connected through a path that contains the root node. Subsequent data calculations are performed on this data structure and assumptions. Figure 3 shows the data processing of this article is divided into two sub-processes, each process is composed of several MapReduce programs. Process one, position entity recognition and adjacent position coding for text data containing position information, to form primary location index, as detailed in the 2.2 section. Process two, classification with item content as the object, to form the the secondary classification index, as detailed in the 2.3 section. The primary location index and the secondary classification index are stored in the position code database.

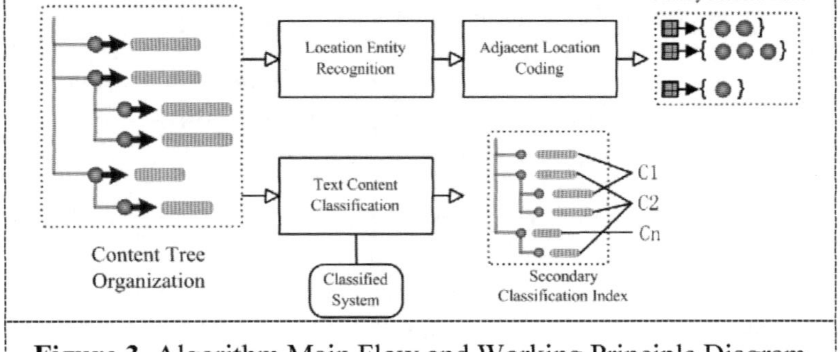

Figure 3. Algorithm Main Flow and Working Principle Diagram.

IOP Conf. Series: Earth and Environmental Science **46** (2016) 012025 doi:10.1088/1755-1315/46/1/012025

3.2. Entity Recognition and Location Coding Technologys

The geographical position obtained by the satellite positioning is the coordinate, and location information in geological data is mostly used in natural language text description. In order to be able to quickly and correctly push the information in the vicinity of the location of the field workers, a method is needed to establish an association from geographic location to geological location and then to geological content. Identifying named entities (geological locations) from text data, this paper uses cascaded hidden Markov algorithm [5] and builds location index by using geohash algorithm. The whole algorithm process is based on the Mapreduce technology framework; the final results are saved in the HBase, then using the geohash code as rowkey, and value column stores geological location-related content referenced in the content repository. Pseudo code:

```
Class Mapper
  //the input of the map function is a content fragment, recorded as :< null, content (i)>
  proceduce Map(null, content(i))
      Use stacked HMM to identify the location entity {POIentire} in content(i).text
      If count(POIentie)>0 then
        for each poi in {POIentire} do
          emit(poi,content(i).ref)
After shuffle and sort, the contents of the same location entity are sent to a reduce program.
Class Reducer
  proceduce Initialize
      // Load position < Body -Coordinate> Map
  proceduce Reduce(poi-entire, Lst(content-ref))
      coords = Map(poi-entire)
      Using coords to generate geohash code of poi-entire
      //take key(geo) as linekey, take the values of poi-entire and Lst(content-ref)
      //as column value write in HBase
      for val in [poi-entire,Lst(conent-ref)] do
        emit(key-geo, val)
```

3.3. Geological content classification based on machine learning

The work of Section 3.1 is to associate a large number of geological content with geological location. So the reasonable classification of geological content is the key technology to realize the correct information pushing, this is the premise of the system to filter out a lot of irrelevant information. As Mentioned above, the geological content is basically recorded in the form of text in the geological survey reports, and a report will contain a large number of different kinds of information, which is often divided in chapters. Due to the huge amount of report, the same kind of geological content point is very dispersive with respect to the overall information in general.

Based on the above analysis, text classification technology is a feasible way to solve this problem. But, if take the document as the smallest level of granularity, the information entropy will be larger, which is very inconvenient for establishing characteristic words and specifying the classified model. This is one of the reasons why we use a paragraph of the document as the smallest level of granularity. According to the modeling method of literature [2], content-item is persistent in HBase, can be directly used as the input of the processing algorithm.

At the same time it should be noted that, if we split each document by paragraph again, the amount of data that needs to be processed will reach one hundred million, so it needs to be done with the MapReduce computing framework. We use machine learning to accomplish the task of mass text classification. Algorithm flow can be divided into the following steps:

1: Stopping word, Chinese word segmentation, count up each sample document frequency, Calculate the cross entropy of each feature word t, and eventually save the largest collection of K feature words *{t1, t2, ..., tk}*.

2: Training corpus, using TF-IDF [6] to build VSM model for each classification, this is a $d \times K$ order matrix, d is the number of samples in each classification, K is the number of feature value selected in step 1， *w (d, k)*is the value of tf-idf.

3: Use MapReduce2 combined with TF-IDF [7,8]to set up the feature vector of the text that to be classified.

4: Use KNN algorithm which is based on MapReduce to complete the classification of text.

The main workload of processing is the establishment of the document VSM model and the iteration of the KNN algorithm， and VSM established mainly relies on cross-entropy and TF-IDF these two methods. This step appear in the training and determine period at the same time, therefore their parallelization will greatly improve the performance.

4. Application Scenarios

The system is deployed in the geological survey cloud platform, by cloud service way to provide functionality for users, the client side use mobile-preferred web front-end development framework, through the browser can use the full functionality of the system. As an application example, this paper uses the real geological survey data as the basic data source.

Somewhere, geological survey workers often focus on the relationship between the various factors and the mineralization in a specific ore-forming geological background. Assumes that field workers engaged in regional geological surveys are located somewhere in Xigaze region of Tibet, and they want to get their location and the surrounding geological information by portable handheld devices. Figure 4 shows the working interface using the system function on a panel computer; system obtains geographical coordinates（90.7031° E， 30.7040° N）of the field workers through the LBS, and through the geological content pushing service to push 5 geological locations of a total of 45 geological contents for the users. System interface is displayed in accordance with the weight of the ranking of the second geological location of the "一、构造地层区划分 等" the 5 geological contents.

Figure 4. The User Interface of System Function.

5. conclusion

Intelligent geological survey system is one of the key tasks of "Geological cloud" construction program China Geological Survey Bureau. In this paper, combined with the actual demand of field geological work, designed geological content pushing system based on location, it is an exploration using big data technology to solve practical problems, its service model is the embodiment of the "recycling field geological survey geographic information intelligent sensing process". The next step, we will proceed with the construction of geo-information in-depth analysis and knowledge discovery using platform， to build the value chain of the geological results data, and launch geosciences' information resource sharing and value-added services, to create favorable conditions for the information integration, resource sharing and knowledge innovation of geological data.

Acknowledgement
Financial support was provided by China geological survey (No. 1212011302660).

References

[1] Li C L, Li J Q, Zhang H C, Gong A H and Wei D Q 2015 Big Data Application Architecture and Key Technologies of Intelligent Geological Survey Geologcal Bulletin of China 34 pp 1288-1299

[2] Wei D Q, Li C L, Wumuti N, Wei J X and Yang J L 2014 Organizing and Storing Method for Large-scale Unstructured Data Set with Complex Content *2014 Fifth International Conference on Computing for Geospatial Research and Application* IEEE 2014 pp 70-76

[3] Wang X, Li Q H and Zhang F J Study on LBS-based Smart Push Technology 2011 *Communications Technology* **44** pp 95-97

[4] Xiao F, Hou Y, Wang L Z and Jia B 2015 Research of Intelligence Information Push Technology Based on LBS *Geomatics&Spatial Information Technology* **38** pp 125-127

[5] Yu H K, Zhang H P, Liu Q, Lv X Q and Shi S C 2006 Chinese named entity identification using cascaded hidden Markov model *Journal on Communications* **27** pp 87-94

[6] Salton G, Buckley C.Term-weighting Approaches in Automatic Text Retrieval 1988 *Information Processing & Management* **24** pp 513-523

[7] Chu C T, Kim S K, Lin Y A, Yu Y Y and Bradski G 2006 Mapreduce for machine learning On multicore *Proceedings of Neural Information Processing Systems Conference19* Vancouver:MIT Press pp 281-288

[8] Fang P L and Yang B R.Study on KNN arithmetic based On cluster 2009 *Computer Engineering and Design* **30** pp 4260-4262

Multi-Scale Visualization Analysis of Bus Flow Average Travel Speed in Qingdao

HAN Yong[1,2], GAO Man[1], ZHANG Xiao-Lei[1], LI Jie[1], CHEN Ge[1,2]

[1]College of Information Science & Engineering, Ocean University of China, Qingdao, China

[2]Laboratory for Regional Oceanography and Numerical Modelling, Qingdao National Laboratory for Marine Science and Technology, China

E-mail: yonghan@ouc.edu.cn

Abstract. Public transportation is a kind of complex spatiotemporal behaviour. The traffic congestion and environmental pollution caused by the increase in private cars is becoming more and more serious in our city. Spatiotemporal data visualization is an effective tool for studying traffic, transforming non-visual data into recognizable images, which can reveal where/when congestion is formed, developed and disappeared in space and time simultaneously. This paper develops a multi-scale visualization of average travel speed derived from floating bus data, to enable congestion on urban bus networks to be shown and analyzed. The techniques of R language, Echarts, WebGL are used to draw statistical pictures and 3D wall map, which show the congestion in Qingdao from the view of space and time. The results are as follows:(1) There is a more severely delay in Shibei and Shinan areas than Licun and Laoshan areas; (2) The high congestion usually occurs on Hong Kong Middle Road, Shandong Road, Nanjing Road, Liaoyang West Road and Taiping Road;(3) There is a similar law from Monday to Sunday that the congestion is severer in the morning and evening rush hours than other hours; (4) On Monday morning the severity of congestion is higher than on Friday morning, and on Friday evening the severity is higher than on Monday evening. The research results will help to improve the public transportation of Qingdao.

1 Introduction

The increasing populations make the traffic congestion increasingly worse at this stage of domestic cities [1]. It is a major source of despairs for road users, affecting individual commuters physically and mentally [2]. To alleviate congestion, researchers have explored various methods and collected various kinds of traffic data, including taxi GPS data, subway IC card data, etc. While data acquisition is not a problem anymore, how to represent, analyze and visualize these large spatiotemporal data effectively presents major challenges to the research community [3].

In this paper, we explore a variety of visualization methods to gain insight into the complex public transportation [4]. Public transportation is the leading of urban transport, which plays a key role in serving people travel [5, 6]. Therefore, multi-scale visualization of bus flow can help us find the hidden problems existing in our city. So we can put forward constructive suggestions to improve the traffic.

The paper is organized as follows: Section 1 briefly introduces the floating bus data and the data processing. In Section 2, multi-scale visualization analyses of bus flow from the perspective of the main district, the four areas, the five business circles and the arterial roads in Qingdao are elaborated.

In Section 3, some factors causing congestion are summarized. Finally, we come to the conclusion of the paper.

2 Data

The data we use comes from Qingdao Public Transportation Group, which has around 5000 buses on the core roads of Qingdao. The data is recorded only when a bus passed through bus stations. There are 1 billion records, from September 2014 to September 2015, as shown in table 1.

Table 1. Data attribute structure of floating bus data

busLineName	busStopName	stationName	actDatetime	stationNum	busId
26	Nanjing Road	The Peri	2014-07-14 20:00:07	19	DD102
28	Taidong One Road	CuoBuling	2014-07-14 20:00:10	7	DD110
11	Polar ocean world	Zhenjiang Road	2014-07-14 20:00:13	14	1413
...	...	...	...	...	...

The average travel speed is retrieved from the above data. It is the division of distance between adjacent stations and the average travel time taken by all buses passing the distance. Figure 1 shows the details of the calculation process.

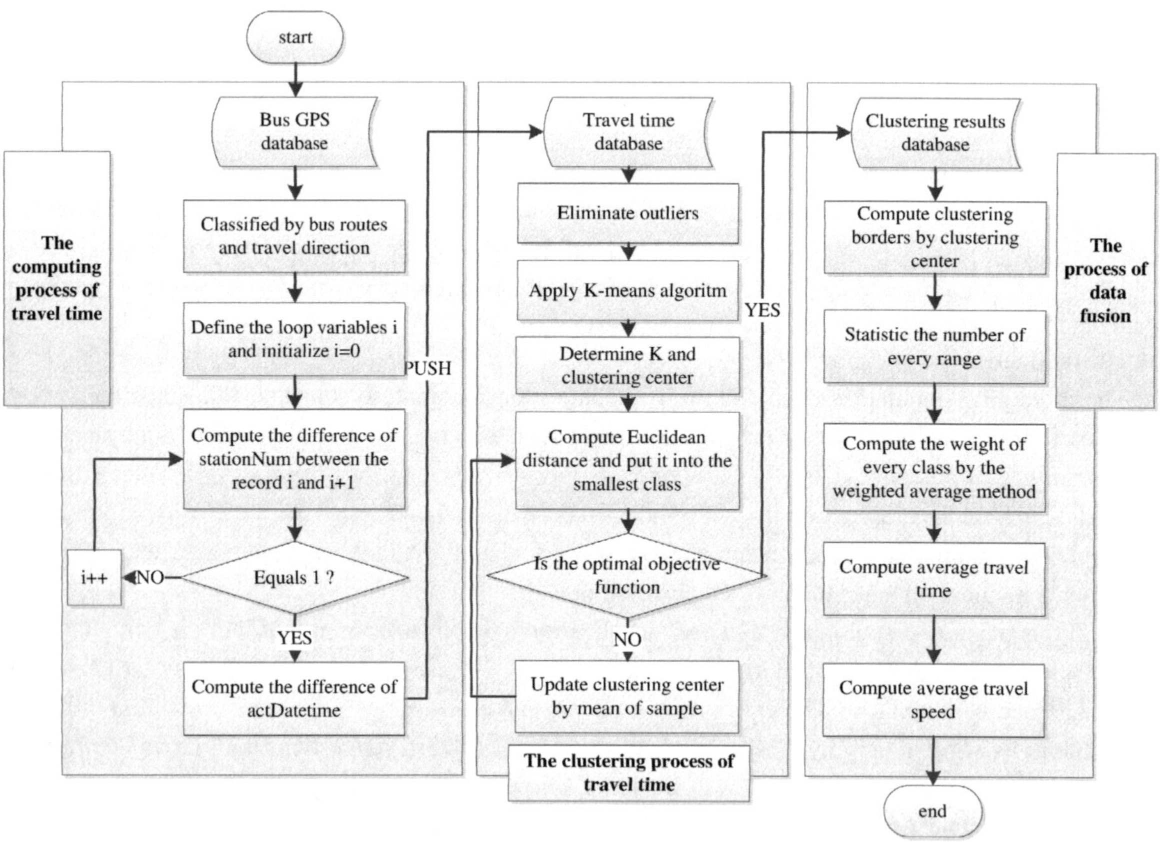

Figure 1. The processing of average travel speed

3 Bus Flow Visualization Analyses in Qingdao

At present, bus travel in Qingdao accounts over 32% among all the travel modes. Buses take more and heavier transportation role in residents' daily lives. Besides, it can help avoid many modern city problems [7]. Full understanding of the space-time characteristics of bus flow can help promote urban road traffic network to maximize its potential effectiveness [8].

3.1 Bus flow visualization analysis in the main district

Bus routes network in the main districts is theory-oriented "the south and north are hydrophobic; the west and east are dredging". It looks like a boot and is in accordance with the structure of road network. The structure determines the mainstream of city buses is the inverted "T" type .

Figure 2 is an isosurface (a three-dimensional contour surface) generated with Msic3D package of R software, which can show where the congestion generates, how it spreads, and where it terminates. An isosurface shows points of equal value in a 3D volume [9].

Figure 2. 3D isosurface of average travel speed in the main district

We can see the dynamic process of congestion from the space and time perspective. Figure 2 shows that congestion starts at 7:00 am, and spreads out during the morning peak hours (7:00 -9:00), and then fades away. A similar trend can be seen in the isosurface during the evening peak hours (17:00-19:00). Moreover, there is longer time of congestion in the evening than in the morning. Many areas experience severe congestion along the coastal edge, such as around May Fourth Square, Ba Da Guan Scenic Area, and Qingdao Railway Station.

3.2 Bus flow visualization analysis of the four areas

The main district is divided into four areas, including Shinan area, Shibei area, Licang area and Laoshan area.

Figure 3 is a K line chart generated by Echarts similarly. It is usually used to draw the direction of stocks. In this paper, we illustrate the traffic instead. In the graph, the line is used to represent the max speed with its top and the min speed with its bottom among all the week. The longer the line is, the greater the change is. We can find there is a bigger change in the morning hours than in the evening hours. So the evening rush hours are relatively stable on the whole week. The box is used to represent the speed of Monday and Friday. If the speed of Monday is lower than that of Friday, the box will be light blue; conversely, the box will be light red. The figure shows an interesting comparison between Monday and Friday in the four areas. On Monday morning the severity of traffic congestion is higher than Friday morning, but On Friday evening the severity is higher than Monday evening. There is a large difference because on Monday morning people just begin their work from outskirts of the city. However, on Friday evening people have finished work for the week, and they may undertake a long journey on weekend, or go out to enjoy the evening. Overall, it is smooth in the Laoshan and Licang areas, compared to the Shinan and Shibei areas.

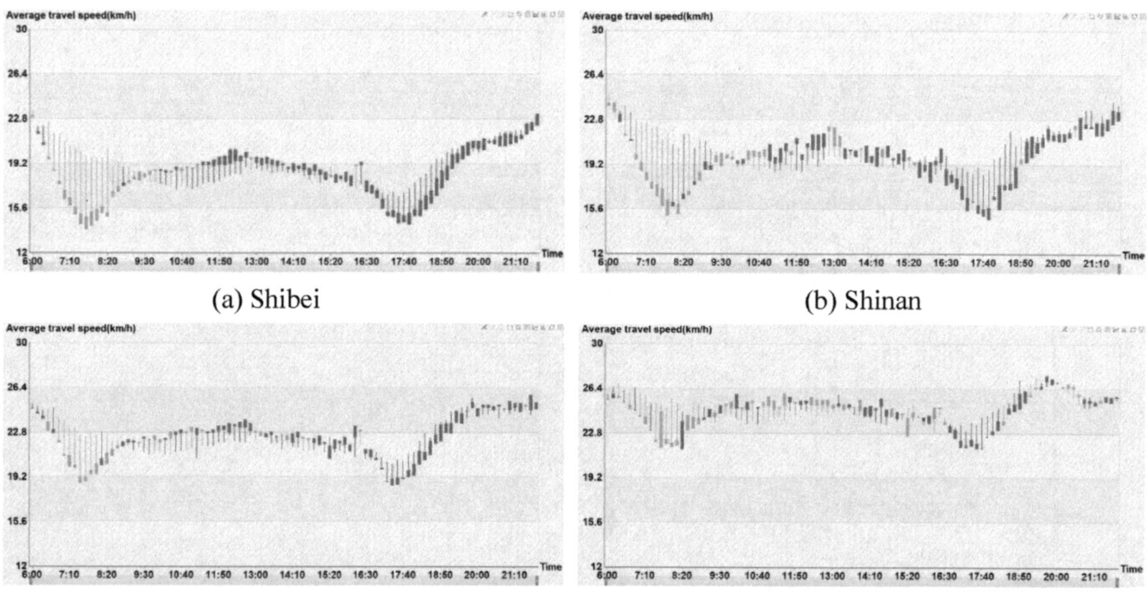

(a) Shibei (b) Shinan

(c) Licang (d) Laoshan

Figure 3. Statistics of average travel speed in the four areas

3.3 Bus flow visualization analysis of the arterial roads

Arterial roads are the framework of urban road network. They are connected to the city's major districts and play key roles in urban transportation. We select six arterial roads to analyze bus flow in Qingdao.

Figure 4 is a 3D wall map made in browser with technologies of WebGL. Under the wall, it is a picture of Baidu map. The figure gives an overall view of the congestion process (accumulating, propagating and subsequently dissipating) in individual links [9].

We can see the high congestion occurs on Hong Kong Middle Road, Shandong Road, Nanjing Road, Liaoyang West Road and Taiping Road. The congestion will spread in space from one section to its adjacent section, which leads to the condition of going slow. In the morning and evening rush hours, congestion is the most severe. It will also spread to the other hours.

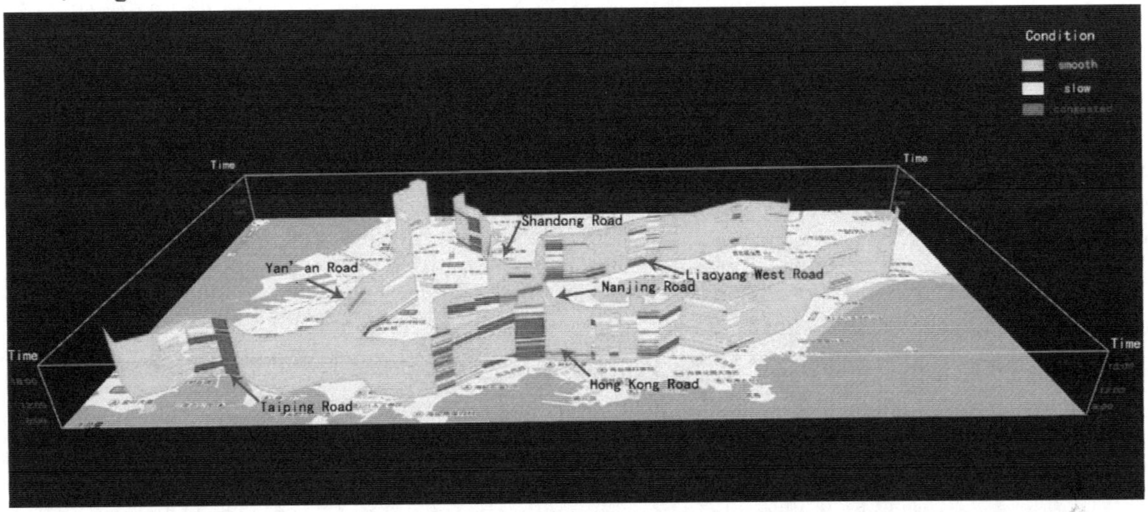

Figure 4. Different traffic condition on the arterial roads

Figure 5 shows the difference between the peak and off-peak period of the arterial roads. Compared to Figure 4, we can find a similar rule.

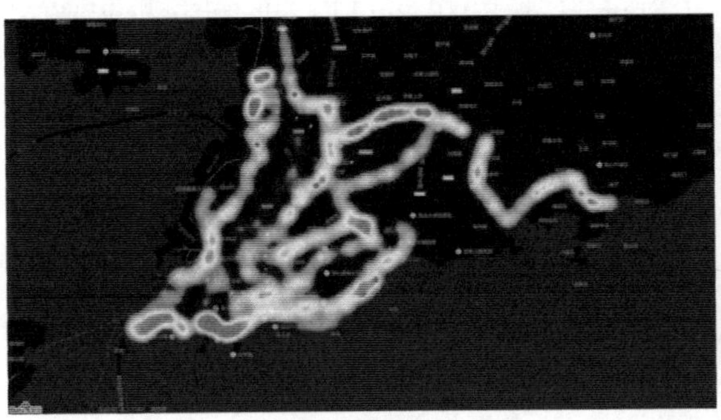

(a) In the rush hours

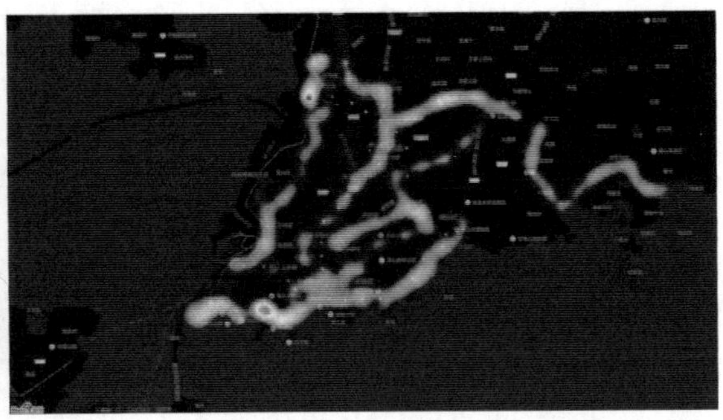

(b) Not in the rush hours

Figure 5. Heat map of average travel speed on the arterial roads

4 Influence Factors Analysis

4.1 The road design of districts.

Shinan and Shibei areas belong to the old districts, with dense road networks and narrow sections. The terrain conditions and historical evolution bring many paths, dead end highways and plug roads, with poor connectivity. Licang and Laoshan areas are built later. Their roads are wide. There is relatively reasonable road design and corresponding daily motor vehicle throughput ratio.

4.2 The spatial layout of the city.

The spatial layout of Qingdao is multi-centric strip shape. The four areas have their business centers (such as shopping plazas, large supermarkets, and offices), entertainment infrastructures (such as parks, playgrounds), and education facilities (such as schools of primary and secondary). Residents' stable patterns of social life, work, school and other activities lead to high congestion in a specific period such as morning and evening rush hours.

4.3 The intensity of population.

In business circles, railway stations, the beaches, intense and flowing population lead to the bad traffic.

5 Conclusions

This paper illustrates the condition of bus flow in Qingdao using different visualization techniques. Large space scale visualization allows us to have a cognitive understanding of congestion from the macroscopic perspective, while the short space scale provides us details of congestion from the microscopic perspective. At last, we summarize some factors which give rise to congestion, including roads, city layout, and population, etc. The analysis of bus flow can promote the traffic administrative department to make a decision to tackle congestion problems.

Acknowledgement

This work was supported by the National High Technology Research and Development Program of China (No. 2014AA09A511), and the Science and Technology Development Plan of Qingdao.

References

[1] Cheng T, Tanaksaraond G, Emmonds A and Sonniki D 2010 Multi-scale visualisation of inbound and outbound traffic delays in London *Cartogr. J.* **47** 323-9.

[2] Hennessy D A, Wiesenthal D L and Kohn P M 2000 The influence of traffic congestion, daily hassles, and trait stress susceptibility on state driver stress: an interactive perspective *J. Appl. Biobehav. Res.* **5** 162-79.

[3] Zhang J Q, Qiu P Y, Duan Y C, Du M Y and Feng L 2015 A space-time visualization analysis method for taxi operation in Beijing *J VIisual Lang Comput* **31** 1-8.

[4] Wang Z C, Ye T Z, Lu M, Yuan X R, Yuan J and Wu Q L 2014 Visual exploration of sparse traffic trajectory data *IEEE T. Vis. Comput. Gr.* **20** 1813-22.

[5] Xu Y H, Feng L G 2010 Situation and strategy planning of public transit priority development J. Traffic Transport. Eng. **10**: 64-9.

[6] Yan W P, Weng J C, Rong J, Ge Yand Chen Z H 2012 Travel time distribution and estimation of urban traffic using vehicle identification data *J. Transport. Sys. Eng. Inform. Technol.* **12** 165-70.

[7] Yu J, Yang X G, Yin R and Xia S G 2007 Public transport planning of developing city based on TOD *J. Traffic Transport. Eng.* **7** 106-11.

[8] Wang Y and Papageorgiou M 2007 Real-time freeway traffic state estimation based on extended kalman filter: A General approach *Transport. Res. B.* **41** 167-81.

[9] Cheng T, Tanaksaraond G, Brunsdon C and Haworth J 2013 Exploratory visualisation of congestion evolutions on urban transport networks *Transport. Res. C.* **36** 296-306.

Estimation of carbon storage and carbon density of forest vegetation in Ili River Valley, Xinjiang

Guo jing[1,2], Zhang renping[3], Wang ranghui[1], Yusupujiang aimaiti[2], Asiyemu tuerdi[2] and Zhang dongya[2]

[1] Nanjing University of Information Science & Technology, Nanjing 210044, China
[2] Xinjiang Academy of Forestry, Urumqi 830000, China
[3]State Key Laboratory of Grassland Agro-ecosystems; College of Pastoral Agriculture Science and Technology, Lanzhou University, Lanzhou 730020, China

corresponding author: Wang ranghui Email: rhwang@nuist.edu.cn

Abstract. Study on the forest carbon storage, carbon density and spatial distribution characteristic are helpful for improving the accuracy of carbon estimation and providing the practical basis for better policy making. In this research, the compiled data of 'Xinjiang Forest Resources Survey Results' in 2011 was used as a source data, by using the biomass-volume regression model and average biomass method, the carbon storage, carbon density and spatial distribution of forest resources in Ili River Valley region were analyzed. Results show that, the total biomass, carbon storage and average carbon density in Ili River valley were 69.647Tg, 34.823Tg and 41.45Mg/hm^2 C respectively. From the aspect of spatial distribution, the northwest region of Ili River Valley has high carbon storage and the southeast region has low carbon storage. The southwest region has low carbon density and the northeast region has high carbon density. The value of forest Carbon storage from high to low was: Arbor > Shrub > Sparse forest > Odd tree > Economic forest > Scattered trees. Mature arbor forest plays an important role in maintaining the balance of carbon dioxide and oxygen in Ili River Valley region.

1. Introduction

Forests compose a major part of terrestrial ecosystems, occupying about 30% of the world's land area [1]. It is estimated that over 80% of global above ground carbon (C) is stored in forest vegetation [2], and the annual C flux between forests and the atmosphere through photosynthesis and respiration is up to 90% of the total annual flux of terrestrial ecosystems [3]. Owing to their huge C pool and high productivity, forest ecosystems play a leading role in the global C cycle [4]. Therefore, estimating carbon storage in forest ecosystems and evaluating their carbon sink capacities have been the focus of numerous studies. For example, the C storage of forest ecosystems has been estimated in the United States [5], Russia [6], China [7], and Europe [8, 9] based on forest inventory data.

Forest inventory data are valuable resources in forest carbon research because they provide true ground-based estimates of C stock and fluxes across heterogeneous regions, and are statistically representative of land-use change and disturbance [10, 11]. Thus, the forest inventory method has been used to evaluate C storage and its dynamics in forest vegetation in China [12-16] summarized C storage estimates for Chinese terrestrial ecosystems in these studies. Overall, the C storage estimation of forests in China differ substantially. The differences among estimates were partially caused by the differences in geography, vegetation, soil type boundaries, data measurement and use [17].

The Ili River valley, where the forest cover rate is 7.9% and the vegetation cover rate of forest and grass is up to 67.7%, is a region with the most humid climate, the most abundant precipitation, and the most developed vegetation and soil in the Xinjiang and Tianshan regions and is regarded as a "wet island" and biotic resource storehouse in an arid region. There are some studies in which data from the Second National Soil Survey was used to estimate the soil organic carbon reserve in the Ili region [18], but there are few studies on the carbon cycle in forest ecosystems in the Ili River valley [19]. Therefore, in this study, based on the Forest Resource Inventory Data of Xinjiang (2011), we aimed to estimate forest vegetation carbon stocks and carbon density in Ili River valley by using the biomass and volume regression model and the average biomass method. In order to lay a foundation for further studies on a regional-scale forest vegetation carbon cycle.

2. Study area

The Ili River valley of Xinjiang, China is located in the westernmost part of the Tianshan Mountains in Xinjiang and on the south western edge of the Jungghar Basin. It is an inland continental river basin in the Central Asia region with fault basins and inter mountain valleys distributed alternately, and is isolated by the southern Tianshan Mountain and the northern Tianshan Mountain. The Ili River valley is located between 80°09′42″E–84°56′56″E and 42°14′16″N–44°50′30″N, its northern, eastern and southern sides are surrounded by high mountains, and the terrain here slopes from east to west and narrows from west to east, like a bell mouth facing westward. The average annual precipitation ranges from 200 mm to 800 mm, and the rainfall mainly occurs in spring and summer, when approximately 60%–70% of the total annual precipitation occurs. The average annual evaporation is 1260–1900 mm. The average annual temperature is 2.9–9.1℃. The frost-free period is 130–180 days. The average annual sunshine duration ranges 2700–3000 hours [18].

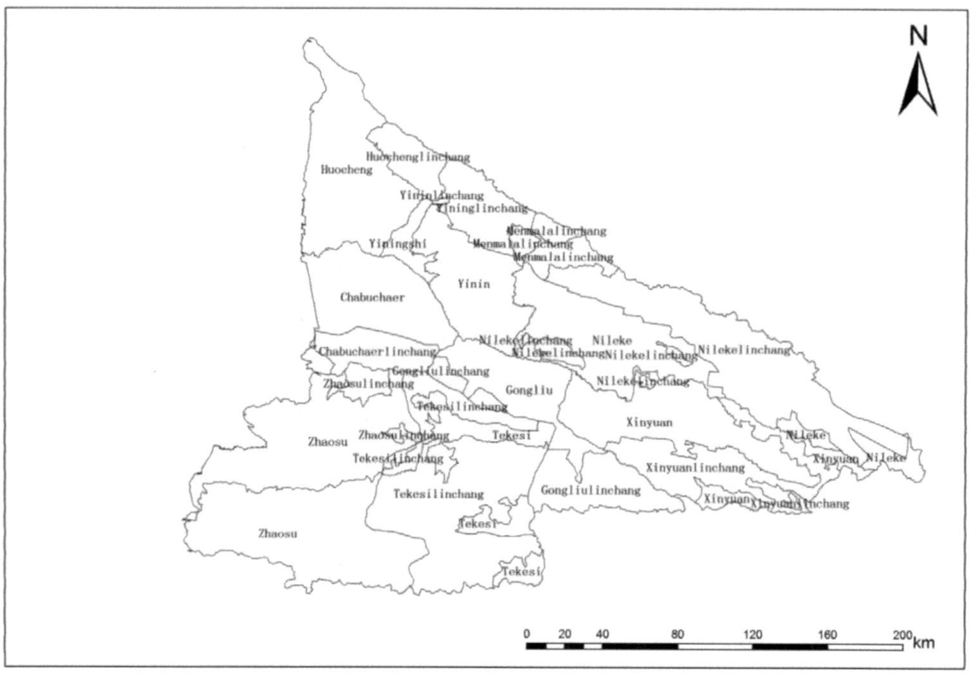

Figure 1. Administration map of the study area

3. Data sources and methods

3.1. Forest inventory data

IOP Conf. Series: Earth and Environmental Science **46** (2016) 012027 doi:10.1088/1755-1315/46/1/012027

The forest inventory data of Ili river valley was parts of the compiled data of 'Xinjiang Forest Resources Survey Results' in 2011, the data sets composed both in county (city) and the forest farm as a unit (including 9 counties (cities), nine forest farm): Ghulja city, Qapqal county, Tokkuztara County, Qorghas county, Nilka county, Tekes county, Kunas county, Mungghulkure county, Ghulja county; Forest farms are include: Kunas forest farm, qapqal forest farm, Tokkuztara forest farm, Qorghas forest farm, Mengmala forest farm, Nilka forest farm, Tekes forest farm, Ili forest farms, Mungghulkure forest farm ,etc. The contents include detailed information relates to the forest type (species and origin), area, volume, extent, age class, etc. Due to the forest inventory data are not include the herb layer, litter layer, soil data, therefore, in this research only calculated the forest vegetation carbon storage.

3.2. Carbon storage calculation

The biomass of forest stands was calculated using the continuous biomass expansion factor (CBEF) method [7, 21]:

$$B = a \bullet V + b \quad (1)$$

Where B is the total stand biomass (Mg ha^{-1}); V is the stand volume (m3ha^{-1}), a and b are the coefficients for a specific forest type. During the calculation, for some specific trees which no significant correspondence between the parameters of the model, parameters were substituted by the approximate tree types in the local area [22]. The specific conversion parameters are shown in Table 1, The fraction of carbon in biomass is assumed to be 0.5 for all forest types, although it varies slightly for different forests [20].

Table 1. Volume-biomass regression parameters of different type

Forest type	Volume-biomass regression parameters	references
Arboreal forest	B =0.5751V+38.706	[21]
Shrubs	B =12.49	[22,23]
Economic forest	B=23.7	[23]
Sparse forest	B =0.5751V+38.706	[21,23]
Mixed broad leaf-conifer forest / Scattered trees	B=0.8119V+12.2799	[21,23]
Poplar /Odd-tree	B =0.6255V1+91.0013	[22]

4. Results and discussion

4.1. Carbon storage of different forest types

Forest vegetation total biomass and carbon storage were 69.647Tg and 34.823Tg C in Ili river valley based on our estimates, and varied largely for different forest types (Table 2). Among all the forest types, the Arboreal forest and Shrubs have the larger proportions, the total biomass were 62.794Tg and 2.813Tg respectively; the carbon storage were 31.397Tg and 1.406Tg respectively. The total biomass and carbon storage Scattered forest have the smaller proportion, about 0.18Tg and 0.09Tg respectively. The proportional order of different forest type's carbon storage were: Arboreal forest >Shrubs> Sparse forest>Odd tree>Economical forest>Scattered forest, accounting for 90.16 %、4.04%、3.64 %、1.44 %、0.46% and 0.26% of the total carbon storage. It can be inferred that, among the forest vegetation in this region, arbor and shrub plays an important role in fixing carbon dioxide and release oxygen in the air, which maintains regional carbon and oxygen balance in this region.

The carbon density of forest vegetation in Ili river valley was 41.45 Mg/hm^2, higher than the average Chinese forest vegetation carbon density 41.32 Mg/hm^2 [23]. Distribution of carbon density among different forest types were uneven, sorting them from the high to low were as follows: arbor > odd trees > scattered tree > sparse forest > economic forest > shrub. The carbon density of arbor and odd were 78.95 Mg/hm^2 and 75.85 Mg/hm^2, shrub and economic forest have the lower carbon density, about 6.24 Mg/hm^2 and 11.87 Mg/hm^2 respectively.

Table 2. Carbon storage and density of different forest types

	Arbor	shrub	economic forest	scattered tree	sparse forest	odd trees	Total
A(hm^2)	397693	225212	13557	2027	40400	6592	685483
P (%)	58.02	32.85	1.98	0.3	5.89	0.96	
B (Tg)	62.794	2.813	0.321	0.18	2.538	1.001	69.647
C (Tg)	31.397	1.406	0.161	0.09	1.269	0.51	34.824
P (%)	90.16	4.04	0.46	0.26	3.64	1.44	
D(Mg/hm^2)	78.95	6.24	11.87	44.40	31.41	75.85	41.45(AV)

Note：A stands for area, B stands for Biomass, C stands for carbon storage, D stands for carbon density, P stands for proportion, AV stands for Average.

4.2. Spatial distribution analysis of carbon storage and density

The distribution of carbon storage and density for different sub-regions in Ili river valley were unevenness. It can be inferred from the Figure 2 that, the Tekes forest farm have the highest carbon storage of 7.613 Tg, accounted for 21.86% of total carbon in the study area; while the Ghulja city have the smallest carbon storage of 0.116 Tg, accounted for 0.33% of total carbon in the study area. The Tokkuztara forest farm have the largest carbon density of 88.8 Mg/hm^2, followed by the Kunas forest farm (63.6 Mg/hm^2), Tekes forest farm (60.4 Mg/hm^2), Mungghulkure county have the smallest carbon density of 19.9 Mg/hm^2 (Figure 3). Carbon density can reflect the quality of the forest, hence the quality of forest resources in Tokkuztara forest farm better than other counties, cities and forest farms in Ili river valley.

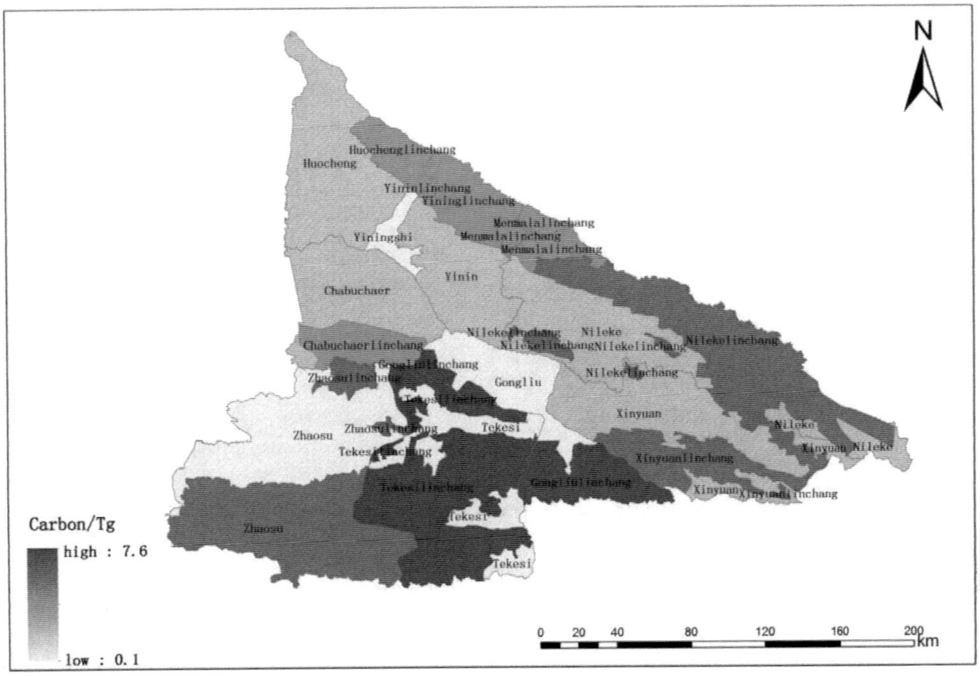

Figure 2. Distribution of carbon storage in Ili river valley

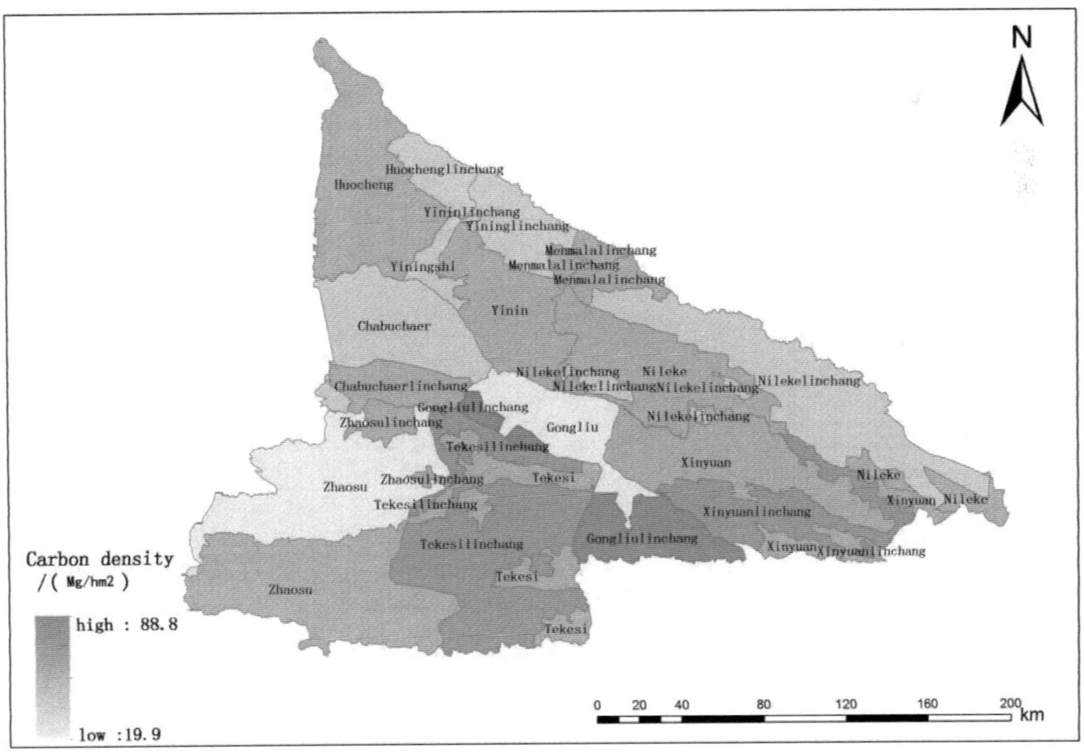

Figure 3. Distribution of carbon density in Ili river valley

5. Conclusion

In this research, based on forest inventory data, the forest carbon storage and carbon density were successfully calculated by using the volume-biomass method. The result shows that total carbon storage and average carbon density of Ili River Valley were 34.823Tg and 41.45Mg/hm^2, respectively. Arbor and shrubs are the main sources of carbon storage which occupies 90.16% and 4.04% of the total carbon storage in this region. Further investigations are needed for calculating the blow ground carbon storage and analyzing the carbon storage of different forest age classes and their spatial distributions.

Acknowledgments
This research was supported by China clean development mechanism (CDM) projects (2013013).

References

[1] Dixon, R. K., S. Brown, R. A. Houghton, A. M. Solomon, M. C. Trexler, and J. Wisniewski,1994. Carbon pools and flux of global forest ecosystems, *Science*, **263**, 185 – 190.

[2] Lieth, H., and R. H. Whittaker,1975. Primary Productivity of the Biosphere, 339 pp., Springer, New York.

[3] Winjum, J. K., R. K. Dixon and P. E. Schroeder,1993. Forest management and carbon storage: An analysis of 12 key forest nations, *Water Air Soil Pollute*, **70**, 239 – 257.

[4] Watson, R. T., I. R. Noble, B. Bolin, N. H. Ravindranath, D. J. Verardo, and D. J. Dokken (Eds.)., 2000. Land Use, land-Use Change, and Forestry— A Special Report of the IPCC, Cambridge Univ. Press, New York.

[5] Turner D P, Koerper GJ, Harmon ME, Lee JJ, 1995. A carbon budget for forests of the conterminous United States. *Ecol Appl*, **5**:421–436

[6] Alexeyev V, Birdsey R, Stakanov V, Korotkov I, 1995. C in vegetation of Russian forests: methods to estimate storage and geographical distribution. *Water Air Soil Pollut* **82**:271–282

[7] Fang J, Chen A, Peng C, Zhao S, Ci L, 2001. Changes in forest biomass C storage in China between 1949 and 1998. *Science* **292**:2320–2322.

[8] Ciais P, Schelhaas MJ, Zaehle S, Piao SL, Cescatti A, Liski J, Luyssaert S, Le-Maire G, Schulze ED, Bouriaud O, Freibauer A, Valentini R, Nabuurs GJ, 2008. Carbon accumulation in European forests. *Nat Geosci* **1**:425–429.

[9] Nabuurs GJ, Hengeveld GM, Van Der Werf DC, Heidema AH, 2010. European forest C balance assessed with inventory based methods-An introduction to a special section. *For Ecol Manage* **260**:239–240

[10] Birdsey, R. A, 1992. Carbon Storage and Accumulation in United States Forest Ecosystems, USDA For Serv. Gen Tech. Rep/ WO-59, p. 51.

[11] Brown, S. L., and Schroeder, P. E, 1999. 'Spatial patterns of above ground production and mortality of wood biomass for eastern U.S. Forests', *Ecol. Appl*, **9**(3), 968–980.

[12] Fang JY, Liu GH, Xu SL,1996. Biomass and net production of forest vegetation in China. *Acta Ecol Sin* **16**:497–508 (in Chinese)

[13] Liu GH, Fu BJ, Fang JY,2000. C dynamics of Chinese forests and its contribution to global C balance. *Acta Ecol Sin*, **20**:733–740 (in Chinese)

[14] Ren Y, Wei XH, Zhang L, Cui SH, Chen F, Xiong YX, Xie PP, 2011. Potential for forest vegetation carbon storage in Fujian Province, China, determined from forest inventories. *Plant Soil* **345**:125–140

[15] Zhao M, Zhou GS,2004. Carbon storage of forest vegetation and its relationship with climatic factors. *Sci Geogr Sin*, **24**:50–54

[16] Yu L, Ren G, 2007. Recent progresses in studies of the terrestrial C storage change for the Past 20 k Year. *Prog Geog* **26**:69–79 (in Chinese)

[17] Hai R, Hua C, Linjun L, Pingheng L, Changmou H, Hongfu W, Qianmei Z, Peixia Z, 2013. Spatial and temporal patterns of carbon storage from 1992 to 2002 in forest ecosystems in Guangdong, Southern China. *Plant Soil*, **363**:123–138

[18] Yang Y, Chen Y, Li W, Chen Y, 2010. Distribution of soil organic carbon under different vegetation zones in the Ili River Valley, Xinjiang. *J. Geogr. Sci*, **20**(5): 729-740.

[19] Sun H, Li W, Yang Y, Yang Y, 2012. Soil Organic Carbon Changing with altitudes on the Ili Mountainous Region. *Scientia Geographica Sinica*, **32**(5): 603-608.(in Chinese).

[20] Chunhua Z, Weimin J, Jing M C, Mei Z, Dengqiu L, Yanlian Z& Xiqun W, 2013. China's forest biomass carbon sink based on seven inventories from 1973 to 2008. *Climatic Change*, **118**:933–948.

[21] Fang JY, Wang GG, Liu GH, Xu SL,1998. Forest biomass of China: an estimate based on the biomass–volume relationship. *Ecol Appl* **8**(4):1084–1091

[22] Fang Y，Liu H，Bai Z,Zang R,Zhang F,Ye G,Han Y,2014. Spatial pattern of carbon storage and carbon density in forest vegetation of the Kanas National Natural Reserve. *Journal of Nanjing Forestry University (Natural Sciences Edition)*, **38**(6):17-22. (in Chinese).

[23] Zhao M, Zhou GS,2004. Carbon Storage of Forest Vegetation and Its Relationship with Climatic Factors. *Scientia Geographica Sinica*, **24**(1):50-54. (in Chinese).

Comparing interpolation techniques for annual temperature mapping across Xinjiang region

Zhang Ren-ping[1,2], Guo Jing[3], Liang Tian-gang[1], Feng Qi-sheng[1], Yusupujiang Aimaiti[3]

[1]State Key Laboratory of Grassland Agro-ecosystems; College of Pastoral Agriculture Science and Technology, Lanzhou University, Lanzhou 730020, China
[2]General Grassland Station of Xinjiang, Urumqi 830049, China
[3] Xinjiang Academy of Forestry, Urumqi 830000, China

corresponding author: Liang Tian-gang Email: tgliang@lzu.edu.cn

Abstract. Interpolating climatic variables such as temperature is challenging due to the highly variable nature of meteorological processes and the difficulty in establishing a representative network of stations. In this paper, based on the monthly temperature data which obtained from the 154 official meteorological stations in the Xinjiang region and surrounding areas, we compared five spatial interpolation techniques: Inverse distance weighting (IDW), Ordinary kriging, Cokriging, thin-plate smoothing splines (ANUSPLIN) and Empirical Bayesian kriging(EBK). Error metrics were used to validate interpolations against independent data. Results indicated that, the ANUSPLIN performed best than the other four interpolation methods.

1. Introduction

Climate variables provide an essential input in global environmental change [1]. Prediction of the impacts of a changing climate on the distribution and functioning of terrestrial ecosystems requires as a first step, the development of reliable, spatially-explicit models of current climate [2]. The development of methods to interpolate climatic data from sparse networks of stations has been a focus of researchers [3-6]. There are several commonly used interpolation methods, such as Inverse distance weighting (IDW), Ordinary kriging, spline-method, etc.[7,8], yet it is critical to choose the optimal interpolation method, due to the variations of used data and the different characters of research areas [9]. Hence,comparative studies have been conducted to determine which method of spatial interpolation is best suited for different contexts, but as of yet, no decisive conclusions have been made [10]. It is important to continue research in this direction in order to gain a better understanding of proper applications of these interpolation techniques [11, 12].

As a typical arid and semi-arid region, there are only 66 meteorological stations in Xinjiang within the scope of 160 * 104 km^2 area, and most stations are distributed within the oasis and unevenly distributed. In actual work, using the meteorological stations data of surrounding areas as an auxiliary data for climatic variations interpolation are considered to be a good solution [13].

In this paper, based on the 66 meteorological stations monthly temperature data and 88 surrounding areas, such as Gansu and Qinghai meteorological station as auxiliary data, we investigate the integration of additional sources of data for comparison of five interpolation methods: IDW, Ordinary kriging, Cokriging, ANUSPLIN and EBK, and validate the interpolation results against independent data. In order to provide a basis for climatic data interpolations in this region and other similar areas.

2. Data sources and methods

2.1. Data sources

The monthly temperature data used in this study were obtained from the 154 official meteorological stations in the Xinjiang region and surrounding areas, among them, the domestic meteorological stations data were acquired from the website of China meteorological science data sharing service (http://cdc.cma.gov.cn/), the meteorological data of outside china were acquired from the US meteorological data sharing website of NOAA (http://www.ncdc.noaa.gov/ghcnm/v3.php). For minimize the edge effect of interpolation, the scope of experimental area was set to be larger than Xinjiang administrational border, approximately cover the area of 70.7-99.20 E and 31.3-51.40 N (Figure 1).

The Digital Elevation Model (SRTM-DEM) data covering the study area with a resolution of 90 m, was received from the website (http://srtm.csi.cgiar.org/),The data was geo-referenced to the Universal Transverse Mercator (UTM) coordinate system, Zone 45 North with World Geodetic System Datum of 1984 (WGS84).

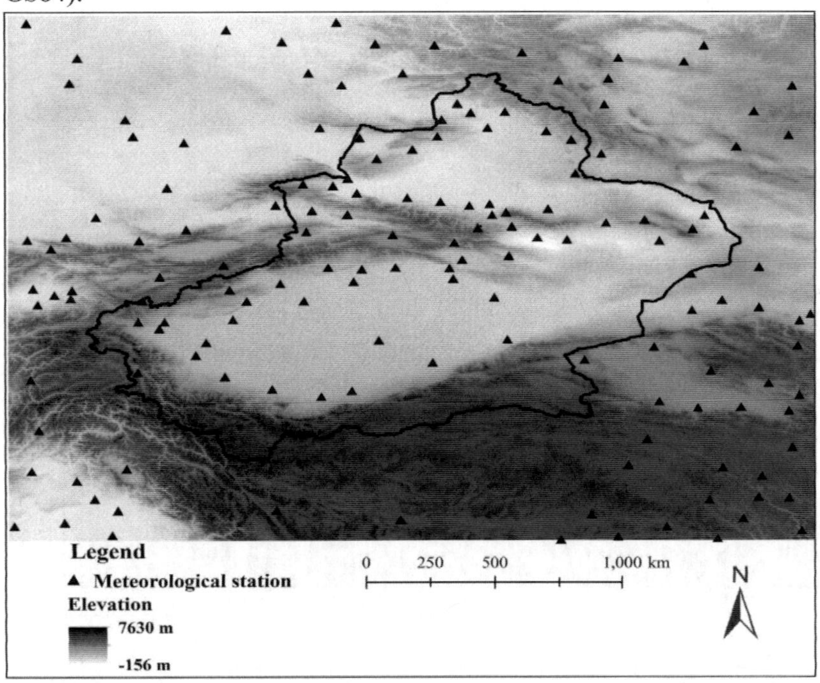

Figure 1. Distribution of meteorological stations in Xinjiang and surrounding areas

2.2. Interpolation methods

Usually Interpolation techniques predict the variable of interest at a specific location by taking known values from the surrounding region into account, and using them to estimate the value at a location where it is unknown. While all of these different types of methods present viable options for interpolation, we only compared five methods. These were chosen based on a number of factors, including computational complexity, ease of implementation, the capabilities of the software being utilized and a thorough review of recent literature. Given these criteria, Inverse distance weighting (IDW), Ordinary kriging, Cokriging, thin -plate smoothing splines (ANUSPLIN) and Empirical Bayesian kriging(EBK) were chosen as the appropriate spatial interpolation techniques.

IOP Conf. Series: Earth and Environmental Science **46** (2016) 012028 doi:10.1088/1755-1315/46/1/012028

2.3. Model evaluation and accuracy assessment

According to the study areas and its current conditions, we choose 15 meteorological stations as validation sites. We assessed the performance accuracy of different interpolation procedures based on the magnitude and distribution of errors between observed and model-predicted values using a series of statistical measures: Mean Absolute Error (MAE), Mean Relative Error (MRE) and Root Mean Squared Error (RMSE). These measures were calculated using the following equations:

$$MAE = \frac{1}{n}\sum_{i=1}^{n}\left|Z_{oi} - Z_{ei}\right| \tag{1}$$

$$MRE = \frac{1}{n}\sum_{i-1}^{n}\left|\frac{Z_{oi} - Z_{ei}}{Z_{oi}}\right| \tag{2}$$

$$RMSE = \sqrt{\frac{1}{n}\sum_{i=1}^{n}(Z_{oi} - Z_{ei})^2} \tag{3}$$

Where Z_{oi}、Z_{ei} *represent the* observed and model-predicted values of temperature, n represent the number of stations.

3. Results and discussion

After interpolating the time series temperature data, we compared the results of 5 interpolation methods in two ways. At first we compared the interpolation results with independent data using the MAE, MRE and RMSE as a criterion. Secondly, the interpolation results are compared on the basis of temperature distribution pattern in Xinjiang. Figure 2 shows the Comparison results of cross-validation MAE, MRE and RMSE for monthly mean temperature during 10years across all stations. The MAE value for annual average temperature are ranged differently: ANUSPLIN< COK < EBP < OK < IDW, their values were 2.17mm，3.11mm，3.43mm，3.73mm，3.74mm respectively. The ANUSPLIN also has the smallest MRE (0.45) and RMSE (2.1mm) values in comparison to other methods, the Cokriging interpolation method has highest RMSE value of 2.69mm. Above all, AUNSPLIN produced the lowest errors when evaluated by all three error metrics (MAE, MRE, RMSE). It can be seen that AUNSPLIN interpolation works best in Xinjiang region.

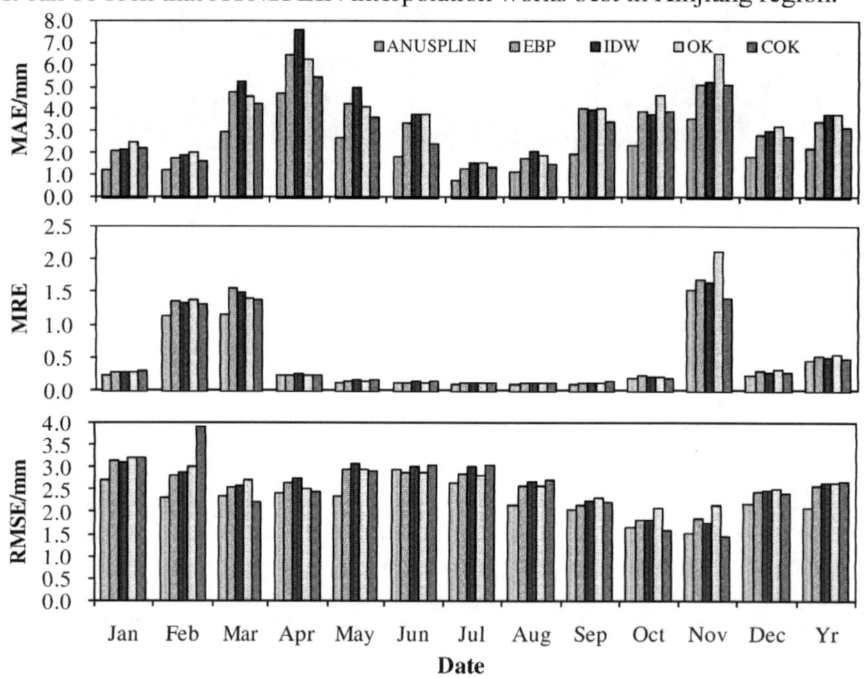

Figure 2. Comparing analysis of cross-validation MAE, MRE and RMSE for monthly mean temperature during 10years across all stations for the six interpolation methods

Even though the interpolation was carried out from 1995-2004, we only selected the interpolation results of average temperature in January 2000 and July 2000. Figure 3 shows the spatial distribution map of interpolation results from 5 methods in January 2000, IDW, EBK and Kriging interpolation results are consistent in spatial distribution, but have certain difference in the maximum and the minimum temperature value. For the Cokriging interpolation method, the lowest temperature are mainly distributed in northern Xinjiang and Hami areas, the temperature is mainly in the range of -21.8~0℃. For the ANUSPLIN interpolation method, the highest temperature is located in the Tarim basin, the lowest temperature in the Altai Mountain and high altitude area of Tianshan Mountain, the temperature decreases with the rise of altitude, and the temperature range in -31.5~ -1.3 ℃.

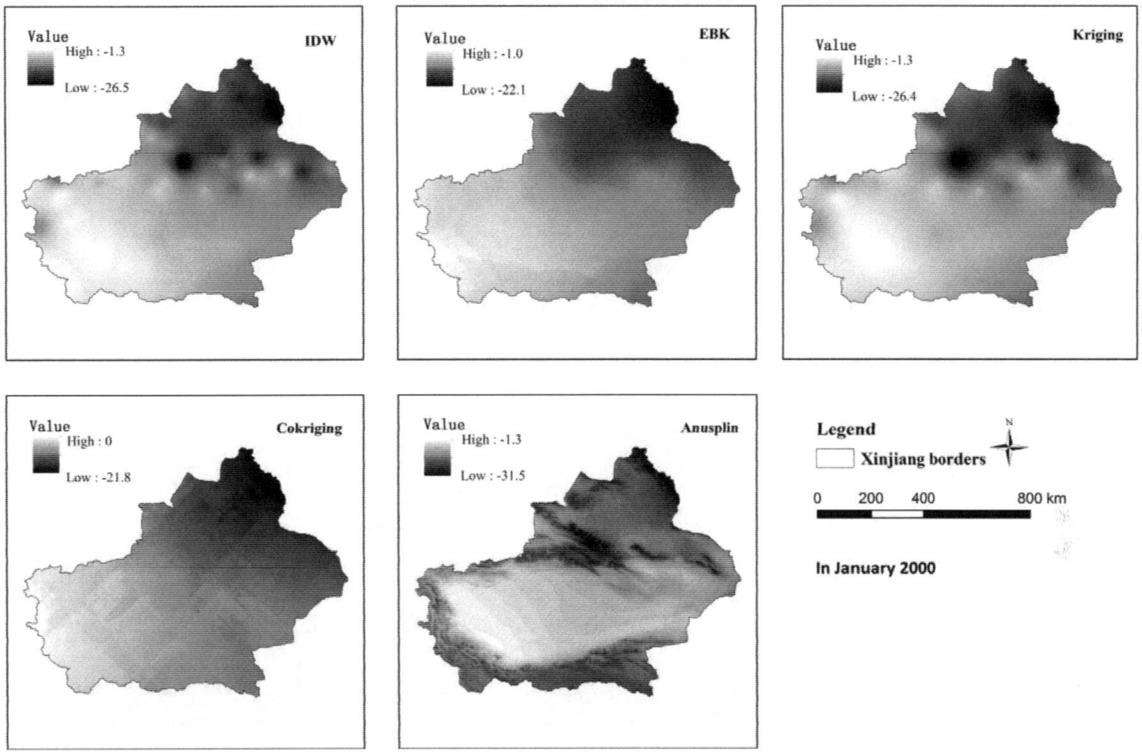

Fig 3. The spatial distribution map of monthly mean temperature on January in Xinjiang

Figure 4 shows the spatial distribution map of interpolation results from 5 methods in July 2000. The spatial distribution of temperature from the IDW, EBK, Kriging, Cokriging and ANUSPLIN interpolation results are consistent, the Tarim basin and Junggar basin have highest temperature, but Altai, Tianshan and Kunlun mountains temperature are lower. In the range of temperature there are certain difference between those five region (Tarim basin, Junggar basin, Altai, Tianshan and Kunlun mountains), their temperature ranged from 6.8~32.7℃, 9.9~31.3℃, 13.6~28℃, 14.5~26.9℃ and 3.8~31.5 ℃ respectively. It can be speculated that ANUSPLIN, which was generally found to perform the best than other methods, and the temperature was evidently affected by altitude that consistent with the study areas climate distribution.

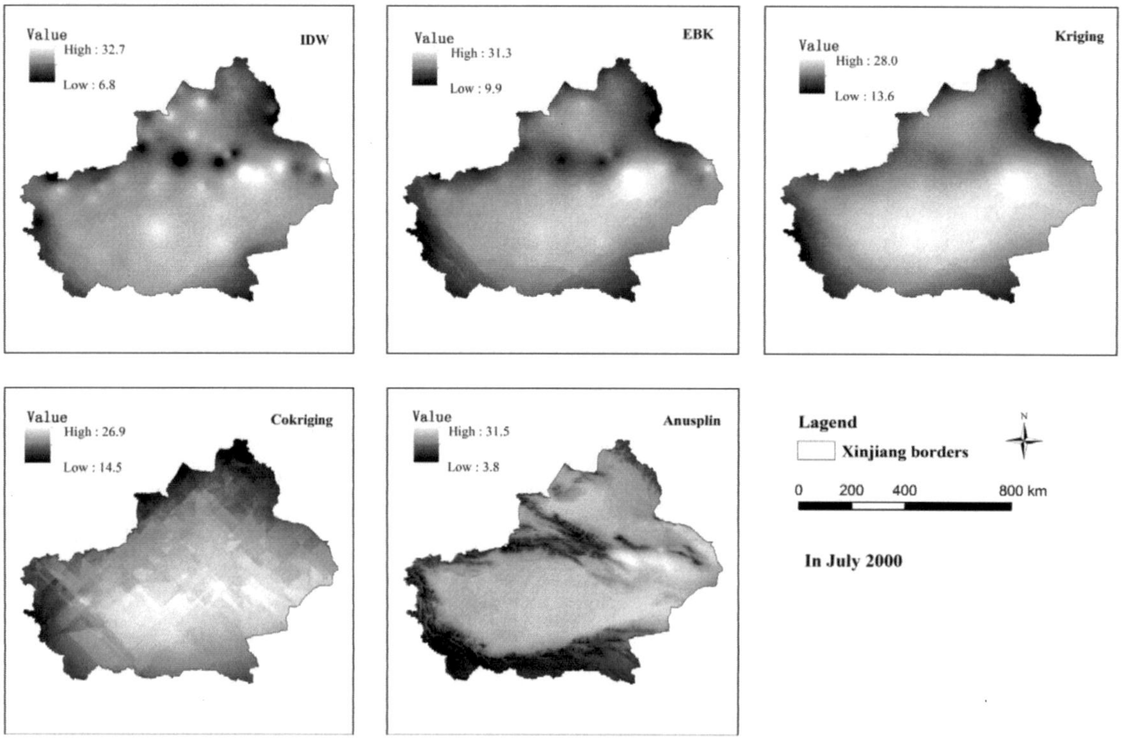

Figure 4. The spatial distribution map of monthly mean temperature on July in Xinjiang

4. Conclusion

This study assessed five different interpolation techniques and their ability to accurately predict monthly mean temperature in Xinjiang. IDW、EBK、Kriging、Cokriging and ANUSPLIN were selected as appropriate methods for interpolating temperature for January to December of 1995 through 2004. Finally, the results were compared of cross-validation MAE, MRE and RMSE for monthly mean temperature during 10years across all stations. The results shows that, ANUSPLIN interpolation method was found to perform the best in spatial distribution and produced low values from all three error metrics.

Interpolating a climatic variable such as temperature is very much dependent on the settings, thus more methods must be tested in order to determine which methods perform well on a consistent basis, and will be taken into account in future research.

Acknowledgments

This study was supported by the National ministry of science and technology's special project for fundamental scientific research(2012FY111900), Chinese academy of engineering research found for major consulting (2012-ZD-7)and the Program for Changjiang Scholars and Innovative Research Team in University (IRT13019). China clean development mechanism (CDM) projects (2013013).

References

[1] James M, Crabbe C., 2008. Climate change, global warming and coral reefs: Modelling the effects of temperature. *Computational Biology and Chemistry*, **32**: 311-314.

[2] David T P, Daniel W M, Ian A N, Michael F H, Jennifer L K., 2000. A comparison of two statistical methods for spatial interpolation of Canadian monthly mean climate data. *Agricultural and Forest Meteorology*, **101**:81–94.

[3] Hughes, D.A., 1982. The relationship between mean annual rainfall and physiographic variables applied to a coastal region of southern Africa. S. *African Geographic J.* **64**, 41–50.

[4] Hutchinson, M.F., 1995. Interpolating mean rainfall using thin plate smoothing splines. *Int. J. GIS* **9**, 385–403

[5] Daly, C., 1994. A statistical-topographic model for mapping climatological precipitation over mountainous terrain. *J. Appl. Meteorol.* **33**, 140–158.

[6] Jayawardene, H., Sonnadara, D.U.J., Jayewardene, D.R., 2005. Spatial interpolation of weekly rainfall depth in the dry zone of Sri Lanka. *Clim. Res.* **29** (3), 223

[7] Zhonghui L, Xingguo M, Hongxuan L., 2002. Comparison of three spatial interpolation methods for climate variables in China.*Acta Geographica Sinica*,**57**(1):47-56. (in Chinese).

[8] Yaozhong P, Daoyi G,Lei D,Jing L,Jing G., 2004. Smart Distance Searching-based and DEM-informed Interpolation of Surface Air Temperature in China. *Acta Geographica Sinica*, **59**(3):366-374.

[9] Willmott C J, Kellji M., 1999. Smart Interpotation of Annually Air Temperature in the United States. *J. Appl. Meteorol*, **59**(3):366-374.

[10] Zimmerman, D., Pavlik, C., Ruggles, A., Armstrong, M.P., 1999. An experimental comparison of ordinary and universal kriging and inverse distance weighting. *Math. Geol.* **31** (4), 375-390.

[11] Robeson S M., 1994. Influence of spatial sampling and interpolation on estimates of air temperature change. *Climate Res*, **4**:119-126.

[12] Ishida T, Kaashima K., 1993. Use of Cokriging to estimate Surface Air Temperature from Elevation. *Thero Appl Climatol*, **47**:147-157.

[13] Willmott C J, Robeson S M, Feddema J J., 1994. Estimating continental and terrestrial precipitation averages from rain-gauge networks. *Int.J.Climatol*.**14**:403-414.

Acknowledgments

This study was supported by the Ministry of Science and Technology Special Basic Work(2012FY111900), the Project of Ministry of Agriculture(201203006), and the Program for Changjiang Scholars and Innovative Research Team in University (IRT13019).

An improved hybrid data-driven model and its application in daily rainfall-runoff simulation

Guangyuan Kan[1, 2], Xiaoyan He[1], Liuqian Ding[1], Jiren Li[1], Tianjie Lei[1], Ke Liang[3], Yang Hong[2, 4]

[1]State Key Laboratory of Simulation and Regulation of Water Cycle in River Basin, Research Center on Flood & Drought Disaster Reduction of the Ministry of Water Resources, China Institute of Water Resources and Hydropower Research, Beijing 100038, China
[2]State Key Laboratory of Hydroscience and Engineering, Department of Hydraulic Engineering, Tsinghua University, Beijing, China
[3]College of Hydrology and Water Resources, Hohai University, Nanjing 210098, China
[4]Department of Civil Engineering and Environmental Science, University of Oklahoma, Norman, OK, United States

E-mail: kanguangyuan@126.com

Abstract. In previous literatures, a coupled data-driven rainfall-runoff (RR) model, NU-PEK, has been proposed and successfully applied in hourly RR simulation task. However, numerical experiments show that its performance for daily RR simulation is unsatisfactory. It is noticed that the poor performance is due to the inability of the original model to capture the much higher non-linear characteristics contained in the daily data. In order to improve the non-linearity simulation capability of the original model, an improved model named NU-PKEK and its calibration methodology are developed in this paper. The improved model is constituted by adding a K-means clustering module and utilizing multiple NU-PEK modules instead of using only one NU-PEK model. This study applies the improved model, the Xinanjiang model, and the original model for daily RR simulation in Chengcun catchment for intercomparison and verification. The simulation results prove that the NU-PKEK performs best, and has better simulation and forecasting capability.

1. Introduction

Data-driven hydrological models have been proved to be effective, simple, and concise methods in recent years (Piotrowski et al., 2013). They have been used widely in hydrological forecasting, water resources management, and environmental simulation (Shouke et al., 2012; Mohammad et al., 2013; Vahid et al., 2013). Applications indicate that the data-driven model is an effective and robust method for the RR simulation and forecasting on the up-dating mode, which can forecast discharge one-time-step ahead. On the other hand, multi-time-step simulation is also studied by some hydrologists (Li et al., 2014; Kan et al., 2015).

The recently proposed NU-PEK model (Kan et al., 2015) aims at the implementation of non-updating RR simulation with higher accuracy and better stability than traditional conceptual models such as the Xinanjiang and the IHACRES models. Nevertheless, numerical experiments show that when applying the NU-PEK model for the daily RR simulation tasks, the model performance is degraded. For the purpose of ameliorating the non-linearity simulation capability of the NU-PEK model and improving model performance in daily RR simulation, an improved model named NU-PKEK and its calibration methodology are developed in this paper. The NU-PKEK model is constituted by adding a K-means clustering module into the original NU-PEK model and utilizing multiple NU-PEK modules instead of using only one NU-PEK model. The input vectors are classified into several categories by the K-means clustering module to weaken the non-linearity contained in the input data. Multiple NU-PEK modules are adopted to simulate their corresponding category of input-output data to generate more precisely forecasted results. This study applies the improved model, the Xinanjiang model, and the original model for daily RR simulation in Chengcun catchment for the purpose of comparison and verification.

2. Study area and data description

We carry on the daily rainfall-runoff simulation in Chengcun catchment (290km^2), which lies in the Qiantang River basin, Anhui province, China. There are ten rainfall gauges located in this area. Observed daily rainfall, evaporation, and average discharges range from 1986 to 1994 were utilized as the calibration data, while data from 1995 to 1999 were utilized as the validation data. The Thiessen polygon map for the Chengcun catchment are shown in Figure 1.

Figure 1. Thiessen polygon map for the Chengcun catchment.

3. Methodology

3.1. PKEK approximator

3.1.1. K-means clustering. The K-means clustering (Grigorios and Aristidis, 2014; Kapageridis, 2015; Rehab and Mohammed, 2015) is a famous and widely used partitioning and clustering method (MATLAB, 2012). The K-means method is usually calibrated or trained by the iterative method.

3.1.2. PEK approximator. The PEK approximator is a hybrid data-driven model and it can be applied for the simulation of the multi-input single-output (MISO) system mapping relationship. Detailed principle of the PEK approximator can be found in the corresponding literatures (Kan et al., 2015).

3.1.3. PKEK approximator. PKEK approximator is an improved version of the previously proposed PEK approximator. It is combined by the PMI (partial mutual information) based IVS (input variable selection) module, the K-means clustering based input variable clustering module, and the ENN (ensemble artificial neural network) and KNN (K-nearest neighbour) simulation module. The PKEK approximator can be seen as a hybrid approximator that composed by a K-means clustering module and multiple PEK modules. The architecture of the PKEK approximator is demonstrated in Figure 2.

The simulation and calibration method of the PKEK approximator is as follows: The input variables are selected by the PMI-based separate IVS scheme to generate the selected input vectors.

After that, each selected input vector is fed into the K-means clustering algorithm to determine which category it belongs to. For each selected input vector, we choose a corresponding PEK module to calculate the output. In the PEK module, the output is estimated by the ENN, and the output error is estimated by the KNN regression. The final simulated output is the sum of the estimated output and output error. The calibration of the PKEK approximator is almost the same with the PEK approximator. The difference is that we need to calibrate the K-means clustering before calibrate PEK modules by an iterative method (Grigorios and Aristidis, 2014; Kapageridis, 2015; Rehab and Mohammed, 2015).

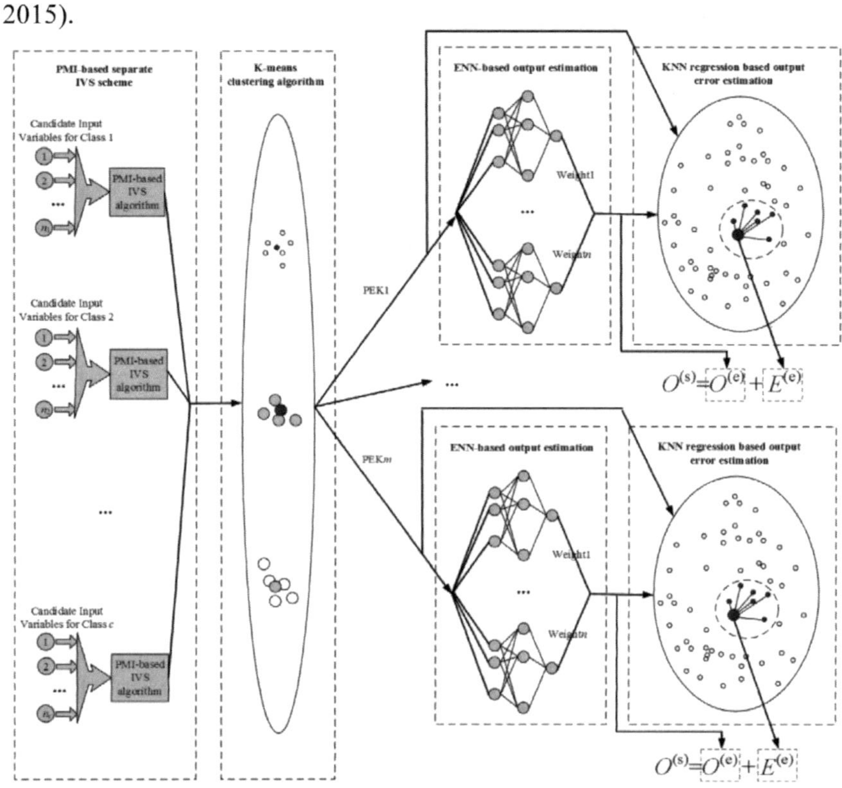

Figure 2. The architecture of the PKEK approximator.

3.2. NU-PKEK module

3.2.1. Non-updating RR simulation of the NU-PKEK model. The non-updating modelling approach of the NU-PKEK model is similar to the previously proposed NU-PEK model. The difference is the addition of the K-means clustering. The modelling approach of the NU-PKEK model is as follows:

$$X_t^{(Q\_\mathrm{SIM})} = \left(Q_{t-1}^{(\mathrm{SIM})},\ Q_{t-2}^{(\mathrm{SIM})},\ \ldots,\ Q_{t-n_Q}^{(\mathrm{SIM})}\right)^{\mathrm{T}} \tag{1}$$

$$X_t^{(SWCR)} = \left(SWCRs_t^{(1)},\ SWCRs_t^{(2)},\ \ldots,\ SWCRs_t^{(n_P)}\right)^{\mathrm{T}} \tag{2}$$

$$X_t^{(\mathrm{S})} = \left(\mathrm{IVS}_{Q\_\mathrm{SIM}}\left(X_t^{(Q\_\mathrm{SIM})}\right),\ \mathrm{IVS}_{SWCR}\left(X_t^{(SWCR)}\right)\right)^{\mathrm{T}} \tag{3}$$

$$Q_t^{(\mathrm{EST})} = \mathrm{F}_{\mathrm{EBPNN}}\left[\mathrm{F}_{K-means}\left(X_t^{(\mathrm{S})}\right)\right] \tag{4}$$

$$E_t^{(\mathrm{EST})} = \mathrm{F}_{\mathrm{KNN}}\left[Q_t^{(\mathrm{EST})}, \mathrm{F}_{K-means}\left(X_t^{(\mathrm{S})}\right)\right] \tag{5}$$

$$Q_t^{(\mathrm{SIM})} = Q_t^{(\mathrm{EST})} + E_t^{(\mathrm{EST})} \tag{6}$$

where $X_t^{(Q\_\mathrm{SIM})}$ denotes the candidate simulated antecedent discharge (SAD) input vector; $X_t^{(SWCR)}$ denotes the candidate sliding window cumulative rainfall (SWCR) input vector; $Q_{t-i}^{(\mathrm{SIM})}$ denotes the SAD;

$SWCRs_t^{(i)}$ denotes the SWCR with sliding window width i; nQ and nP denote the order of the SAD and the SWCR respectively; IVSQ_SIM denotes PMI-based IVS for candidate SAD input vector; IVS_SWCR denotes PMI-based IVS for candidate SWCR input vector; $X_t^{(S)}$ denotes the selected input vector; $Q_t^{(EST)}$ denotes the estimated output discharge associated with $X_t^{(S)}$; $E_t^{(EST)}$ denotes the estimated output discharge error associated with $X_t^{(S)}$; $Q_t^{(SIM)}$ denotes the simulated output discharge and it is the final prediction of the output discharge; FEBPNN denotes the EBPNN discharge estimation; FKNN denotes the KNN discharge error estimation; FK-means denotes the K-means clustering. The non-updating simulation procedure of the NU-PKEK model for each flood event is almost the same with the NU-PEK model. The difference is that after the PMI-based separate IVS, the K-means clustering is executed to classify the input vector and fed it to different PEK module to calculate the forecasted discharge value.

3.2.2. Calibration of the NU-PKEK model. The calibration of the NU-PKEK model is almost the same with the NU-PEK model. The difference is the addition of the calibration of the K-means clustering. The K-means algorithm is calibrated by the iterative method and this calibration process is repeated for 100 times with different initial parameters to avoid the local minimum issue. The optimal number of classes is determined by the Silhouette value. In this paper, approximately 70% of the flood events are utilized to create the calibration data and the remaining events are utilized to create the validation data.

3.3. Xinanjiang model

The Xinanjiang model (Zhao et al., 1980, 1984, 1992) is a conceptual RR model which has been successfully and widely used in China. The flowchart of the Xinanjiang model is demonstrated in Figure 3 (Yao et al., 2009).

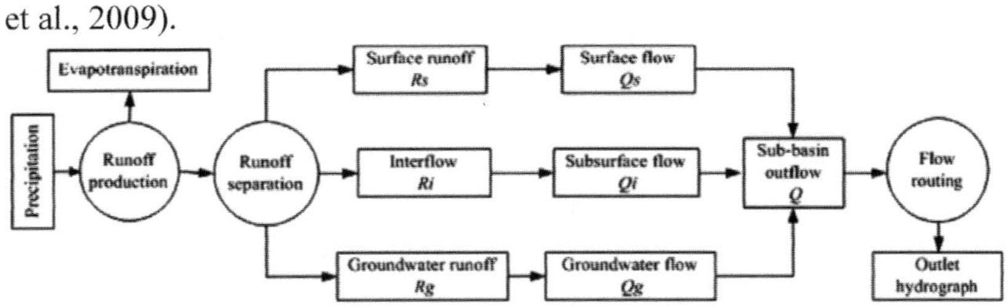

Figure 3. Xinanjiang model flow chart (Yao et al., 2009).

4. Results and discussion

4.1. Calibration of the NU-PKEK and NU-PEK models

4.1.1. Input variable selection. Considered that we are studying the daily simulation, we set the order of SWCR and SAD (i.e. nP and nQ) equal to 10, and 10 is large enough and rational for daily RR simulations. This means that the rainfall will becomes runoff with a lag at most 10 days. After the IVS we have found that the lag time of most selected SADs are less than 10. Therefore, this result conform that the order of 10 is large enough to describe the mapping relationship between the daily rainfall and runoff data of the Chengcun catchment.

4.1.2. K-means clustering module. We use the Silhouette value to determine the optimal cluster number. The Silhouette refers to a method of interpretation and validation of consistency with clusters of data. The technique provides a succinct graphical representation of how well each object lies within its cluster (Rousseeuw, 1987). The classification results of the K-means clustering algorithm is shown in Figure 4. Figure 4 (A) demonstrates the relationship between the cluster number and the Silhouette value. It can be seen from the figure, when we dividing the input samples into 3 clusters, the

IOP Conf. Series: Earth and Environmental Science **46** (2016) 012029 doi:10.1088/1755-1315/46/1/012029

Silhouette value reaches the maximum value. Therefore, we divide the input sample into 3 clusters. Figure 4 (B) shows the Silhouette value distribution when the input samples are divided into 3 clusters. We can see that the Silhouette value distributes evenly and the clustering results is rational good and acceptable.

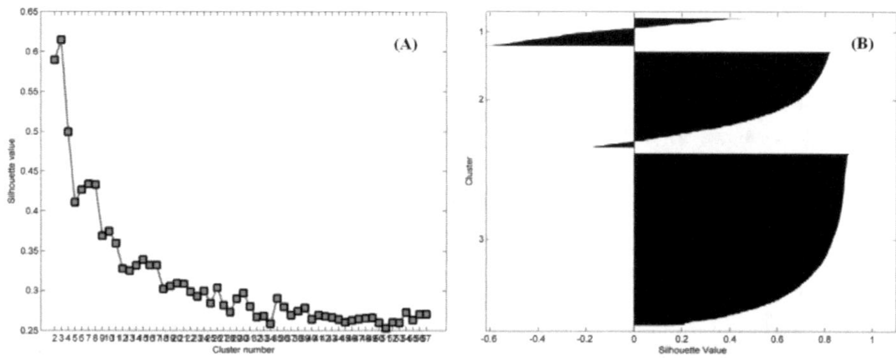

Figure 4. The classification results of the K-means clustering algorithm. (A) Relationship between the cluster number and the Silhouette value; (B) The Silhouette value distribution when the samples are divided into 3 clusters.

4.1.3. The PEK module. The PEK module is calibrated by a NSGA-II, Levenberg-Marquardt (LM), and leave-one-out cross-validation coupled method. The arithmetic parameters of the NSGA-II algorithm are set as: population size pop=90, evolution generation total number gen=1000, crossover probability pc=0.85, and mutation probability pm=0.15. The arithmetic parameters of the LM method are set as follows: minimum performance gradient min_grad=1-10, initial mu=0.001, mu decrease factor mu_dec=0.1, mu increase factor mu_inc=10, and maximum mu mu_max=110. The lower and upper boundaries of K for the KNN algorithm are set to 1 and 300.

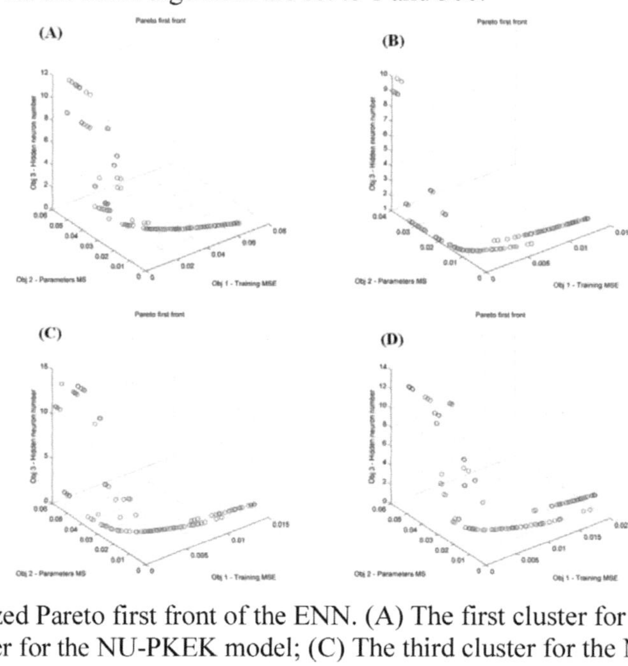

Figure 5. The optimized Pareto first front of the ENN. (A) The first cluster for the NU-PKEK model; (B) The second cluster for the NU-PKEK model; (C) The third cluster for the NU-PKEK model; (D) The NU-PEK model.

The optimized Pareto first front of the ENNs for the NU-PKEK and NU-PEK models are shown in Figure 5. Because we divide the input samples into 3 clusters, we construct three PEK modules for the NU-PKEK model. Figure 5 (A), (B), and (C) demonstrate the Pareto first front for the first, second, and third cluster. Figure 5 (D) demonstrates the Pareto first front for the NU-PEK model. The objective function used for the NSGA-II algorithm is mean squared error (MSE), mean squared value of the network parameters (MS), and hidden layer neuron number. As we can see, all of the four Pareto fronts distribute evenly which means that the optimization results is reasonable. It can be seen that the configuration of the Pareto fronts in (D) seems like a combination of Pareto fronts in (A), (B), and (C). This fact implies that the improved model ((A) + (B) + (C)) contains the simulation capability of the original model (only (D)) and can simulate more precisely by utilizing three Pareto fronts.

4.2. Calibration of the Xinanjiang model

The Xinanjiang model parameters are optimized by the SCE-UA (Duan et al., 1992, 1993, 1994) method. The algorithmic parameters are set as: p (number of complexes)=15, maxn (maximum number of objective function evaluations)=100000, kstop (number of shuffling loops in which the objective function must improve by the specified percentage pcento)=10, pcento (percentage by which the objective function must change in the specified number of shuffling loops kstop)=0.1, and peps (minimum parameter space allowed)=0.001, respectively.

4.3. Model performance comparison

We use the scatter plots of the observed and simulated discharges and the regression R values to inspect the overall performance of the three models (see Figure 6).

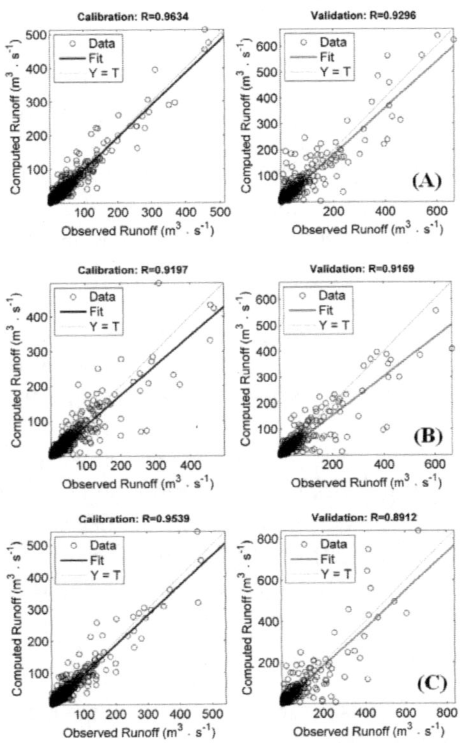

Figure 6. The scatter plots of the observed and simulated discharges and the regression R values: (A) NU-PKEK model, (B) Xinanjiang model, (C) NU-PEK model.

As demonstrated in Figure 6, as for the calibration period, the NU-PKEK model obtains the best result (R = 0.9634), the NU-PEK model obtains the moderate result (R = 0.9539), and the Xinanjiang model obtains the worst result (R = 0.9197). It can be noticed that the data scatters of the NU-PKEK model intensively lie close to the 45 degree line. The distribution of the data scatters of the NU-PKEK model is relatively more uniform for small, middle, and large discharge values. These results indicate that the NU-PKEK model generates the best result and obtains the most stable performance in the calibration period.

As for the validation period, the NU-PKEK model obtains the best result (R = 0.9296), the Xinanjiang model obtains the moderate result (R = 0.9169), and the NU-PEK model obtains the worst result (R = 0.8912). As demonstrated in the Figure 6, the data scatters of the NU-PKEK model distributed normally around the 45 degree line. This result indicates that the simulation results of the NU-PKEK model do not have bias to be larger or smaller and shows a very stable property. However, the simulated discharges of the Xinanjiang model is smaller than the observed values, while the simulated discharges of the NU-PEK model is larger than the observed values. This phenomenon becomes more obvious especially for large discharge values.

After analyzing the accuracy in calibration and validation period, the accuracy declination is also compared. The Xinanjiang model obtains the best accuracy declination value (0.9169/0.9197 = 0.9970), the NU-PKEK model obtains the moderate accuracy declination value (0.9296/0.9634 = 0.9649), and the NU-PEK model obtains the worst accuracy declination value (0.8912/0.9539 = 0.9343). The results show that although the NU-PKEK model obtains worse accuracy declination value than the Xinanjiang model, it outperforms the original NU-PEK model and generates better forecasted results.

The analysis of the three models shows that the NU-PKEK model can generate better overall results than the original NU-PEK model and the Xinanjiang model. Especially for peak values, the NU-PKEK model is much better than other two models. The comparison results indicate that the NU-PKEK model outperforms other models in Chengcun catchment and prove its satisfactory accuracy and stability.

5. Conclusions

An improved hybrid data-driven model, named NU-PKEK, is proposed in this paper to overcome the disadvantages of the previously proposed NU-PEK model in daily RR simulation. For the purpose of comparing performances of different models, the NU-PKEK, Xinanjiang, and NU-PEK models are applied to daily RR simulation in Chengcun catchment. It can be remarked that the NU-PKEK model is able to simulate generally much better runoff hydrograph than other models. From the application and analysis of the NU-PKEK model, we can conclude that:

(1) The NU-PKEK model have all the advantages of the previously proposed NU-PEK model, including better calibration and validation non-updating simulation accuracy, automatic calibration without much impact from human experiences, good compromise between the sufficiency and parsimony of the input data, and good compromise between the simulation accuracy and topology complexity, and so on.

(2) Compared with the original NU-PEK model, the performance of the NU-PKEK model becomes much better for the daily RR simulation. The improvement in the goodness-of-fit to the observed discharge of the NU-PKEK model is owed to the addition of the K-means clustering method combined with the utilization of multiple PEK modules. With the clustering method and multiple PEK modules, the NU-PKEK model can simulate the characteristics of the RR relationship much precisely and it's forecasting capability and stability becomes better.

References

[1] Grigorios T., Aristidis L., 2014. The MinMax k-means clustering algorithm. Pattern Recognition. 47 (7), 2505-2516.

[2] Kan G., Yao C., Li Q., Li Z., Yu Z., Liu Z., Ding L., He X., Liang K., 2015. Improving event-based rainfall-runoff simulation using an ensemble artificial neural network based hybrid data-driven model. Stochastic Environmental Research and Risk Assessment. DOI: 10.1007/s00477-015-1040-6.

[3] Kapageridis I.K., 2015. Variable lag variography using K-means clustering. Computers & Geosciences. In Press, Corrected Proof, Available online 13 April 2015.

[4] Li Z., Kan G., Yao C., Liu Z., Li Q., Yu S., 2014. An improved neural network model and its application in hydrological simulation. Journal of Hydrologic Engineering. 19 (10),

04014019-1 – 04014019-17.
[5] MATLAB, 2012. User's manual. MATLAB documentation.
[6] Mohammad T. Dastorani, Jamile Salimi Koochi, Hamed Sharifi Darani, Ali Talebi and M. H. Rahimian, 2013. River instantaneous peak flow estimation using daily flow data and machine-learning-based models. J. Hydroinform. 245, 1089-1098.
[7] Piotrowski, A.P., Napiorkowski, J.J., 2013. A comparison of methods to avoid overfitting in neural networks training in the case of catchment runoff modeling. J. Hydrol. 476, 97-111.
[8] Rehab D., Mohammed A., 2015. A novel approach for initializing the spherical K-means clustering algorithm. Simulation Modelling Practice and Theory. 54, 49-63.
[9] Rousseeuw, P.J., 1987. Silhouettes: a graphical aid to the interpretation and validation of cluster analysis. Computational and Applied Mathematics. 20, 53-65.
[10] Shouke Wei, Depeng Zuo and Jinxi Song, 2012. Improving prediction accuracy of river discharge time series using a Wavelet-NAR artificial neural network. J. Hydroinform. 143, 974-991.
[11] Vahid Nourani and Masoumeh Parhizkar, 2013. Conjunction of SOM-based feature extraction method and hybrid wavelet-ANN approach for rainfall–runoff modeling. J. Hydroinform. 141, 829-848.
[12] Yao, C., Li, Z., Bao, H., Yu, Z., 2009. Application of a developed Grid-Xinanjiang model to Chinese watersheds for flood forecasting purpose. Journal of Hydrologic Engineering, 14 (9), 923-934.
[13] Zhao, R., Zhang, Y., Fang, L., Liu, X., Zhang, Q., 1980. The Xinanjiang model. In: Hydrological Forecasting Proceedings Oxford Symposium. IAHS. pp. 351-356.
[14] Zhao, R., 1984. Watershed Hydrological Modeling - Xinanjiang Model and Northern Shaanxi Model. China Water Power Press, Beijing, China, pp. 126-127 (in Chinese).
[15] Zhao, R., 1992. The Xinanjiang model applied in China. Journal of Hydrology 135, 371-381.

Acknowledgments
This research was funded by the IWHR Research & Development Support Program (JZ0145B052016) IWHR Scientific Research Projects of Outstanding Young Scientists "Research and application on the fast global optimization method for the Xinanjiang model parameters based on the high performance heterogeneous computing" (No. KY1605), Specific Research of China Institute of Water Resources and Hydropower Research (Grant Nos. Fangji 1240), the Third Sub-Project: Flood Forecasting, Controlling and Flood Prevention Aided Software Development - Flood Control Early Warning Communication System and Flood Forecasting, Controlling and Flood Prevention Aided Software Development for Poyang Lake Area of Jiangxi Province (0628-136006104242, JZ0205A432013, SLXMB200902), Study of distributed flood risk forecast model and technology based on multi-source data integration and hydrometeorological coupling system (2013CB036400), and the IWHR application project of multi-source precipitation fusion and soil moisture remote sensing assimilation. We gratefully acknowledge the support of NVIDIA Corporation with the donation of the Tesla K40 GPU used for this research.

IOP Conf. Series: Earth and Environmental Science **46** (2016) 012030 doi:10.1088/1755-1315/46/1/012030

Study on analysis from sources of error for Airborne LIDAR

H C Ren[1,2], Q Yan[1], Z J Liu[1], Z Q Zuo[1], Q Q Xu [1,2], F F Li [1,3]and C Song[1,4]

[1]Institute of Photogrammetry and Remote Sensing, Chinese Academy of Surveying and Mapping, Beijing 100830, China
[2]Lanzhou Jiaotong University, Gansu Provincial Engineering Laboratory for National Geographic State Monitoring, Lanzhou, 730070, China
[3]School of Geomatics Liaoning Project Technology University, Fuxin 12300, China
[4]China University of Geosciences, Information Engineering Institute, Wuhan, 430074, China

E-mail: renhaicheng2016@163.com

Abstract: With the advancement of Aerial Photogrammetry, it appears that to obtain geo-spatial information of high spatial and temporal resolution provides a new technical means for Airborne LIDAR measurement techniques, with unique advantages and broad application prospects. Airborne LIDAR is increasingly becoming a new kind of space for earth observation technology, which is mounted by launching platform for aviation, accepting laser pulses to get high-precision, high-density three-dimensional coordinate point cloud data and intensity information. In this paper, we briefly demonstrates Airborne laser radar systems, and that some errors about Airborne LIDAR data sources are analyzed in detail, so the corresponding methods is put forwarded to avoid or eliminate it. Taking into account the practical application of engineering, some recommendations were developed for these designs, which has crucial theoretical and practical significance in Airborne LIDAR data processing fields.

1. Introduction

In recent decades, with the development of Aerial Photography, Airborne LIDAR has attracted the mapping industry a lot. Airborne Light Detection and Ranging (LIDAR) technology that is an emerging technology for earth observation, is mainly made up of the Laser Range Finder, Inertial Navigation System (INS), Global Navigation Satellite System (GNSS) and High-resolution digital cameras and other equipment. Pulsed laser range finder can take advantage of information behind to get the real target detection directly, steep slope, the roughness coefficient case, even reflectivity information, yet digital cameras can record real-time image information of the target, through a particular process to produce ground real three-dimensional and three-dimensional imaging results from surface area and coordinate information of sampling points, a certain strip, such as high-density point cloud data, high-resolution orthophoto DOM, a variety of Digital Terrain Model(DTM) and large scale topographic maps, and digital surface models (DSM),etc. Lidar scanner data through post-processing software can be widely used in land resources, urban planning and construction management, digital three-dimensional modeling, to be applied to an electronic map [1].

Airborne LIDAR measurement technology, its' height accuracy may be up to 13cm, or horizontal accuracy may be up to 20 centimeter in the open air. To improve the accuracy of Airborne LIDAR Systems and access to high quality point cloud data, some scholars conducted extensive studies. Domestic and foreign scholars have conducted in-depth study of the accuracy. Domestic scholars such as Shukai Li, Shaochuang Liu, especially Hong You in construction and other red from photogrammetry collinear equations deduced Airborne 3D imager system geometry, and affect the positioning accuracy error sources qualitative analysis [2]. Xiaohong Zhang (2007) studied the Airborne LIDAR' accuracy from elevation or horizon [7]. Foreign scholars, Schenk (2001) devoted much time to studying the sources of error on LIDAR System [3]. However, there are numerous foreign scholars exploring on the impact of Airborne Laser scanning measurement accuracy systematic error analyzes [4, 5]. In this paper, considering the research status of domestic and foreign scholars, we demonstrates common error sources were elaborated or analyzed from Airborne LIDAR, yet the corresponding method to avoid or eliminate will be enforced to improve accuracy based on practical experience of great significance for subsequent processing point cloud data.

2. Airborne LIDAR error source

Airborne LIDAR System is a complex integrated system, resulting in its accuracy affected by the interaction of the various components of the system. From Airborne LIDAR 'error sources being various even very complex, they can be divided into two major aspects: single error and integrated error, as shown in Figure 1.

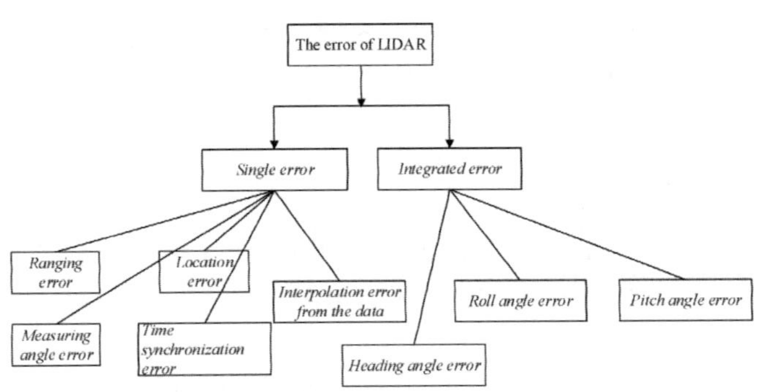

Figure.1 Error Category

2.1 Single error

2.1.1 Ranging error. The core equipment used to be rangefinder about Airborne LIDAR System. All factors can affect the precision of laser pin point coordinates, but the ranging accuracy, which is the most complex. According to Schenk (2001), described the mathematical model, says that LIDAR point clouds no error geolocation equation:

$$P_w = P_{GPS} + R_w R_{GEO} R_{INS} (R_{lu} R_{LB} s + l_0) \quad (1)$$

P_w: Laser target point coordinates in WGS84 ; P_{GPS} : GPS antenna phase center coordinates in WGS84; R_W : Local Ellipsoid System to the WGS-84 coordinate system transformation matrix; R_{GEO} : Navigation frame transformation matrix to local ellipsoid system; R_{INS} : IMU attitude where the carrier to coordinate the navigation frame transformation matrix; R_{LU}: Laser scanners IMU coordinate system to the coordinate transformation matrix carrier; R_{LB} : The instantaneous position of the laser beam to the laser coordinate system coordinate system; s : Laser spot position coordinates of the laser beam coordinate system; l_0 : The eccentric amount of the emitted GPS antenna phase center to the center of the laser emission.

Be taken into account when R_W, R_{GEO} value is relatively small, LIDAR point cloud positioning equation can be rewritten as:

$$P^*_w = \Delta P_{GPS} + P_{GPS} + \Delta R_{INS} (\Delta R_{lu} R_{lu} \Delta R_{LB} R_{LB} (\Delta s + s) + \Delta l_0 + l_0) + \Delta p_s \quad (2)$$

ΔP_{GPS} : the GPS positioning error; ΔR_{INS}: Attitude IMU error; ΔR_{LU} and ΔR_{LB} :placed foot error and laser scanning angle error; Δs: Laser point ranging error; ΔP_S : Error Laser point with GPS and INS; After the adoption of (1) and (2) added to the system error correction, Laser ranging' error is between 2 centimeter and 4 centimeter [6]. Ranging error elimination can be carried out by various experiments in the room to eliminate, the residual error after calibration can reach centimeter level [7].

2.1.2Measuring angle error. Angle measurement error is attitude angle error and scan angle error from the GPS and INS integrated. Attitude (IMU) error is mainly horizontal error except for laser gyro sight deviation, acceleration allowed timing, complex measurement noise, the presence of non-orthogonal between the bearing and the like, mainly affecting roll angle, pitch angle, flight angle, which can by flying at low altitude to reduce the influence. Due to the installation, design and other reasons such scanning system axis direction and original state with a certain bias, leading to the starting angle scanning angle greater than 0 degrees, named scanning angle error, can be measured at the factory as long as the instruments prescribed time calibration to, but the actual work do not consider such errors generally [6].

2.1.3 Location error. The positioning errors of GPS are all kinds ,including satellite orbit error, satellite clock error, the receiver clock error, multipath, phase center of instability, and there is a constellation of satellites, observation noise, integer ambiguity resolution is whether is correct or not, etc. (Zhang Xiao Hong, 2001). GPS positioning uses a dynamic difference method, while recording the GPS carrier phase and pseudo distance from ground-based reference stations from the data in conjunction with data from the satellite carrier phase and pseudo-together after the flight to use Novatel Inertial Explorer8.6 (IE) solver, less than 50km in baseline length can reach between 23cm and 32 cm accuracy[8]. Some scholars use differential GPS real-time dynamic three-dimensional coordinate measurement target, up to centimeter-level accuracy (Cong hui Feng, 2007).

2.1.4 Time synchronization error .There are Laser rangefinder, GPS differential positioning, IMU attitude, which interfere with each other, working independently of each other, with having their own time recording apparatus. In order to get a correct three-dimensional coordinates of the laser point in space, we must ensure that the laser ranging value at launch, and the attitude of observational position are at the same time, so they need to be imputed to the respective time uniform time UTC or GPS time. If the time is not, there will be positioning errors [9]. Interpolation can be linear or low-order polynomial fitting method to eliminate the time offset, but some tests shows that if the data updated rate GPS signal receiver is more than 2 times per second, two adjacent epoch GPS antenna location linear interpolated between two adjacent epochs GPS antenna position at a time, you can reduce the impact of the time synchronization error (Da hai Guo etc., 2006).

2.1.5 Interpolation error from the data. Interpolation error is due to various airborne LIDAR system having different sampling frequencies caused. At present, the frequency of the laser ranging system can be up to 20 kHz, however, IMU data sampling frequency is generally 200Hz, while the GPS data sampling frequency is only 20Hz. Therefore, in order to get the correct position and attitude information of each of the laser spot, it is necessary for GPS and IMU data interpolation processing to eliminate certain outside influences. Domestic scholars, Shukai Li [10] , made full use of pulse time difference 5s, namely track data fitting processing, accuracy up to centimeter level. Generally, the interpolation error depends on weather changes airflow during flight, the error of the time offset impact is between 3 centimeter and 5 centimeter [11].

2.2 Integrated error

We need to consider the relationship between the axes of the reference coordinate system and the scanner IMU reference coordinate system when each of system is integrated. In general, the installation integrates axes must parallel to each other, but it does not guarantee parallel to each other between themselves in the actual installation process, resulting in the flight direction around the roll angle (Roll) error, about the direction of the wing pitch angle (Pitch) error angle about a vertical navigation (Heading) error. Roll angle error is mainly reflected in the plane of the scan lines having a certain inclination; flight angle error can cause ranging inaccurate, mainly in track route deviation, distorted, broken phenomenon [12]. Haikun Yu, Peng Li, Nan Liu [13] and other studies of airborne LIDAR collimation calibration method, application of new technology in the Leica Gallery measurement are discussed in detail so that both sides of the roll angle error route inconsistent level phenomenon; pitch angle error will cause the laser plane position data before and after the dislocation phenomena; flight angle error will cause the strip to the right of the point cloud is shifted forward, the left back offset. In the actual operation of the integrated error elimination generally do calibration experiments can eliminate the corresponding error.

3. Route design

Based on the above sources of error analysis, the paper puts forward the design of some suggestions, to a certain extent, the elimination of error plays a role of great importance for subsequent processing point cloud data. Laser radar equipment and high-resolution digital camera parameters employed, by means of a number of surface features characteristic to develop a reasonable rule of Airborne LIDAR flight program, considering the point cloud density point cloud accuracy, attitude accuracy, overlap and other issues [13], Following routes laid:

- The best test area is more typical spire angular roof or gable roof, mutually perpendicular to the road;
- Endurance test area is best taken by the forward and reverse flight alternate, also jumper aerial;
- Next to the route to the best degree of overlap in the 15% -35%;

After the route design, we also need to work to improve data quality secondary post-processing, mainly on the ground measuring points laid condition. There are two types: static and linear measuring point measuring point. Linear measuring points can be laid flat along the road near the route laid, namely RTK measurement, at regular intervals a point about 5 meter, a length of approximately 4 kilometer, accuracy should be at least within 5 centimeter, this method is cumbersome, the workload is relatively large, but the actual work is not taken. While static measuring point layout, can be good to overcome this problem, but also improve accuracy. To choose a certain geographical constraints, the general choice of four weeks without cover, the terrain is more open, less water area, if there is terrain depressions, preferably with a tripod stand up for a long time measurement accuracy can be improved, accuracy up to 2 centimeter [14] .

4. Conclusion

The paper discourses Airborne LIDAR measurement system based on the principles of composition, from the time the sensors integrated error generated by a detailed discussion on the causes and effects of error generation mechanism, was analyzed, and a variety of error processing were discussed, summed up the corresponding elimination method, the accuracy of the evaluation were analyzed. Route planning and design of the measuring point programs can reduce the error object to meet specific project needs. In the coming years, Airborne LIDAR mapping technology will become the focus of research in the field, with high precision, high density, high efficiency, product rich features obvious advantages, will gradually replace the traditional mapping technology. In some areas, it already has applications such as emergency mapping for geological disasters, earthquakes, landslides, mudslides and other automatic tracking and dynamic monitoring, real-time assessment of post-disaster situations; remote sensing applications for feature identification and classification of auxiliary geography Census; high-voltage transmission, power line for the safety distance detection.

References

[1] Ye S, Cao B and Zhang Y 2010 *Surveying Mapping* **3** 68-69
[2] Li S k, Liu S C and You H J 2000 *The Efficient 3D Remote Sensing System Integration Technology* (Beijing: Science Press)
[3] Schenk T 2001 *Proc. OEEPE Workshop Airborne Laser Scanning Interferometer SAR Detailed Dig. Elevation Models* (Amsterdam: Elsevier) pp 40-48
[4] Toni S 2001 *OEEPE Workshop Airborne Laser Scanning Interferometer SAR Detailed Dig. Elevation Models* (Amsterdam: Elsevier) p 78
[5] Ackermann F 1999 *ISPRS* **54** 64-67
[6] Lai X D 2009 *Airborne Laser Radar Fundamentals and Applications* (Beijing: Electronic Industry

Press)
[7] Zhang X H 2007 *Airborne Laser Radar Measurement Technology Theory and Method* (Wuhan: Wuhan University Press)
[8] Jiang Y S 2001 *Remote Sens.* **5** 241-6
[9] Zhang Y T 2012 *Airborne LIDAR systems integration and precision calibration error assessment* (Qingdao: Shandong University of Science and Technology)
[10] Li S K 2003 *Temporal Remote Sensing Information Integration Technology and Application* (Beijing: Science Press)
[11] Hebel M and Silla U 2012 *IEEE Tran. Geoscience Remote Sens.* **50** 2364-79
[12] Xiong A W and Yang M M 2014 *Surveying Mapping* **3** 75-78
[13] Sun L M, Deng Q J and Ling X C 2015 *Electric Power Survey Design* **2** 15-18
[14] Zhu Y 2012 *Railway Survey Design* **3** 23-25

Acknowledgments
This work was supported by the China Southern Power Grid key R & D projects for mapping geographic information nonprofit industry (approved No. GD-KJXM-20150903) .

IOP Conf. Series: Earth and Environmental Science **46** (2016) 012031 doi:10.1088/1755-1315/46/1/012031

Comparing NetCDF and SciDB on managing and querying 5D hydrologic dataset

Haicheng Liu[1] and Xiao Xiao[1]

[1]Changjiang River Scientific Research Institute, Changjiang River Water Resource Commission, Wuhan, China

E-mail: liuhaicheng@mail.crsri.cn

Abstract. Efficiently extracting information from high dimensional hydro-meteorological modelling datasets requires smart solutions. Traditional methods are mostly based on files, which can be edited and accessed handily. But they have problems of efficiency due to contiguous storage structure. Others propose databases as an alternative for advantages such as native functionalities for manipulating multidimensional (MD) arrays, smart caching strategy and scalability. In this research, NetCDF file based solutions and the multidimensional array database management system (DBMS) SciDB applying chunked storage structure are benchmarked to determine the best solution for storing and querying 5D large hydrologic modelling dataset. The effect of data storage configurations including chunk size, dimension order and compression on query performance is explored. Results indicate that dimension order to organize storage of 5D data has significant influence on query performance if chunk size is very large. But the effect becomes insignificant when chunk size is properly set. Compression of SciDB mostly has negative influence on query performance. Caching is an advantage but may be influenced by execution of different query processes. On the whole, NetCDF solution without compression is in general more efficient than the SciDB DBMS.

1. Introduction

In hydrologic domain, all kinds of simulation models never stop running to produce essential results for decision making. These modelling datasets are normally featured by high dimensionality, large data size and diverse storage formats, which brings inconvenience for professionals to extract effective information for certain applications efficiently. One of the most frequently used format is NetCDF which is notable for its simple data model, ease of use, portability, and strong user support infrastructure [1].

However, according to practical experience, traditional NetCDF solutions perform inefficiently in retrieving information from large spatio-temporal datasets for certain queries. This is caused by the way it stores variable values, which is known as contiguous storage structure. Basically, for a grid full of variable values in a certain spatial area, NetCDF stores values into a one-dimensional array according to a row-majored order. And in this way, to query the value in a particular cell, calculating position of the cell in the one-dimensional array is needed. Extraction of a time series becomes more expensive due to accessing required cell values, which are widely spread over the disk in a one-dimensional array. Alternatively, it is possible to store time series of every location as a one-dimensional variable in NetCDF, but then retrieving the complete grid at a single time step becomes the problem.

Chunked storage structure is introduced to the later version of NetCDF, i.e. NetCDF-4. The chunked storage structure divides a whole dataset into separate chunks with specified chunk sizes [2,3]. Based on this storage structure, specific array addressing and relative offset calculation are then applied to index values, which is proved to be of high query efficiency [4]. Within DBMS scope, MD array DBMS also employs chunked storage structure. Additionally, it is optimized further for MD array data management such as plenty native functionalities to manipulate arrays, smart caching algorithms, etc [2]. It also supports efficient storage and query features like versioning and overlapping [4,5].

This research investigates whether the MD array DBMS SciDB can achieve better performance in processing queries on high dimensional hydrologic simulation datasets than classic non-chunked NetCDF solution, and if SciDB has competitive performance when compared to chunked NetCDF-4 file based solution. To address the main research question, benchmark tests are executed. Several steps should be performed: define benchmark, create benchmark test environment, execute benchmarks and analyse results.

2. Data and query

Queries and datasets frequently used by water experts should be collected for devising a scientific benchmark. Therefore, several experts were first interviewed. Next a representative dataset and set of queries were specified for the benchmark.

Global Ensemble Forecast System (GEFS) dataset (table 1; figure 1) is calculated from a global forecast model. It contains 5 dimensions. The forecast dimension refers to the time steps simulated while model run represents the time to run the forecast model in reality. Increments for both dimensions are 6 hours, i.e. every 6 hours the model will be run and forecast temporal interval is also 6 hours. Ensemble represents initial condition and 20 ensembles simulate 20 different initial conditions as the input for the forecast model. Only one sample is applied for benchmarking and the model run time is 6:00 15-05-2014. This is because, for forecast dataset, it is always the latest forecast data that are intensively concerned.

Table 1. Details of GEFS dataset

Variables	Temperature 2m above ground; Maximum temperature 2m above ground; Minimum temperature 2m above ground; Relative humidity 2m above ground; Total precipitation; Total Cloud Cover; U-Component of Wind 10m above ground; V-Component of Wind 10m above ground
Dimension count	5
Dimension Span	Longitude, latitude, forecast, ensemble, model run; (360,181,40,20,1)
Temporal resolution	6 hours
Spatial resolution and coverage	1 degree (111 km); Global
Single file size	1.55 GB
Data format	NetCDF 64-bit offset

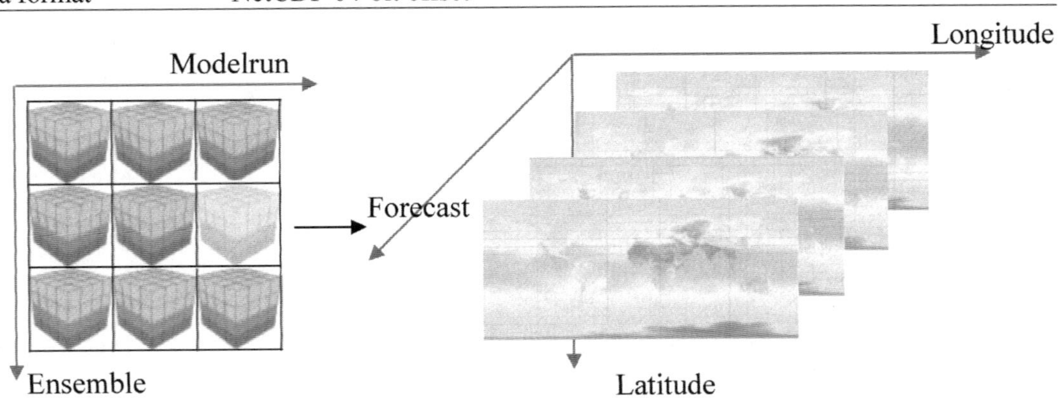

Figure 1. Visualization of 5D GEFS data.

According to most important classes of query types from expert interview and the dataset selected, conceptual queries for benchmarking are designed as follows:

- (Q1) All ensembles for Delft at one time step: Select total precipitation for all ensembles at (4, 52) for one forecast step.

- (Q2) Forecast time series for Delft in 10 days: Select total precipitation for all ensembles at (4, 52) for 40 forecast steps.
- (Q3) 80th percentile for Delft in 10 days: Select the 80th percentile at (4, 52) from all ensembles for 40 forecast steps.
- (Q4) Ensemble mean for the Netherlands in 10 days: Select mean of 20 ensembles of total precipitation for 40 forecast steps in the bounding box (3, 51) (7, 54).

3. Benchmark environment

The benchmark test environment is constructed on the HydroNET-4 system [6] which is running on Microsoft Windows. However SciDB (version 14.3) is only available on Linux. Thus a SciDB connector is developed such that the query performance of both NetCDF and SciDB can be assessed in the same HydroNET frontend system (figure 2).

As is shown in figure 2, a query starts at a web client which sends an HTTP POST request with a JSON object containing essential parameters for executing the query. The data access API receives the request and parses it. Then the API communicates with the catalog DBMS to retrieve the data from the specific data storage system through connectors. After data extraction, the result is returned as a binary HydroNET in-memory object to HydroNET-4. NetCDF connector to files in 64-bit offset format is based on HydroNetCDF library developed and optimized by the company. NetCDF-4 format connector is constructed on the standard library NetCDF-4.1.3. SciDB connector is constructed on shim, a super-basic SciDB client that exposes limited SciDB functionality through a simple HTTP API. All connectors can communicate with the 'Processor' component. For benchmarking, the Processor is utilized to realize complex queries such as aggregation and percentile calculation for the NetCDF solutions, while SciDB processes all calculations itself and communicates with the API directly. The total time spent for executing a query, i.e. the benchmark result, can be measured in the HydroNET-4 frontend. The whole benchmark environment is installed on a Dell Inc. OptiPlex 745 server. The sever has one Intel processor with 2 cores, 6600 at 2.4 GHz, 4 x 2 GB DDR2 RAM, 3 TB SATA 5400 rpm Western Digital hard disk.

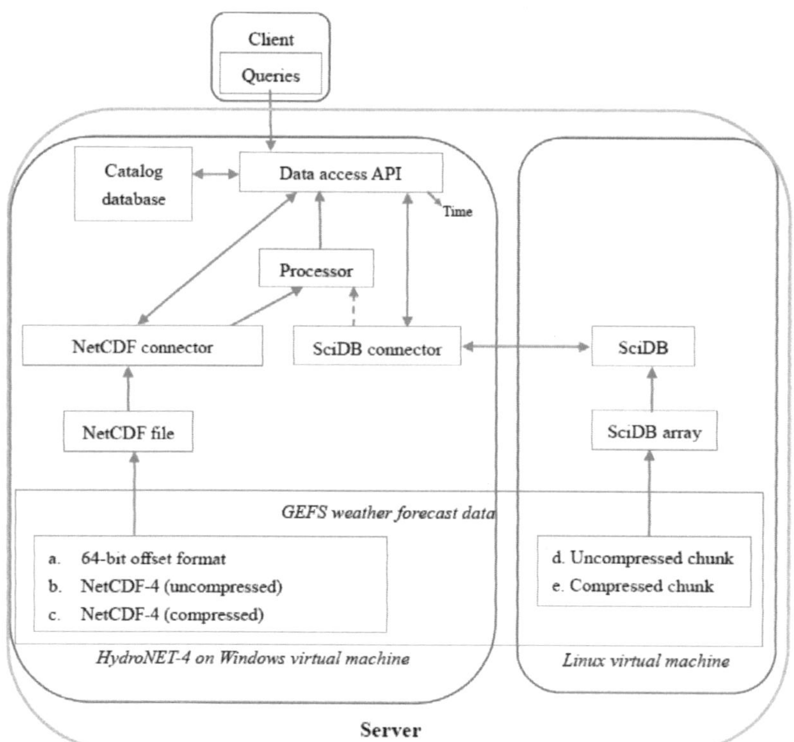

Figure 2. Benchmark architecture.

4. Results

Data used for benchmarking are stored in NetCDF 64-bit offset format, NetCDF-4 format and SciDB separately (table 2). In table 2, the last dimension in the second column is the outer most dimension to organize the storage of data. S1, S2, S3 and S4 represent 4 data schemes, i.e. structures for GEFS sample storage caused by modifying the order of dimensions. The chunk sizes of these 4 arrays keep at the same modest level, i.e. 1 value in the model run and forecast dimension, 20 values in the ensemble dimension, 181 values in the Y dimension and 360 values in the X dimension. To fully investigate whether data organization inside the chunk has an influence on query performance, GEFS sample is additionally stored into one chunk but with 4 different dimension orders, i.e. C1, C2, C3 and C4. The compression refers to DEFLATE method. All arrays store the same 8 attributes (table 1). In S5 scheme, one chunk only contains a spatial grid, which is a decision comes from recommendations of earlier researches [3].

Table 2. Storage information of NetCDF solutions and SciDB.

Array name	Dimension part in the schema	Chunk size	Chunk count	Average chunk storage size	Total storage size
NetCDF_64bit_offset	-	-	-	-	1.55 GB
NetCDF4_S3	(X=360, Y=181, Forecast=1,Ensembe=20,Modelrun=1)	360 x 181 x 1 x 20 x 1	40	40 MB	1.55 GB
NetCDF4_S3_C (Compression)	(X=360, Y=181, Forecast=1,Ensembe=20,Modelrun=1)	360 x 181 x 1 x 20 x 1	40	16.4 MB	654 MB
NetCDF4_S5	(X=360, Y=181, Forecast=1,Ensembe=1,Modelrun=1)	360 x 181 x 1 x 1 x 1	800	2 MB	1.55 GB
NetCDF4_S5_C (Compression)	(X=360, Y=181, Forecast=1,Ensembe=1,Modelrun=1)	360 x 181 x 1 x 1 x 1	800	0.7 MB	561 MB
SciDB_S1	[M_idx=0:675,1,0,E_idx=0:19,20,0,F_idx=0:39,1,0, Y_idx=0:180, 181,0,X_idx=0:359,360,0]	1 x 20 x 1 x 181 x 360	40	6.7 MB	268 MB
SciDB_S1_C (Compression)	[M_idx=0:675,1,0,E_idx=0:19,20,0,F_idx=0:39,1,0, Y_idx=0:180, 181,0,X_idx=0:359,360,0]	1 x 20 x 1 x 181 x 360	40	2.16 MB	86.4 MB
SciDB_S2	[M_idx=0:675,1,0,F_idx=0:39,1,0,Y_idx=0:180,181, 0,X_idx=0:359,360,0,E_idx=0:19,20,0]	1 x 1 x 181 x 360 x 20	40	6.2 MB	247 MB
SciDB_S2_C (Compression)	[M_idx=0:675,1,0,F_idx=0:39,1,0,Y_idx=0:180,181, 0,X_idx=0:359,360,0,E_idx=0:19,20,0]	1 x 1 x 181 x 360 x 20	40	2 MB	79 MB
SciDB_S3	[X_idx=0:359,360,0,Y_idx=0:180,181,0,F_idx=0:39, 1,0,E_idx=0:19,20,0,M_idx=0:675,1,0]	360 x 181 x 1 x 20 x 1	40	6.2 MB	246.5 MB
SciDB_S4	[E_idx=0:19,20,0,X_idx=0:359,360,0,Y_idx=0:180,1 81,0,F_idx= 0:39,40,0,M_idx=0:675,1,0]	20 x 360 x 181 x 1 x 1	40	5.8 MB	232.4 MB
SciDB_C1	[M_idx=0:675,1,0,E_idx=0:19,20,0,F_idx=0:39,40,0, Y_idx=0:180, 181,0,X_idx=0:359,360,0]	1 x 20 x 40 x 181 x 360	1	268 MB	268 MB
SciDB_C2	[M_idx=0:675,1,0,F_idx=0:39,40,0,Y_idx=0:180,181 ,0,X_idx=0:359,360,0,E_idx=0:19,20,0]	1 x 40 x 181 x 360 x 20	1	246.7 MB	246.7 MB
SciDB_C3	[X_idx=0:359,360,0,Y_idx=0:180,181,0,F_idx=0:39, 40,0,E_idx= 0:19,20,0,M_idx=0:675,1,0]	360 x 181 x 40 x 20 x 1	1	246.5 MB	246.5 MB
SciDB_C4	[E_idx=0:19,20,0,X_idx=0:359,360,0,Y_idx=0:180,1 81,0,F_idx= 0:39,40,0,M_idx=0:675,1,0]	20 x 360 x 181 x 40 x 1	1	250 MB	250 MB

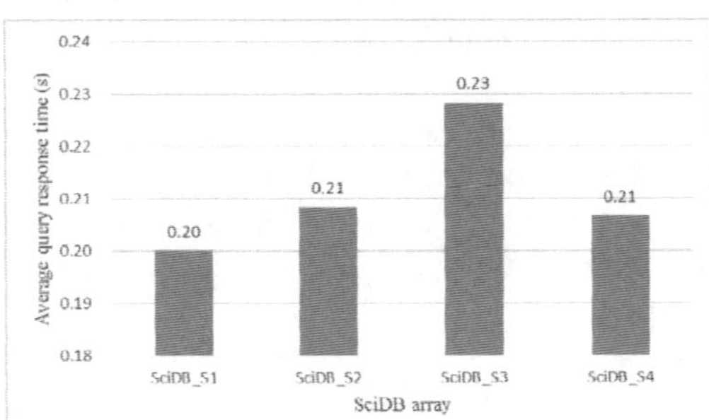

(a)

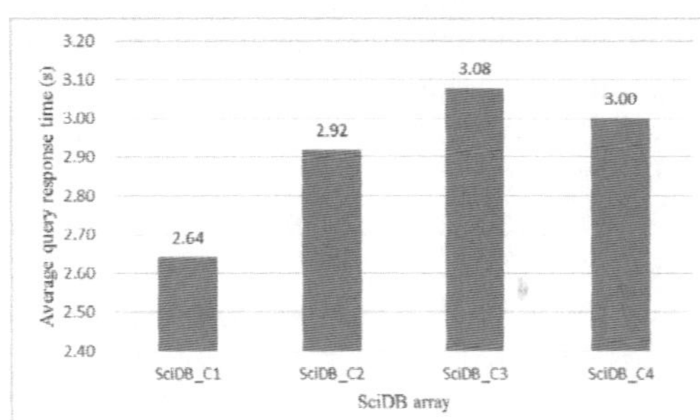

(b)

Figure 3. Performance of Q1, i.e. selecting all ensembles of total precipitation in Delft at one forecast step, (a) each array has 40 chunks (b) each array has only one chunk.

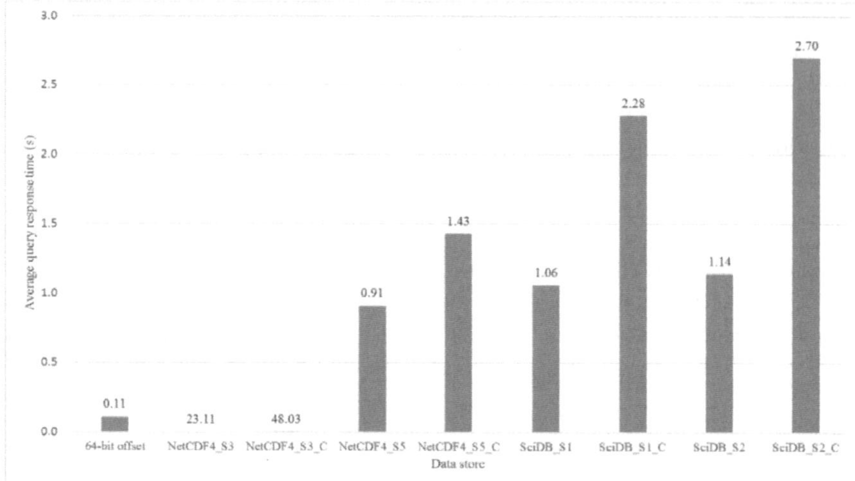

Figure 4. Performance of Q2, i.e. extracting forecast time series of total precipitation.

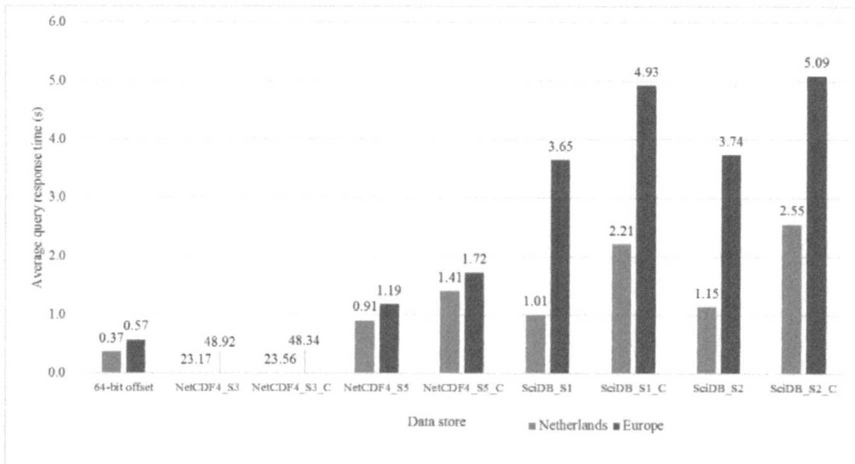

Figure 5. Performance of Q4, i.e. calculating ensemble mean of total precipitation at Netherlands and Europe scale.

Regarding benchmarking, a query is executed 20 times alternating between different storage solutions. The final number for query response time is the average of the middle 12 time records with the largest 4 and the smallest 4 values removed for each solution. Benchmark performance for different solutions is provided in figure 3, 4 and 5.

It is commonly believed that retrieving values which are stored close to each other on the disk should be faster than values stored discontinuously. So considering dimension order, SciDB_S1 should take less time to execute Q1 than SciDB_S2 while SciDB_S4 should response faster than SciDB_S3. Besides, SciDB_S1 and SciDB_S4 should both response more quickly than either SciDB_S2 or SciDB_S3. This is verified in figure 3. Referring to query execution records, SciDB_S1 outperforms SciDB_S2 by 8 ms, while SciDB_S3 is slower than SciDB_S4 by 21 ms. However, the gap is not significant due to the short query execution time. To exclude the effect of chunk size on performance, GEFS dataset is managed into 4 other arrays. Then, effect of dimensions order on query performance becomes more significant (figure 3b). However, an exception appears that SciDB_C2 in which ensemble values in a single location at one forecast step are stored discontinuously responses faster than SciDB_C4 where ensemble values are stored successively. High dimensionality of GEFS dataset adds much complexity to the mapping from dimensions order to data organization on disk and this patterns is still not fully understandable.

Nevertheless, from complementary queries tested, dimension order can indeed influence query performance to a certain extent.

In figure 4, performance of NetCDF-4 solutions with schema S3 are not visualized for their high values. On the whole, 64-bit offset solution performs the best and is 9 time faster than NetCDF-4 with smaller chunk size which comes the second. SciDB arrays without compression take more time to execute the query than 64bit offset solution, but the noise included in the measurements cannot be neglected (refer to section 3). SciDB_S1 takes less time than SciDB_S2 to process the query while the gap is insignificant. For their compressed versions, i.e. SciDB_S1_C and SciDB_S2_C, the difference becomes more obvious. The compression of SciDB arrays doubles the query response time compared with non-compressed arrays. This is also the case for NetCDF4_S3_C. The compression on NetCDF4 S5 solution however does not cause so much degradation. The high response time of NetCDF-4 S3 stores is due to little caching. Specifically, the buffered data of NetCDF4 S3 stores are flushed by caching of other NetCDF stores. While NetCDF-4 with smaller chunk size, i.e. scheme S5 can benefit from caching as presented in the figure, a possible explanation might be the chunk size of NetCDF-4 can influence the behaviour of caching as well. But more tests need to be implemented to verify it. Result of Q3 is not shown as it presents analogous patterns as Q2, but as more computation is involved, performance of SciDB experiences much degradation.

Regarding Q4, to investigate the scalability of different solutions, two locations are used for precipitation ensemble mean calculation at all 40 forecast steps (figure 5). One is the Netherlands containing 20 cells (5 x 4) and the other Europe scope has 3888 cells (72 x 54) in GEFS dataset. For the calculation of either query area, the pattern of performance keep similarity to Q2 and Q3. The effect of dimensions order is insignificant for SciDB arrays. Thanks to caching, compression of NetCDF4 S5 store cause less negative impact on query performance compared with SciDB arrays. NetCDF-4 solutions with larger chunk size, i.e. S3 do not benefit from caching which is undermined by query execution on other data stores. In general, NetCDF4 S5 data stores present the best scalability and with the query area increases 200 percent, the query response time only grows 20 to 30 percent. The query processing time of 64-bit offset solution experiences a 54 percent increase. With little caching, NetCDF-4 S3 data stores cost twice the time to on the Europe scope compared to the Netherlands scope. The scalability of SciDB solutions rank the last. This is mostly because the realization of ensemble mean calculation in SciDB connector is a hybrid function using two operators, i.e. "aggregate" and "between", which adds more complexity to query execution.

5. Conclusions and future work

Query results show that exchanging the order of dimensions inside each chunk plays significant role in query performance. But when chunk size reduces to a modest level, the impact of dimension order decreases as well. It is thus suggested that the research emphasis on chunked storage structure should move to designing appropriate chunk sizes or smartly indexing chunks instead of data structure and indexing inside chunks. Compression in most cases causes degradation especially for NetCDF-4 solutions with large chunk sizes. Caching can definitely improve performance, but the positive effect may be insignificant as cached data can be flushed by other query processes. Realization of hybrid function is a drawback for SciDB solutions, and a possible solution is to move complex calculation to the Processor, i.e. computation on top of SciDB, a hybrid solution. On the whole, NetCDF 64-bit offset solution presents the highest efficiency.

In regard to existing flaws as well as more interests inspired by the research, extension work can be conducted. Typical directions are:

- Chunk model. Previous researches [3] and this study demonstrates that the chunk size is perhaps the most crucial factor for efficiency of chunked storage structure. On the other hand, the effect of chunk size is unpredictable and for example, at which chunk size a specific query performance experience significant tuning point. Regarding this, intensive experiments can draw specific conclusions while more generic understanding relies on theoretical support. A possible way is to propose a model concerning important factors such as hardware configuration, data size and query habits to predict the performance of chunks. After this, the model should be verified by comprehensive benchmark tests.

- "Hot" and "cold" query tests. Normal approaches used to measure the average query response time include two types. "Hot" test means that the same query is executed on each data store consecutively several times and then average time is calculated. Caching effect is elaborated. As the opposite, "cold" test purges the cache every time before executing each query. Lower boundary of query response time comes from "hot" test while upper boundary depends on "cold" test. The method used in this research is a state close to "hot", i.e. cache is not cleaned for a specific data store but can be flushed by caching of other data stores. Honestly, none of these 3 approaches reflect the reality. A more scientific way might be through analysing query logs recorded to learn what users query in reality and try to simulate those scenarios for testing.
- Parallelization. Parallel query processing of SciDB is not explored in the research. Also the research demonstrates the favorable capability of NetCDF solutions without parallelization. Then a potential topic in the future is to compare parallel query performance between SciDB and NetCDF solutions. Besides, as data loading is indeed a problem for SciDB, another direction can be parallel loading techniques.

References

[1] Rew R, Hartnett E and Caron J 2006 NetCDF-4: Software implementing an enhanced data model for the geosciences *22nd Int. Conf. on Interactive Information Processing Systems for Meteorology Oceanograph and Hydrology* (Atlanta GA: USA)

[2] Brown P G 2010 Overview of SciDB: large scale array storage, processing and analysis *Proc. of the 2010 ACM SIGMOD Int. Conf. on Management of Data* (Indianapolis IN: USA/ACM SIGMOD Record, ACM Press) p 963-68

[3] Lee C, Yang M and Aydt R 2008 NetCDF-4 Performance Report Retrieved from http://www.hdfgroup.org/pubs/papers/2008-06_netcdf4_perf_report.pdf

[4] Colliat G 1996 OLAP, relational, and multidimensional database systems *ACM SIGMOD Record* **25** p 64-9

[5] Seering A, Cudre-Mauroux P, Madden S and Stonebraker M 2012 Efficient versioning for scientific array databases *Data Engineering (ICDE) 2012 IEEE 28th Int. Conf.* (Arlington VA: USA) p 1013-24

[6] Reichard L, Lobbrecht A, Clark S, Catalano C, Tate B andCox D 2014 Supporting water managers making effective decisions by using HydroNET *35th Hydrology and Water Resources Symp.* (Australia) p 710-17

Acknowledgments

Financial supports for this research from the National Soft Science Research Program (2012GXS2B008), the Fundamental Research Funds for Central Public Welfare Research Institutes (CKSF2015018/KJ, CKSF2016263/KJ) and the Grant Projects of China Clean Development Mechanism Fund (2013015) are gratefully acknowledged.

IOP Conf. Series: Earth and Environmental Science **46** (2016) 012032 doi:10.1088/1755-1315/46/1/012032

Remote sensing-based analysis on temporal and spatial changes about environmental elements in the northwest of Junggar basin, China

Liu Yang[1] , Zhang Nannan[1] , Dong Wentong[1], Zou Liqun[1] , Huang Shanghong[2]

[1] Institute of Petroleum Exploration & Development, PetroChina, Beijing, China
[2] PetroChina Exploration & Production Company, Beijing, China

E-mail: liuyang_rs@petrochina.com.cn

Abstract. This paper presents a study of revealing the environmental elements change during the process of local industrialization based on remote sensing technique in the western part of China. Spatio-temporal evolution of vegetation cover derived from NDVI and land surface water distribution was analyzed by time-series analysis of MSS and Landsat data from 1977 to 2011. Results show that remote sensing provide a way for monitoring the influence of local industrialization on regional environment elements in gobi region.

1. Introduction

The western China covers vast territory with the typical landform of gobi desert, exists a mass of energy enterprises. Energy industry has great significance on promoting local industrialization, especially in the northwest of Junggar basin, China. Population growth and economic development occurred simultaneously with increasingly rapid development of energy production. Regional environmental elements are affected by human activity and undergoing continuous change. Due to the large geography extent, inaccessible parts and low population density, it is essential for environment supervision to timely acquire the distribution information of base environmental elements and track their evolution during the industrialization process. Periodic monitoring is required to control the possible and potential environment problems.

Vegetation and land surface water are two fundamental elements of ecological environment. It is important for characterizing the regional environment state by obtaining the overview of their spatial and dynamics. Remote sensing provides a complement way to accomplish the monitoring on regional vegetation cover and land surface water distribution[1-6]. This is beneficial to get better understanding of local environment feature and its evolution.

With a large coverage, short repeatability, convenient acquisition and long term data series, Landsat TM imagery has been proven to be a useful tool for field environmental monitoring. The aim of this study was to reveal the spatio-temporal evolution of regional environment based on vegetation cover and land surface water during the industrialization process by remote sensed technique.

In this paper, we present results obtained from time-series analysis performed over selected Landsat imageries. The paper is organized as follows. Section 2 focused on the description of the research conducted. Section 3 illustrated the results and discussion. Remarks and conclusions were summarized in section 4.

2. Materials and methodology

2.1. Study and dataset

The study area is located in the northwest margin of Junggar basin in the west of China, with the typical landform of gobi desert, between longitudes 84° 30′ E and 85° 30′ E, latitudes 45° N and 46° N. After the discovery of China's first large scale oil field in Karamay in 1955, oil industry has been the main industry of local economic development. This area consists of a wide range of ecosystem types, including desert communities, wetland, grazed grassland, irrigated agriculture and urban land cover.

In this study, the image time series of the MSS and Landsat between 1977 and 2011 was exploited. In order to analyze the trend of environment variation, the time series covers the same month of September except MSS image for the reason of poor data quality. Five different temporary of satellite images were selected which acquired on 11 July 1977, 2 September 1998, 21 September 2002, 11 September 2007 and 6 September 2011 respectively.

The image processing was performed by the following steps. Firstly, the reflectance-based multispectral image was derived based on image. Then, Vegetation cover extraction was performed by the thresholding setting based on NDVI index. Finally, water extraction was performed by the combined application of NDVI and NDWI. MSS image was not used to extract water information for the haze condition and low resolution.

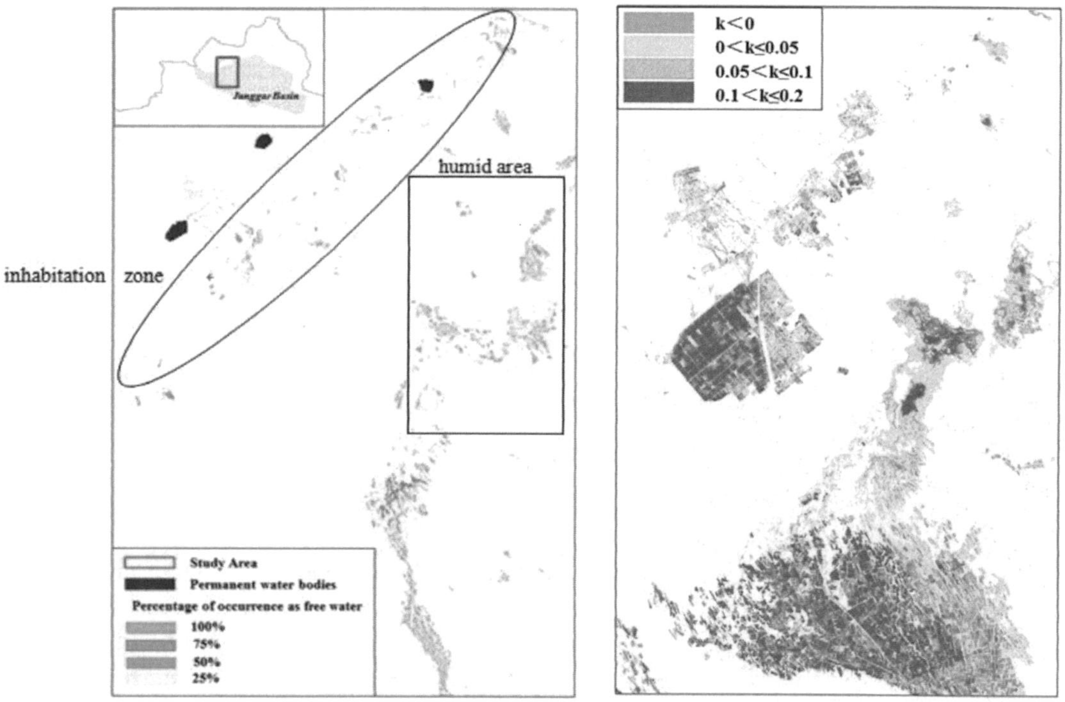

a b

Figure 1. Mapping of results derived by time-series analysis
-a. Spatial distribution of surface water recurrence in September from 1998 to 2011;
-b. Spatial distribution of vegetation cover variation tendency from 1977 to 2011.

2.2. Methodology

2.2.1. Time series analysis on land surface water

In the research, land surface water is classified into two categories of permanent water body and free water body[7]. Permanent water body is mainly the stable water like the water in the reservoir, while free water body is mainly the natural water like river and so on. The analysis of the time series is concentrated on free water.

Percentage of occurrence as free water (O_{fw}) is used to analyze the regional land surface water environment. O_{fw} is defined as the following:

$$O_{fw} = n_w / n_{nw} \tag{1}$$

where n_w is the number of the pixel detected as free water, and n_{nw} is the number of the time-series image used. The map of regional surface water distribution state was obtained by thresholding setting of O_{fw}.

2.2.2 Time series analysis on vegetation cover

For vegetation cover data derived from NDVI, simple linear regression method[8-9] was used to make time-series analysis on the trend of spatial distribution of vegetation. The slope k of the regression equation was defined as:

$$k = \frac{\sum_{i=1}^{n}(x_i - \bar{x})(y_i - \bar{y})}{\sum_{i=1}^{n}(x_i - \bar{x})^2} \tag{2}$$

where x_i and y_i are acquisition year of image and NDVI value of vegetation cover respectively, $\bar{x}$ and $\bar{y}$ are mean value of x_i and y_i of time-series data, i is the order of satellite image sequence, and n is the number of images. The slope of k reflects the trend of vegetation cover variation over years. The value of k indicates the degree of vegetation cover improvement when $k > 0$ and the degree of vegetation cover degradation when $k < 0$. The map of vegetation cover distribution variation tendency was obtained by thresholding setting of k.

3. Results and discussion

Figure 1.a shows the state of surface water distribution by time-series analysis of Landsat data in September from 1998 to 2011 within the study area. Permanent water is the water in the reservoir which one of them existed in 1998 and the other two were built after 2000 along the inhabitation communities. Spatial distribution of actual surface water occurrence exhibits slow increasing of permanent water around inhabitation area and the seasonal recurrence of water bodies in humid regions.

Figure 1.b shows the distribution of vegetation cover evolution from 1977 to 2011. The statistical results of vegetation cover percentage with different variation degree are summarized in table 1. Degradation percentage of vegetation cover is only 3.19%. It mainly occurred along the inhabitation zone. Vegetation cover presents a significant and long-lasting increasing over 34 years. Given the increasing geographic extent of vegetation cover throughout region, the expanding trend shows the correlation with the distribution of inhabitation communities and irrigated agricultural region related to the activities of regional industrialization.

Table 1. Statistics of vegetation cover variation tendency

k	variation	pixels	percentage
k<0	Degradation	36087	3.19
0<k≤0.05	Slight improvement	421800	37.25
0.05<k≤0.1	Middle improvement	306116	27.04
0.1<k≤0.2	Remarkable improvement	368288	32.53

In the research, satellite image sequence includes five temporal data over 34 years. The impact of uneven time interval of the sequence and inadequate images on the time-series analysis should be taken into account. The accuracy of spatio-temporal evolution analysis depends to a great extent on the design of a scientific sequence with consideration of factors corresponding to the specific objective.

4. Conclusions

By using MSS and Landsat data, Spatio-temporal evolution of regional environment was analyzed based on elements of vegetation cover and land surface water with time-series analysis. The temporal consistency between variations in vegetation cover and land surface water distribution implies the environment improvement during the industrialization process within study area. Research demonstrated that big data in remote sensing can be of great help in monitoring the influence of local industrialization on regional environment elements in the western part of China by time-series analysis.

References

[1] Alsdorf ,D.E, Rodriguez.,Lettenmaier,D.P. Measuring surface water from space. *Reviews of Geophysics.* 45 (RG2002).

[2] Caplar R and Kulisic P 2009 Selective logging changes forest phenology in the Brazilian Amazon: Evidence from MODIS image time series analysis *Remote Sensing of Environment* 113(11) p 2431-2440

[3] Clinton N E, Potter C, Crabtree B, Genovese V and Gross P 2010 Remote Sensing-Based Time-Series Analysis of Cheatgrass (Bromus tectorum L.) Phenology *Journal of Environmental Quality* 39(39) p 955-963

[4] Fang Jingyun, Piao Shilong, He Jinsheng, et al. In recent 20 years the activities of vegetation in China is enhanced. *Science in China(Series C)*33(6)p554-565

[5] Frans Van Den Bergh, Konrad J. Wessels, Simeon Miteff, Terence L. Van Zyl and Albert D. Gazendam 2012 HiTempo: a platform for time-series analysis of remote-sensing satellite data in a high-performance computing environment *International Journal of Remote Sensing* 33(15) p 4720-4740

[6] Bhandari S, Phinn S and Gill T 2012 Preparing Landsat Image Time Series (LITS) for Monitoring Changes in Vegetation Phenology in Queensland, Australia *Remote Sensing* 4(12) p 1856–1886

[7] Haas E M, Bartholome E and Combal B 2009 Time series analysis of optical remote sensing data for the mapping of temporary surface water bodies in sub-Saharan western Africa *Journal of Hydrology* 370(1) p 52-53

[8] Han Xiuzhen, Li Sanmei, Luo Jingning, Ji Xiang. Study on Spatiotemporal Change of Vegetation in China since 20 years. *Arid Zone Research* 25(6)p 753-759

[9] Wang Dianlai, Liu Wenping, Huang Xinyuan. Trend analysis in vegetation cover in Beijing based on Sen+Mann-Kendall method. *Computer Engineering and Applications.*

Wave height possibility distribution characteristics of significant wave height in China Sea based on multi-satellite grid data

W Han and J Yang

The First Institute of Oceanography, SOA, Qingdao, China
E-mail: yangjg@fio.org.cn

Abstract. This paper discusses the group of wave height possibility distribution characteristics of significant wave height in China Sea based on multi-satellite grid data, the grid SWH data merges six satellites (TOPEX/Poseidon, Jason-1/2, ENVISAT, Cryosat-2, HY-2A) corrected satellite altimeter data into the global SWH grid data in 2000~2015 using Inverse Distance Weighting Method. Comparing the difference of wave height possibility distribution of two schemes that scheme two includes all of 6 satellite data and scheme one includes all of other 5 satellite data except HY-2A in two wave height interval, the first interval is [0,25）m, the second interval is [4,25）m, finding that two schemes have close wave height probability distribution and the probability change trend, there are difference only in interval [0.4, 1.8）m and the possibility in this interval occupies over 70%; then mainly discussing scheme two, finding that the interval of greatest wave height possibility is [0.6, 3）m, and the wave height possibility that the SWH is greater than 4m is less than 0.18%.

Keywords: Significant wave height; China Sea; Satellite grid data; Possibility distribution

1. Introduction

Significant wave height (SWH), defined as the crest-to-trough height of the 1/3 highest waves in the sea wave field of view, is an important parameter of ocean research and marine environment forecast. SWH measurements are used for a variety of purposes, these studies improve the understanding of the physical processes responsible for wind-wave evolution to the validation and calibration of models for waves and other ocean processes as well as wave climate investigations, which have implications for shipping and offshore engineering projects[1].

Wave data over periods of many years are necessary for estimating the wave climate all over the world, traditionally, data coverage over the ocean has been poor, and wave observations have come from ships and moored buoys clustered around coastlines. During the 1970s, global monitoring of SWH became a reality and resulted in the first global atlas of SWH data derived from satellite observation [6]. Since then, the SWH data from several different satellites have become available, including SEASAT (1978), GEOSAT (1985-1990), TOPEX/POSEIDON (1992-2006), ERS-1 (1991-2000), ERS-2 (1995-2011), ENVISAT (2002-2012), Jason-1 (2002-2013)[7], Jason-2 (2008-present), Cryosat-2 (2010-present), HY-2 (2011-present), Saral/Altika (2013-present), Jason-3(2016.01.17), SENTINEL-3A (2016.02.16)[2].

In this paper, first, introducing the satellite altimeter data and multi-satellite grid data, next, representing wave height possibility statistics analysis method, then, comparing the difference of wave height possibility distribution of two schemes and mainly analyzing the wave height possibility of scheme two, finally, the last part states some conclusions.

2. Data

Merging SWH data comes from 6 satellite altimeter data that are distributed by the various space agencies, the detailed information of satellite altimeter is given in Table1[1]. SWH data from these satellite altimeters is corrected through regression equations, merging corrected SWH data into global grid SWH data using Inverse Distance Weighting(IDW) method[3,4], the grid data starts Jan. 1, 2000, ends to Sep. 1, 2015, the spatial resolution is 0.25° × 0.25° [5], temporal resolution is one day. Selecting the grid data in China Sea that is 100~130° E, 0~40° N.

Table 1. Basic Information for satellite radar altimeter[2]

Satellite	T/P	Jason-1	ENVISAT	Jason-2	Cryosat-2	HY-2A
Operational Period	1992.8.10-2005.10.18	2001.12.7-2013.07.03	2002.03.01-2012.04.08	2008.06.20-	2010.04.08-	2011.08.16-
Data Period	1992.09.25-2005.0924	2002.01.15-2012.12.11	2002.05.17-2012.04.08	2008.07.04-2015.09.01	2010.07.16-2014.09.12	2011.10.01-2015.08.16
Orbital Altitude/km	1336	1336	799.8/782.4	1336	717	971/973
Inclination /°	66	66	98.5	66	92.03	99.34
Spacing at equator /km	315	315	80	315	7.5	207.6/18.3
Repeat /days	9.9156	9.9156	35/30	9.9156	369	14/168
Institute	NASA,CNES	NASA,CNES	ESA	NASA,CNES	ESA	NSOAS

3. Methods

In order to possibility statistics analysis[8], first defining statistics parameters, assume M_h is the total number that SWH belongs to [h, 25) m, $[SWH_{h,i-1}, SWH_{h,i})$, i=1,2,..., I_h is SWH the interval [h,25) m(the maximum measurement value of satellite altimeter SWH is 25m), I_h is the total number of interval [h, 25) m, $SWH_{h,i-1}$ and $SWH_{h,i}$ are the left and right SWH threshold value in the group i in the interval [h, 25) m, $N_{h,i}$ is the number of SWH in the group i in interval [h, 25) m, $P_{h,i}$ is the frequency of SWH in the group i in interval [h, 25) m , $F_{h.,i}$ is the accumulative frequency of SWH in the group i in interval [h, 25) m, the expressions are as follows:

$$N_{hi} = 0 + \sum_{j=1}^{M_{hi}} 1, SWH_{hi-1} \le SWH_{hj} < SWH_{hi} \tag{1}$$

$$P_{hi} = \frac{N_{hi}}{M_{hi}} \tag{2}$$

$$F_{hi} = \sum_{i=1}^{I_h} P_{hi} \tag{3}$$

using above statistical models discuss probability distribution of merging grid SWH, the internal [0,25) m and [4,25) m are selected, that is $h_1=0$, $h_2=4$.

4. Results and discussions

Using IDW method merges 6 satellite altimeters according to two schemes, one scheme is foreign satellite altimeter data including T/P, Jason-1/2, ENVISAT and Cryosat-2, other scheme adds HY-2A base on scheme one.

Selecting grid SWH data that starts Oct.1, 2011 due to HY-2A data starts from Oct. 1, 2011. Here h=0, the interval is [0, 25) m, grouping is that using interval range 0.2m divide the interval [0,4.8) m into 24 groups, the interval [4.8,25)m is the 25th group. Calculating the wave height possibility of two schemes and its possibility difference, it is showed in figure 1, it reveal two schemes have the close wave height probability distribution and the probability change trend, there are difference only in interval [0.4,1.8) m, the possibility in this interval occupies over its' total 70%, the wave height possibility of scheme two is less than the wave height possibility of scheme one in this interval, the result is opposite in interval [1.0,1.8) m, this phenomenon demonstrates that the wave height probability in interval [0.4,1.0) m is becoming lower and is becoming higher in interval [1.0,1.8) m.

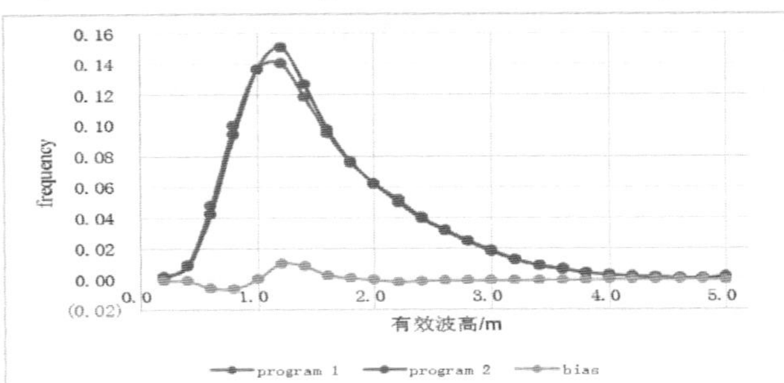

Figure 1. wave height possibility distribution of two schemes

Analyzing the SWH that the wave height is greater than 4m, here h=4, the group interval is [4, 25) m, the grouping is that using interval range 1.0m divide the interval [4, 25) m into 21 groups, calculating the wave height possibility of two schemes and its possibility difference as show in Figure 2, there are difference in the interval [4, 7) m between two schemes, this interval is the interval of higher SWH, the wave height possibility of scheme two is greater than it of scheme one in interval [4, 5) m, whereas, scheme two is becoming smaller in interval [5, 7) m. Above all, the wave height possibility is greater in interval [4, 5) m after adding HY-2A and is becoming smaller in other interval.

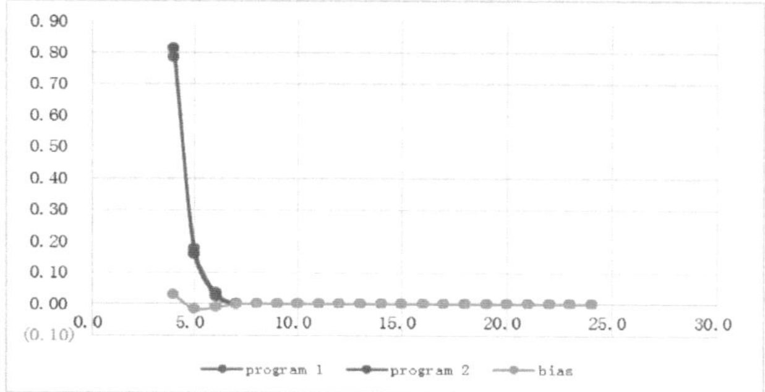

Figure 2. wave height possibility distribution of two schemes in interval [4,25)m

Above the discussion between two schemes could find that the difference of SWH in the interval [0, 2) m after adding HY-2A, the change trend of bias like Sine function. New HY-2A data could illustrate wave height possibility, so, it is essential to analyze the long term wave height possibility distribution of scheme two. The wave height possibility of SWH shows in Figure 3, data starts from Oct. 1, 2000 to Sep. 12, 2015 is close 16 years, the SWH mainly distributes the interval [0.6, 3) m, and the sub-peak distribution interval is [0, 0.2) m, the greatest wave height possibility is greater than 14%, the interval is [1,1.2) m .

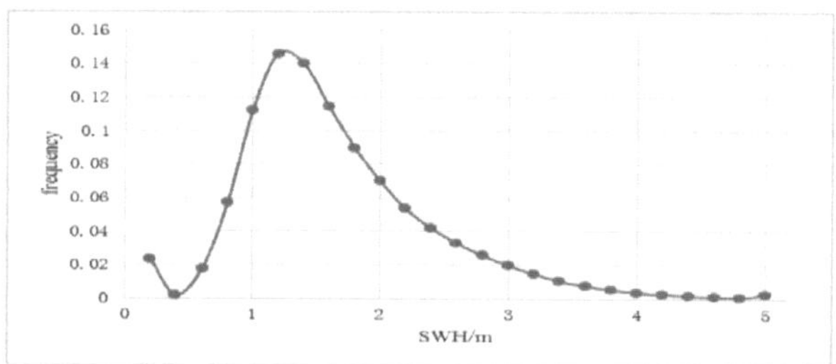

Figure 3. wave height possibility distribution in the interval [0,5)m in 2000~2015

Figure 4 mainly discusses that the wave height possibility of every interval is less than 0.18% while SWH is greater than 4m, and is almost 0 while SWH is greater than 8m, so, in general, we always focus on the SWH that is lower than 8m, above is the reason.

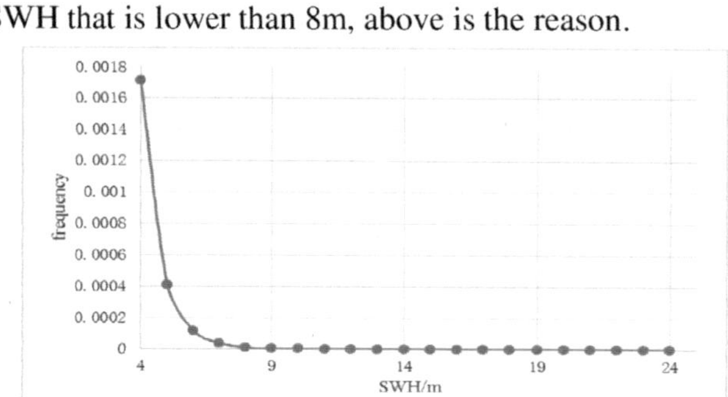

Figure 4. wave height possibility distribution in the interval [4,25) m in 2000~2015

5. Conclusions

This paper discusses the wave height possibility of two SWH grid schemes in two intervals, finding that the interval of greatest wave height possibility is [0.6, 3) m, and the wave height possibility that the SWH is greater than 4m is less than 0.18%. Above two schemes have the close probability distribution and the probability change trend, there are difference only in interval [0.4, 1.8) m, the possibility in this interval occupies over 70%; the possibility is greater in interval [4, 5) m and smaller in other interval after adding HY-2A; main distribution interval of SWH is [0.6, 3) m and possibility is less than 0.18% in interval [4, 7) m.

References

[1] Birol F Roblou L Lyard F et al 2006 Towards using satellite altimetry for the observation of coastal dynamics *Proceedings of Symposium on 15 Years of Progress in Radar altimetry* 614.

[2] Li J and Jin T 2013 On the main progress of satellite altimetry and its applications *Journal of Geomatics* **38(4)** 1-8

[3] Queffeulou P Croizé-Fillon D Queffeulou P et al 2010 Global altimeter swh data set Version 7 *Biochemistry & Oral Biology* **100(1)** 1-4

[4] Li X Yuan F Jia C et al 2012 Contrastive study of s-a multifractal filtering method based on inverse distance weighted and kriging interpolation *Science of Surveying & Mapping*

[5] Shanas P Kumar V Hithin N 2014 Comparison of gridded multi-mission and along-track mono-mission satellite altimetry wave heights with in situ near-shore buoy data *Ocean Engineering* **83(2)** 24-35

[6] Mcmillan J 1981 A global atlas of geos-3 significant wave height data and comparison of the data with national buoy data

[7] Durrant T Greenslade D Simmonds I 2009 Validation of jason-1 and envisat remotely sensed wave heights *Journal of Atmospheric & Oceanic Technology* **26(1)** 123-134

[8] Shanas P Kumar V Hithin N 2014 Comparison of gridded multi-mission and along-track mono-mission satellite altimetry wave heights with in situ near-shore buoy data *Ocean Engineering* **83(2)** 24-35

Acknowledgements

The authors thank the European Space Agency for providing ENVISAT RA-2 altimeter data, National Aeronautics and Space Administration (NASA) for providing T/P, Jason-1/2 data, National Data Buoy Center (NDBC) in the National Oceanic and Atmospheric Administration (NOAA) for providing NDBC buoys and National Satellite Ocean Application Service of State Oceanic Administration for providing HY-2 data. This study is supported by National 863 Program (No.2013AA122803), the National Marine Public Welfare Industry Special Funding Scientific Research Program (No.201305032) and Chinese National Natural Science Foundation (No.41406007).

Growth pattern research on the modern deposition of Ganjiang delta in Poyang lake basin by spatio-temporal remote sensing images

Hongying Zhou*, Xuanjun Yuan, Youyan Zhang, Wentong Dong, and Song Liu

Institute of Petroleum Exploration & Development, PetroChina, Beijing 100083, China

E-mail: zhouhys@petrochina.com.cn

Abstract. It is of great importance for petroleum exploration to study the sedimentary features and the growth pattern of shoal water deltas in lake basins. Taking spatio-temporal remote sensing images as the principal data source, combined with field sedimentation survey, a quantitative research on the modern deposition of Ganjiang delta in the Poyang Lake Basin is described in this paper. Using 76 multi-temporal and multi-type remote sensing images acquired from 1973 to 2015, combined with field sedimentation survey, remote sensing interpretation analysis was conducted on the sedimentary facies of the Ganjiang delta. It is found that that the current Poyang Lake mainly consists of three types of sand body deposits including deltaic deposit, overflow channel deposit, and aeolian deposit, and the distribution of sand bodies was affected by the above three types of depositions jointly. The mid-branch channels of the Ganjiang delta increased on an exponential growth rhythm. The main growth pattern of the Ganjiang delta is dendritic and reticular, and the distributary channel mostly arborizes at lake inlet and was reworked to be reticulatus at late stage.

Key words: Remote sensing, Modern deposition, Growth pattern, Ganjiang delta, Poyang Lake.

1. Introduction

Over recent years, the petroleum exploration in continental basins of China has revealed that typical fluvial dominated shoal water deltas is developed in large depression lake basins. The researches on modern deposition of shoal water deltas can build petroleum exploration depositional pattern with more comparable bases. With respect to its advantages (such as wide imaging scope and fine details), remote sensing is taken as an important base for modern deltaic deposit survey. According to its multi-temporal and multi-stage characteristics like historical filing and repeated monitoring, remote sensing is also utilized to quantitatively and dynamically analyze the sedimentary features and growth patterns of modern deltas.

2. Data source

Ganjiang delta spreads in a fan shape horizontally. Its apex is located in the vicinity of Bayi Bridge, Nanchang City, Jiangxi Province, whereas its front presents an irregular arc, with two end points at Wucheng Town and Sanjiangkou Town. The lobe at front of the mid-branch is a delta lobe growing the fastest and preserved the most intact in recent years, therefore, it is selected as the key study area in this paper.

A total of 76 cloud-free images were collected from 1973 to 2015: including 17 MSS images, 2 SPOT images, 1 ALOS image, 1 QuickBird image, 2 GF1 images, 2 HJ images, 5 ZY images, 44 TM/ETM images and 2 ASTER images.

Table 1. Different types of Satellite images in Ganjiang Delta

Satellite Sensor	Temporal List
MSS (17Images)	19731224; 19751030; 19760427; 19761006; 19770703; 19770826; 19771106; 19790403; 19790702; 19790912; 19791105; 19810507; 19810928; 19830128; 19831011; 19831128; 19841106;
TM/ETM (44Images)	19890213; 19890715; 19930312; 19950326; 19950902; 19950918; 19951105; 19951207; 19960209; 19960904; 19961022; 19991210; 20000923; 20010606; 20010910; 20020929; 20030220; 20030308; 20041012; 20050929; 20060612; 20061103; 20070106; 20070802; 20080516; 20081210; 20090409; 20090511; 20090604; 20091026; 20091103; 20091221; 20100223; 20100615; 20100802; 20100818; 20100919; 20101005; 20101021; 20101106; 20131122; 20140109; 20150213; 20151011;
Others (15Images)	20010129ASTER; 20010214 ASTER ; 20071122ALOS; 20080327SPOT; 20111204QB; 20111224HJ; 20120324ZY3; 20120506HJ; 20130102ZY02C; 20130305SPOT6; 20131104ZY3; 20140220GF1; 20141014ZY3; 20141019ZY3; 20141208GF1;

how to extract useful information from these large data sets has become a difficult problem. data sieving, thematic information extracting and spatio-temporal rules mining are key research contents.

3. Methodology

Based on the sedimentary features of shoal water lake basin and Ganjiang delta of the Poyang

Lake, combined with the characteristics of remote sensing data, an analysis process of remote sensing survey technique for shoal water delta of modern lake basins is proposed in this paper. The whole process consists of four steps, including sieving and processing of remote sensing data, remote sensing interpretation of lake basin delta, remote sensing quantitative statistical analysis of delta growth element and analysis, and summing-up of delta growth pattern, as shown in Fig.1. In the first step, on the basis of sieving and preprocessing of image data, combined with field verification on sedimentation, remote sensing interpretation of lake basin delta is conducted. In the second step, based on the interpretation vectors of multi-temporal delta images, quantitative statistical analysis on delta growth elements, and quantitative superposition analysis on object oriented sedimentary facies are conducted. In the third step, based on quantitative statistical analysis, the growth and depositional patterns of delta are described.

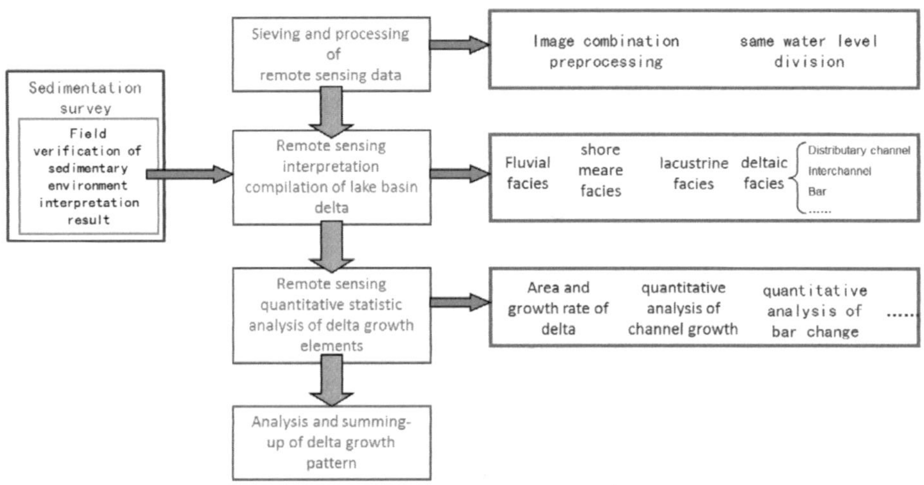

Fig.1 Analysis process of remote sensing survey technique for modern deposition of shoal water deltas in lake basins

3.1 Optimization and Preprocessing of Remote Sensing Data

(1) Image combination preprocessing

For the sake of studying the dynamic growth pattern of delta, it needs to try to collect different types of remote sensing images with the maximum time span. In this study, we collect and sort out 60 moderate resolution and 6 high resolution remote sensing images of Ganjiang delta in Poyang Lake basin got in 1973-2013. The moderate resolution landsat images are adopted for macro study, and the high resolution satellite images are used for local growth study. The data type of remote sensing image is primarily optical image, and the wave band scope covers visible light and near-infrared band.

Because there are differences of spectral signatures and spatial coordinates among remote sensing data of different satellite sensors, and they have larger influence to the quantitative analysis and superposition analysis on the images, a preprocessing method combining image fusion, geometric fine correction and image enhancement is adopted to reduce these discrepancies. The image fusion is accomplished using the IHS transformation fusion method, and the geometric fine correction is finished using the polynomial method. The combination preprocessing of images strengthens the resolution of remote sensing images, ensures the accurate spatial coordinate corresponding relationship among multi-temporal images, and makes the results of subsequent image interpretation and growth change quantitative analysis more reliable.

(2) Image water level gradation and optimization The water-level fluctuation of Poyang Lake is very large all the year round; the imaging time of different remote sensing images is different, the corresponding water level is different, and the exposure status of delta front is also different; as a result, dynamic quantitative correlation and analysis cannot be conducted directly, and it needs to conduct gradation and optimization on the water level of images. When the hydrologic data related to images are available, the real water level data should be used as the water level gradation criteria of all images. However, if the hydrologic data related to them are not available, it should, based on the water-level fluctuation status of Ganjiang delta of Poyang Lake all the year round, select relatively stable reference object to conduct water level gradation of images. The reference objects selected may include: ① the maximum lakeshore (embankment); ② delta front lobe; and ③ dragon's pool (exposure of dragon's head and tail). Based on the exposure status of reference objects on different images, the relative water levels of the images can be indirectly judged, and then, the water levels of the images are divided into different grades.

The same water level grade remote sensing images with constant duration within the scope of the maximum time span, including 5 moderate resolution (December 24, 1973; November 28, 1983; December 7, 1995; November 3, 2006 and November 22, 2013) and 3 high resolution (November 22, 2007; December 4, 2011 and March 5, 2013) images, are selected as the basic image data for follow-up analysis.

3.2 Remote Sensing Identification Chart for Sedimentary Microfacies

The Ganjiang delta of Poyang Lake is a typical open flow shoal water lake basin delta. Based on the sedimentary system of lacustrine delta, combined with the morphological characteristics spread of Ganjiang delta images, after having conducted preliminary analysis, it is believed that sedimentary microfacies like channel, interchannel, natural levee and crevasse splay are mainly developed in Ganjiang delta. All sorts of sedimentary microfacies have obviously different reflections on the remote sensing images. Fine correction and enhancement processing are conducted for different types and different-temporal remote sensing images, and systematic observation and description are implemented on the remote sensing image characteristics of all sorts of microfacies; furthermore, sedimentation survey of Ganjiang delta is carried out at the same water level episode of images, and the remote sensing image data of the corresponding periods are acquired. Based on the correspondence of image characteristics and surface features, the remote sensing identification chart of sedimentary microfacies (channel, interchannel, channel bar, distributary bar, crevasse splay, natural levee and abandoned channel) of Ganjiang delta is built (Fig.2), and the image characteristics of various sedimentary microfacies are summed up as the following:

(1) Channel： water is dusty blue, low curvature ribbon channel predominates;

(2) Interchannel: low-lying shoal water swamp between channels, yellow green, overlain by water locally;

(3) Distributary bar: in the vicinity of lake inlet of channel, mostly yellow-gray, having fusoid and crescent bottomlands;

(4) Channel bar: shoal exposed in dry season in the centre of channel, resulted from waster accumulation, elongated;

(5) Crevasse splay: fan shape spread waster accumulation body in interchannel, resulted from pour of flood after it has made a breach in natural levee;

(6) Natural levee: tint ribbons distributed at both sides of channel, and extended parallel to it;

(7) Abandoned channel: in interchannel, dark ribbon with channel shape.

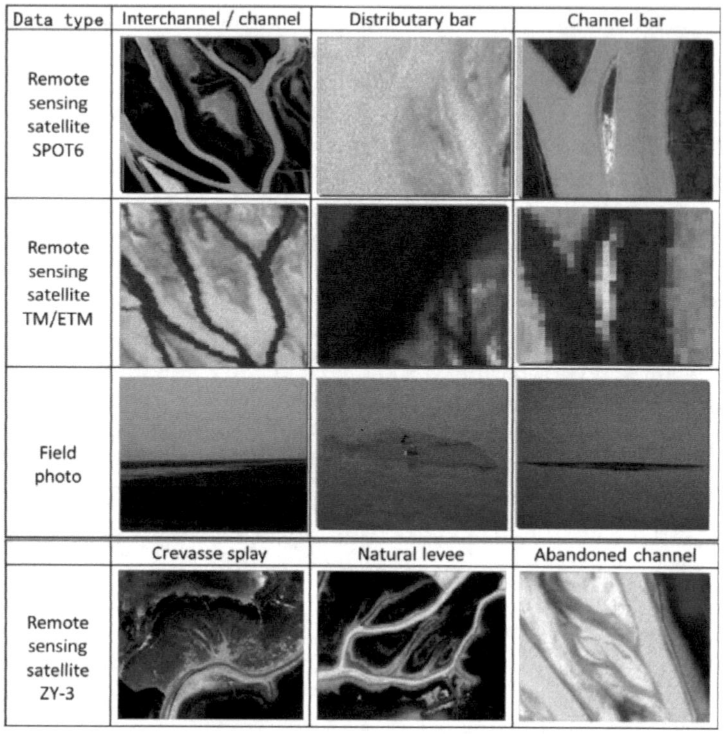

Fig.2 Remote sensing identification chart for sedimentary microfacies of Ganjiang delta of Poyang Lake

3.3 Sedimentary facies remote sensing interpretation

Based on multi-temporal remote sensing analysis, the sedimentary facies of the whole Poyang Lake basin are interpreted by medium resolution images(Fig.3).

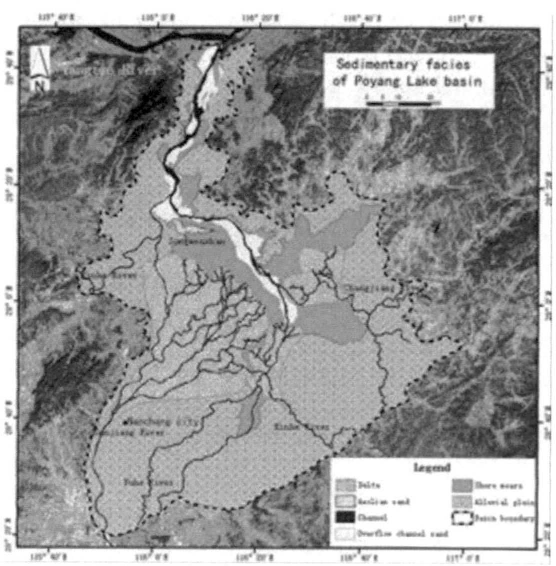

Fig.3 Remote sensing interpretation map of sedimentary facies in the Poyang Lake Basin

Based on remote sensing identification chart, the sedimentary facies of mid-branch of Ganjiang delta are interpreted by high resolution image, including channel, inter channel, distributary bar and channel bar(Fig.4).

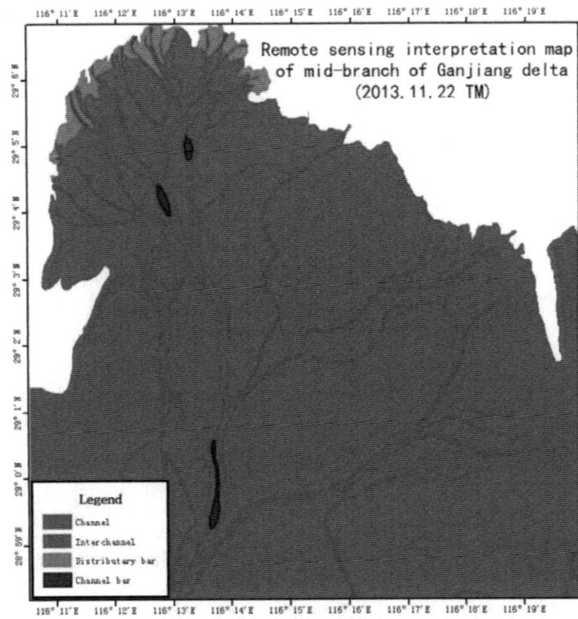

Fig.4 Remote sensing interpretation map of mid-branch of Ganjiang delta

4. Results

Based on interpretation analysis of remote sensing images and verification of field sedimentation survey, there are three types of sand body deposits in the center of the Poyang Lake Basin: deltaic deposit, overflow channel deposit and aeolian deposit.

The overflow channel of the Poyang Lake starts from Sanjiangkou in the southern lake region, runs across the north of Songmenshan beach bar, continually flows northward and mingles into the Yangtse River from lake outlet (Fig.3). Under the coaction of inflow-outflow current and tractional current, following the once swinging of the overflow channel, overflow channel sand body parallel to the channel is formed. The overflow channel is 2-3 km wide, while the sand bodies at both sides of it can be as wide as 4.5 km maximally, with an area about 400 km^2.

Linear deflation geomorphology is developed at lake outlet section of the Poyang Lake; furthermore, the linear deflation trend is consistent with the dominant wind power of the Poyang Lake, i.e., they both come from the NNE direction. The sands at lake neck were raised by the NNE wind, and extensively deposited in the Poyang Lake region; among which, the coarser sands were firstly deposited, and resulted in the formation of Songmenshan beach bar in the southern lake neck. The Songmenshan beach bar is a typical aeolian deposit, where medium sands predominate; the beach bar sand body is about 14 km long, its maximum width is 2.6 km, and its maximum thickness exceeds 70 m, and the area of aeolian sand is 68.5 km^2 (Fig.3).

The analytic results of sedimentary facies interpretation show that: from 1973 to 2013, the mid-branch front sand body had grown the most rapidly in the Ganjiang delta, and its area had increased from 6 km^2 to 25 km^2 ; among which, the right wing of mid-branch of Ganjiang delta was the sub-branch that had grown most rapidly during 2007-2013, advanced for more than 600 m towards lake, and became the key growth point of the Ganjiang delta at next stage.

Based on integrated analysis on channel growth of the Ganjiang delta, the growth rhythm of channel front of mid-branch of the Ganjiang delta is summed up as exponential type (Fig.5), and the growth pattern of the Ganjiang delta is furthered summed up as dendritic and reticular.

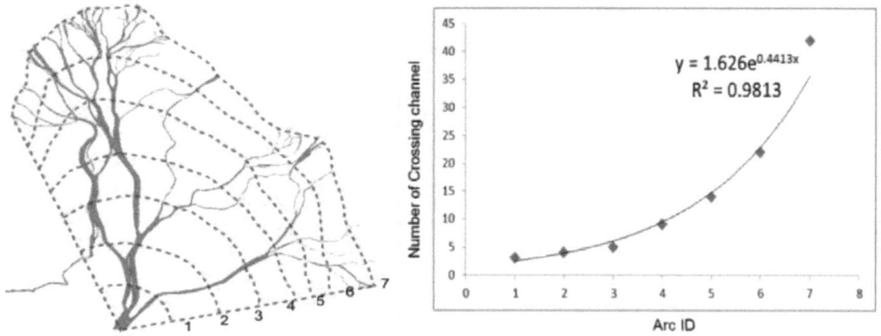

Fig.5 The mid-branch channels of the Ganjiang delta increased on an exponential growth rhythm

In spatial dimension, the distributary channels at west of the SN trend main channel of Ganjiang River flow into the lake in NW trend, and exhibit a dendritic distribution on the whole and reticular distribution locally; whereas those at east of it flow into the lake in NE and NEE trends, and exhibit a distribution pattern to the contrary. In temporal dimension, the distributary channels mostly arborize at lake inlet and were reworked to be reticulatus at late stage (Fig.6). The Ganjiang delta presents a dendritic growth in spatial and temporal dimensions, accompanied by the enclosure and mergence of distributary channels, which make the channel sands present a dendritic and reticular distribution, and it represents the distribution and growth pattern of two kinds of typical sand bodies of the Ganjiang delta.

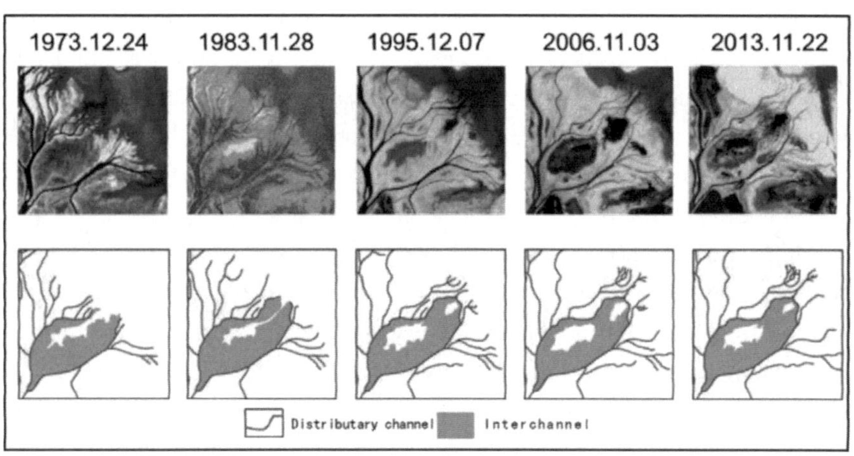

Fig.6 Growth pattern of transformation from dendritic deposit to reticular deposit

5. Conclusions

Based on quantitative research on modern deposition of the Gangjiang delta in the Poyang Lake

Basin by remote sensing technique, some conclusions are obtained as the following:

(1) Remote sensing is an important advanced technical means for modern deposition survey.

By means of unique advantages like global imaging and detail highlighted, historical filing and repeated monitoring, global scope and unrestricted, the remote sensing technique can be used to study the depositional pattern of shoal water deltas in lake basins, and it is an important advanced technical means for modern deposition survey.

(2) Multiple types of sand body deposits are developed in continental lake basin.

Based on remote sensing image analysis, combined with field survey verification, it is believed that three types of sand body deposits are mainly developed in the Poyang Lake Basin: deltaic deposit, overflow channel deposit and aeolian deposit; and the distribution of sand bodies in the lake basin is affected by the above three types of depositions jointly.

(3) Multi-temporal remote sensing dynamic quantitative analysis reveals the growth pattern of delta.

Based on multi-temporal remote sensing dynamic quantitative analysis, the growth rhythm of channel front of mid-branch of the Ganjiang delta is exponential type, the growth pattern of the Ganjiang delta is dendritic and reticular, and the distributary channel mostly arborizes at lake inlet and was reworked to be reticulatus at late stage.

Reference

[1] Cornel Olariu, Janok P. Bhattacharya, "Terminal distributary channels and delta front architecture of river-dominated delta systems,"Journal of Sedimentary Research, vol.76, pp. 212–233, 2006.

[2] Adrian J. Hartley, Gary S. Weissmann, Gary J. Nichols, "Large distributive fluvial systems: characteristics, distribution, and controls on development,"Journal of Sedimentary Research, vol. 80, pp.167–183, 2010.

[3] Zou Caineng, Zhao Wenzhi, Zhang Xingyang, "Formation and distribution of shallow—water deltas and central—basin sandbodies in large open depression lake basins," Acta Geologica Sinica, vol. 82, no. 6, pp. 813-825, 2008.

[4] Dang Shuqing, Han Zhiyong, Li Xusheng, "Evolution mode of linear wind eroded landforms on sand hills on the shore of the Poyang lake," Quaternary Sciences, vol. 33, no. 2, pp. 388-397, 2013.

[5] Wang Jiangong, Wang Tianqi, Wei Pingsheng, "Sedimentary mode of shallow lacustrine delta of large continental basin-An example from Putaohua Formation in northern Songliao Basin," Lithologic Reservoirs, vol. 19, no.2, pp. 28-34, 2007.

Coastline detection in SAR images using discriminant cuts segmentation

Xianwen Ding[1*], Xiaolin Zou[2] and Tan Yu[1]

[1]College of Marine Sciences, Shanghai Ocean University Shanghai, 201306, China
[2] School of Mathematics and Statistics, Zhaoqing University, Zhaoqing 526061, China

[*] Email: xwding@shou.edu.cn

Abstract. The discriminant cut algorithm is used to detect coastlines in synthetic aperture radar (SAR) images. The proposed approach is a region-based one, which is able to capture and utilize spatial information in the image. The real SAR images, e.g. ALOS-1 PALSAR and COSMO-SkyMed SAR images, together with in-situ GPS data were collected and used to validate the performance of the proposed approach for coastline detection in SAR images. The accuracy is better than 4 times the image resolution. The efficiency is also tested.

1. Introduction

Coastline detection is important for land/water resources management, safe navigation, coastal erosion monitoring, among others. In recent years, coastline detection from spaceborne synthetic aperture radar (SAR) images is a fast growing development field due to SAR's wide spatial coverage, high resolution, and all-day/all-weather imaging capabilities. Automatic coastline extraction from a SAR image is not a trivial task due to both the presence of the speckle noise and the non-uniform characteristics of the sea surface roughness caused by different mechanisms including wind, wave, and their interactions [1].

Many studies have devoted to develop semi-automatic or automatic algorithms for coastline detection in SAR images. Among them, many researches [2-4] on coastline detection in SAR images are pixel-based. Therefore they are not able to capture and utilize spatial information in the image data. In this study, we proposed a new region-based approach for coastline detection in SAR images, which is based on discriminant cuts segmentation [5]. To validate this approach, the ALOS-1 PALSAR and COSMO-SkyMed (CSK) SAR images, together with in-situ GPS data were collected. The experimental results show that the proposed approach is effective and efficient in coastline detection in SAR images.

2. Methodology

Discriminant cut is a graph-based spectral cluster algorithm. It emphasizes the general criterion of the cluster algorithms: maximizing the within-cluster similarities while minimizing between-cluster associations [5]. We introduce the discriminant cut algorithm to detect coastlines in SAR images.

The proposed approach includes three image processing steps: pre-processing, land-sea segmentation, and post-processing. In the pre-processing step, speckle reduction, geometric correction and anisotropic diffusion filtering [6-7] are performed. In the land/sea segmentation step, the divide-

and-combine approach and discriminant cuts segmentation are combined to derive the land/sea boundary of SAR images. For divide-and-combine approach, the image is first divided into many regular sub-blocks (a few tens of pixels by a few tens of pixels) for further processing. By doing so, local characteristics in the image may have due care or attention; and the processing speed may be enhanced due to smaller size of sub-blocks. And then each block is partitioned into regions using discriminant cuts segmentation. After that, all the regions are combined. Lastly, a threshold calculated automatically (e.g., Otsu's method [8]) is applied to partition the combined image into land and sea classification, with a binary image output. In the post-processing step, some objects whose boundaries are not the coastline are removed by filtering out isolated targets that consist of less than a predefined T pixels from the binary image. And then the exterior boundaries of land are traced automatically. Finally, the detected boundaries are converted to a vector coastline file.

3. Experimental results

Two experiments are undertaken. The first experiment is on the application of the proposed approach in Japanese ALOS-1 PALSAR images. The second experiment is on the application of the proposed approach in Italian COSMO-SkyMed SAR images.

3.1. Experiment I

Two Japanese ALOS-1 PALSAR FBS images covered the coast of Shanghai, China on January 15, 2010 were collected. Table 1 shows their relevant information. Their spatial resolution is 12.5 m.

Table 1. Summary of the ALOS-1 PALSAR images

Sensor	Beam	Orbit	Date (dd/mm/yy)	Acquisition time (UT)
PALSAR	FBS	21188	15012010	14:25:10
PALSAR	FBS	21188	15012010	14:25:18

The ALOS-1 PALSAR images were first calibrated, re-projected (UTM), and resampled to 6.25 m pixel spacing. As the study area is covered by these two ALOS-1 PALSAR scenes: south and north, then they were mosaicked and the image of the study area was clipped. The clipped image size is 5112 × 3582 pixels. We detected the coastline in the clipped image by using the proposed approach. In the pre-processing step, the parameters in anisotropic diffusion filtering are set as follows: number of iterations is set as 15; integration constant is set as 1/7; the gradient modulus threshold that controls the conduction is set as 30. In the land/sea segmentation step, the size of sub-blocks is set as 50 pixels. Each block is partitioned into 5 regions using discriminant cuts segmentation. In the post-processing step, T is set as 5000. For comparison, we also detected the coastline in the clipped image by using global thresholding approach. The threshold value used in global thresholding approach is the same as that used in the proposed approach.

To evaluate the effectiveness of the proposed approach on coastline detection in ALOS-1 PALSAR images, we compared the detected coastlines by using the proposed approach with that by using global thresholding approach. The comparison result is shown in figure 1. The detected coastline by using the proposed approach is shown in green; that by using global thresholding approach is shown in red. To show more details, we enlarge some areas (indicated by small blue box) by 2.5 times (indicated by large blue box). From figure 1, the detected coastline in the ALOS-1 PALSAR image by using the proposed approach fits that by using global thresholding approach well in most areas. By visual inspection, we find the detected coastline by using the proposed approach differs from that by using global thresholding approach in some areas, e.g. area A ~ area D. In area A, both approaches fail to detect coastlines accurately due to low land/sea contrast ratio for this area is wet. In area B, the

IOP Conf. Series: Earth and Environmental Science 46 (2016) 012035 doi:10.1088/1755-1315/46/1/012035

proposed approach succeeds but global thresholding approach fails. The reason is that the proposed approach is able to capture and utilize spatial information while global thresholding approach is unable. In area C, both approaches work not well for this area is complex. In area D, both approaches work not well either due to interference from inland water. In summary, the proposed approach performs well under mid-high land/sea contrast ratio in coastline detection in the ALOS-1 PALSAR image. And it is able to capture and utilize spatial information in the ALOS-1 PALSAR image.

Figure 1. The detected coastlines in ALOS-1 PALSAR image on January 15, 2010. The detected coastline by using the proposed approach is shown in green; that by using global thresholding approach is shown in red. The large rectangle blue box is 2.5 times the length of the small rectangle blue box. Letters 'A', 'B', 'C', and 'D' represent the names of the small boxes.

3.2. Experiment II

An Italian CSK SAR Stripmap PingPong mode HH/VV polarized data covered the coast of Shanghai, China on December 3, 2012 was collected. Table 2 shows its relevant information. Its spatial resolution is 15 m. The tide level is middle in the acquisition time.

Table 2. Summary of the PingPong mode COSMO-SkyMed SAR data and ancillary tide information

Sensor	Polarizations	Date (dd/mm/yy)	Acquisition time (UT)	Tide level (m)
COSMO-SkyMed	HH/VV	03122012	02:18	1.42 M*

Note: 'M' indicates middle tide

The HH channel image was used to test the proposed approach. It was first calibrated, re-projected (UTM), and resampled to 7.5 m pixel spacing. Then the image of the study area was clipped. The clipped image size is 2714 × 2985 pixels. We detected the coastline in the clipped image by using the proposed approach and by using global thresholding approach respectively. The parameters used in the proposed approach and global thresholding approach are the same as those in experiment I.

To evaluate the performance of the proposed approach on coastline detection in the COSMO-SkyMed SAR image, we compared the detected coastlines by using the proposed approach with that by using global thresholding approach. The comparison result is shown in figure 2. The detected coastline by using the proposed approach is shown in green; that by using global thresholding approach is shown in red. To show more details, we enlarge some areas (indicated by small blue box) by 2.5 times (indicated by large blue box). From figure 2, the detected coastline in the ALOS-1 PALSAR image by using the proposed approach fits that by using global thresholding approach well in most areas. By visual inspection, we find the detected coastline by using the proposed approach differs from that by using global thresholding approach in some areas, e.g. area A ~ area D. To quantitatively evaluate the accuracy of the proposed approach, we collected in situ GPS information along the coastline under middle tide condition on July 4, 2013 and July 16, 2013. The accuracy is 46 m. It is better than 4 times the image resolution. For more information on validation of coastline detection approach using SAR images, readers can refer to [9].

Figure 2. The detected coastlines in COSMO-SkyMed SAR image on December 3, 2012. The detected coastline by using the proposed approach is shown in green; that by using global thresholding approach is shown in red. The large rectangle blue box is 2.5 times the length of the small rectangle blue box. Letters 'A', 'B', 'C', and 'D' represent the names of the small boxes.

The efficiency is also tested. In experiment I, the run time is about 26 minutes on a personal workstation. In experiment II, the run time is about 16 minutes on the same personal workstation.

4. Conclusions

A novel approach of coastline detection in SAR images based on the discriminant cuts segmentation is proposed. The real ALOS-1 PALSAR and COSMO-SkyMed SAR images over the coast of Shanghai, China, together with in-situ GPS data were collected and used to validate the performance of the proposed approach. The experimental results show that the proposed approach is able to capture and utilize spatial information in SAR images and the accuracy of the proposed approach is better than 4 times the image resolution.

References

[1] Mason D C and Davenport I J 1996 Accurate and efficient determination of the shoreline in ERS-1 SAR images *IEEE Geosci. Remote Sens.* **34** 1243-53

[2] Baselice F and Ferraioli G 2013 Unsupervised coastal line extraction from SAR images *IEEE Geosci. Remote Sens. Lett.* **10** 1350–4

[3] Liu H and Jezek K C 2004 Automated extraction of coastline from satellite imagery by integrating canny edge detection and locally adaptive thresholding methods *Int. J. Remote Sens.* **25** 937-58

[4] Dellepiane S, De Laurentiis R, and Giordano F 2004 Coastline extraction from SAR images and a method for the evaluation of the coastline precision *Pattern Recogn. Lett.* **25** 1461-70

[5] Chen W, Feng G, and Liu Z 2011 Discriminant cuts for data clustering and analysis *ACPR2011* **32** 120-4

[6] Perona P and Malik J 1990 Scale–space and edge detection using anisotropic diffusion *IEEE Trans. Pattern Anal. Mach. Intell.* **12** 629-39

[7] Saint-Mark P, Chen J, and Medioni G 1991 Adaptive smoothing: a general tool for early vision *IEEE Trans. Pattern Anal. Mach. Intell.* **13** 514-29

[8] Otsu N 1979 A threshold selection method from gray-level histograms *IEEE Trans. Sys., Man., Cyber.* **9** 62–6

[9] Ding X, Nunziata F, Li X, and Migliaccio M 2015 Performance analysis and validation of waterline extraction approaches using single- and dual-polarimetric SAR data *IEEE J. Sel. Top. Appl. Earth Observ. Rem. Sens.* **8** 1019-27

Acknowledgements

This work is supported by a grant from the National Natural Science Foundation of China (No. 41306184).

IOP Conf. Series: Earth and Environmental Science **46** (2016) 012036 doi:10.1088/1755-1315/46/1/012036

Database assessment of CMIP5 and hydrological models to determine flood risk areas

Ponthip Limlahapun[1] and Hiromichi Fukui[2]

[1] Fellow of the Japan Society for the Promotion of Science hosted by Chubu University, 1200 Matsumoto-cho, Research Center, Kasugai, Aichi, Japan

[2] Director of the International Digital Earth Applied Science Research Center, Chubu University, 1200 Matsumoto-cho, Research Center, Kasugai, Aichi, Japan

E-mail: thip.limlahapun@gmail.com

Abstract. Solutions for water-related disasters may not be solved with a single scientific method. Based on this premise, we involved logic conceptions, associate sequential result amongst models, and database applications attempting to analyse historical and future scenarios in the context of flooding. The three main models used in this study are (1) the fifth phase of the Coupled Model Intercomparison Project (CMIP5) to derive precipitation; (2) the Integrated Flood Analysis System (IFAS) to extract amount of discharge; and (3) the Hydrologic Engineering Center (HEC) model to generate inundated areas. This research notably focused on integrating data regardless of system-design complexity, and database approaches are significantly flexible, manageable, and well-supported for system data transfer, which makes them suitable for monitoring a flood. The outcome of flood map together with real-time stream data can help local communities identify areas at-risk of flooding in advance.

1. Introduction

Climate change has been a significant research topic in recent years. Although climate change is largely considered harmful to human habitation, it may also deliver positive impacts to some regions. For example, an ADB economics working paper reported that future projections on grain products (corn and soybean) in China will increase significantly by 2030 [1]. However, the negative consequences of climate change seem more likely, as nature and society are affected worldwide, with the negative impacts on population, urbanization, and even industrial and economic developments continuing to increase every year. These changes could influence environmental and social vulnerabilities regardless of geographical frontiers. There is a growing scientific consensus that inconsistent weather patterns (i.e., extreme wet and stormy conditions) has the potential to intensify the hydrological model [2], leading to flood risk expansion, intensification, and significant societal implications for activities of daily living and quality of life. Locating flood prone areas where serious or frequent flooding occurs could make communities more aware of their risks in certain areas and perhaps take steps to prepare ahead and aid recovery afterwards.

The Coupled Model Intercomparison Project (CMIP) was established by the working group under the World Climate Research Programme (WCRP) since 1995. The CMIP delivers a standard experimental framework for studying the output of the Global Climate Models (GCMs) [3] and supplies infrastructure to support a large number of climate models, including analysis, validation, inter-comparison, and data access for international climate communities to understand, analyze and facilitate model improvement [4]. Recently, more than 60 CMIP5 models [5] distributed from about 30 various agencies have been assembled and publicized to support climate studies and the estimation of future climates [3]. The large volume of CMIP5 model simulations in capacity of terra bytes has required the assessment of data sharing across research community. Taylor et al. (2012) noted that the CMIP5 output will be archived in data nodes distributed at modeling centers and data centers near where the model output is produced. The nodes will be linked together and the model output will be freely accessible through data portals (or gateways) integrated in a way that retains much of the convenience of a single repository [6]. This worldwide climate research has generated large-scale models and shared results to increase the accuracy of various models working with big datasets [7].

Big data refers to a vast number of datasets, heterogeneous, complex, and difficult to store, analyze, process, and visualize [8]. In other words, big data is often defined by its characteristics of increased data capacity; receiving and processing speed; and a variety of data types, sources, and formats as 3V volume, variety, and velocity [9], which require new forms of processing to enable enhanced decision making, insight discovery, and process [10]. Recently, big data has been involved in many studies (i.e., hydrological, natural, environmental, engineering, and computer sciences studies). The term "big data" may appear in social media like Google or Twitter or in scientific formats (like climate model research, distributed sensors, and satellite observation data) that assist in tracking a dynamic situation, observe patterns, or witness near-real-time events [11].

One challenge of this paper is to obtain information by accessing big data (meteorological CMIP5) and integrate hydrological models to identify inundation areas using a database approach. An intelligent alternative in a proper system will provide a faster analytical data transfer, increased usability, consistency, and better sharing. Nonetheless, model selection from a number of available models is an essential requirement intimately tied to the hydrological models structures, input requirements, and results.

2. Methodology

2.1. Data source

CMIP5 model datasets were obtained from the Data Integration and Analysis System (DIAS) provided coordinating support to the University of Tokyo. We acknowledge the World Climate Research Programme's Working Group on Couple Modelling, which is responsible for CMIP, and we thank the climate modeling groups (listed in Table 1) for producing and making available their model output.

2.2. Approach

2.2.1. Analyzing precipitation

a. Model selection

The model selection approach was evaluated from 20 CMIP5 models based on parameters (precipitation, air temperature, sea surface temperature, sea level pressure and zonal wind). Figure 1 shows model

Table 1. Selected modelling group

Modeling Center (or Group)	Institute ID	Model Name
Beijing Climate Center, China Meteorological Administration	BCC	BCC-CSM1.1 BCC-CSM1.1(m)
Japan Agency for Marine-Earth Science and Technology, Atmosphere and Ocean Research Institute (The University of Tokyo), and National Institute for Environmental Studies	MIROC	MIROC-ESM MIROC-ESM-CHEM
Atmosphere and Ocean Research Institute (The University of Tokyo), National Institute for Environmental Studies, and Japan Agency for Marine-Earth Science and Technology	MIROC	MIROC4h MIROC5
Max Planck Institute for Meteorology	MPI-M	MPI-ESM
Meteorological Research Institute	MRI	MRI-CGCM3 MRI-ESM1

Source: WCRP [5].

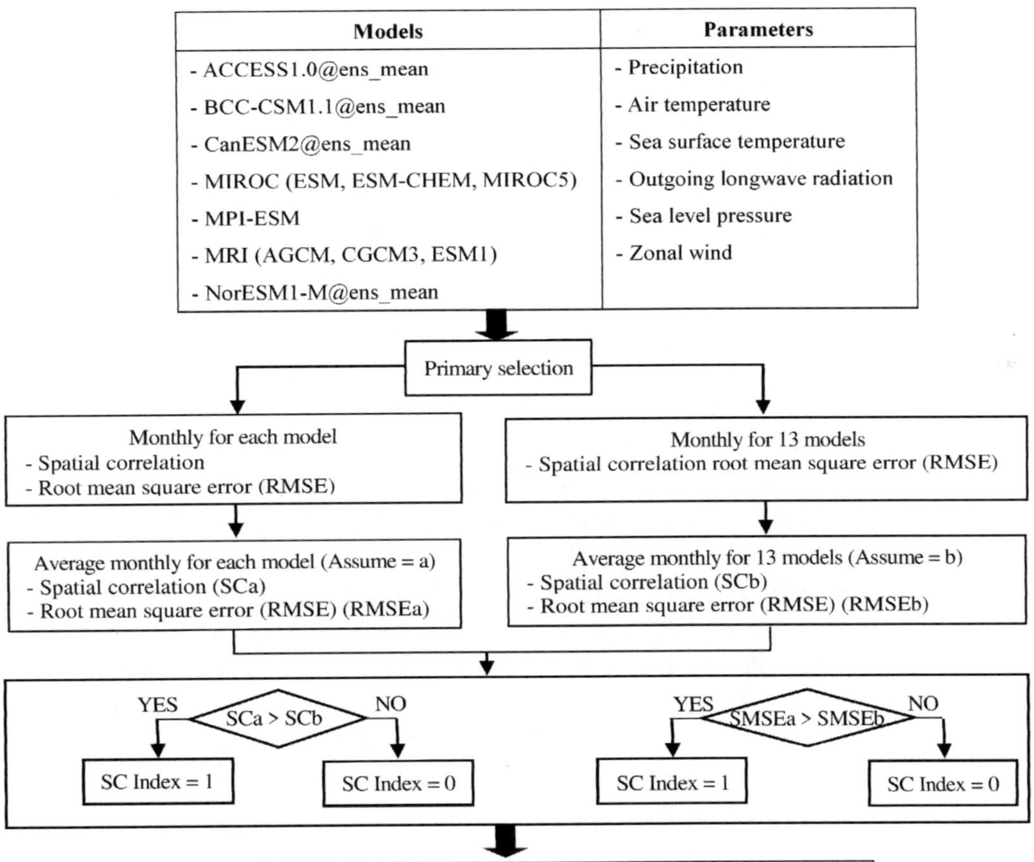

SC index	RMSE index	Overall index (SC and RMSE)
1	1	1
1	0	0
0	1	0
0	0	-1

Figure 1. Model selection procedure

selection methods. Out of these, seven models (MIROC-ESM-CHEM, MIROC5, MPI-ESM-MR-LR, MRI-CGCM3-ESM1) that returned positive overall index were selected for bias correction over Shonai River in Kasugai.

b. Precipitation extraction

The model selection assembled precipitation information and the degree of spread in future climate projections [12] from multiple GCMs via the CMIP5. In this study, the historical data was taken from 1981-2000 based on a number of models. The observed gridded precipitation data (0.05°×0.05° grid) was derived and accessed from the DIAS, which used JRA55 and other reanalysis data as reference data for comparison with CMIP5 data provided by the University of Tokyo [13].

2.2.2. Examining discharge with flood analysis system model

The Public Works Research Institute (PWRI), the International Centre for Water Hazard and Risk Management (ICHARM), under the auspices of UNESCO, developed the Integrated Flood Analysis System (IFAS), a satellite rainfall based simulation model for determining rainfall-runoff distribution [14]. The data required to calculate discharge includes elevation, land cover, soil, watershed boundary, and daily rainfall. Digital Elevation Model (DEM) 5-metre grid data was download from Geospatial Information Authority of Japan (GSI). The process of filling DEM; generating water flow; flow accumulation to lower cell and direct to outlet cell; and output as watershed boundary, and daily rainfall. Digital Elevation Model (DEM) 5-metre grid data was download from Geospatial Information Authority of Japan (GSI). IFAS is an executed installation program. Users proceed step-by-step through graphical interfaces provided by the system. To reduce processing time, we adapted the program by running it from a command line. The output from IFAS was graph and table, which could be plotted for a hydrograph (rainfall-discharge) using original IFAS program based on a grid-cell or a MS Excel spreadsheet.

2.2.3. Determining an inundation map with a hydrological model

The Hydrologic Engineering Center River Analysis System (HEC-RAS) is a hydraulic model to execute a steady or unsteady flow calculation [15]. The data preparation was created from HEC-GeoRas on ArcGIS extension. The Hec-GeoRAS interface allows the preparation ofgeometric data in so-called RAS layers (streamline, river-bank, cross-sectional, and floodplain boundary) for importing and processing simulation results from HEC-RAS [15]. Data requires for creating an import file includes the UTM projected digital terrain model (TIN format) and background Landsat imagery. Once the RAS layers were created, attribute data could be assigned using the GeoRas tools (river and reach name to stream network, channel-left-right to rivers, overbanks, and flow paths). The layers were then ready to written out in spatial data file (SDF) format to an extensible markup language (XML) file, but it is practical to validate data and attributes before importing to the HEC-RAS geospatial data exchange. The final outcome is to create the inundation and velocity spatial distribution of flooding and identify affected areas.

3. Results

3.1. Results of precipitation analysis derived from CMIP5

The historical data from each model was verified similarly to the references in Asian Precipitation-Highly Resolved Observational Data (APHRODITE). Based on the literature and reviewed reports, the length and intensity of rainfall caused flooded along Shonai River in 2000. The Tropical Rainfall

Measuring Mission (TRMM) dataset was also collected and displayed to verify rain-patterned and intensity performed in the areas. The future climate prediction shows a large spread in projections, and the shifting rain pattern is obvious. This result occurs due to variations in the climate sensitivity that determines the influence magnitude of the average global response, and large variations in spatial patterns of change, particularly for precipitation [16].

3.2. Results of discharge analysis using flood analysis model

The precipitation extracted from DIAS system was then formatted for use in IFAS. The data were structured into daily records and simulations were conducted. The result calculated based on the year 2000 data average discharge amount of 2,300 m^3/s, showed good agreement with the JMA records of approximately 2,500 m^3/s. The same procedure for deriving precipitation data from the CMIP5 model was applied to estimate future flood peak. Additionally, the near-real-time observation rain gauge and water level were collected at 10-minute intervals to observe stream flow. The thresholds were used for three level of alert system (warning, evacuation, and flooding). The verification was crossed checked with the observation data. These results show that the IFAS model could be a useful tool for the overflow estimates and accessible CMIP5 data for prediction of flood inundation process in any area of interest. The TRMM satellite earth observation and near-real-time system assist to monitor rainfall influences to surrounding to global climate and the current situation, accordingly.

3.3. Results of inundation map extracting from hydrological model

The final process to generate an inundation map was derived using HEC, based on the discharge result and other structure (i.e., bridge, land use, and so on). The HEC estimates water flows through system and compute water surfaces and evaluates flood encroachments.

The serial result data input to and output from each process can be transferred among modeling and simulation techniques. This database approach is recognized for its speed and flexibility while also providing the ability to manage with amounts of data. Databases are designed to store data, so data coming into the system must be cautiously considered.

4. Conclusion

Empirical relationships between rainfall rates generated from the big data model over the past 20 years and the characteristics of the actual flood have been formulated. The changes in rainfall intensity and its behavior have been forecasted for the 20-year period from 2045 to 2065. The result shows an increase in the amount and duration of rainfall, which may lead to significant future flooding, escalation in the flood peak, and total flood volume. In the context of disaster preparedness, the results may be shocking, but it is objectively apparent that communities may find themselves at considerable risk and such models afford them the opportunity to take specific actions to reduce their risks before it is too late.

A complex issue like being prepared for a natural disaster may not be solvable with a simple solution or the application of one particular model. A selection of appropriate models from a number of available standards is challenging due to the variability of spatial and temporal aspects. This paper proposed alternative ways to approach results to select models, worked backwards and illustrated outcomes we need as a basic notion. Then we craft solutions to the underlying issues, factors, and resources to solve these problems. Three selection models were applied, using one model's output values to serve as another model's input, ultimately producing a delineated inundation map. By developing a database to act as the assembly tool for each component, receive rainfall intensity, transfer parameters of flows in channel network, and storing forecast datasets that are capable of generating the distributed and predicting inundated areas, it has been found that a database approach is an efficient methodology.

References

[1] Tun L, Xiaoyun L, Guanghua W, Xian X, and Yongsheng Z 2011 Impacts on climate change on the people's Republic of China's grain output—Regional and crop perspective. *ADB economics working paper series*. **243**.

[2] Guoyi Z et al. 2011 Quantifying the hydrological responses to climate change in an intact forested small watershed in Southern China. *Global Change Biology*. **17** pp. 3736-3746.

[3] Guo Y, Dong W J, Ren F M, Zhao Z C, and Huang J B 2013 Surface Air Temperature Simulations over China with CMIP5 and CMIP3. *Advances in climate change research*. **4** No. 3 pp. 145-152.

[4] David A R and Richard A W Climate Models and Their Evaluation. https://www.ipcc.ch/pdf/assessment-report/ar4/wg1/ar4-wg1-chapter8.pdf

[5] WCRP (World Climate Research Programme), CMIP Coupled Model Intercomparison Project, http://cmip-pcmdi.llnl.gov/index.html?submenuheader=0

[6] Taylor K E, Ronald J S, and Gerald A M 2012 An Overview of CMIP5 and the Experiment Design. *American Methodological Society*. pp. 485-498.

[7] Jon B 2014 Weathering the flood of Big data in climate research. *ESnet News*, [Accessed on March 2, 2016, https://www.es.net/news-and-publications/esnet-news/2014/weathering-the-flood-of-big-data-in-climate-research/

[8] Akhtar U and Mehdi H 2015 Big Data Mining Based on Computational Intelligence and Fuzzy Clustering. *Handbook of Research on Trends and Future Directions in Big Data and Web Intelligence*. 130-148.

[9] Siddesh G M, Srinidhi H and Srinivasa K G 2014 Driving Big Data with Hadoop Technologies. *Handbook of Research on Cloud Infrastructures for Big Data Analytics*. 232-262.

[10] Laney D 2001 3D Data Management. *Controlling Data Volume, Velocity and Variety*.

[11] Tsou M H 2015 Mapping human dynamics with social media for disaster alerts. *Proceedings of ESRI users conference July 20-24, 2015, San Diego, California, U.S.A.*

[12] Carol F M, Richard G J 2016 How representative is the spread of climate projections from the 5 CMIP5 GCMs used in ISI-MIP? *Climate Services*. doi: 10.1016/j.cliser.2016.02.001

[13] DIAS, Data Integration and Analysis System, The University of Tokyo, http://www.dias.nii.ac.jp/modelvis/cmip5/

[14] (The) International Centre for Water Hazard and Risk Management (ICHARM) 2014 IFAS ver.2.0 technical manual. *Public Works Research Institute*. Japan, p.259.

[15] OPW, The Office of Public Work 2009 The planning system and flood risk management. *Guidelines for planning Authorities*. p.73.

[16] Guojun G and Robert F A 2015 Spatial Patterns of Global Precipitation Change and Variability during 1901-2010. *Journal of Climate*. **28** No. 11.

Acknowledgement

This project and publication were funded by the Japan Society for the Promotion of Science (JSPS) under a Postdoctoral Fellowships for Overseas Researchers and Grants-in-Aid for Scientific Research (KAKENHI) grant. Any opinions, findings, and conclusions expressed in this publication are those of the authors and do not necessarily reflect the views of the JSPS.

Design and realization of RS application system for earthquake emergency based on digital earth

Xiaoxiang Yuan[1] Xiaoqing Wang[1] Jianxing Guo[2] Aixia Dou[1] Xiang Ding[1]

[1] Institute of Earthquake Science, CEA. Beijing, China
[2] Earthquake Administration of Shannxi Province, Xi'an, China
E-mail: yuanxx1983@163.com

Abstract. The current RS-based earthquake emergency system is mainly based on stand-alone software which cannot meet the requirements of massive remote sensing data and parallel seismic damage information extraction after a devastating earthquake. Taking Shaanxi Province as an example, this paper explored firstly the network-based working mode of seismic damage information extraction and data management strategy for multi-user cooperative operation based on analysing work flow of the RS application to earthquake emergency. Then, using WorldWind java SDK, the RS application system for earthquake emergency based on digital earth platform was brought out in CS architecture. Finally, spatial data tables of classification and grade of seismic damage were designed and the system was developed. This system realized functions including 3D display, management of seismic RS image and GIS data obtained before and after earthquake for different user levels and cooperative extraction and publish of such seismic information as building damage, traffic damage and seismo-geological disasters caused by earthquake in real time. Some application to earthquake cases such as 2014 M_s6.5 Ludian earthquake show that this system can improve the efficiency of seismic damage information interpretation and data sharing, and provide import disaster information for decision making of earthquake emergency rescue and disaster relief.

1. Introduction

As the vigorous development of domestic high satellite and UAV remote sensing technology, the remote sensing data source for earthquake emergency is more and more rich after destructive earthquakes occur. Therefore, how to efficiently use multi-source remote sensing data for earthquake emergency services becomes the one of the problems that need to be solved. There are some scholars began to research and develop earthquake disaster management and recognition system for rapid earthquake remote sensing information extraction. Turkera M and Sumerb E [1] developed the seismic damage evaluation system based on buildings by Matlab and proposed segmentation algorithm based on watershed to detect intact buildings and damaged building. Xiaoqing Wang, etc. [2] developed earthquake disaster emergency and damage assessment system using ENVI/IDL which realized building seismic damage quickly identify and model management based on aerial and satellite images, and completed the development of earthquake emergency remote sensing (EERS) analysis and processing system on the basis of research of key techniques such as earthquake emergency disaster information extraction and damage assessment in 2008. After that, considering the characteristics of urban application, Aixia Dou [3] developed Tianjin Remote Sensing Earthquake Damage Analysis and Processing System (TJ-RSEDAPS), which realized many practical functions including image management, damage degree identification of buildings, key-objects and major facilities, losses assessment of life and economic, damage distribution mapping, and etc.

With the development of 3D GIS technology, research on earthquake disaster management and recognition technique based on 3D digital earth is becoming a hot spot. Chongjun Yang [4] developed 3D geographic information system on wenchuan earthquake on account of GeoBeans platform, and realized the functions including 3D display of multi-source remote sensing damage images, seismic road interpretation, and etc. Based on discussing the technical methods on application of multi-sensor remote sensing technology for earthquake disaster management, Jixian Zhang, etc. [5] built remote sensing monitoring of the Wenchuan earthquake disaster situation and the information service system which provide functions including disaster information management, visualization and statistical analysis. Jieping Zhou, etc. [6] implemented a 3D visualization management system for UAV remote sensing images based on the VGE-3DGlobeEarth platform, which provided the quad-tree pyramid and LOD model for managing large volume remote sensing images from earthquake zone. There are many advantages on 3D digital earth system such as intuitive and convenient data sharing. However, the currently existing system focuses on the 3D data management and display, and earthquake disaster information extraction and evaluation function is weak. Based on the earthquake emergency prototype system of remote sensing digital earth [7], this paper analyse work flow of remote sensing earthquake emergency using network, and discusses the multi-user collaborative damage extraction mode on the basis of the network and data management strategy, proposed CS architecture of earthquake emergency digital earth platform system of RS applications based on Worldwind.

2. The overall design of the system

Usually EERS work based on single version software, the specific process includes pre-earthquake background remote sensing data preparation, extraction of seismic damage, seismic damage and loss estimation of earthquake damages and earthquake damage products production and display [3].In this situation, collaborative work and how to share the background data and the real-time extracted information of earthquake are questions need to be resolved.

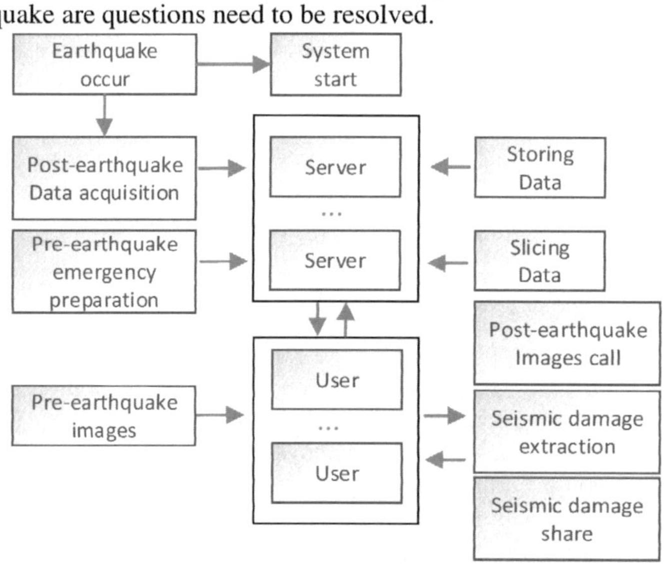

Figure 1. Network-based workflow of remote sensing application to earthquake emergency.

In this paper, we based on the WorldWind digital earth platform to make sure full use of the network environment (the specific work process is shown in figure1), including network-based data preparation for the earthquake emergency response and the post-earthquake emergency work. Network-based data preparation before earthquake mainly contains slicing and storing of fundamental geographic information, elevation data, background of remote sensing images, population, and economic data through distributed server providing data support for earthquake emergency. After the earthquake, on the one hand, the server slice and import multi-source data from different stages. On the other hand, different users from different computers can use the earthquake emergency client calls for post-earthquake and pre-earthquake disaster images to be displayed in 3D mode, then they can analyse and extract the seismic disaster information such as the building damage, road damage, geological disaster,

at the same time the system will share the extracted information shows to different users in real-time (figure 1 shows the overall workflow).
According to the business workflow above, the system is divided into the presentation tier, business logic tier and data service tier. The first tier is the presentation tier, whose mainly function is the system access for the three kinds of earthquake emergency users such as administrators, professional users, and emergency decision makers by the client. The second tier's function is logic control by the client-side, including web through the network environment, invoking geography information. The third tier is composed of spatial database and remote sensing database, storage the data produced in the process of earthquake emergency, and basic data provide requirements for server and client calls, as shown in figure 2.

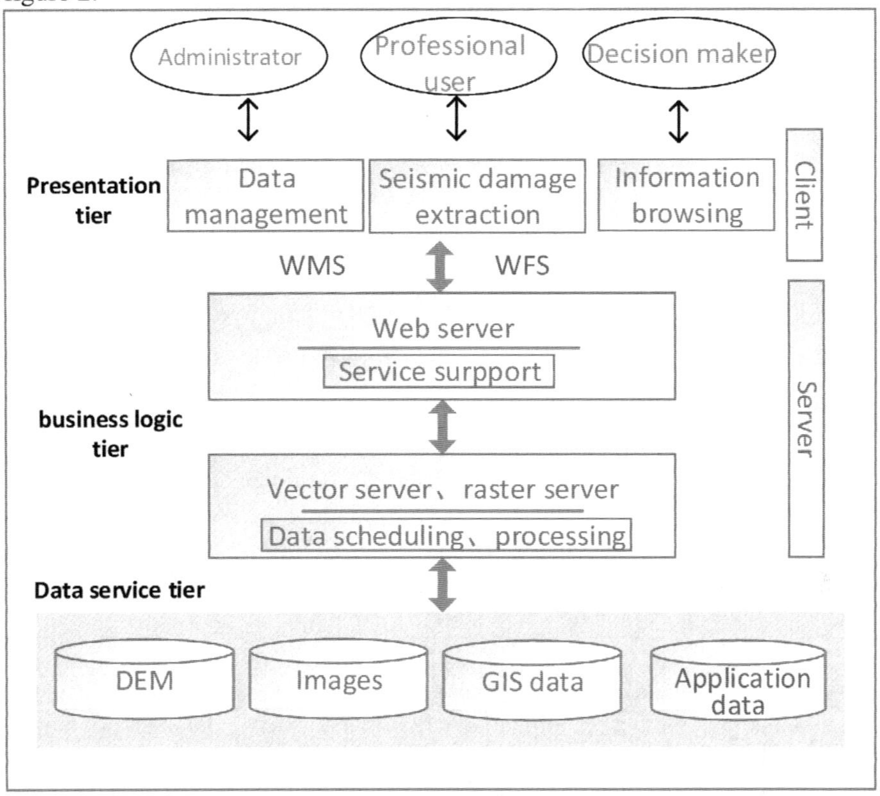

Figure 2. Overview of an RS application system for collaborative seismic damage information extraction and visualization based on 3D digital earth.

3. The design of remote sensing earthquake emergency database

In order to make the management of EERS data standard, efficient, easy to use, database design (figure 3) can be classified into spatial database, remote sensing image database, earthquake professional database and system management database. According to the specific function, different database stores corresponding earthquake-related information. Details of the design are as follows.

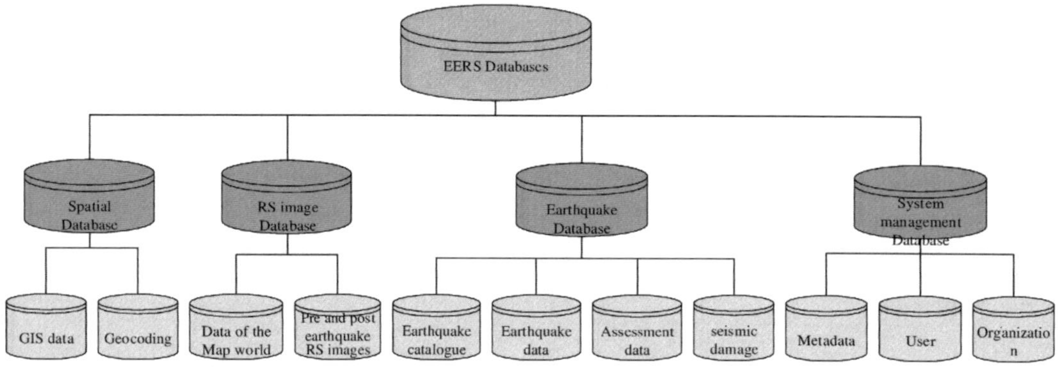

Figure 3. Overview of database structure of RS application system.

(1) Spatial database. It stores the emergency basic spatial data of the residential area, the key objects, administrative divisions and related provinces, cities, counties, townships and attributes of the standard geographic coding data in demonstration zone.

(2) Remote sensing image database. It stores the pre-earthquake background RS data of the Map world and the medium and high resolution in demonstration area, pre-earthquake high resolution data of key area and multi-source RS images acquired after earthquake such as UAV images and airborne high resolution images.

(3) Earthquake professional database. It stores the historical earthquake, basic information of current earthquake, rapid assessment of earthquake, seismic intensity map, classification of buildings and seismic geological hazard information. Table 1 shows the design of earthquake damage levels of different objects from RS images, which is a classification standard for different users sharing and updating the visual interpretation results to the server.

(4) System management database. It stores the data of system operation related to the user, the organization and related metadata, etc.

Table 1. The designation of earthquake damage levels of different objects from RS images.

Disaster type	Feature type	Classification	Level
Building damage	Single(point), Group(polygon)	High-rise building, Multi-storey building, Ordinary house, Industrial plant, Large open house, Other house	Collapse, Partial collapse, not collapsed with obvious damage, not collapsed without obvious damage
Road damage	Point, Line	Motorway, First class road, Second class road, Third class road and others	Completely damaged, Severely damaged, Moderately damaged, Slightly damaged, Intact
Seismic geological damage	Point, Line, Polygon	Earthquake fault, Ground fissure, Water and sand erupt, Landslide, Rockfall, Debris flow, Other	Severe, Moderate, Slightly
Other Secondary disaster	Point, Line, Polygon		Severe, Moderate, Slightly

4. The main functions and applications of the system

The system is CS architecture containing clients and servers. The Client developed by WoldWind SDK, whose integrated environment menus is shown in figure4. The server integrate Geoserver and Arcgis Server, as is shown in figure 5. After destructive earthquake occur, the client of the system (as is shown in figure 6) automatically load seismic RS images of arranged region from distributed servers, which provide 3D display of disasters for professional users. The extracted result can be stored automatically to servers for other users' reference. The main functions are introduced as below.

(1) Data management function. On the server side, it mainly contains users' data management of adding, deleting, modifying users of different roles, earthquake-related metadata management, slicing and updating of background RS images, DEM. On the client side, the user can upload and update specific earthquake-related seismic disaster images, vector data, GDP and population data for key areas.

(2) 3D browsing thematic information function. According to different requirements, the system carries out corresponding 3D rendering to the background information of the earthquake, damage images and thematic information, and elevation exaggeration rendering for special interpretation needs. For the key object in database, it can be displayed by serial number or name.

(3) Real-time damage interpretation and publishing function. For professional users, according to the needs of multi-source remote sensing data such as the domestic satellites and UAV, the client load the corresponding data for extraction of building damage, road damage and geological disasters caused by earthquake in real time. After extracting the damage vector information is automatically stored in the database through the server, another user on demand, automatic access to available earthquake damage information as a basis for collaborative task and working. In the case of M_s6.5 Ludian earthquake occurred at August 5 2014, figure 7-9 show the different disaster extraction interface.

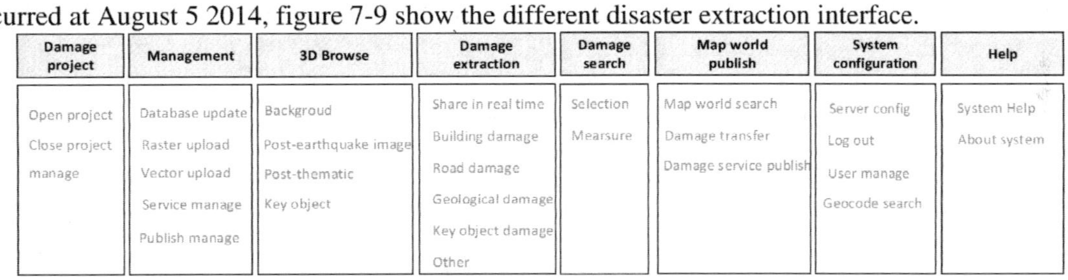

Figure 4. The integrate environment menus of client.

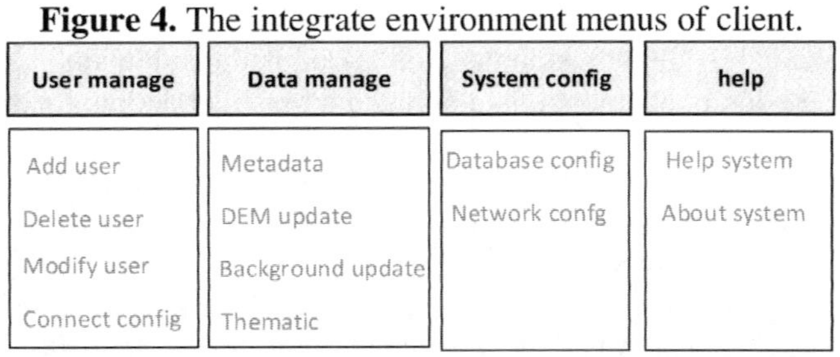

Figure 5. The integrated environment menus of server.

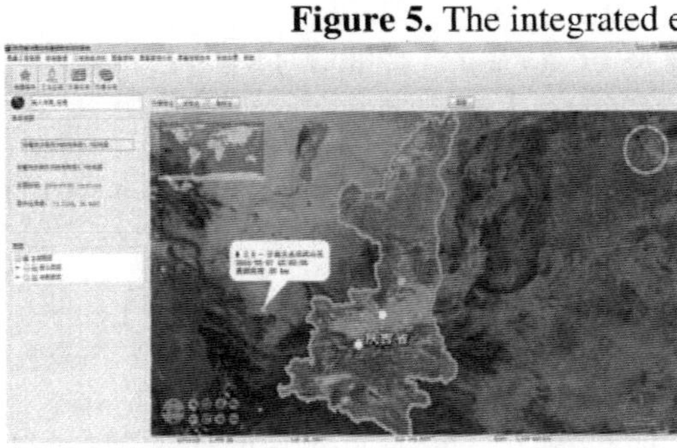

Figure 6. Main interface of client.

Figure 7. Single building damage extraction.

Figure 8. Road damage extraction.

Figure 9. Seismic geological disaster extraction.

5. Discussion and Conclusion

3D GIS is intuitive and realistic, which provides a good environment for the damage extraction users. With the rapid development of remote sensing technology and network technology, multi-user collaborative disaster information extraction based on digital earth is an important direction of development of remote sensing.

Analysing workflow of EERS using network, this paper realized CS architecture of RS application system based on WorldWind SDK. This system realized 3D display and management of remote sensing image and geographic information obtained before and after earthquake for different user levels. The system also implemented the functions of the cooperative extraction and seismic information publishing such as building damage, road damage and geological disasters caused by earthquake in real-time. Some applications to earthquake cases such as 2014 M_S6.5 Ludian earthquake show that this system can improve the efficiency of seismic damage information interpretation and data sharing, and provide import disaster information for decision making of earthquake emergency rescue and disaster relief.

6. References

[1] Turker M and Sumer E 2008 Building-based damage detection due to earthquake using the watershed segmentation of the post-event aerial images *International journal of remote sensing* **29** pp 3073–89

[2] Wang X Ding X and Dou A 2004 Study on model base system of earthquake damage extraction from RS in ENVI/IDL environment *Geoscience and Remote Sensing Symposium (IGARSS) IEEE International* pp 2252–55

[3] Dou A X Wang X Q and Ding X 2012 The development of Tianjin Remote Sensing Processing System Earthquake Damage Analysis *Geoscience and Remote Sensing Symposium (IGARSS), IEEE International* pp 2237–40

[4] Yang C J Zhang F Q Wu S Ren Y C Zhu H Y Wang G Liu D L Guo H T Fu L Liu Y L et al. 2008 3D Geographic Information System on Wenchuan Earthquake *Journal of Remote Sensing* **12**(6) pp 893–99 (in Chinese)

[5] Zhang J X Liu Z J and Liu J P 2008 Remote Sensing Monitoring of the Wenchuan Earthquake Disaster Situation and the Information Service System *Journal of Remote Sensing* **12**(6) 871–76(in Chinese)

[6] Zhou J P Gong J H Wang T 2008 Study on UAV Remote Sensing Image Acquiring and Visualization Management System for the Area Affected by 5•12 Wenchuan Earthquake *Journal of Remote Sensing* **12**(6) pp 877–86

[7] Guo J X Wang X Q Dou A X Yuan X X and Zhang W 2013 Remote sensing data dissemination system for earthquake based on OpenGIS and digital earthquake platform *Earthquake* **33**(2) pp 123–31 (in Chinese)

Acknowledgments

The research was supported by the Fundamental Research Funds of Institute of Earthquake Science, China Earthquake Administration (Grant No. 2013IES0204).

Open Standards in Practice: An OGC China Forum Initiative

Peng Yue[1], Mingda Zhang[2], Trevor Taylor[3], Jibo Xie[4], Hongping Zhang[5], Xiaochong Tong[6], Jinsongdi Yu[7], and Juntao Huang[8]

[1] School of Remote Sensing and Information Engineering, Wuhan University
[2] State Key Laboratory of Information Engineering in Surveying, Mapping and Remote Sensing (LIESMARS), Wuhan University
[3] Open Geospatial Consortium
[4] Institute of Remote Sensing & Digital Earth, Chinese Academy of Sciences
[5] National Geomatics Center of China
[6] Zhengzhou Institute of Surveying & Mapping
[7] Fuzhou University
[8] Wuda Geo Information Engineering Technology Co. Ltd.

Email: pyue@whu.edu.cn

Abstract. Open standards like OGC standards can be used to improve interoperability and support machine-to-machine interaction over the Web. In the Big Data era, standard-based data and processing services from various vendors could be combined to automate the extraction of information and knowledge from heterogeneous and large volumes of geospatial data. This paper introduces an ongoing OGC China forum initiative, which will demonstrate how OGC standards can benefit the interaction among multiple organizations in China. The ability to share data and processing functions across organizations using standard services could change traditional manual interactions in their business processes, and provide on-demand decision support results by on-line service integration. In the initiative, six organizations are involved in two "MashUp" scenarios on disaster management. One "MashUp" is to derive flood maps in the Poyang Lake, Jiangxi. And the other one is to generate turbidity maps on demand in the East Lake, Wuhan, China. The two scenarios engage different organizations from the Chinese community by integrating sensor observations, data, and processing services from them, and improve the automation of data analysis process using open standards.

1. Introduction

With the advancement of sensor technologies, large volumes of geospatial data are collected every day [1]. Various tools and platforms are developed to access and process these data. The data and software usually have their own proprietary features or operations. For example, it is often a challenge to combine processing functions from different tools/platforms to work collaboratively. Open standards bring an interoperate approach to address such an issue, which allows heterogeneous data and software to work together in an interoperable way. For example, the data service standards like the Open Geospatial Consortium (OGC) Web Map Service (WMS), Web Feature Service (WFS), and Web Coverage Service (WCS) are a set of standard interfaces for finding and accessing data in data archives of varied sizes and sources. The OGC Web Processing Service (WPS) specifies a standard interface and protocol for discovering and executing distributed geoprocessing processes. And the Open Modelling Interface (OpenMI) provides standard interfaces to support data exchange at runtime [2], which enables interoperability across modelling components. The main geospatial standards are provided by ISO/TC

211 and OGC. ISO/TC 211 [3] is a standard technical committee in International Organization for Standardization (ISO), which is responsible for ISO series of standards in geographic information. In the geospatial Web services area, OGC is the main organization working on developing geospatial services standards by adapting or extending common Web service standards.

Service Oriented Architecture (SOA) is increasingly popular in recent years, and a large amount of geospatial data and processing functions are available in the form of online Web services. OGC has developed a set of OGC Web Service (OWS) testbeds, through which a set of open standards are developed, such as WMS, WFS, WCS, WPS, Web Coverage Processing Service (WCPS), Web Map Tile Service (WMTS), and Catalogue Service for Web (CSW). WMS provides HTTP interfaces for clients to access to geo-registered map images from distributed geospatial databases on web map server. WFS provides feature operations for clients to communicate with servers on features and feature properties. WCS defines a standard for sharing and interoperability of coverage data. The WCPS defines a protocol-independent language for the extraction, processing, and analysis of multi-dimensional coverage data. WMTS provides an interface for accessing to some map service. The OGC WPS specification makes traditional analysis functions accessible through standard interfaces on the Web. The OGC CSW standard specifies interfaces, bindings, and a framework for metadata catalogues to publish and discover geospatial data, services and related resource information. OGC Sensor Web Enablement (SWE) [4] standards provide a framework for sensor web interoperability. For example, Sensor Observation Service (SOS) specification provides standard interfaces for accessing sensor observations. The Sensor Event Service (SES) performs filtering of sensor data (streams) and send notifications of events [5].

The distributed services following OGC standards could be combined together as workflows to automate the extraction of information and knowledge from heterogeneous geospatial data. This paper will present an on-going China "MashUp" project that adopts OGC standards to facilitate information interaction among multiple organizations in China. The project is being coordinated by the OGC China Forum. It will provide a demonstration/mashup by organizations in China designed to show exactly how OGC services work using Chinese scenarios and Chinese data. Two scenarios are used to demonstrate the advantages of using standard services and workflow technologies.

The remainder of the article is organized as follows. Section 2 introduces the role of OGC-China Forum. Section 3 presents the OGC China forum initiative. Conclusions and future work are given in Section 4.

2. The role of OGC-China Forum

The OGC-China Forum can be traced back to October 2007, when the OGC China Symposium was held in Wuhan, China. The symposium topics include the OGC standards process and infrastructure, the business and technology environment for OGC standards, overview of OGC standards architecture from business and development perspectives, OGC standards for Location Services, Sensor Webs and spatial-temporal apps. [6]. The goals of the symposium aim to introduce OGC to the geospatial community in China and initiate effective communications concerning geospatial standards development. This symposium improved the cooperation between China and OGC. Later, the Asia Forum of the OGC was established in October 2011. China is one of the founding members. The OGC Asia Forum aimed to address OGC outreach in the Asia region [7].

The OGC-China Forum could help improve awareness of geospatial interoperability standards in China and assist in the definition of a set of standards to meet requirements of China. In addtion, OGC–China Forum also provides a platform for Chinese developers and users to access the latest products and services of OGC.

3. An OGC-China forum initiative

The "MashUp" project involves six organizations from the Chinese community:1) Wuhan University (abr. WHU); 2) National Geomatics Center of China (abr. NGCC); 3) Institute of Remote Sensing & Digital Earth, Chinese Academy of Sciences (abr. RADI); 4) Fuzhou University (abr. FZU); 5)

Zhengzhou Institute of Surveying & Mapping (abr. ZISM); and 6) Wuda Geo Information Engineering Technology Co. Ltd. (abr. GEOSTAR). The organizations involved include both OGC members (e.g., WHU, RADI, FZU, ZISM) and non-members (e.g., NGCC, GEOSTAR). They will join the two scenarios on disaster management. In the scenarios, they will work together to integrate sensor observations, data, and processing services, and improve the automation of data analysis process using OGC standards.

3.1. Scenarios

Two motivating scenarios are used to demonstrate how OGC standards benefit the interaction among multiple organizations and how the automation of data analysis process works. The first scenario is to derive flood maps in the Poyang Lake, Jiangxi. And the other one is to generate turbidity maps on demand in the East Lake, Wuhan, China.

(1) Scenario 1

Supposing Mr. Li is a staff member in a disaster monitoring department in China, and he wants to know the situations of flood inundation in a region around the Poyang Lake in China in 2010. One satisfactory result would be a thematic image for the flood area, which renders regions that have different flood situations using different colors.

Mr. Li would like to formulate a geoprocessing task on the flood analysis, which can generate such an image. He finds that a geoprocessing model can perform the flood analysis and output the image (figure 1). The model uses input images in different periods. In each period, there are two images in the first and second wavelength bands of MODIS data respectively. Each group of images goes through Normalized Difference Vegetation Index (NDVI) calculation, binarization, and rendering processes to gain the range of the flood in the same place in each period. Then, the resultant images can be mixed by a blend process to generate the required flood thematic image.

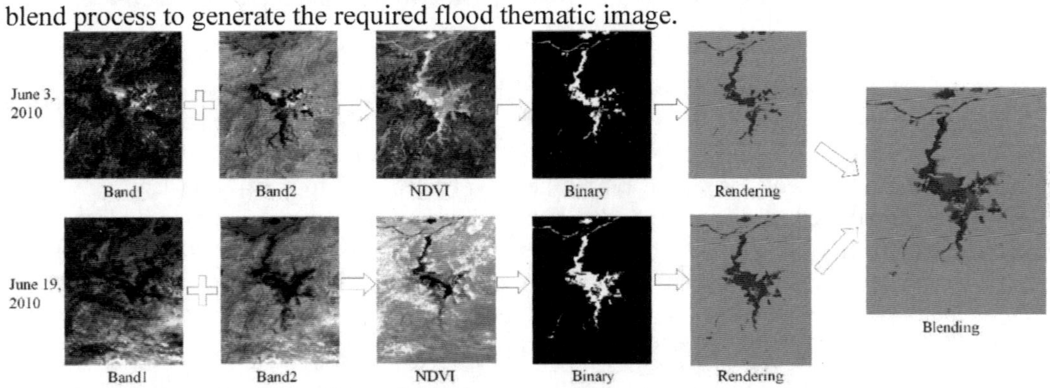

Figure 1. A scientific workflow for change extraction of water coverage area from remote sensing imagery.

In this scenario, the images could be provided by data services following Web Coverage Service (WCS) specification, while the processes could be accessed via OGC WPS interfaces. It is then possible to integrate data and computational resources in a distributed information infrastructure using workflow technologies. In case users want to generate another flood thematic image in other periods, users just need to adjust input parameters. Then time-series maps on demand are produced.

(2) Scenario 2

Suppose Mr. Wang is an employee of the National Disaster Reduction Center (NDRC) in China. He is responsible for monitoring the water quality of East Lake in Wuhan City, Hubei Province. He asked the GEOSTAR company to place a number of water quality sensors in East Lake, which can continuously collect chlorophyll concentration and water turbidity data. When detecting anomaly (e.g., Water turbidity exceeds the threshold), Mr. Wang will collect suitable remote sensing observation covering East Lake, process the data based on a turbidity extraction processing flow (figure 2). The observation imagery will go through orthorectification, radiometric calibration, atmospheric correction, clipping, Normal Differential Water Index (NDWI) calculation, mask building, silt inversion, and mapping processes to provide decision support on whether pollution happens or the range of pollution.

Traditionally, experts have to check in-situ observation data regularly, and if anomaly is detected, find remote sensing imagery and process the data step by step using an image processing software. With the standard Sensor Web technologies, in-situ observation data is sent to the data center, accessed through the SOS interface, and checked automatically against the threshold. Once an event is detected, the turbidity extraction processes could be triggered automatically. Thus interoperable services and the workflow approach improve the automation of the environment monitoring.

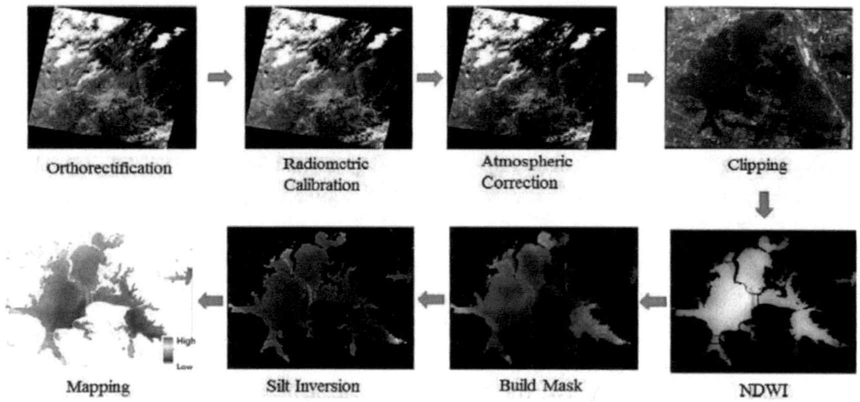

Figure 2. A scientific workflow for turbidity extraction from remote sensing imagery.

3.2. Architecture

The architecture used in the scenarios follows the publish-find-bind paradigm (shown in figure 3). The various kinds of services (e.g., WCS, WFS) provided by different Chinese organizations are published in a registry (e.g., CSW). The requestors are able to find appropriate services to consume through the registry. After the service discovery, the result services will be bound to specific tasks. For complex spatial problems, distributed services need to work together as service chains, and workflow technologies are employed to accomplish this. The integration environment will provide a user-friendly tool for service composition to support on-demand generation of geospatial data products.

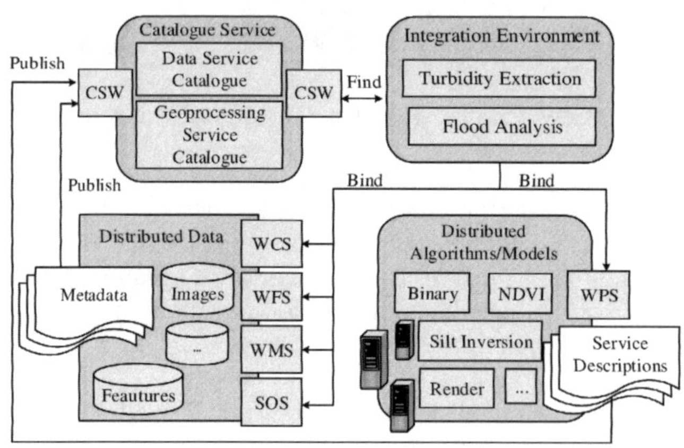

Figure 3. Service-oriented architecture using OGC standards.

In scenarios, distributed algorithms are exposed as OGC WPS interfaces, and distributed data (e.g., images, features, observations) are accessible through OGC data service standards (e.g., WCS, WFS,

SOS). All services are published into an OGC CSW. The scenarios adopt the scientific workflow approach for turbidity extraction and flood analysis, in which the workflow-based service chaining is used.

3.3. Implementations

All developed services follow the OGC standards. And, the service integration environment is implemented by extending an existing geoprocessing workflow tool, named GeoJModelBuilder [8].

Table 1 shows services provided by the six organizations. RADI provides processing functions, such as binarization and rendering used in Scenario 1, which follow the WPS specification. The processing functions used in scenario 2 are provided by WHU. The CSW is offered by WHU. FZU provides remote sensing data services. NGCC operates a national geographic information service named the MapWorld, which provides base maps. The feature service is offered by ZISM, which provides the area of interest. GEOSTAR provides a sensor observation service which follow the SOS specification.

Table 1. Services provided by six organizations following OGC standards.

Data/Processing functions	Type of services	Organizations
silt inversion, mask building, clipping, etc.	WPS	WHU
Service registry	CSW	WHU
NDWI, Binarization, Rendering	WPS	RADI
Images	WCS	FZU
Features	WFS	ZISM
MapWorld	WMTS	NGCC
Observations	SOS	GEOSTAR

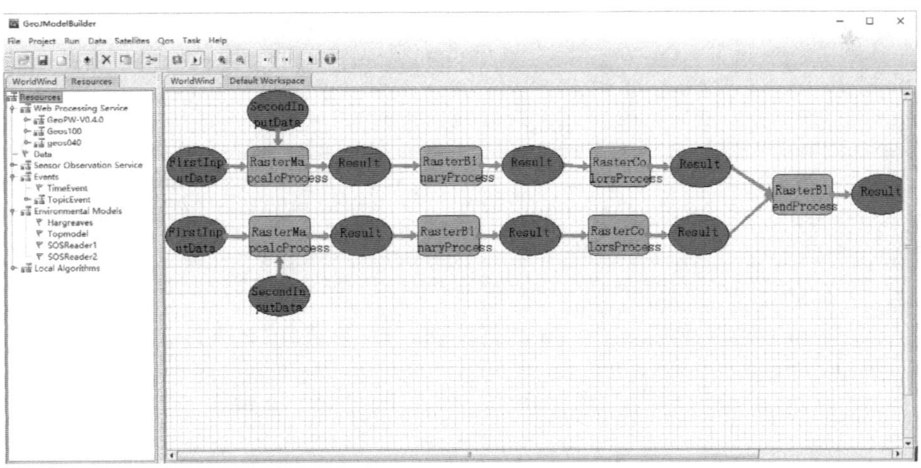

Figure 4. Flood analysis in GeoJModelBuilder.

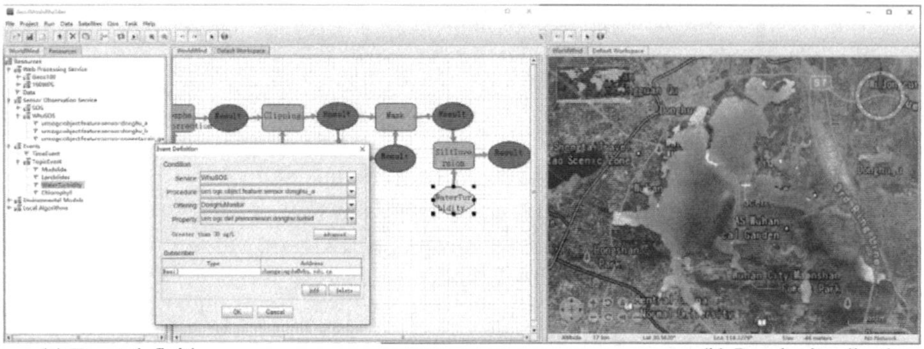

(a) Event definition (b) Result visualization

Figure 5. Turbidity extraction in GeoJModelBuilder.

GeoJModelBuilder is a geoprocessing workflow system for environmental monitoring and integrated modelling, which could couple sensor web, geoprocessing services, and data services using workflow technologies. The tool is extended to support the ability to discover appropriate services from a service registry following the OGC CSW specification. Figure 4 shows that the services are integrated for the flood analysis in the GeoJModelBuilder. In the figure, rectangles stand for services and ellipses stand for input or output. When users double click on the rectangle, a dialog is popped up to set parameter values. Users could use the *Search* button to search for input data from a service registry. Figure 5 shows that GeoJModelBuilder is used for turbidity extraction. Users could specify the threshold of observations to define an event. Once the event happens, the turbidity extraction process will be triggered automatically. The result also could be visualized in the GeoJModelBuilder.

3.4. Result analysis

In the two scenarios, various geospatial data and processing functions from six organizations are coupled into a workflow in a flexible, reusable, and interoperable way using OGC standard services. The first case derives flood maps in the Poyang Lake in China. Once MODIS images in different periods are provide, a thematic map showing the change of water coverage area will be produced. In the second case, SOS and SES are used to support "active" environmental monitoring. When an event is triggered, the turbidity extraction process could be executed automatically. Users don't need to extract the turbidity maps step by step. The ability to share data and processing functions across organizations using standard services could change traditional manual interactions in their business processes, and provide on-demand decision support results by on-line service integration.

4. Conclusions and future work

In this paper, an ongoing OGC China Forum initiative is introduced. OGC standards are used to share data and processing functions in an interoperate way over the Web. Geospatial services from various organizations could be integrated for the automation of information and knowledge extraction. Two scenarios are developed to demonstrate the advantages of using standard services and workflow technologies to provide on-demand decision support. Six organizations in China are involved in the "Mashup" project. The work is still going on. For example, since the Donghu lake is located inside the city of Wuhan, the second scenario could be extended to support some cases on smart cities. We will have a special OGC-China Forum session to make further planning in the annual conference of Chinese GIS society, which will be held in Shenzhen University, Shenzhen, China in September 2016.

Acknowledgements
The work was funded partly by National Natural Science Foundation of China (41271397), Hubei Science and Technology Support Program (2014BAA087), and Program for New Century Excellent Talents in University (NCET-13-0435).

References

[1] Tatem AJ Goetz SJ Hay SI 2008 Fifty years of earth observation satellites. *American Scientist*, vol 96 no 5 pp 390-398.

[2] Gregersen J, Gijsbers P and Westen S 2007 OpenMI: Open modelling interface *Journal of Hydroinformatics* vol 9 pp 175-191

[3] ISO/TC 211 Geographic information/Geomatics *http://www.isotc211.org/*

[4] Botts M Percivall G Reed C and Davidson J 2008 OGC® sensor web enablement: Overview and high level architecture *GeoSensor networks* pp 175-190.

[5] Echterhoff J and Everding T 2011 OGC Event service-review and current state *Open Geospatial Consortium. Inc*

[6] The OGC China Symposium Agenda *http://www.ogcnetwork.net/node/324*

[7] OGC Asia Forum *http://www.opengeospatial.org/ogc/regions/AsiaForum*

[8] Yue P Zhang M and Tan Z 2015 A geoprocessing workflow system for environmental monitoring and integrated modelling *Environmental Modelling & Software* vol 69 pp 128-140

Big Data Cognition for City Emergency Rescue

Xin Zhang[1] Yongxin Chen[1] and Weisheng Wang[2]

[1] Institute of Remote Sensing and Digital Earth, Chinese Academy of Sciences, Beijing, China 100101;

[2] Xinjiang Institute of Ecology and Geography, Chinese Academy of Sciences, 818 South Beijing Road, Urumqi, Xinjiang, China :830011)

E-mail: zhangxin@radi.ac.cn

Abstract. There are many kinds of data produced in the city daily life, which operates as an elementary component of the citizen life support system. The city unexpected incidents occurs in a seemingly unpredictable patterns. With the Big Data analysis the emergency rescue can be carried out efficiently. In this paper, the Big Data cognition for city emergency rescue is studied from four perspectives. From the data volume perspective, the spatial data analysis technology is divided into two parts, the indoor data and the outdoor data. From the data velocity perspective, the big data is collected from the eyes in the sky and objects on-the-ground networks, together with demographic data. From the data variety analysis perspective, the population distribution data, the socio-economic data and model estimates are included. From the data value mining perspective, the crime model estimates are studied. In the end, the application in the big public venues emergency rescue is introduced, which is located in Urumqi, Xinjiang, China.

Massive amounts of data are produced in the process of urban development, such as urban services data, environmental protection data, and public safety data and so on. With the constant development of information technology, the collection, analysis, integration of these vast amounts of information become a reality. City unexpected incidents often have great destructive power, which is difficult to predict and cope with.[1] However, through the integration of diverse source information and collaboration between different departments in the city, we can make use of resources and response the emergency incidents timelier. Big data has changed people's cognitive pattern. It not only refers to large data quantification, diverse data and large data analysis and utilization, but also contains a thinking of big data which means considering the whole as target sample, focusing on the concept of relevance, and a cognitive pattern that try to process, analysis mass data, and thereby participate in the decision.[2]

The comprehensive ability of high resolution remote sensing or aerial imagery, multi-level large-scale topographic maps, 3D spatial information and other sources of information need to be developed. The timeliness and authenticity of urban environment simulation are also hard to meet the need of intelligence decision. In this paper, the Big Data cognition for city emergency rescue is studied from four perspectives: data volume, data velocity, data variety analysis, and data value mining.[3] Finally, this paper introduces a big public venues emergency rescue application cases in Urumqi, Xinjiang China.

1. The characteristics of big data

1.1 The Large Data Volume

The most important feature of big data is the massive data volume. The International Documentation Center (IDC) has issued a report named with the "digital universe" in 2020 which predicts that by 2020 the global digital universe will swell to 40000 exabytes, between each person is 5200 GB or more. Take the location data which includes the indoor and outdoor location as an example, the amount of data is far beyond the traditional survey methods that can be acquired. The accumulation of record involves progress and time. The amount of data is so big that we can can't fully analysis and utilization them.[4]

1.2. The Velocity of Data

The application of big data usually request collect, analysis, react to the data rapidly and accurately. For example, the emergency rescue need the data which is collected from the eyes in the sky and objects on-the-ground networks, together with demographic data. An efficient multi-source heterogeneous data engines that can operate the massive amount of data is needed.

1.3 Data variety analysis

A challenge in the use of big data is that the data is more and more unstructured and complicated. The data needed in the emergency rescue contain real scene pictures / Street View, 3D models, the population distribution data, the socio-economic data and so on.[5] Spatial position sensor, different application, gathering content vary widely, which leads that the data from the format to the data storage pattern are significantly different.

1.4 Data mining

Big Data is not just for information extraction but it's bound to combine with data mining, especially mining implicit, non-obvious patterns, rules and knowledge. According to the different application, the big data cognition require that analysis the relevant data in detail. In this study, a lot of crime data is used to build a emergency model which could determine position with a high incidence of crime or emergency. With the emergency model, we could forecast key areas that emergencies might occur, guide people emergency exercises, and develop emergency plans.[6]

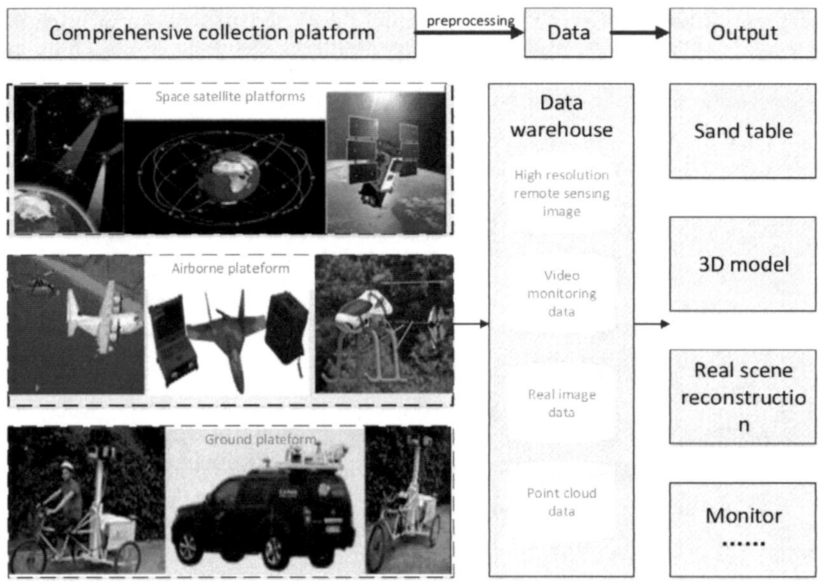

Figure 1. Big data

2. Stage of big data application

Emergency management can be divided into three stages: monitoring and preparing phase, facing-it phase and recovery and rehabilitation phase according to the sequence of the emergency.

In the phase of monitoring and preparing, configure the software and hardware equipment and collect relevant information and data as detailed as possible. The big data take the whole as the sample. At the same time, pay attention to the personnel training.[7]

In the phase of facing-it, big data means that we could integration the advanced technical support and the wisdom of lots of people conveniently. The big data bring great for collaboration and also make the efficiency of the emergency disposal high.[8]

In the phase of recovery and rehabilitation, big data can gather many scattered forces on the Internet altogether and conclude the optimal solution. For example, with the application of big data, we can analyze where area is the most serious or design optimal rescue route in order to prevent the waste of resource.

3. Application

Due to the situation of indoor evacuation system is complicated and hard to assess authentic, there is no effective method to test whether the operational emergency plan pre-established policies or security is valid and feasible, and policymakers cannot get a way to gain experience in dealing with large-scale events or disasters. So it's hard for them to make the right decisions in times of emergency.

Figure 2. Emergency simulation

This paper introduce an application of big data in which a large-scale events venue 3D simulation scenarios is created to simulate emergency including fire and blasts and the emergency response process after the event simulation in China - Asia-Europe Expo Urban Activity. The large venues in the city always gather large numbers of people and traffic in short term. Rapid evacuation should be implemented when emergency occurs. We build the virtual environment of interactive 3D model, integrated with video monitoring data, personnel deployment data, radio wave data, basic geographic data, social-economic data and so on. So, we can utilize the system to simulate a variety of emergency situations.

4、Conclusion and Future work

Big data technology can help to improve the auxiliary decision-making ability when facing the emergency. The government, enterprises and individuals should continue to support the research of big data technology. Encourage the application of emergency management of large data help to improve the cities' service level. Stress the cultivation of talent team

However, the application of big data in emergency rescue still have a lot of room to improve. (1)The management of huge amounts of data efficiently still need improving. (2) The research of high performance priority algorithm to find out useful information from huge amounts data still need to be continued. (3) The use of virtual reality technology to realize the dynamic simulation of emergency monitoring ability needs to be improved in the future.

Reference

[1] Manyika J, Chui M, Brown B, et al. Big Data: The Next Frontier For Innovation, Competition, And Productivity[J]. Analytics, 2011.

[2] Li D, Wang S, Li D. Theories and Applications of Spatial Data mining[J]. 2015.

[3] Andres Sevtsuk, Carlo Ratti. Does Urban Mobility Have a Daily Routine? Learning from the Aggregate Data of Mobile Networks[J]. Journal of Urban Technology, 2010, 17(1):41-60.

[4] Sagl G, Delmelle E, Delmelle E. Mapping collective human activity in an urban environment based on mobile phone data[J]. Cartography & Geographic Information Science, 2014, 41(3):272-285.

[5] C. Shupeng, J. van Genderen. Digital Earth in support of global change research[J]. International Journal of Digital Earth, 2008, 1(1):43-65.

[6] Ding Liang, Niu Xinyi, Song Xiaodong. Urban Spatial Studies with Big Data of Mobile Location: A

Progress Review[J]. Urban Planning International, 2015, 30(4):53-58.
[7] Li Qingquan, Li Deren. Big Data GIS[J]. Journal of wuhan university: information science, 2014, 39(6):641-644.
[8] A. Gruen. Reality-based generation of virtual environments for digital earth[J]. International Journal of Digital Earth, 2008, 1(1):88-106.

Acknowledgements
The authors would like to thank anonymous reviewers for their constructive comments and suggestions. The study is funded by the Free Exploring Program of the State Key Laboratory of Remote Sensing Science of China(No. 14ZY-03), the National Science & Technology Pillar Program （2015BAJ02B01, 2015BAJ02B02） the Special Research Project for the Commonweal of the Ministry of Water Resources of the People's Republic of China (grant no. 201201092),the National Natural Science Foundation of China (grant no. 61473286 and 61375002).

Three-dimensional Modelling Technology for City Indoor Positioning and Navigation Applications

Xin Zhang[1] Yongxin Chen[1] and Weisheng Wang[2]

[1] State Key Laboratory of Remote Sensing Science, Institute of Remote Sensing and Digital Earth, Chinese Academy of Sciences, Beijing, China 100101;
[2] Xinjiang Institute of Ecology and Geography, Chinese Academy of Sciences, 818 South Beijing Road, Urumqi, Xinjiang, China :830011)

E-mail: chenyx@radi.ac.cn; zhangxin@radi.ac.cn

Abstract. For city indoor positioning and navigation applications, there are two technical problems should be solved, which are the modelling efficiency and the validity and accuracy of the spatial models. In this paper, the quick modelling technology is introduced which uses the multi-angle remote sensing based on the unmanned aerial vehicle measurement. To enhance the validity and accuracy of the spatial models, we proposed an algorithm to remove the line style and planar style foreground occlusions before reconstructing backgrounds. The three-dimensional models can only provide the spatial framework for the city indoor positioning. Furthermore, the simple indoor three-dimensional modelling technology which is based on the building design drawings. In the end, the application in the public safety emergency rescue is introduced.

Under the guidance of the concept of smart city, the future urban development proposes higher demand of city public security. A thorough indoor navigation technology is without fail the essential guarantee condition and an important technical part of the modern city safety control system. There are two key issues in the indoor positioning and navigation modeling process: fast three-dimensional modeling and spatial model accuracy problems. However, the function and the structure of modern city tends to be diverse and complex. The data sources and spatial accuracy of existing underlying database cannot meet the rapid, accurate modeling requirements. The acquisition method of city multidimensional data need to be strengthened [1].

To solve the above problem, this paper proposes a method to build a multi-angle and multi-level three-dimensional city foundation database based through the integration of data processing and rapid three-dimensional urban modeling technology. In order to improve the accuracy of the spatial model, this paper describes an algorithm to remove the line style and planar style foreground occlusions before reconstructing backgrounds. This algorithm can realize the modeling of buildings and streetscape rapidly. The indoor modeling of the building can be realized through building design drawings and modeling software. Finally, this paper introduce an application in the public safety emergency rescue.

1. Current status and progress

1.1 The platform of spatial information retrieve

With the rapid development of computer technology, communications network technology, the improvement of the unmanned aerial vehicle (UAV) performance and continuous availability to a variety of digital, light weight, small size, and high precision detection sensors, spatial data collection technology has achieved rapid development. The traditional ways of conventional ground survey, such as astronomical geodesy, total station, GPS receivers, are not suitable for collecting spatial data. The way of collecting spatial coordinates and attributes by-point or by-sectional surface based on laser radar is out of date. Instead, space and aerial remote sensing is a feasible way of acquiring high resolution image in a large area, which can be further extracted geometric and physical characteristics to express the subtle surface features. During the research and development of sensors, only Charge-coupled Device (CCD) cameras is used at first. But now the CCD and the laser scanners can be used together to acquire planar information and three-dimensional information of the city at the same time. The user can use close-range aerial photogrammetry survey to meet the demand of high precision spatial data.

UAV remote sensing status and progress can be summarized as the following characteristics: (1) Using plane array camera to collect data. (2) The integration of low-altitude remote sensing data acquisition and post-processing. (3) The widespread use of high-precision navigation positioning and fixed point exposure technology. (4) Processing data automatically.

The characteristics of oblique photography can be summarized as following: (1) Acquiring abundant space images. (2) The capability of serving for mass levels of application. (3) The outstanding advantages in terms of three-dimensional modeling.

The integrated application of Global Positioning System (GPS) and inertial navigation system (IMU) technology promote the development of airborne, automotive and other three-dimensional data acquisition system [2]. There are many urban ground vehicle data acquisition system, such as with the current Optech Lynx laser collection vehicle, Earthmine panoramic collection vehicles, Avideh Zakhor dual laser radar scanning system and so on. These three-dimensional data acquisition system equip the CCD cameras and laser scanners simultaneously. So they can settle the problem of loss information caused by the obstruction from high buildings.

1.2 Review on city 3D modeling method

Three-dimensional modeling of surface feature is based on geometric approach originally. The model is entirely generated by geometric modeling software manually. The complex and precise object model requires experienced operators to spend a long time. At the next stage, modeling based on the image is developed. The method use a camera to capture a set of objects in the image, and then reconstruct the geometry of the object. The relatively new 3D modeling method is to obtain three-dimensional model of the surface features through various optical equipment. The method is widely used in vehicle and airborne 3D data acquisition system. In addition, using a laser rangefinder instrument to extract the depth information of a number of probing points which cover the entire surface of the object is a hot new technology of three-dimensional modeling.

In recent years, some new 3D modeling of buildings reconstruction methods have been proposed. The trend is embodied in three aspects: (1) the integration of diverse data. (2) Introducing the theory of artificial intelligence in the process of identification matching. (3) The diversification of building model, such as parametric model, prism model, polyhedral model, CSG models and so on [3].

There are some developed reconstruction systems which include TotalCalib system developed by the Bougnoux from French INRIA, 3D surface automatic generation system developed by the Pollefeys etc. from Belgium KULeuven University and the PhotoBuilder system developed by the University of Cambridge Computer Vision Research Group.

2. Research method

2.1 Data source for modeling

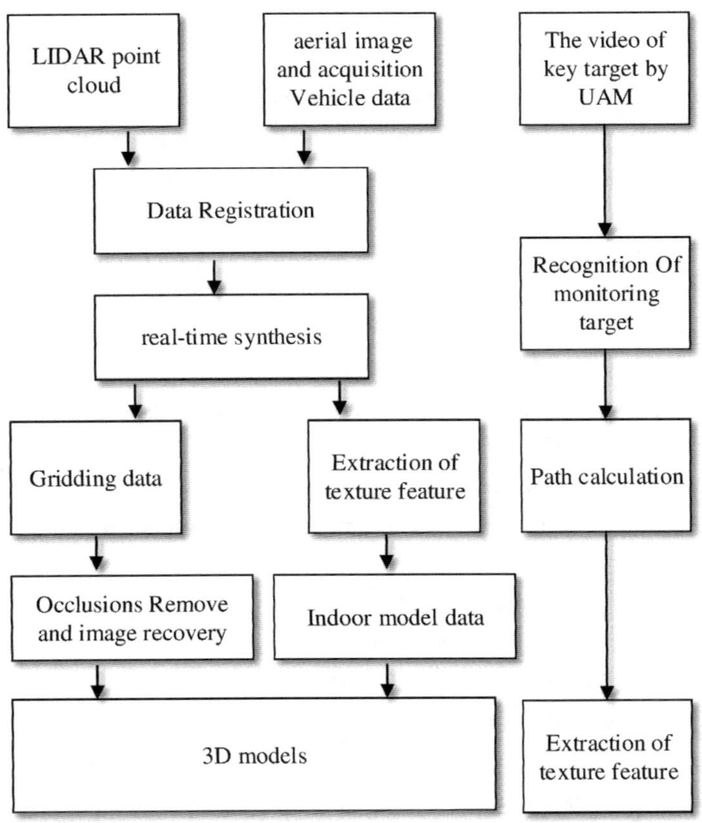

Figure 1. Brief technical roadmap

This research mainly relying on the high-resolution Earth observation system as well as some close-range data acquisition devices. Based on the think of a multi-platform system integration, we build a three-dimensional data acquisition and surveillance platforms which integrate space, air, land data. And the research proposes a method to build a multi-angle and multi-level three-dimensional city

IOP Conf. Series: Earth and Environmental Science **46** (2016) 012040 doi:10.1088/1755-1315/46/1/012040

foundation database based through the integration of data processing and rapid three-dimensional urban modeling technology [4]. In the aspect of software systems development, we use the indoor and outdoor positioning integration technology for the establishment of three-dimensional virtual city scene and network application service system. On the network, we set up the integration process of urban three-dimensional information monitoring / control, the reproduction of three-dimensional scene dynamically, and intelligence service. In the same time, the system can visualize the city static and dynamic models in one scene and develop an interactive virtual urban environment to achieve dynamic intelligence, simulation and contingency plans and emergency maintenance applications.

2.2 Occlusions Remove algorithm

Removing occlusions on the foreground image will be based on abundant linear information straight of the building which can be used to split planar obstruction. The algorithm adopts edge protection template matching method to recover the block texture. Edge detection and tracking method is applied to extract the image on the blocked area for the simple linear vertical obstruction. The texture recovery can be realized by a one-dimensional mean value interpolation and bulk filling way. Experiments show that the method has good application effect.[5]

This research make full use of the structure information of three-dimensional buildings and obstructions to separate and track the blocked area on image. The texture of blocking area only correlate with its neighboring image pixel. The sequence images with overlapping degree acquired from the photogrammetric way can fully meet the requirements of image number.[6] The obstruction is classified for two types according to its shape. One obstruction type is planar images such as a large pedestrian occupied area, car parked on the street; the other is linear objects, such as the street lights, flagpoles. The linear objects showed a narrow strip area on the image. The algorithm adopted different methods for planar and linear block objects removal.

The brief introduction of occlusions remove algorithm is introduced as follows:

Divide the image by a certain size of a rectangular grid, and count the parallel lines density of the grid. If grid density is less than a given threshold, it is considered in the shelter area, otherwise, it is not blocked area. The results from above operation will contain misnomer segmentation block grids area which is generally considered containing less linear features or being covered by substantial advertising signs.

(2) Eliminate false segmentation grid

Compare the segmented block grid with the corresponding grid in the adjacent image from image sequence each grid to judge the degree of similarity to determine whether the adjacent grid is incorrect segmentation grid. If the two grid have a high degree of similarity, segmented block grid can be considered wrong, and removed the grid block. To avoid the situation that the grid area cannot block the entire block, expand a grid width along the border after removing the incorrect segmentation block grid.

(3) Template matching method based on edge preservation

With the right range blocked area, texture can be acquired from the adjacent image. This paper adopts the template matching algorithms based on edges preservation to recover the texture of the blocked area. This method references the priority edge ideas in image patching algorithm and the principle of optimal pixels local texture [7].

Obtain the rough linear edge by edge detection and tracking method. If the edge width and around texture satisfy some certain conditions, use the one-dimensional average interpolation for recovery, otherwise use the affine transform.

3. The application

Study of dynamic spatial evolution of spatial objects on fire, sudden mass emergency and other serious situation help to improve emergency response capacity. Combined with the emergency response process of emergency management department after the event, setting up the three-dimensional model of unexpected events and rules of the sport corresponding dynamic model can provide support for urban public safety in an emergency situation emergency treatment.[8]

Take the based geographic data of a city in Xinjiang as test data. Combined with the convention center, stadiums and other venues, facilities and space data related municipal infrastructure projects. In the same time, collect the real-time data of the stadiums environment in and around dynamically. Establish the large-scale 3D simulation scenarios of events venue.

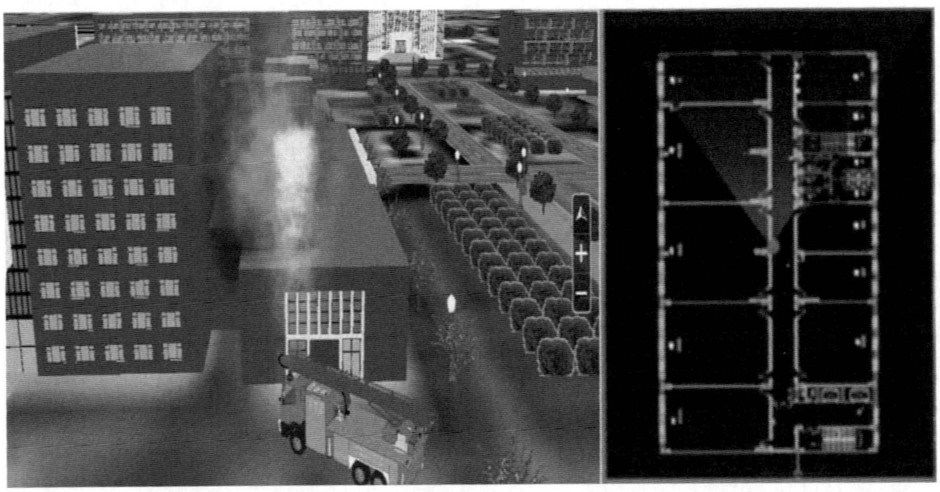

Figure 2. A sketch of emergency rescue and indoor navigation

In the process of simulation, add the static 3D model such as building, road, streetlight. Add the car model and route information to make the car moving on the road. Add water pumps, fire, fire engine model to simulate the fire. The focus is on the integration of indoor and outdoor positioning information, analog of building and interior firefighters, and rescue personnel process.

References

[1] Feng Mei. The study of 3D modeling method based on the LiDAR and aerial images[J]. Bulletin of Surveying and Mapping, 2011, 1(12): 16-18.

[2] J. Gong, X. Yang, F. Su, et al. An approach to extracting information of residential areas from Beijing-1 image based on Gabor texture segmentation[J]. International Journal of Digital Earth, 2009, 2(2):186-193.

[3] J. Gong, X. Yang, F. Su, et al. An approach to extracting information of residential areas from Beijing-1 image based on Gabor texture segmentation[J]. International Journal of Digital Earth, 2009,

2(2):186-193.
[4] Mcardle G, Tahir A, Bertolotto M. Interpreting map usage patterns using geovisual analytics and spatio-temporal clustering[J]. International Journal of Digital Earth, 2014, 8(8):1-24.
[5] Zheng Shunyi, Wang Zhihe, Yang Fan. The method of fast automatic 3D modeling of city based on visual multi-image[J]. INFRARED, 2007, 28(4):18-22.
[6] Wang wei, Huang Wenwen, Zhen Jiao. Pictometry Oblique photography and its application in 3D modeling of city [J]. Geomatics & Spatial Information Technology, 2011, 34(3):181-183.
[7] Jiang Julang. Research and Realization of Texture Mapping Techniques[D]. Hefei University of Technology, 2003.
[8] Tan Yong. The Study on Selection of Rescue Vehicle Routing Base onEmergent Events about City [D]. Lanzhou Jiaotong University, 2014.

Acknowledgements
The authors would like to thank anonymous reviewers for their constructive comments and suggestions. The study is funded by the Free Exploring Program of the State Key Laboratory of Remote Sensing Science of China(No. 14ZY-03), the National Science & Technology Pillar Program （2015BAJ02B01, 2015BAJ02B02） the Special Research Project for the Commonweal of the Ministry of Water Resources of the People's Republic of China (grant no. 201201092),the National Natural Science Foundation of China (grant no. 61473286 and 61375002).

Extraction of cross-sea bridges from GF-2 PMS satellite images using mathematical morphology

Chao Chen[1], Xinxin Sui[2], Guangwei Zhen[3], Biyun Guo[1] and Xiaowei Chen[1]

[1] Marine Science and Technology College, Zhejiang Ocean University, Zhoushan, Zhejiang 316022, China
[2] China Aero-geophysical Survey and Remote Sensing Center for Land Resources, Beijing 100083, China
[3] Sinomine Resource Exploration Company Limited, Beijing 100089, China

E-mail: sxx5666@163.com

Abstract: Cross-sea bridges are typical target of natural sceneries. Extracting cross-sea bridges from remote sensing images appears significant. In this study, for the complexity of ground objects in GF-2 PMS satellite images, we have selected direction-augmented linear structuring elements to be used for extraction of cross-sea bridges. Firstly, pre-processing of original image has been carried out; secondly, calculate NDVI and extract water bodies; thirdly, according to the direction of water bodies, select appropriate direction-augmented linear structuring elements, proceed mathematical morphology operations on the water bodies, and assisted by prior knowledge of the bridge, extract the bridge object; finally, select an experimental area to verify the effectiveness and applicability of the method. Our work shows that the direction-augmented linear structuring elements are effective and efficient for extraction of bridges with different directions in GF-2 PMS satellite images, and the approach in this study is important to expand the application areas of domestic high-resolution remote sensing images.

1. Introduction

Information extraction is the hotspot and difficult point in remote sensing image processing on current. As cross-sea bridge is an important artificial building and transportation hub. Due to this particularity of its location, accurately recognition and precise positioning of the bridge target is beneficial not only for the timely update of geographic information data, but also for the disaster prevention and reduction, and the function as a prior referential knowledge for other objects.

In recent years, many methods of bridge extraction have been proposed using remote sensing images. In summary, the methods can be roughly divided into three categories: the method based on difference image, edge detection, and the method based on road information. These three methods have achieved a certain extent of success in some applications. The method based on different image has problems such as hard control on the times of erosion and dilation, and unicity of structural elements despite the fast operation speed. The method based on edge detection picks up bridge targets only by the feature of parallels, not considering the location relationship between the bridge and the river, leading to more production of false information and the uncertainty of the Hough transform parameters. The method based on road information requires the knowledge known of road information, at the same time road information extraction is a difficult point in remote sensing image understanding, which means this kind of method has some theoretical importance but bigger difficulty in practice.

Considering the complexity and particularity of ground objects in remote sensing images, aiming at the flaws such as geometric form's unicity, and weak directivity of structuring elements in mathematical morphology, in this paper, direction-augmented linear structuring elements are selected, and be used for extraction of cross-sea bridges with different directions in remote sensing images, to solve problems like incompleteness of extraction in utilizing common structuring elements to extract the same category ground objects with different directions in one remote sensing image.

2. Method

The specific flowchart is shown in figure 1. The main steps are divided into image pre-processing, separation of land and sea, extraction of cross-sea bridges and precision evaluation.

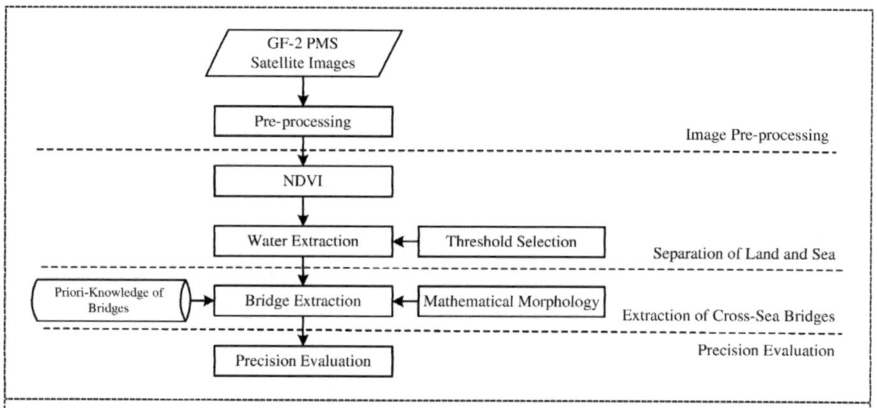

Figure 1. Flow chart of extraction of cross-sea bridge from GF-2 PMS satellite images based on mathematical morphology.

2.1 Image pre-processing

During imaging, due to the influence of atmosphere and flight attitude, the quality of the original image can't meet the requirement of target recognition. So, the image needs a pretreatment to improve the quality and enhance the accuracy of target recognition.

2.2 Separation of land and sea

Water extraction plays a decisive role in the correct recognition of bridge target. In this paper, the

NDVI is calculated, and a threshold value is selected for extraction of water bodies.

2.3 Extraction of Cross-sea bridges

On the basis of water extraction, according to the directions of the water bodies, pick the corresponding direction-augmented linear structuring elements (as shown in figure 2), respectively, and carry out mathematical morphological operations, to connect the broken water bodies because of the bridges. Then, conduct overlay analysis with the result of water extraction, potential bridge targets can be obtained. After that, based on the prior knowledge of the bridges (rectangle fit, minimal area, etc.), verify the potential bridge targets to wipe out invalid information and obtain initial bridge.

To obtain more accurate bridge targets, post-treatment is necessary on the initial bridge targets. The opening operation has the functions of eliminate tiny objects, can smooth the image, eliminate the tiny objects in highlights (e.g. the burr on the edge and isolated spots), and the closing operation has the effect of filtering on the image, able to fill the small holes and cracks of the image. So, the opening and closing operation are needed for the initial bridge targets to acquire more accurate results.

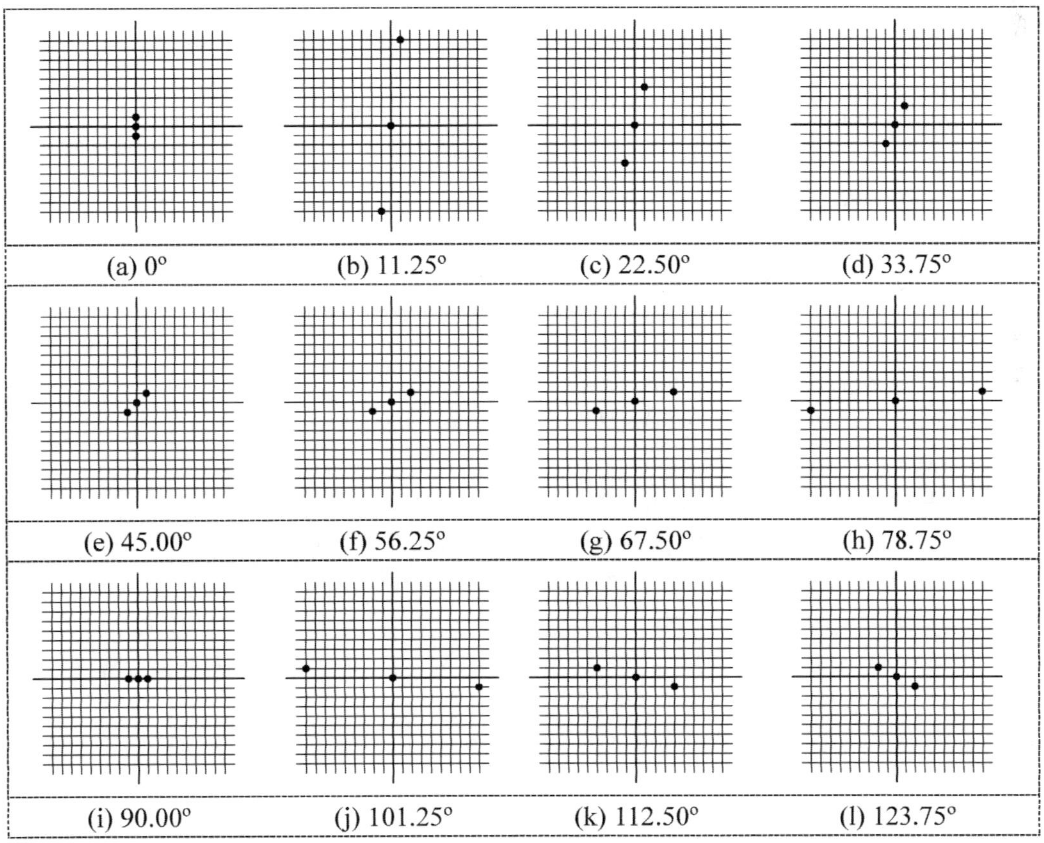

(a) 0° (b) 11.25° (c) 22.50° (d) 33.75°
(e) 45.00° (f) 56.25° (g) 67.50° (h) 78.75°
(i) 90.00° (j) 101.25° (k) 112.50° (l) 123.75°

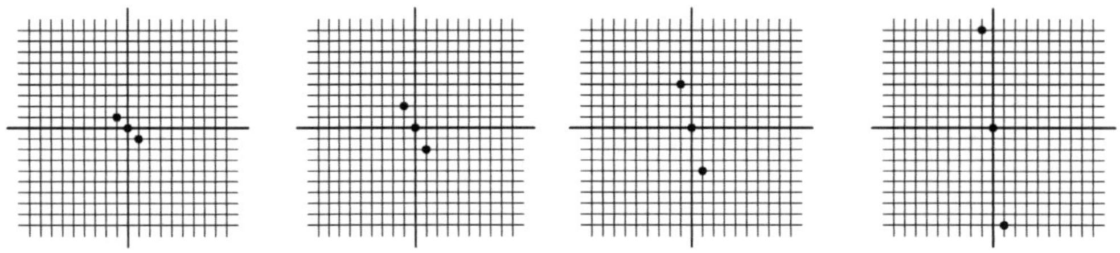

(m) 135.00° (n) 146.25 ° (o) 157.50° (p) 168.75°

Figure 2. The direction-augmented linear structuring elements.

2.4 Precision evaluation

Precision evaluation is not only the standard for judging the evaluation method and the influence on following applications, but also an important basis for evaluating the method performance, adjusting model parameters, optimizing extraction process. This article verifies the results from two aspects of qualitative and quantitative evaluation. By definition, qualitative evaluation is to judge the extraction results by visual observation with the support of prior knowledge by the observer; quantitative evaluation is to judge the extraction results using statistical parameters. For the bridge targets, the former mainly includes the quantity, location, shape and so on, whereas the latter mainly includes the length, the width and the area.

3. Experiment and analysis

3.1 Data

The GF-2 satellite is a Chinese sun-synchronous, high-resolution satellite. It was launched on 19 August 2014. The GF-2 satellite is equipped with two cameras of 0.8 m panchromatic and 3.2 m multispectral resolution with 23 km width.

As a case study, a part of an multispectral image with the size of 3023×2538 pixels from Zhoushan in China is chosen, which is taken in February, 2016, with the spatial resolution of 3.2 m. Figure 3 shows study area and the original image.

It can be seen that, this remote sensing image includes ground objects like bridges, roads, forest and water bodies. The bridge across the sea and links with the road. Affected by the spatial resolution, lanes interval lines on the road are very clear. Affected by the flow of the water and the layout of the road, the directions of four bridges are not the same. Therefore, the algorithm is required to accurately recognition and precise positioning every bridge.

IOP Conf. Series: Earth and Environmental Science **46** (2016) 012041 doi:10.1088/1755-1315/46/1/012041

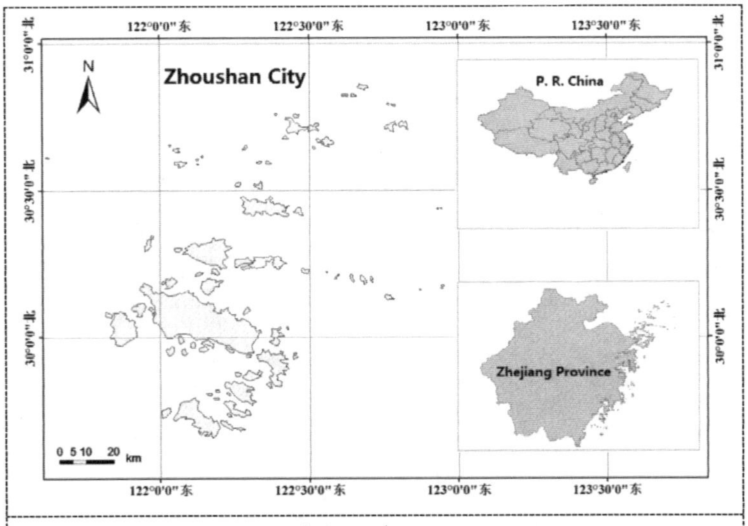

(a) study area

(b) original image

Figure 3. Study area and original image.

3.2 Experiment and results

The images are clear and high-quality because of good weather conditions, the water bodies have a sharp contrast with other objects. The NDVI threshold value of water bodies information is *0*. The result, shown in figure 4, indicates that the water bodies extracted from the image are complete, and the boundary of which is sharp.

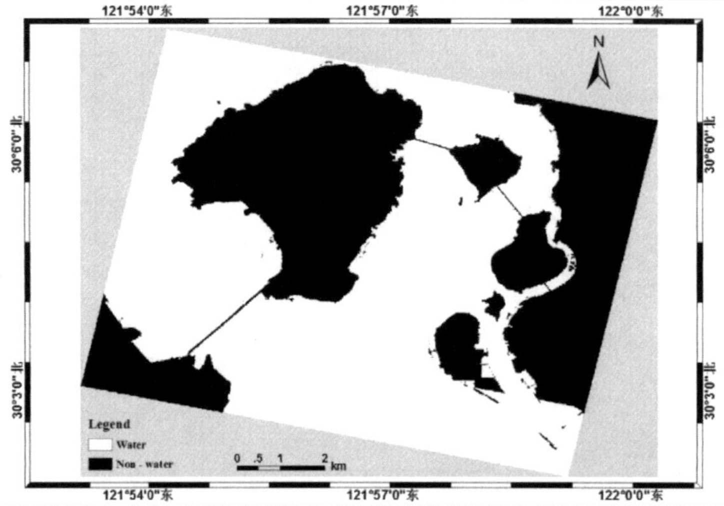

Figure 4. Water bodies.

When the water bodies are extracted accurately, the direction-augmented linear structuring elements at four different directions (33.75°, 45.00°, 56.25° and 135.00°) are employed to perform mathematical morphology operations on it to identify the potential bridges. The threshold values of the rectangle degree and the area are *0.50~1.0* and *500~20000* pixels respectively, which are selected to screen out the false bridges and obtain the initial bridges. Finally, the initial bridges are executed in post-processing by opening and closing operation of mathematical morphology to determine the true bridges. The overlap image is generated by result image and original image, as the figure 5 shows. It is obviously observed that the four bridges can be identified exactly. Comparing with the real bridges, the position is accurate and the shape is complete of bridges extraction in our study.

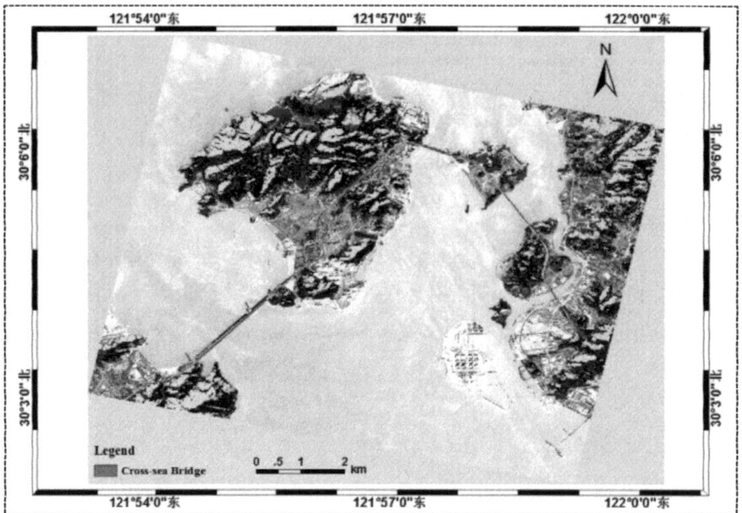

Figure 5. Results of extraction of cross-sea bridges.

3.3 Precision evaluation

3.3.1 Qualitative evaluation. By comparing the visual interpretation result of the original image and the extraction result, we can get the following conclusion. As to quantity, there are four bridges in the original image, the same as the extracted bridges. As to location, the extracted bridges match with the real ones very well. As to shape, the other three bridges are rectangular as same as the real bridges, expert Xiangjiaomen bridges.

3.3.2 Quantitative evaluation. We calculate the length (in direction of the water body), the width (in direction of the road) and the area of every single bridge. See details listed in table 1.

Table.1 Quantitative evaluation of bridge targets (the unit is pixel)

	real	**extracted**	**deviation (%)**	**real**	**extracted**	**deviation (%)**
		Xihoumen bridge			***Taoyaomen bridge***	
length	*517*	*514*	*0.58*	*229*	*231*	*0.87*
width	*12*	*12*	*0*	*12*	*12*	*0*
area	*9160*	*8112*	*11.44*	*2324*	*2187*	*5.89*
		Xiangjiaomen bridge			***Cengang bridge***	
length	*194*	*197*	*1.55*	*49*	*49*	*0*
width	*12*	*12*	*0*	*12*	*12*	*0*
area	*2785*	*1912*	*31.35*	*633*	*541*	*14.53*

The quantitative evaluation result is shown in table 1. When it comes to length and width, extracted bridges match well with true bridges, the max deviations between the extracted bridges and the true values is 1.55%. For area, because of large content of suspended sediment, the water extraction is not complete, so the extracted bridges is not completely, the maximum error for the Xiangjiaomen bridge is 31.35%. In the real world, a real bridge usually has a part located on the land, but all parts of extracted bridges are over rivers because the extraction process is based on the water extraction.

4. Conclusion and Discussion

For the defects of mathematical morphology that the geometric shapes of common structuring elements are too simple and unidirectional, direction-augmented linear structuring elements with directions are constructed and applied in the extraction of cross-sea bridges in GF-2 PMS satellite images. Our work shows that the direction-augmented linear structuring elements presented in this paper can relatively correctly extract bridges with different directions and basically satisfy the need of actual application.

But, in this study, the bridge extraction is performed on the basis of extracted water bodies, the recognition accuracy and positioning preciseness of bridge depend on the effect of water extraction. So, develop more effective method of water extraction in order to extract water body more accurately, get real width of bridge (in the direction of the road) using more prior referential knowledge, and get the direction-augmented linear structuring elements with direction to extract cross-sea bridges are our future work.

Acknowledgments

This word is supported by Zhejiang Provincial Natural Science Foundation of China (Grant NO.

LY16F010011); the National Natural Science Foundation of China (Grant No. 51479179); the Science Foundation of Zhejiang Ocean University (Grant NO. Q1502); the Grants from Key Laboratory of Oceanographic Big Data Mining and Application of Zhejiang Province (Grant NO. OBDMA201510).

References

[1] Catherine P, Dennis P L, Filipe A and Fabrice P 2016 Toward a high-resolution monitoring of continental surface water extent and dynamics, at global scale: from GIEMS (Global Inundation Extent from Multi-Satellites) to SWOT (Surface Water Ocean Topography) *Remote Sensing and Water Resources* **55** 149-165

[2] Chaudhuri D and Samal A 2008 An automatic bridge detection technique for multispectral images *IEEE Transactions on Geoscience and Remote Sensing* **46** 2720-2727

[3] Chen C, Qin Q M, Chen L, Wang J L, Liu M C and Wen, Q 2013 Extraction of bridge over water from high-resolution remote sensing images based on spectral characteristics of ground objects *Spectroscopy and Spectral Analysis* **33** 718-722

[4] Chen C, Qin Q, Zhang N, Li J, Chen L, Wang J, Qin X and Yang X 2014 Extraction of bridges over water from high-resolution optical remote sensing images based on mathematical morphology International *Journal of Remote Sensing* **35** 3664-3682

[5] Cheng G and Han J W A survey on object detection in optical remote sensing images *ISPRS Journal of Photogrammetry and Remote Sensing* **117** 11-28

[6] Claire T, Thierry R, Lucien W and Jocelyn C 2008 Synthesis of multispectral images to high spatial resolution: A critical review of fusion methods based on remote sensing physics. *IEEE Transactions on Geoscience and Remote Sensing* **46** 1301-1312

[7] Luo J, Ming D, Liu W, Shen Z, Wang M and Sheng H 2007 Extraction of bridges over water from IKONOS panchromatic data *International Journal of Remote Sensing* **28** 3633-3648

[8] Luo J, Ming D, Shen Z, Wang M and Sheng H 2007 Multi-scale information extraction from high resolution remote sensing imagery and region partition methods based on GMRF-SVM *International Journal of Remote Sensing* **28** 3395-3412

[9] Uddin M N, Anwar M F, Rahman M T and Mobin M N 2014 An investigation on the pattern of land use change in Dhaka city using remote sensing and GIS application *Journal of Environmental Science and Natural Resources* **7** 105-109

[10] Zhu H C, Cai L J, Liu H Y and Huang W 2016 Information extraction of high resolution remote sensing images based on the calculation of optimal segmentation parameters *PLoS ONE* **11** 1-15

IOP Conf. Series: Earth and Environmental Science **46** (2016) 012042 doi:10.1088/1755-1315/46/1/012042

Responses of water resource of the Yarlung Zangbo River Basin to climate changes and glacier-snow fluctuations in recent years

N Nie[1, 2], W C Zhang[2, 3] and C Deng[1, 2]

[1] State Key Laboratory of Pollution Control & Resource Reuse, School of the Environment, Nanjing University, Nanjing 210023, China.
[2] Key Laboratory of Digital Earth Science, Institute of Remote Sensing and Digital Earth, Chinese Academy of Sciences, Beijing 100094, China.

E-mail: zhangwc@radi.ac.cn

Abstract. Responses of river water resource of the Yarlung Zangbo River (YR) Basin to climate change and glacier-snow fluctuations in recent years were systematically studied by taking the best use of meteorological data and remote sensing images. Climate variation trends over the YR basin, mainly focusing on temperature, precipitation and actual evapotranspiration during 1979-2006, were examined by means of the GIS spatial interpolation techniques, meteorological statistics analyses, Mann-Kendall nonparametric test, etc. Based on the daily snow depth dataset derived from SMMR and SSM/I, annual snow-cover days and average snow depth (maximum and mean) were collected and analysed in the YR basin during 1979-2006. According to the water budget for a closed watershed, in association with the relevant achievements obtained from pioneer studies, the response of river water resources to the variation of climate, glaciers and snow cover in the YR basin was systematically studied. The results suggest that, under the background of the increasing temperature, precipitation, actual evapotranspiration and glacier melting together with the decreasing snow cover over the basin, river runoff presented the increasing trend to respond the climate and glacier changes at the speed of 18.7 $m^3s^{-1}a^{-1}$ with the significant level of $\alpha<0.1$ from 1979 to 2006.

1. Introduction

As the highest great river in the world with a mean elevation of over 4000 m [1], the Yarlung Zangbo River (YR) Basin is also one of the major maritime glacier developed regions with abundant solid (glaciers and snow) and liquid (stream flow) water resource in China. The water resources in this region are of great importance for both China and India. Prior studies have shown that the climate, glacier, snow cover and water resources over the YR basin have changed under the background of global change. However, due to the special geographical location of the YR basin, many difficulties existed for logistic support of field observations in the region, and comprehensive and systematical study on the climate change, glacier-snow fluctuations and the corresponding response of river water resource over the whole basin was difficult to conduct. This study is attempted to take the best use of meteorological data, basic geographic information, and daily snow cover datasets derived from

[3] To whom any correspondence should be addressed.

SMMR and SSM/I collected in the YR Basin, in association with the relevant achievements obtained from pioneer studies in the region, under the principle of water budget for a closed watershed, to explore the responses of river water resource to climatic and glacier-snow changes of the whole basin.

2. Data and methods

2.1. Data

Climatic data from meteorological stations located over the YR basin obtained from China Meteorological Data Sharing Service System were analyzed. Climatic variation trends over the YR basin from 1979 to 2006 for air temperature and precipitation collected from 27 meteorological stations as the geographic locations illustrated in figure 1 were explored.

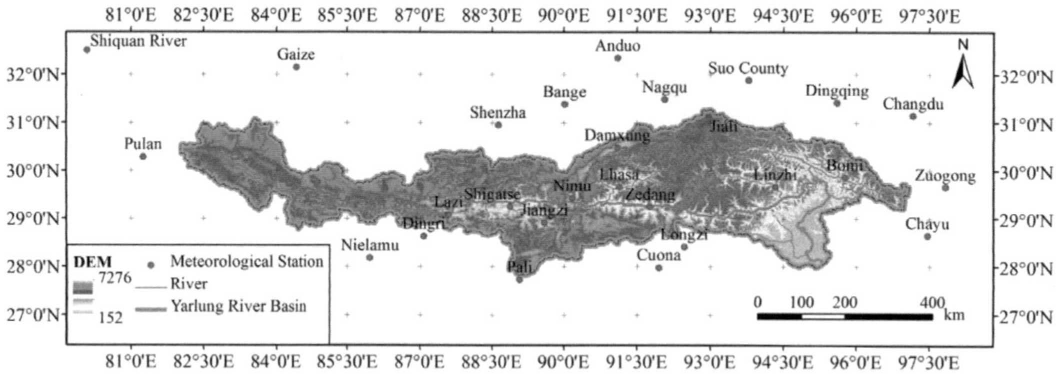

Figure 1. Map showing fundamental geographic information of the YR Basin.

Studies suggested that Takahashi Formula estimated actual evapotranspiration had large bias compared with the observations over the Qinghai Tibet Plateau because of the effects of frozen soil and snow thaw on the Plateau [2]. By introducing a coefficient on the precipitation part without changing the original form of Takahashi Formula, Fu et al.[2] amended the formula as shown in equation (1) for accurate estimation of actual evapotranspiration in Lhasa.

$$E = \frac{3100(P+X)}{3100 + 1.8P^2 \exp\left(\dfrac{-34.47T}{235+T}\right)} \qquad \begin{array}{ll} T \le 0°C,\ X = 5 & \\ 0°C < T \le 5°C,\ X = 10 & 10°C < T \le 15°C,\ X = 15 \\ 5°C < T \le 10°C,\ X = 20 & T > 15°C,\ X = 80 \end{array} \quad (1)$$

Here E indicates monthly actual evapotranspiration, P and T represents monthly precipitation and air temperature respectively. In this study, we selected equation (1) to calculate monthly actual evapotranspiration over the YR Basin. Annual actual evapotranspiration was obtained by adding monthly result of a year.

With a 7-channel 4-frequency radiometer, SMMR and SSM/I have been utilized for retrieval of various snow parameters. The daily snow depth derived from SMMR (from 1979 to 1987) and SSM/I (from 1987 to 2006) radiometric data for China can be downloaded from http://westdc.westgis.ac.cn. The algorithms and computation flow for this dataset can be found in the literature of [3] and [4]. Based on the daily snow depth dataset, annual snow-cover days and average snow depth (maximum and mean) were analyzed by using RS and GIS techniques with special attentions being paid on snow covers in the YR basin during 1979-2006.

2.2. Methodology

Common interpolation methods, such as Inverse Distance Weighting (IDW), Kriging and Spline, were applied to obtain the spatial distribution of climatic elements (temperature and precipitation), among which IDW was demonstrated the best in the YR basin [5]. Therefore, IDW was applied to obtain the spatial distribution of climatic elements in this study. Linear regression analysis (least squares method)

and Mann-Kendall nonparametric test method were used to detect possible long-term trends and significant level of variations in time series of metrological parameters. In the Mann–Kendall test, null hypothesis H_0 is that the data (X_1, X_2 ... X_n) is a sample of n independent and identically distributed random variables. The alternative hypothesis H_1 of a two-sided test states that the distributions of X_k and X_j are not identical for all k, j. The Mann–Kendall test may be simply stated as follows:
Null hypothesis: H_0. Significance level: α. Test statistics: Z_c.
Rejected H_0: $|Z_c| > |Z_{1-\alpha/2}|$, in which $Z_{1-\alpha/2}$ is the standard normal deviation and α represents the significance level of the test. The principle and the calculation steps of this method can be found in the literature of [6] and [7].

3. Results and discussion

3.1. Climate change

Time series of spatial annual mean temperature, precipitation and actual evapotranspiration during 1979-2006 over the YR basin were derived with the methodology as described in the previous section. Linear regression analysis (least squares method) and Mann-Kendall nonparametric test method were adopted to detect possible long-term trends and significant levels of variations on these meteorological parameters, and the results were shown in figure 2.

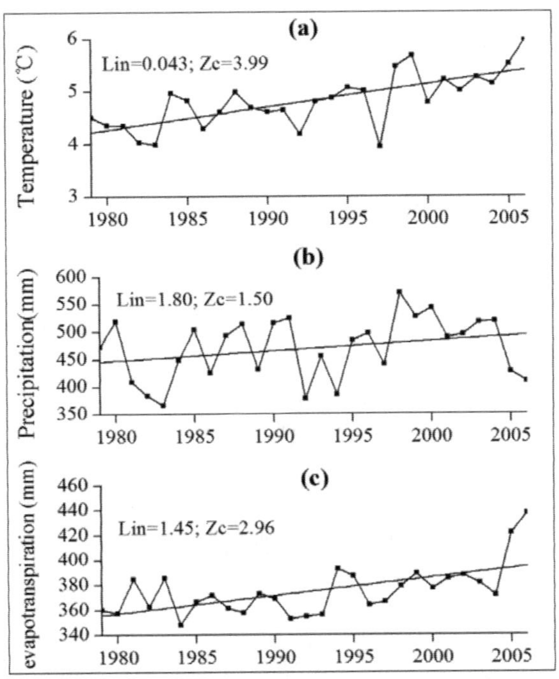

Figure 2. Variation trends of annual mean air temperature (a), precipitation (b) and actual evapotranspiration (c) over the YR basin in period of 1979–2006. The straight line represents linear trends, and *Lin* stands for linear-change-rate and Z_c for Mann–Kendall test statistics.

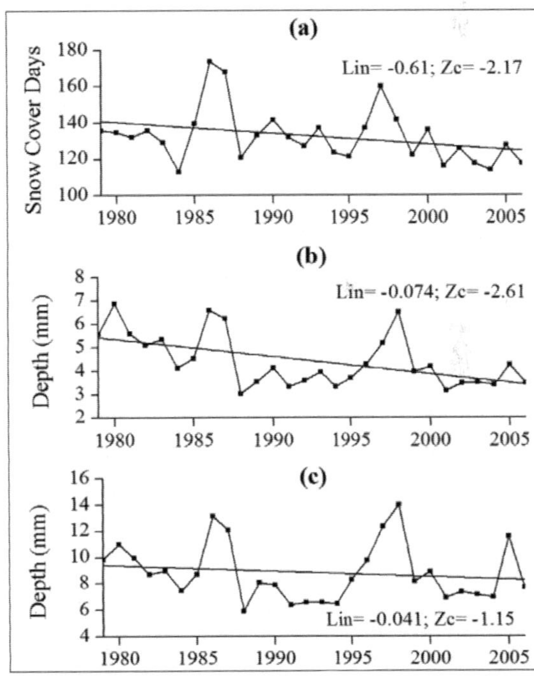

Figure 3. Same as figure 2 but for snow cover days (a), mean snow depth (b) and maximum snow depth (c) averaged over the YR basin from 1979 to 2006.

From figure 2a, it can be seen that annual air temperature for the YR basin as a whole during 1979-2006 had increased about 1.2℃ at a speed of 0.43℃/10a. The warming trend was very obvious with the significant level of $\alpha<0.001$. Yang et al. [8] indicated that during the period of 1971-2004, the linear increasing rate of air temperature averaged in China and globally was about 0.226℃ and

0.148℃/decade respectively. The linear increasing rate of air temperature averaged in the YR Basin was significantly higher than those for China and for the world in the same period. Statistical analyses on annual mean precipitation averaged for the YR basin suggested a slightly increasing period from 1992 to 1999 with a speed of 21.2mm/a and a slowly decreasing period from 2000 to 2006 with a speed of -17.7mm/a (figure 2b). In general, annual mean precipitation averaged in the basin exhibited a non-significant increasing trend of 18.0 mm/decade with the significant level of $\alpha > 0.1$ during the period 1979-2006. Annual mean actual evapotranspiration experienced a decreasing trend during the period 1979-1993 and then exhibited an increasing trend afterwards (figure 2c). Generally, annual mean actual evapotranspiration averaged over the basin from 1979 to 2006 showed a significant increasing trend with speed of 14.5 mm/decade under the significant level of $\alpha < 0.01$.

3.2. *Variation of snow cover*
Annual snow-cover days and maximum and mean snow depth, derived from SMMR and SSM/I data by means of GIS spatial analysis techniques, averaged over the basin during 1979-2006 were illustrated in figure 3 for variation trends analysis. According to figure 3a, we can find that there were more than 110 days snowpack covered on average over the YR basin. The linear decreasing trend of snow-cover days for the YR basin during the period 1979-2006 can be detected at speed of -6 days/10a with the significant level of $\alpha < 0.05$. Meanwhile, mean and maximum snow depth became thinner at the speed of -0.74mm/10a (with the significant level of $\alpha < 0.01$) and -0.41mm/10a, respectively (figures 3b and 3c). As a consequence, it is believed that snow meltwater probably decreased over the YR basin during the past 30 years.

Table 1. Changes of glaciers in representative regions in YR Basin during the past few decades.

Study Area	Period /year	Number of Glaciers Studied	Area Change / km^2	Proportion of Area Change / %	Source of Data	Notes
Gangrigabu Range	1980-2001	88	-2.02	-0.25	[9]	24 glaciers over the Ranwu Lake Basin and 64 glaciers over the Danlongqu Basin; Maritime Glaciers
Ranwu Lake Basin	1980-2005	All Glaciers over This Basin	-29.7	-5.98	[10]	Maritime Glaciers
Nianchuhe River Basin	1980-2005	83	-13.41	-7.3	[11]	Continental Glaciers
YR Basin	1976-2005	21	-17.9	-1.25	[12]	Large Maritime Glaciers
Jiema Yangzong Glacier	1974-2010	1	-1.11	-5.10	[13]	The source of Yarlung Zangbo River; Continental glacier on the east side of ridge, and maritime glacier on the west side of ridge
Zhadang Glacier	2005-2008	1	Annual glacier mass balance was -1143, -838, 208 mm.w.e. from 2005 to 2008, respectively.		[14]	Field Measurements; Continental Glacier

3.3. *Glacier fluctuations*
Table 1 summarized the glacier mass balance measurement results reported in different literatures concerning with the YR basin glacier studies. Glaciers listed in Table 1 of which total area accounting for about 21% of that in the whole basin in 1976, are different in the size (big or small), characteristic (continental glacier or maritime glacier) and geographical location (upstream, midstream and downstream). Therefore, trends for variation of these glaciers were representative enough for that of

the glaciers over the whole YR basin in past decades. According to these previous studies, we can find that most glaciers in the YR basin tended to retreat in general since the late 1970s. As a consequence, it was believed that glacial meltwater might increase over the YR basin in past decades.

3.4. Response of river water resource to climate and glacier-snow changes over the region

Theoretically, if the water consumptions for human activities are less enough to be neglected, the multi-year water balance equation for YR basin can be expressed as [15]:

$$P + S + G - E = R \tag{2}$$

in which *P*, *S*, *G*, *E* and *R* are precipitation, snow meltwater, glacier meltwater, actual evapotranspiration and runoff during a period, respectively.

Considering the glacier meltwater directly influenced by climate factors such as temperature, precipitation and so on, in association with Equation (2), changes of temperature, precipitation, actual evapotranspiration and glacier-snow were directly related to the variation of watershed runoff. Correlation analysis between runoff and these factors during 1979-2006 was shown in Table 2. The observed runoff data of Nuxia hydrometric station (29.14°N, 94.34°E) located in the downstream of YR basin was obtained from literature [16]. Variation of the runoff measured by Nuxia hydrometric station of which controlling area accounts for 70% of the whole watershed area can represent the change trend of the river water resource.

Table 2. Correlation analysis between runoff and relevant factors (temperature, precipitation, actual evapotranspiration and snow cover).

	P	*T*	*E*	Snowcover Day	Maximum Snowdepth	Mean Snowdepth
Correlation Coefficient	0.871	0.444	0.006	-0.077	0.138	0.029
Significance	$\alpha<0.01$	$\alpha<0.05$	/	/	/	/

Correlation coefficient of precipitation-runoff and temperature-runoff were 0.871 and 0.444 with the significant level of $\alpha<0.01$ and $\alpha<0.05$, respectively (Table 2). However, river runoff has the poor correlations with actual evapotranspiration and snow cover. It suggested that the precipitation and glacier meltwater were the main sources of YR water resource supplies. Although watershed evapotranspiration increased and snow cover decreased in recent years, the river runoff increased at a speed of 18.7 $m^3s^{-1}a^{-1}$ during 1979-2006 with the significant level of $\alpha<0.1$ under the background of the increasing precipitation, temperature and glacier meltwater.

4. Conclusions

According to previous studies plus present result obtained in this study, the response of river water resources to the climate, glaciers and snow cover changes in the YR basin was systematically studied. From 1979 to 2006, under the background of the increasing temperature, precipitation, actual evapotranspiration and glacier meltwater and decreasing snow cover over the basin, river runoff presented the increasing trend to respond the climate and glacier changes at the speed of 18.7 $m^3s^{-1}a^{-1}$ with the significant level of $\alpha<0.1$.

Acknowledgements

This work was financially supported by the National Key Technologies R&D Program of China (No. 2016YFA0602302 and No. 2016YFB0502502), and CRSRI Open Research Program (CKWV2015226/KY), respectively.

References

[1] You Q, Kang S, Wu Y and Yan Y 2007 Climate change over the Yarlung Zangbo River Basin

during 1961-2005 *J. Geogr. Sci.* **17** 409-20

[2] Fu J, Fan G Z and Zhou D W 2012 The applicability and modification of Takahashi formula for evaporation estimation in Lhasa *Journal of Applied Meteorological Science* **23** 231-7

[3] Che T, Li X, Jin R, Armstrong R and Zhang T 2008 Snow depth derived from passive microwave remote-sensing data in China *Ann. Glaciol.* **49** 145-54

[4] Bi X and Zhang W 2009 Long term analyses on spatial and temporal evaluation of snow covers in northern China by using RS and meteorological data. In: *2008 International Workshop on Education Technology and Training and 2008 International Workshop on Geoscience and Remote Sensing, Vol 2, Proceedings,* ed L Qi p 368-71

[5] Nie N, Zhang W and Deng C 2012 Spatial and temporal climate variations from 1978 to 2009 and their trend projection over the Yarlung Zangbo River Basin *Journal of Glaciology and Geocryology* **34** 64-71

[6] Belle G V and Hughes J P 1984 Nonparametric tests for trend in water quality *Water Resour. Res.* 127-36

[7] Wei F 2007 *Modern Climatic Statistical Diagnosis and Prediction Technology* (Beijing: China Meteorological Press)

[8] Yang X, Zhang Y, Zhang W, Yan Y, Wang Z, Ding M and Chu D 2006 Climate change in Mt.Qomolangma region since 1971 *J. Geogr. Sci.* **16** 326-36

[9] Liu S, Shangguan D, Ding Y, Han H, Zhang Y, Wang J, Xie C, Ding L and Li G 2005 Glacier variations since the early 20th century in the Gangrigabu range, southeast Tibetan Plateau *Journal of Glaciology and Geocryology* **27** 55-63

[10] Xin X, Yao T, Ye Q, Guo L and Yang W 2009 Study of the fluctuations of glaciers and lakes around the Ranwu Lake of southeast Tibetan Plateau using remote sensing *Journal of Glaciology and Geocryology* **31** 19-26

[11] Li Z, Yao T, Ye Q, Tian L and Wang W 2010 Glaciers in the upstream Manla Reservoir in the Nianchu river basin, Tibet:shrinkage and impact *Journal of Glaciology and Geocryology* **32** 650-8

[12] Nie N, Zhang Z, Zhang W and Deng C 2013 Analysis of the characteristics of the glaicer change based on remote sensing in the Yarlung Zangbo River basin during the past 30 years *Journal of Glaciology and Geocryology* **35** 541-52

[13] Liu X and Xiao C 2011 Preliminary study of the Jiemayangzong Glacier and lake variations in the source regions of the Yarlung Zangbo River in 19742010 *Journal of Glaciology and Geocryology* **33** 488-96

[14] Yao T D, Li Z G, Yang W, Guo X J, Zhu L P, Kang S C, Wu Y H and Yu W S 2010 Glacial distribution and mass balance in the Yarlung Zangbo River and its influence on lakes *Chin. Sci. Bull.* **55** 2072-8

[15] Rui X F 2004 *The Principle of Hydrology* (Beijing: China Water & Power Press)

[16] Luozhu Nima, Wang J and Xu X 2011 Variation characteristics and trend analysis of runoff over the Yarlung Zangbo River Basin *Journal of China Hydrology* **31** 76-9

Spatial-temporal variations and their dynamics of the saline lakes in the Qaidam Basin over the past 40 years

S L Zhou[1], W C Zhang[1,3] and F Wang[2]

[1] Academy Key Laboratory of Digital Earth, Institute of Remote Sensing and Digital Earth, CAS, Beijing 100094, China
[2] China Institute of Water Resources and Hydropower Research, Beijing 100044, China

E-mail: zhangwc@radi.ac.cn

Abstract. The saline lakes in the Qaidam basin, providing numerous saline mineral resources to the development of China's western region have changed greatly in recent decades. Understanding the spatial-temporal variations of the saline lakes for exploring the dynamics of those changes is not only essential for sustainable development of the region, but also is of significance for scientific studies. By using multi-temporal Landsat and HJ-1A/1B remote sensing data, this study focused on detecting the spatial-temporal variations of the saline lakes concerning with their extent changes with the influence of climate and human activities in the Qaidam basin from 1973 to 2014 based on the normalized difference water index (NDWI) and maximum likelihood supervised classification techniques. The results indicated that most of the saline lakes studied in this paper across the study region had experienced great changes over the past 40 years, and each saline lake differed considerably on extent. The runoff supplies from the surrounding rivers and the large-scale saline resource exploitation were considered as direct reasons for the saline lake extent changes, in which climatic changes characterized by the increase of temperature and slight decrease of precipitation over the region had significant impacts on the lake area variations directly or indirectly.

1. Introduction

As one of the most important lake types, saline lakes, rich in special saline mineral resources such as potassium salt, magnesium salt, boron salt, and lithium salt, are an important strategic resource for the development of countries [1]. Exploitation and resource management of saline lakes require high-precision spatial-temporal variation information of saline lakes concerning with their extents, which are usually difficult to be investigated in China by the conventional means due to the harsh environmental conditions of most saline lakes. Images from earth observation satellites can provide unique data sources rapidly and efficiently for the investigation of saline lakes, and the development of various remote sensing image classification methods also contributes to the application of remote sensing on saline lake researches [2].

Lakes, which constitute the essential components of the hydrological and biogeochemical cycles, are a sensitive indicator of global change [3-7]. An increasing number of studies on the changes of saline lakes in the Qaidam basin where a number of main saline lakes distributed in China had been

[3] To whom any correspondence should be addressed.

reported [8-10]. The variations of temperature and precipitation and the large-scale saline resource exploitation were regarded as the main dynamics for the various changes of the lakes. Most of such studies focused on the relations of lake area with temperature and precipitation across the Qaidam basin, while the analysis on seasonal extent changes as well as the variations of the saline lakes with runoff supplies was relatively few.

By means of multi-temporal HJ-1A/1B and Landsat satellite images acquired from 1973 to 2014, spatial-temporal extent variations of the saline lakes in the Qaidam basin for the last decades were investigated, annual and seasonal variations on individual saline lake extent with air temperature, precipitation were analysed and meanwhile the driving dynamics for those changes were discussed.

2. Materials and methods

2.1. Study region

Located in Qinghai Province, close to the northern margin of the Qinghai–Xizang Plateau, the Qaidam basin covers an area of 120,000 km^2, situated between 90° 09' - 99° 16' E and 34° 41' - 39° 20' N (figure 1). The Qaidam basin is one of the regions significantly affected by the global change in the Qinghai–Xizang Plateau, and the saline resource exploitation here persistently enhances with the development of the country, especially after 2000. Under the circumstances, some saline lakes there have taken on great changes in the last decades and some of them, as shown in figure 1, were the main research objects in this study.

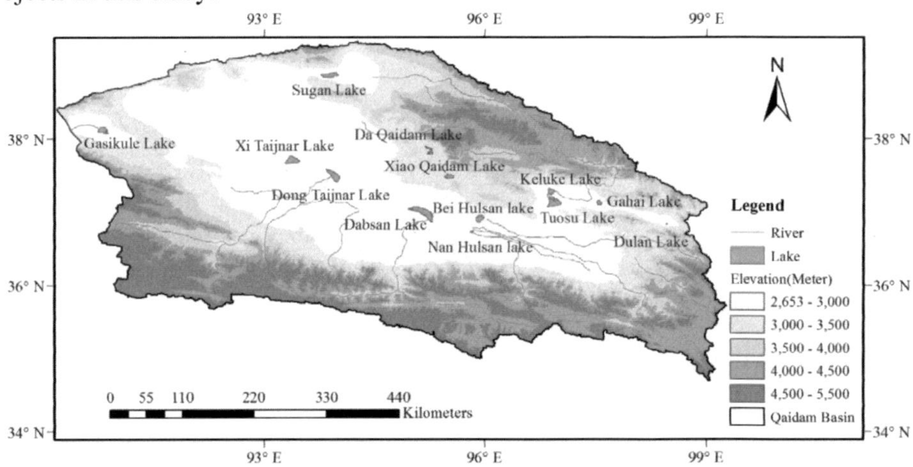

Figure 1. Study region.

2.2. Datasets

Landsat MSS, TM, ETM+ imagery are the most frequently applied remote sensing data in the lake variation detection researches [11], which were also the main satellite imagery sources for this study. Besides, some images from HJ-1A/1B satellites were also used to provide supplementary information. Cloud-free or nearly cloud-free, calibrated and geometrically corrected using ENVI software Landsat images from 1973 to 2014 and HJ-1A/1B images in 2013 and 2014, downloaded from USGS GLOVIS (http://glovis.usgs.gov) and China Centre for Resources Satellite Data and Application (http://cresda.com/en/gywm/zxgk/index.shtml), were applied in this study.

In order to analyse the dynamics of the lake extent variations, annual precipitation and average temperature data during 1973–2013 were collected from China Meteorological Data Network (http://data.cma.cn/). Annual river runoff data in the Qaidam basin were also used in this paper to study the relations of lake changes to runoff. In addition, the 90m-resolution DEM data from Shuttle Radar Topography Mission Digital Elevation Model (SRTM-DEM) and the boundary vector of the Qaidam basin (after slight adjustment by reference to the literatures) from National Science &

Technology Infrastructure of China, Data Sharing Infrastructure of Earth System Science-Lake-Watershed Science Data Center (http://lake.geodata.cn) were obtained to provide the basic geographic information of the study region.

2.3. Water body identification

By means of multi-temporal Landsat and HJ-1A/1B images, the extraction of the saline lakes and retrieval of lake surface area were conducted. Based on the spectral absorption characteristics of water body that differ very much from the other ground objects in visible and infrared spectral bands, several water indices were developed in recent years by combining two or more bands for water body identification, among which the normalised difference water index (NDWI) [12] was a frequently used one, and can be expressed as in equation (1) according to Mcfeeters (1996):

$$NDWI = \frac{(Green - NIR)}{(Green + NIR)} \tag{1}$$

Where Green is the reflectance of green band and NIR is that of near-infrared band.

In this paper, the extent of saline lakes was derived mainly by NDWI methodology, and the maximum likelihood supervised classification method was also applied to the supplementary of variation information. A suitable cut-off threshold was needed when water index methods were used to detect water body. Generally pixels with the water index value greater than 0 represented water bodies. But due to the water index value varied for different lakes caused by environmental factors or lake features, it's necessary to calibrate the thresholds in order to make the lake delineation more accurate and efficient. Hence, the water index values of several sample points classified as water body by the visual interpretation are calculated to adjust the thresholds. It was found that 0.3 was most suitable for water body detection in this study, although slight adjustments in the practical applications were necessary for each image. In this paper, the visual interpretation results referenced with lake area data derived from past literatures were used to evaluate the accuracy of detection and identify the errors in the estimation. Tests showed that the water body identification results matched the visual interpretation perfectly, and also comparatively consistent with the reference data derived from published literatures for the same period.

3. Results and discussions

3.1. Variations of saline lake surface area in the Qaidam Basin

The dynamic changes of saline lakes in the Qaidam basin from 1973 to 2014 were presented in table 1. Because of the lack of suitable images for some lakes in 1973, the 1970s surface area data were derived from images in 1973 or 1976. This table indicated that some saline lakes in the Qaidam basin expanded greatly during the past 40 years, such as Dulan Lake. However, there are also some lakes, like Xi Taijnar Lake, in contrast decreased a lot. Dong Taijnar Lake and Bei Hulsan Lake can't even be detected using satellite images in 2014 probably because their water levels were too low to be detected or they had disappeared at that time. Though the total changes of the saline lakes differed considerably in the past 40 years, their change trends during the period 1970s-2000 and the period 2000-2014 are relatively similar. Most of their area decreased from 1970s to 2000, and increased from 2000 to 2014. As an example, the spatial-temporal changes of Da Qaidam Lake are shown in figure 2.

Table 2 gives the seasonal changes of some saline lakes in the Qaidam basin from Jan 2013 to July 2014, from which it can be found that these lakes didn't change greatly during different seasons compared with the inter-annual changes.

3.2. Dynamics of the variations of saline lake surface area

3.2.1 Climate changes in the Qaidam Basin. The river runoff feeding by the glacial and snow melting, as well as precipitation, is the main water supply of saline lakes in the Qaidam basin [10], so the temperature and precipitation have significantly influences on the surface area of the saline lakes. However, the relations of the saline lake changes to the temperature and precipitation were not so direct according to the study results. Figure 3 shows the inter-annual variations of the average data of precipitation and average temperature derived from six meteorological stations across the study area. Compared with precipitation, the temperature had an obvious increasing trend, which may accelerate glacier and snow melting to provide more water resource to the lakes. However, while studying on the area changes and temperature in specific years, it was found that there was no strong correlation between the variations of temperature and the extent changes of saline lakes. It was the same with precipitation. In addition, the relationship between lake area and climate factors took on strong regional features when using the nearest meteorological station data to analyze the variations of specific lakes, which was also presented by Wu (2014). For example, compared with Da Qaidam Lake, the variations of Gasikule Lake area had a stronger correlation with the changes of temperature.

Table 1. The area changes of the saline lakes in the Qaidam basin over the past 40 years.

Saline Lake	Area（km^2）			Change（%）		
	1970s	2000	2014	1970s-2000	2000-2014	1970s-2014
Gahai Lake	32.17	28.32	33.05	-11.98%	16.71%	2.73%
Tuosu Lake	157.15	133.56	147.54	-15.01%	10.46%	-6.12%
Da Qaidam Lake	42.01	21.23	34.12	-49.47%	60.73%	-18.79%
Xiao Qaidam Lake	44.06	51.34	81.51	16.50%	58.77%	84.97%
Gasikule Lake	81.03	118.30	109.49	46.00%	-7.45%	35.13%
Sugan Lake	105.16	102.03	104.53	-2.98%	2.45%	-0.60%
Dong Taijnar Lake[a]	133.04	202.19		51.97%		
Xi Taijnar Lake	113.04	18.79	33.62	-83.38%	78.93%	-70.26%
Nan Hulsan Lake	3.37	0.51	11.33	-84.87%	2120.28%	235.89%
Bei Hulsan Lake[a]	88.21	52.55		-40.43%		
Dulan Lake	7.94	1.84	22.44	-76.83%	1119.67%	182.65%
Dabsan Lake	238.36	167.13	226.15	-29.88%	35.31%	-5.12%

[a] The lakes can't be detected using satellite images in 2014.

Table 2. The seasonal area changes of the saline lakes in the Qaidam basin from Jan 2013 to July 2014.

Saline Lake	Area(km^2)						
	Jan-13	Apr-13	Aug-13	Oct-13	Feb-14	May-14	Jul-14
Tuosu Lake	151.97	149.27	149.11	149.31	153.91	149.29	147.54
Gahai Lake	35.91	34.98	33.95	34.02	36.09	34.69	33.05
Keluke Lake	54.87	53.03	51.35	52.48	56.05	53.03	50.51
Xiao Qaidam Lake	99.23	91.77	88.77	87.77	95.97	85.59	81.51
Dabsan Lake	305.10	321.93	235.88	194.09	198.08	226.15	157.06
Nan Hulsan lake	44.40	63.88	32.58	27.44	25.71	91.94	11.33

To analyze the relationship between the lake area and river runoff, according to the annual runoff data of the main supply rivers of these saline lakes acquired from seven hydrological stations during the period 1956-2012, the annual runoff design values of typical hydrological years were calculated using the annual runoff frequency curve means, where the runoff values at the frequencies 25%,50%,75%,90% were chosen as the design values of wet year, normal year, dry year and extra ordinary dry year. By the calculated annual runoff design values, the typical or nearly typical hydrological year from 1973 to 2002 (due to the large scale saline recourse exploitation started from 2000, the area variations before 2000 are more related to climate changes) were chosen and the lakes area were derived from satellite images in these years, presented in table 3. The table shows that compared with temperature and precipitation, the variations of lakes seemed more directly relative to the runoff supplies from the surrounding rivers, probably due to the situation that the influences of temperature and precipitation on the lake area in the Qaidam basin were always cross-regional, lagged and more complex, not as obvious as that of the river runoff. In spite of that, temperature and precipitation were still ones of the primary dynamics of the saline lake variations in the Qaidam basin, because the river runoff was also significantly affected by them.

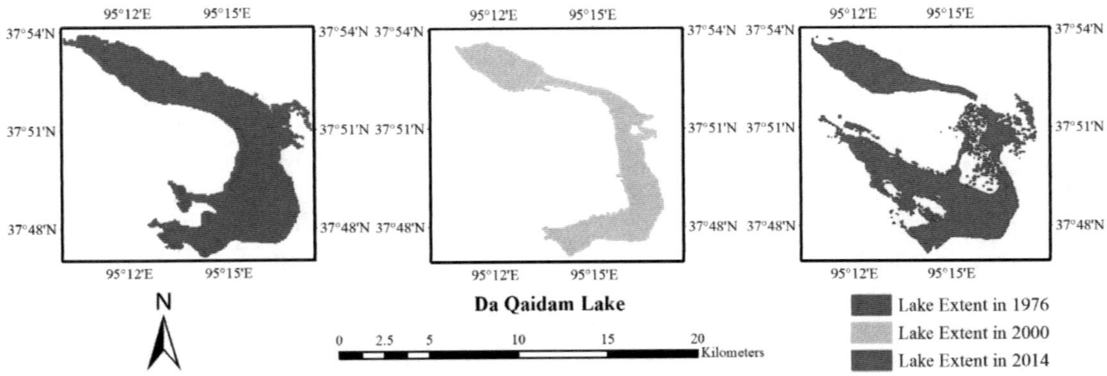

Figure 2. The spatial-temporal changes of Da Qaidam Lake from 1976 to 2014.

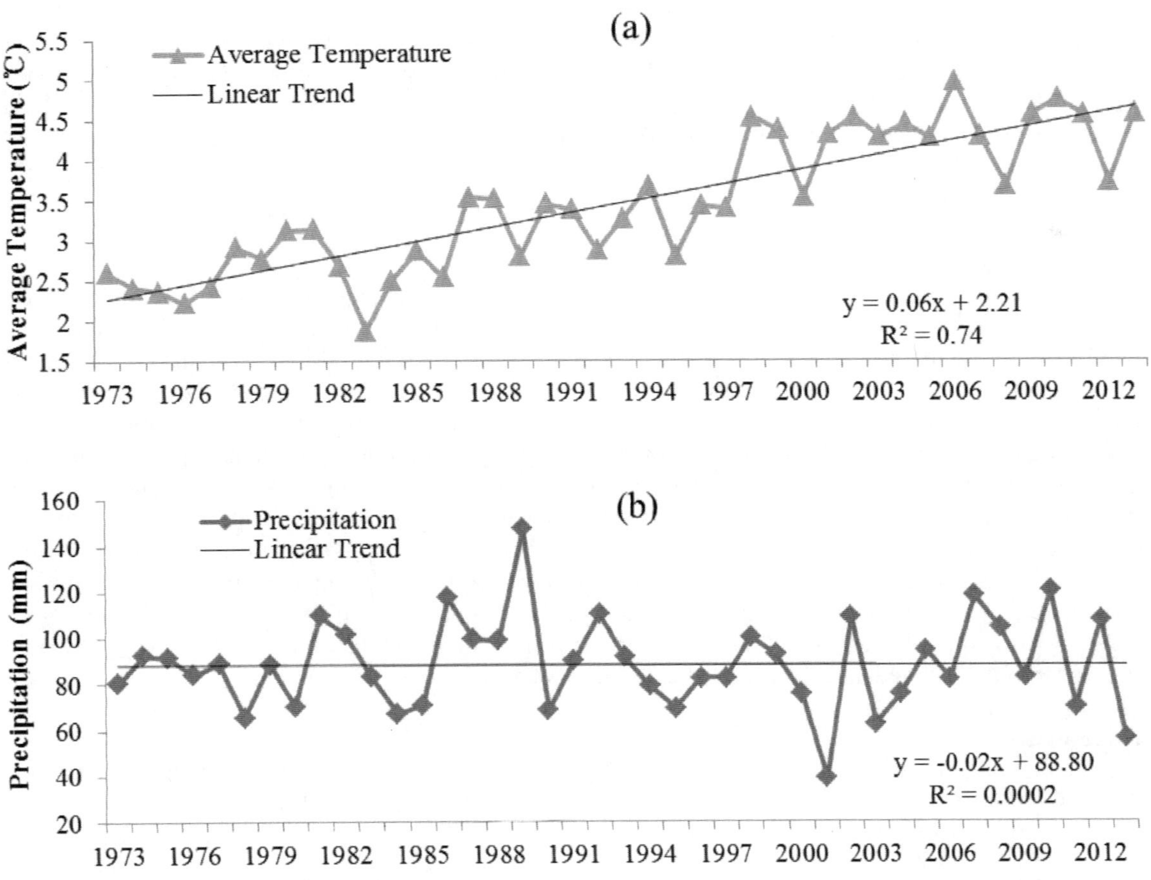

Figure 3. The variations of average temperature (a) and precipitation (b) in the study region from 1973 to 2013.

Table 3. The area changes of the saline lakes in the Qaidam basin in typical hydrological years.

Saline Lake	Wet year		Normal year		Dry year		Extra ordinary dry year	
	Year	Area (km^2)	Year	Area (km^2)	Year	Area (km^2)	Year	Area (km^2)
Dabsan Lake	1976	238.36	1987	212.03	1991	263.92	1998	121.71
Gahai Lake	1987	28.41	1991	29.71	2000	28.32	2001	27.97
Tuosu Lake	1987	147.13	1991	146.04	2000	133.56	2001	132.32
Da Qaidam Lake	1987	34.47	1988	33.88	2000	21.23	2001	20.57
Xiao Qaidam Lake	1987	55.17	1988	50.68	2000	51.34	2001	45.19
Gasikule Lake	2000	118.30	1992	115.13	1990	115.70	1973	81.03
Sugan Lake	2000	102.03	1998	101.72	1987	99.89	1990	101.19
Dong Taijnar Lake	2000	202.19	1992		1990	164.71	1973	133.04
Xi Taijnar Lake	2000	18.79	1992	127.16	1990	129.42	1973	113.04
Nan Hulsan Lake	1976	3.37	1988	6.41	2000	0.51	2002	7.59
Bei Hulsan Lake	1976	88.21	1988	69.79	2000	52.55	2002	62.95
Dulan Lake	1991	11.50	1995	1.30	1999	1.37	2000	1.84

3.2.2 Impact of human activities on changes in lake area. The large-scale human activities, especially the saline resource exploitation, have considerable impacts on the area changes of these saline lakes. As shown in table 2, Dong Taijnar Lake had decreased greatly after 2000. From the satellite images, it was found that Dong Taijnar Lake had a large area in 1970s and 2000, but with the salt pans building nearby, the area of the lake decreased rapidly. In 2014, the main water body of the lake can't be even detected based on remote sensing, and instead the salt pans were still obvious on the images and had increased compared with before. The expanding of Dabsan Lake and Nan Hulsan Lake after 2000 was also thought to be related to the 400-ton potash fertilizer product project in Qinghai Province by Lu (2014).

4. Conclusion

The present study showed that most of the saline lakes studied in the Qaidam basin had great changes in the past 40 years, and the changes of the saline lakes differed considerably. Half of them had significantly decreased in surface area, while there are meanwhile several lakes expanding considerably during the period. The trends of the lake changes were distinct across different regions, and the relations of the changes to temperature and precipitation also took on strong regional features. The results illustrated that although temperature and precipitation had significant impacts on the lake area directly or indirectly, the runoff supplies from the surrounding rivers were more directly correlated to the area of these lakes, the variations of which may be one of the main dynamics of the lake changes. The analysis based on remote sensing images in this paper also indicated that human activities especially the saline resource exploitation, had considerable impacts on the area changes of these saline lakes, which were considered as one of the main reasons of the decrease of Dong Taijnar Lake. Accordingly, it is important and necessary to control the scale of saline resource exploitation for the protection of saline lake resource in the Qaidam basin.

Acknowledgments

This work was financially supported by the National Key Technologies R&D Program of China (No. 2016YFA0602302 and No. 2016YFB0502502), and CRSRI Open Research Program (CKWV2015226/KY), respectively.

References

[1] Zhou L G, Feng X Z, Xiao P F, Xie S P and Wang D Y 2009 Advance and Prospection of Remote Sensing Application to Salt Lakes *Advances In Earth Science* **24** 141-9

[2] Wang Y F, Chen R B, Bai Z J, Fang H B, Zuo A P and Liu C L 2004 Remote Sensing Analysis of Tibetan Salt Lakes by TM Imaging *Journal Of Salt Lake Research* **12** 1-7
[3] Mason I, Guzkowska M, Rapley C and Street-Perrott F 1994 The response of lake levels and areas to climatic change *Climatic change* **27** 161-97
[4] Bai J, Chen X, Li J, Yang L and Fang H 2011 Changes in the area of inland lakes in arid regions of central Asia during the past 30 years *Environmental monitoring and assessment* **178** 247-56
[5] Chai H, Cheng W, Zhou C, Zhao S and Liu H 2013 Climate effects on an inland alpine lake in Xinjiang, China over the past 40 years *Journal of Arid Land* **5** 188-98
[6] Emmer A, Merkl S and Mergili M 2015 Spatiotemporal patterns of high-mountain lakes and related hazards in western Austria *Geomorphology* **246** 602-16
[7] Yao X, Liu S, Long L I, Sun M and Luo J 2014 Spatial-temporal characteristics of lake area variations in Hoh Xil region from 1970 to 2011 *Journal of Geographical Sciences* **24** 689-702
[8] Yan L J, Zheng M P and Yuan Z J 2014 Influence of climate change on salt lakes in Qinghai Province and their mineral resources exploitation in the past forty years: A case study of Xiao Qaidam Lake *Mineral Deposits* **33** 921-9
[9] Wu J J 2014 Analysis On The Changing And Reason Of The Salt Lake surface water area In The Qaidam Basin Since Recent 20 Years (Qinghai Institute of Salt Lakes, Chinese Academy of Sciences)
[10] Lu N 2014 Changes of lake area in Qaidam basin and the influence factors *Journal of Arid Land Resources and Environment* **8** 015
[11] Ma M, Wang X, Veroustraete F and Dong L 2007 Change in area of Ebinur Lake during the 1998–2005 period *International Journal of Remote Sensing* **28** 5523-33
[12] McFeeters S 1996 The use of the Normalized Difference Water Index (NDWI) in the delineation of open water features *International journal of remote sensing* **17** 1425-32

Research on algae blooms forecasting based on the multivariate data driven method: a case study of the Chaohu Lake

Xiang bo[1], Zhao Yanchuang[1, 2], Wang Xinyuan[1], Zong Xin[1, 2]

[1] Key Laboratory of Digital Earth Science, Institute of Remote Sensing and Digital Earth, Chinese Academy of Sciences, 100094 Beijing, China
[2]University of Chinese Academy of Sciences, 100049 Beijing, China

E-mail: xiangbo@radi.ac.cn

Abstract. The data driven is one of the main methods for forecasting algae bloom, which requires lots of continuous and accurate monitoring data. It is an effective way to increase sample data size by combining in-situ, and remote sensing data. The Chaohu Lake was taken as the case study. Based on water quality data (TLI), meteorological data (sunshine duration, temperature, wind speed, wind direction) and bloom grade data, provided respectively by remote sensing and in-situ monitoring, an artificial neural network was employed to build empirical data-driven models. The model accuracy was evaluated by algae bloom grade recognition rate and bloom trend recognition rate. The results showed that the bloom grade recognition rate of model driven by remote sensing data was better than others. Bloom trend recognition rate of model driven by in-situ data is higher than others. These results provide some insights for algae bloom forecasting.

1. Introduction

Algae bloom is one of the most widespread environmental problems of inland water in the world. It is important to develop methods for forecasting algae blooms, which is useful for water management and protection. There are mainly two kinds of methods for forecasting algae bloom: mechanical models and data driven models. The mechanical methods predict algae blooms by modeling the processes of the hydrological, hydrodynamic and algae growth. However, the mechanical methods, based on the processes of hydrological, hydrodynamic and algae growth, need a number of parameters, such as initial conditions and ecological variables. It is very complex to get all data, which limits the application of model driven methods on forecasting algae blooms.

The data driven approaches for forecasting algae blooms by constructing empirical-statistical models based on multi-source heterogeneous data, such as in-situ monitoring data, automatic monitoring data, remote sensing data. ANN (artificial neural network), Decision Tree, GA (genetic algorithm), SVM (support vector machine) are common methods to perform data driven processes. Wei et al. (2001) simulated the relationship between algae density and multi water quality factors, through ANN and predicted the outbreak of algae bloom. Lee et al. (2003) forecasted water quality factors and algae bloom of coastal waters in Hong Kong. Zeng et al. (2007) combined decision tree modeling and non-linear regression, to predict algae bloom of 'six lakes' in Beijing. Muttil et al. (2005) predicted algae bloom in Causeway Bay, Hong Kong by using GA. SVM can solve the problem of the empirical trial resulting from ANN. Lu et al. (2006) established the water safety warning evaluation model by SVM.

However, in order to ensure the accuracy and reliability of the data driven models, lots of continuous and accurate monitoring data are required. Datasets in-situ is often used to establish the data driven model. This type of data is usually intuitive and with high accuracy, but the collection of these data is higher time-consuming, laborious and costly. Besides, the information obtained by in-situ monitoring are usually at station scale. Remote sensing data can provide global detail information about eutrophication and algae blooms, quickly at different scales with low cost. Thus, it is an effective way to increase sample data size by combining in-situ data and remote sensing data. Here, we took Chaohu Lake as the case of study. Based on water quality data (TLI), meteorological data (sunshine duration, temperature, wind speed, wind direction) and bloom grade data, provided respectively by remote sensing and in-situ monitoring. An artificial neural network method was employed to build empirical data driven models. Then algae bloom grade recognition rate and bloom trend recognition rate were used to evaluate the model accuracy. We hope this study can provide some insights for algae bloom forecasting.

2. Study area

Figure 1. Location of Chaohu Lake

The Chaohu Lake is located between the Yangtze and the Huaihe rivers. It belongs to the north drainage of the Yangtze River downstream and is one of the five largest freshwater lakes in China. Fig. 1 shows the location of the Chaolu Lake. The study area is located in the humid subtropical monsoon climate zone with abundant rainfall. The annual average temperature is 15-16℃. The main wind direction in winter is northeast and southeast in summer. The average annual precipitation is about 1100mm. The mean water depth is 2.69 m covering an area of 780 km^2. Phosphorus-rich strata are widely distributed in the north shore of the basin and phosphorus is input into the lake along runoff. It is not only the main drinking water source for Hefei and Chaohu cities, but also has the function of shipping, fishing, tourism, etc. the water quality of the Chaohu Lake is of the utmost importance.

3. Materials and methods

3.1 Data source

Water quality data from 12 monitoring sites (Figure 1) from 2013 to 2015 were provided by Chaohu Authority. Chlorophyll a (Chl–a), total phosphorus (TP), total nitrogen (TN), transparency (SD), chemical oxygen demand (COD), and algae density were included. The monitoring frequency is once a day from May to October each year and once a week for the rest of the months. The meteorological data were obtained from the National Meteorological Information Center, China Meteorological data network. The accuracy of sunshine duration, average temperature, and wind speed are 0.1h, 0.1 °C and 0.1m/s respectively. The monitoring frequency is once a day. The algae bloom coverage data between 2013 and 2015 were obtained from Chaohu cyanobacteria bloom inversion product datasets, which were provided by the National Science and Technology Platform - National Earth system data sharing platform - Lake - Watershed Science Data Center. MODIS surface reflectance product (MOD09) was obtained from the National Aeronautics and Space Administration, (NASA) website.

3.2 Data preprocessing

3.2.1 Grade of algae bloom feature. In agreement with the actual monitoring of Chaohu, the method grading algae bloom feature in this study was according to that employed by the China Environmental Monitoring Center and the Anhui Environmental Monitoring Center Station. As shown in Table 1, the algae feature in the Chaohu Lake was divided into five grades. The algae density threshold is employed by in-situ data and the algae coverage threshold are employed by the remote sensing data. In order to determine the algae bloom coverage for each site, the averaged value of 3*3 grid, fixing the station as center, were calculated.

Table 1. Thresholds used in the segmentation of thermal patches

Algae bloom feature	Algae density (10^4 /L)	Algae coverage (%)
No bloom	<200	≤5
Sporadic bloom	≥200	≤10
Local bloom	≥500	≤40
Regional bloom	≥1000	≤60
Global bloom	≥10000	>60

3.2.2 Trophic Level Index (TLI). The Trophic Level Index is a general indicator to evaluate the eutrophication status of inland water. It is also one of the most important quality factors for algae bloom prediction model inputs. TLI was calculated based on in-situ data according to the following formulas (Wang et al, 2002; Jin et al., 1990).

$$TLI(\Sigma) = \sum_{j=1}^{m} W_j \times TLI(j) \quad (1); \quad W_j = r_{ij}^2 / \sum_{j=1}^{m} r_{ij}^2 \tag{1}$$

TLI(j) is the *j*th composite indicator with the corresponding weight W_j. The r_{ij} value is the correlation coefficient for the relationship between the reference chlorophyll concentration and each indicator (Table 2). The TLI of all the observation points was calculated from 5 components, including chemical oxygen demand (COD), total phosphorus (TP), total nitrogen (TN), chlorophyll a (Chl–a), and Secchi depth (SD). Formulas for the TLI of each component are given below:

$$TLI(chl - a) = 10(2.5 + 1.086 \ln(chl - a)) \tag{2}$$

$$TLI(TP) = 10(9.436 + 1.624 \ln(TP)) \tag{3}$$

$$TLI(TN) = 10(5.453 + 1.694 \ln(TN)) \tag{4}$$

$$TLI(SD) = 10(5.118 + 1.94 \ln(SD)) \tag{5}$$

$$TLI(COD) = 10(0.109 + 2.661 \ln(COD)) \tag{6}$$

Equations 2 to 6 are empirical regression equations based on a survey of eutrophication levels from more than 20 lakes in China (Jin et al., 1990).

Table 2. Correlation between Chl-a, and other substances influencing TLI

	Chl-a (mg/m³)	**TP (mg/ℓ)**	**TN (mg/ℓ)**	**SD (m)**	**COD (mg/ℓ)**
r_{ij}	1	0.84	0.82	−0.83	0.83
r_{ij}^2	1	0.7056	0.6724	0.6889	0.6889

Please refer to our previous research work about the procedure of TLI calculation based on remote sensing data (Xiang et al, 2015), which will not be described in detail here.

3.3 Multivariate data driven method

Based on water quality data (TLI), meteorological data (sunshine duration, temperature, wind speed, wind direction) and bloom grade data; an artificial neural network was employed to build an empirical model. Meteorological data and bloom grade data were used as the input and output layer respectively. Neural networks are a machine learning algorithm that imitates brain processes and was originally developed to solve non-linear problems like over-fitting, pattern recognition, clustering, and time-series prediction (Mitchell, 1997). They have been successfully applied in environmental sciences (Sattari et al., 2012; Krasnopolsky and Chevallier, 2003; Krasnopolsky and Schiller, 2003; Song et al., 2014a). In this study, the neural network was designed as a feed-forward type and divided into three layers (input layer, hidden layer, and output layer). Back propagation algorithm was used to determine the parameters in the network. In-situ data, remote sensing data and their combination between 2013 and 2014 were selected to perform the driven process. In order to prevent over-fitting, the number of nodes in hidden layer was adjusted based on the results of verification.

3.4 Evaluation of model accuracy

In-situ data, remote sensing data, and their combination in 2015 were selected to evaluate the model accuracy respectively. Bloom grade recognition rate and bloom trend recognition rate were used to assess the predictive effect of different types of data driven models. To calculate bloom trend recognition rate, assuming that it is correct when the difference between forecast bloom grade and the true grade is limited in one grade.

4 Results and analysis

Fig. 2 shows the accuracy of model driven by different types of data (in-situ data, remote sensing data as well as their combination). The blue, red and green solid lines, are respectively the bloom grade recognition rate of remote sensing data (RS) driven model, the in-suit data drive model, and their combination data driven model. The blue dotted, red dashed and green dotted lines are respectively corresponding to blooms trend recognition rate.

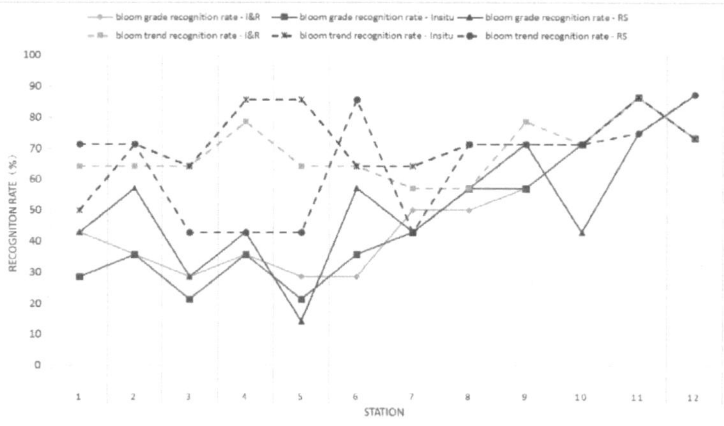

Figure 2. Accuracy of model driven by different types of data.

As can be seen in Fig. 2, there are totally 9 stations whose bloom grade recognition rate of remote sensing data (RS) driven model is not lower than that of in-situ data. 5 stations are in the western half lake and the others are in the eastern half lake. There are only 2 stations whose bloom grade recognition rate of collection data driven model is higher than that of the single dataset. 5 stations are in the western half lake and the others are in the eastern half lake. The in-situ data have more effect on the blooming trend recognition rate than the remote sensing data and the collection data. The bloom trend recognition rate of model driven by collection data fluctuates between that of the single data.

Table 3. Statistical results for accuracy of models driven by different types of data

Type of data	Bloom grade recognition rate（%）	Standard deviation（%）	Bloom trend recognition rate（%）	Standard deviation（%）
In-situ	47.26	21.56	71.67	10.68
RS	51.64	20.34	64.73	17.06
I&R	49.05	19.60	68.69	9.10

The statistical results for accuracy of models driven by different types of data based on 12 stations are shown in Table 3. For bloom grade recognition rate, the accuracy of model driven by remote sensing data is better than others. According to bloom trend recognition rate, the accuracy of model driven by in-situ data is higher than others. The reasons are that the model driven by remote sensing data, forecast algae bloom grade by algae coverage and the model driven by in-situ data, forecast algae bloom grade by algae density.

Algae coverage reflects the extent of the lake surface covered by algae blooms, which is a visible trait; while algae density reflects the degree of enrichment of algae in the lake, which is a latent trait. Visible traits characterize the current situation of algae bloom and are more suitable to forecast bloom grade. Latent traits characterize the possibility of algae bloom and are more suitable to forecast bloom trends. The algae bloom grade in the western half lake is more than that in the eastern half lake; the difference between visible and latent features is more than that the in eastern half lake. Thus, the difference of effect of data type on model accuracy is more significant in the western half lake.

5. Conclusion

The data driven is one of the main methods to forecast algae bloom. The approach requires lots of continuous and accurate monitoring data. It is an effective way to increase sample data size by combining in-situ and remote sensing data. The Chaohu Lake was the case of study. Based on water quality data (TLI), meteorological data (sunshine duration, temperature, wind speed, wind direction) and bloom grade data, provided respectively by remote sensing and in-situ monitoring. An artificial neural network was employed to build empirical data driven models. The model accuracy was evaluated by algae bloom grade recognition rate and bloom trend recognition rate. The results showed that the bloom grade recognition rate of model driven by remote sensing data was better than others. Bloom trend recognition rate of model driven by in-situ data is higher than others. These results provide some insights for algae bloom forecasting.

6. References

[1] Wei B, Sugiura N and Maekawa T 2001 *Water Res.* **35(8)** 2022–2028.

[2] Lee JHW, Huang Y, Dickman M, et a1. 2003 *Ecol. Model.* **159(2-3)**: 179–201.

[3] Zeng Y, Yang Z and Liu J 2007 *Adv. Water Sci.* **18(1)**: 79-85.

[4] Muttil N and Lee JHW 2005 *Ecol. Model.* **189(3-4)**: 363–376.

[5] Lu M, Zhang Z and Feng B 2006 *Comput. Eng.* **32 (15)**: 44-45.

[6] Wang M, Liu X and Zhang J 2002 *Environ. Monit. China* **18(5)**: 47–49.

[7] Jin X, Tu Q, Zhang Z, Jiang X, Wang Y, Zhu X, Shu J, Xu N, Huang C and Xu R 1990 *Specifications for Lake Eutrophication Surve.* 2nd Edition. Beijing: China Environment Science Press: 291.

[8] Xiang B, Song J, Wang X and Zhen J 2015 *Water SA* **5**: 753-761.

Precipitation Interpolation by Multivariate Bayesian Maximum Entropy Based on Meteorological Data in Yun-Gui-Guang region, Mainland China

Chaolin Wang[1], **Shaobo Zhong**[1*], **Fushen Zhang**[1], **Quanyi Huang**[1]

[1]Department of Engineering Physics/ Institute of Public Safety Research, Tsinghua University, Beijing 100084, China

[*]Corresponding author: zhongshaobo@tsinghua.edu.cn

Abstract. Precipitation interpolation has been a hot area of research for many years. It had close relation to meteorological factors. In this paper, precipitation from 91 meteorological stations located in and around Yunnan, Guizhou and Guangxi Zhuang provinces (or autonomous region), Mainland China was taken into consideration for spatial interpolation. Multivariate Bayesian maximum entropy (BME) method with auxiliary variables, including mean relative humidity, water vapour pressure, mean temperature, mean wind speed and terrain elevation, was used to get more accurate regional distribution of annual precipitation. The means, standard deviations, skewness and kurtosis of meteorological factors were calculated. Variogram and cross-variogram were fitted between precipitation and auxiliary variables. The results showed that the multivariate BME method was precise with hard and soft data, probability density function. Annual mean precipitation was positively correlated with mean relative humidity, mean water vapour pressure, mean temperature and mean wind speed, negatively correlated with terrain elevation. The results are supposed to provide substantial reference for research of drought and waterlog in the region.

1. Introduction

Precipitation has close relation to some meteorological and environmental factors. Humidity, water vapor pressure, temperature, wind and terrain elevation etc. are potential meteorological factors that affect the spatial distribution of precipitation. However, the observations of meteorological stations are discrete and sparse in time and space. Hence, spatial interpolation for meteorological factors is necessary to obtain regionalized distribution of annul mean precipitation [1]. Precipitation interpolation has been a hot area of research for many years. Bayat et al. [2] compared Bayesian maximum entropy (BME) and ordinary kriging methods for spatial and spatiotemporal variations of annual precipitation prediction, and cross-validation was used to assess their performances. Shi et al. [3] employed BME model taking TRMM satellite data as soft data, observations of meteorological stations as hard data, to estimate the precipitation of Fujian Province, China. Many researches have demonstrated that an appropriate covariate can improve the estimation accuracy largely [4, 5]. Zhang et al. [6] compared four BME models to estimate the spatial distribution of mean annual precipitation using meteorological, elevation and the nearest distance to coastline variables. The results were assessed by cross validation and proved that multivariate BME method was optimal. Most methods and models were only based on indexes using hard data. However, soft data not only can provide some useful information, but also can save calculation cost and improve the predictions. The application of spatial distribution of precipitation was helpful to risk assessment of drought and inundation, which could map disaster risk accurately combined with disaster index (say, drought index).

This paper took precipitation data as main variable, mean relative humidity, water vapor pressure, mean temperature, mean wind speed and terrain elevation as auxiliary variables to study the spatial interpolation of precipitation in Yunnan, Guizhou provinces and Guangxi Zhuang autonomous region during 1964-2013. The means, standard deviations, skewness and kurtosis of meteorological factors were calculated. Variogram and cross-variogram were fitted between precipitation and auxiliary variables. The results are supposed to provide substantial reference for research of drought and waterlog in the region.

2. Materials and methods

2.1. Study area

The study area is Yunnan, Guizhou provinces and Guangxi Zhuang autonomous region (Yun-Gui-Guang region, 97°E-113°E, 20°N-29°N), which is located in the Yunnan-Guizhou Plateau, southwest of Mainland China, eastern along the border of China, Vietnam, Laos and Myanmar and coastline of South China Sea (Figure 1). Mean terrain elevation of the region is 1000~2000 meters. This region covers an area of 795,300km^2, and contains 39 big cities, such as Kunming, Guiyang and Nanning. The climate of this region is subtropical monsoon climate. Drought happens easily in spring because of poor rainfall and strong evaporation. There are 129.77 million resident people in the region. Guizhou and Guangxi are relatively richer in water resource than Yunnan. And all this region is undeveloped provinces compared to east China coastal area.

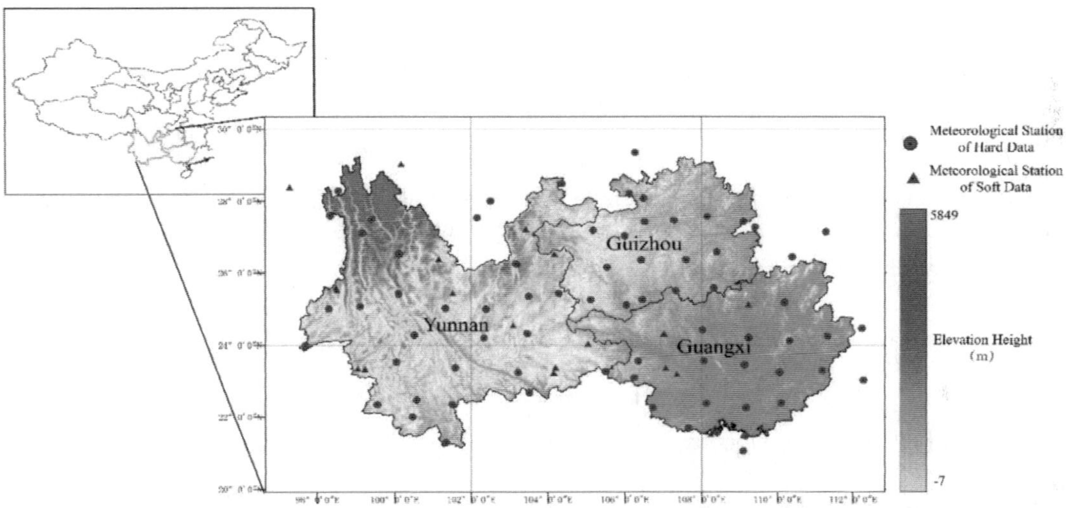

Figure 1. The terrain of Yun-Gui-Guang region and distribution of meteorological stations.

2.2. Data

Data used in this paper was acquired from the annual China's terrestrial climate data set, which was published by China Meteorological Data Sharing Service System (http://cdc.nmic.cn). We took 50 years (1964-2013) of annual precipitation as main variable, mean relative humidity, water vapor pressure, mean temperature and mean wind speed as covariates to study the spatial distribution of precipitation from 91 meteorological stations in or around the study area. However, Observations are not recorded completely at 19 meteorological stations which are replaced by -32766 or 32766. In this paper, they were replaced by soft data, precipitation probability density function(pdf) which was matched by kernel density estimation method by MATLAB software using precipitation data set from 1951 to 2015. These 19 stations were called meteorological station of soft data; other 72 stations were called meteorological stations of hard data. All stations were marked in map in Figure 1. Digital Elevation Model(DEM) data were also used in this paper, called ASTER GDEM v2(Figure 1). The spatial resolution is 0.1°.

2.3. BME method

Christakos (1990) established the BME method. BME method takes many types of data and different types of knowledge bases into spatial interpolation. These data and information are divided into general knowledge(K_G) and site-specific knowledge(K_S) [7]. The steps for calculating BME are as follows.

Prior stage. The aim of prior stage is to figure out the pdf, based on general knowledge K_G, called prior pdf. Information entropy is defined as equation (1) by Shannon information measure [8]:

$$Info_G\left(Z_{map}\right)=-\log f_G\left(Z_{map}\right) \tag{1}$$

Where, Z_{map} is the stochastic variable in the study area, $Z_{map}=\left(Z_{hard},Z_{soft},Z_0\right)$, Z_{hard}, Z_{soft} and Z_0 indicate the value of hard data, soft data and location for estimating, respectively. $f_G\left(Z_{map}\right)$ indicates the pdf based on general knowledge K_G. Based on these constraints and Lagrange multiplier approach, we can get the prior pdf:

$$f_G\left(Z_{map}\right)=A^{-1}\exp\left(\sum_{\alpha=1}^{N_C}\mu_\alpha g_\alpha\left(Z_{map}\right)\right) \tag{2}$$

Where, μ_α indicates Lagrange's multiplier, $g_\alpha(Z_{map})$ is the known function associated with Z_{map} based on K_G, A indicates normalization coefficient:

$$A=\int\exp\left(\sum_{\alpha=1}^{N_C}\mu_\alpha g_\alpha\left(Z_{map}\right)\right)dZ_{map} \tag{3}$$

Pre-posterior stage. The aim of pre-posterior stage is to collect and organize additional auxiliary information in appropriate forms to produce site-specific knowledge K_S. And then they will be used in BME model. Hard data has been incorporated into prior stage indirectly, and will be used directly at this stage.

Posterior stage. The aim of posterior stage is to update prior pdf based on Bayesian conditional probability theorem and K_S. So we can get posterior pdf. When the distribution of hard data and soft data is certain, posterior pdf of spatial variable Z at location x_0 is:

$$f_K\left(Z_0\right)=f_G\left(Z_0\,|\,Z_{hard},Z_{soft}\right)=f_G\left(Z_0,Z_{hard},Z_{soft}\right)\times\left(f_G\left(Z_{hard},Z_{soft}\right)\right)^{-1} \tag{4}$$

Where, $Z_{hard}=\left[x_1,\ldots,x_n\right]'$, $Z_{soft}=\left[x_{n+1},\ldots,x_m\right]'$, n, m indicate the number of hard data and soft data within the scope of maximum distance $d_{\max}$ to estimation point, respectively.

3. Numerical results and plots

Table 1 shows the means, standard deviations, skewness and kurtosis of meteorological factors during 1964 to 2013 in Yun-Gui-Guang region. Figure 2 and figure 3 show the pdf and the cumulative density function(CDF) of meteorological factors in study area, respectively.

Table 1. The basic statistical indicates of meteorological factors during 1964 to 2013 in Yun-Gui-Guang region.

	Annual Mean Precipitation(0.1mm)	Mean Relative Humidity(%)	Water Vapor Pressure(0.1hPa)	Annual Mean Temperature(0.1C°)	Annual Mean Wind Speed(0.1m/s)
Means	12974	95.07	184.78	194.30	98.73
Standard Deviations	1.91×107	5.94×105	5.92×105	5.91×105	2.68×106
Skewness	1.1716	42.382	42.215	42.234	19.922
Kurtosis	2.76	1.79×103	1.78×103	1.79×103	394.87

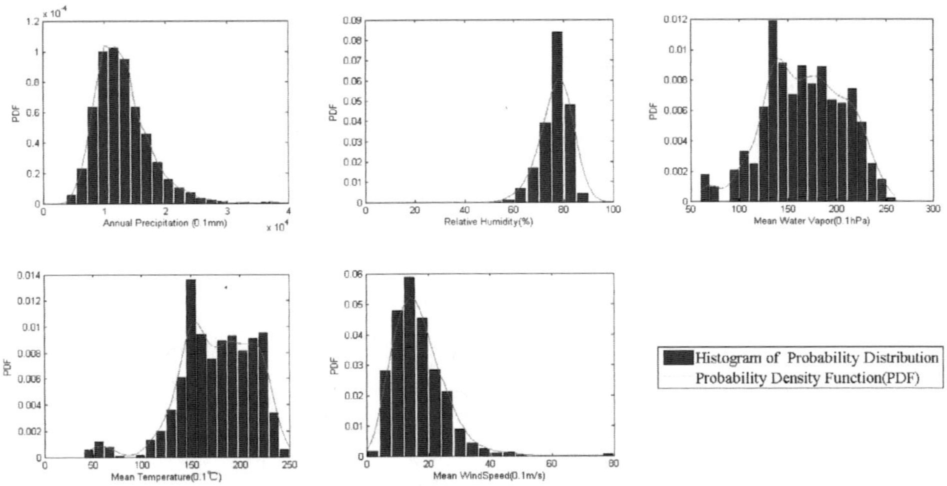

Figure 2. The PDF of meteorological factors in study area.

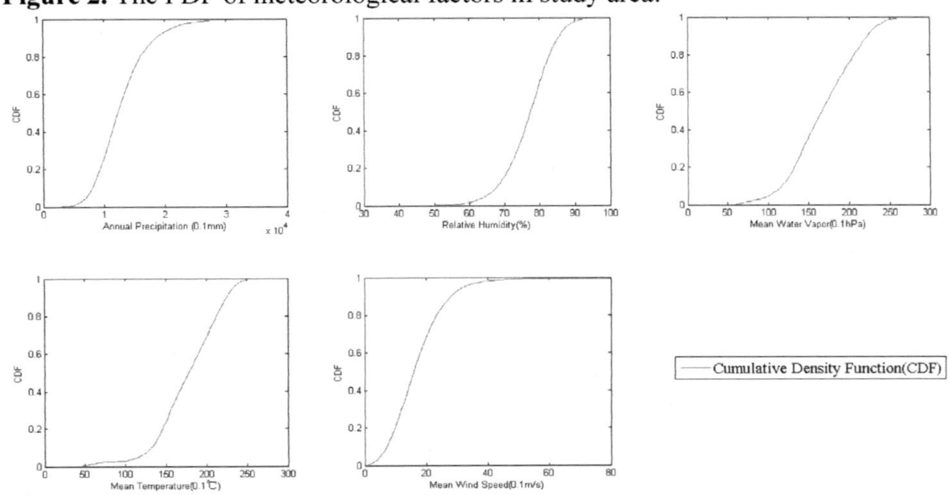

Figure 3. The CDF of meteorological factors in study area.

The constitutive property and randomness of annual mean precipitation can be described by variogram. We set distance lag as 1° to compute variogram value in interval (0, 10). The directional variograms of precipitation along the N-S (stars), the W-E (circles) directions and omnidirectional variogram of precipitation (red circles) are computed and shown in Figure 4. The calculated variogram value is shown by blue line in Figure 4. Then Nested model is defined with Nugget model and Gaussian model. Then we can get the equation after fitted by least square method:

$$\gamma(h)=1.115\times10^{7}+1.501\times10^{7}\times\left[1-\exp(-3h^{2}/10^{2})\right],\quad h>0 \tag{5}$$

Where, first half part of the equation is Nugget model, and the other is Gaussian model. Fitted variogram curve is shown by red line in Figure 4. It can be seen that the fitted curve coincides with the calculated value of variogram well. Increasing of variogram with distance lag indicates that the correlation of annual precipitation increases with distance lag.

In this paper, we take mean relative humidity, water vapor pressure, mean temperature, mean wind speed and terrain elevation as cross variates. Cross-variograms between annual precipitation and cross variates describe the correlation of spatial distribution. The nested models are defined same to variogram. Cross-variograms between any two variates are fitted and are shown in Figure 5. As we can see, the correlations of cross-variogram between mean relative humidity, water vapor pressure, mean temperature and annual precipitation are positive, terrain elevation is negative. And mean wind speed is unstable.

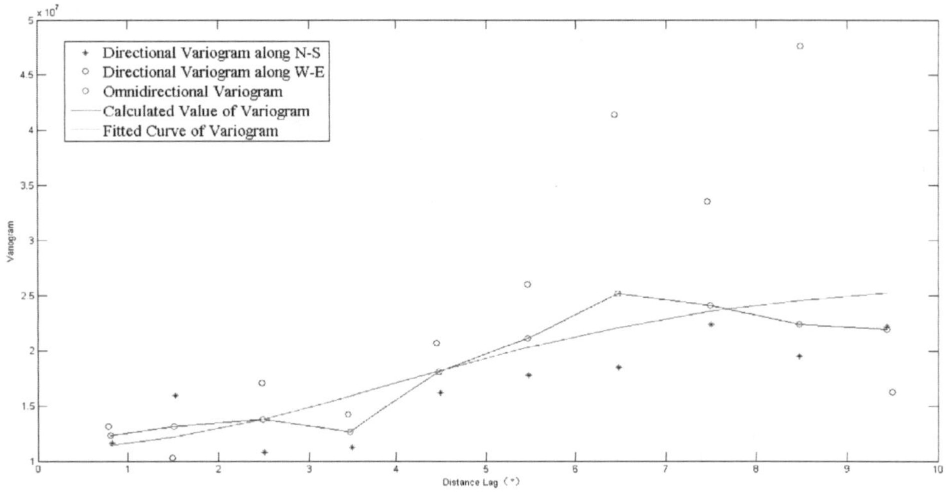

Figure 4. Variogram of annual precipitation.

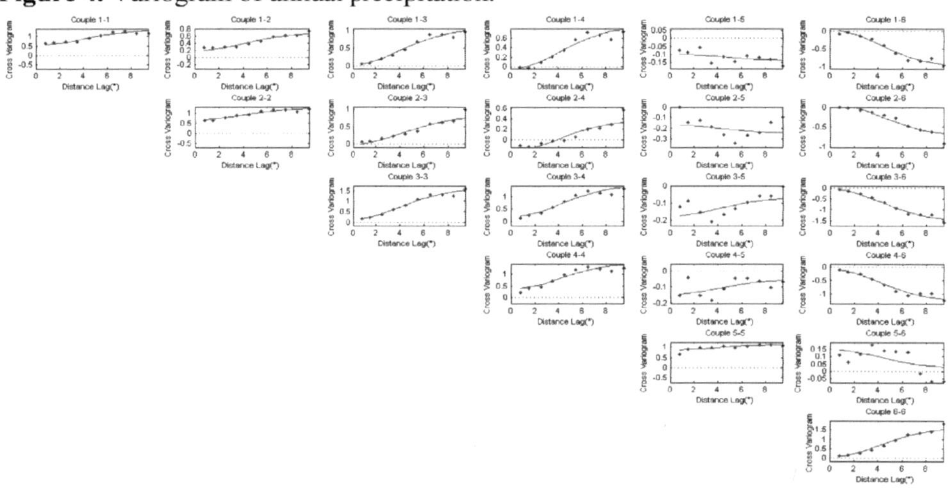

Figure 5. Cross-variogram of Variates ($h>0$). (1) Annual mean precipitation; (2) Mean relative humidity; (3) Mean water vapor pressure; (4) Mean temperature; (5) Mean wind speed; (6) Terrain elevation.

Based on the achievement in the past, multivariate BME method were employed to estimate the spatial distribution of annual precipitation in Yun-Gui-Guang region. We studied year by year from 1964 to 2013 at a 0.05° spatial resolution. Then the annual precipitation distribution images were mapped in ARCGIS 10.2.2 software. Figure 6 shows the interpolation annual mean precipitation distribution maps of Yun-Gui-Guang region. Overall, there is more rain in the south and southeast because of being adjacent to South China Sea, and it is rainless in the north-west where is deep in Mainland China. In addition, there is more rain in south than north of the study area, east than west.

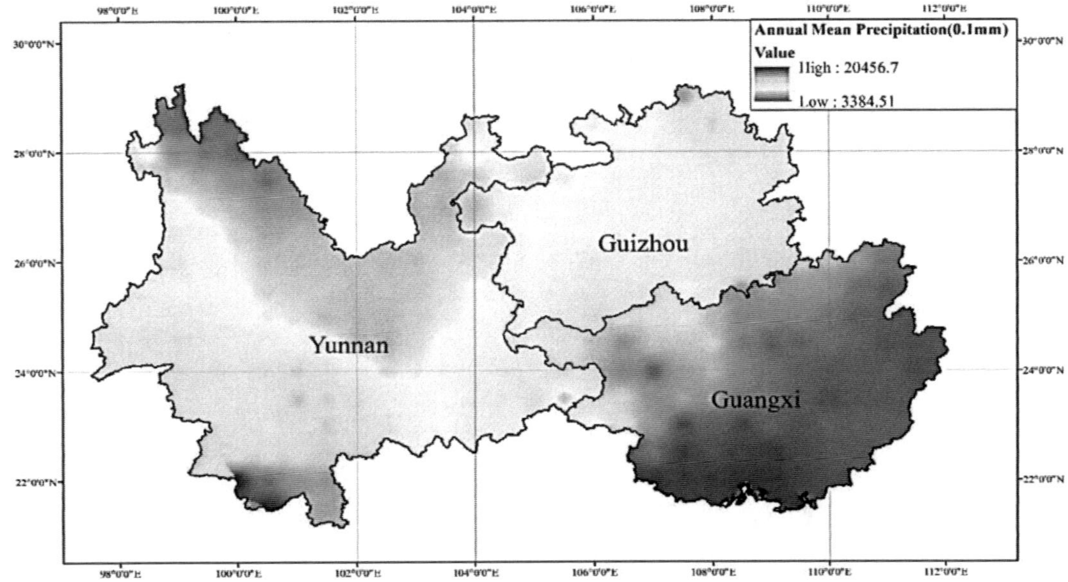

Figure 6. The Annual Mean Precipitation Distribution Map from 1964 to 2013 in Yun-Gui-Guang Region.

4. Conclusions

This paper takes annual precipitation with auxiliary variables, mean relative humidity, water vapor pressure, mean temperature, mean wind speed and terrain elevation into consideration for spatial interpolation. The PDF and CDF are analyzed for multivariate BME method, which are used to get more accurate regional distribution of annual mean precipitation. The variogram of annual precipitation and cross-variogram of mean relative humidity, water vapor pressure, mean temperature, mean wind speed and terrain elevation are studied to get the correlation between mean annual precipitation and covariates. Finally, the spatial distribution of annual precipitation was mapped in Yun-Gui-Guang region. The results show that:

(1) Annual mean precipitation was positively correlated with mean relative humidity, mean water vapor pressure, mean temperature and mean wind speed, negatively correlated with terrain elevation.

(2) There is more rain in the south and southeast, and it has a trend that there is more rain in south than north of the study area, east than west.

Some natural disasters, such as drought, waterlog and so on, have close relation to precipitation. The precipitation map is obtained in this paper by BME method of the study area of interest. The results are supposed to provide substantial reference for disaster prevention and mitigation of drought and waterlog in the region. However, the mapping procedure of precipitation is also affected by the factor of time. The space-time estimation method of precipitation should be paid more attention in further research.

Acknowledgments

The authors would like to thank the support of the National Natural Science Foundation of China (Grant No. 91224004), the Project in the National Science & Technology Pillar Program during the Twelfth Five-year Plan Period (Grant No.2015BAK12B03). The authors appreciate support for this paper by the Collaborative Innovation Center of Public Safety.

References

[1] Yu, H., Chen, J., Christakos, G. and Jerrett, M. 2009 Environ Health Perspect 117 537-544.
[2] Bayat, B., Zahraie, B., Taghavi, F. and Nasseri, M. 2013 Theoretical and applied climatology, 113 429-444.
[3] Shi, T., Yang, X., Zhang, T., Liu, L. and Tian, W. 2014 Journal of Geo-information Science 16 470-481.

[4] Yu, X., Ma, J., Zhu, J. and Sun, S. 2011 Science of Surveying and Mapping 36 83-85+55.
[5] Jost, G., Heuvelink, G. B. M. and Papritz, A. 2005 Geoderma 128 258-273.
[6] Zhang, F., Yang, Z., Zhong, S. and Huang, Q. 2016 Advances in Meteorology 2016 1-13.
[7] Douaik, A., Van Meirvenne, M. and Tóth, T. 2005 Geoderma 128 234-248.
[8] Shannon, C. E. 1948 Bell System technical Journal 27 379-423 and 623–656.

IOP Conf. Series: Earth and Environmental Science **46** (2016) 012046 doi:10.1088/1755-1315/46/1/012046

Experiences of using UAVs for monitoring levee breaches

J Brauneck[1], R Pohl[2] , R Juepner[1]

[1] Hydraulic Engineering and Water Management, Department of Civil Engineering, TU Kaiserslautern/Germany

[2] Institute of Hydraulic Engineering and Technical Hydromechanics, TU Dresden/Germany

E-mail: jens.brauneck@bauing.uni-kl.de

Abstract. During floods technical protection facilities are subjected to high loads and might fail as several examplės have shown in the past. During the major 2002 and 2013 floods in the catchment area of the Elbe River (Germany), some breaching levees caused large inundations in the hinterland. In such situations the emergency forces need comprehensive and reliable real-time information about the situation, especially the breach enlargement and discharge, the spatial and temporal development of the inundation and the damages. After an impressive progress meanwhile unmanned aerial vehicles (UAV) also called remotely piloted aircraft systems (RPAS) are highly capable to collect and transmit precise information from not accessible areas to the task force very quickly. Using the example of the Breitenhagen levee failure near the Saale-Elbe junction in Germany in June 2013 the processing steps will be explained that are needed to come from the visual UAV-flight information to a hydronumeric model. Modelling of the breach was implemented using photogrammetric ranging methods, such as structure from motion and dense image matching. These methods utilize conventional digital multiple view images or videos recorded by either a moving aerial platform or terrestrial photography and allow the construction of 3D point clouds, digital surface models and orthophotos. At Breitenhagen, a UAV recorded the beginning of the levee failure. Due to the dynamic character of the breach and the moving areal platform, 4 different surface models show valid data with extrapolated breach widths of 9 to 40 meters. By means of these calculations the flow rate through the breach has been determined. In addition the procedure has been tested in a physical model, whose results will be presented too.

1. Background/Introduction

During a flood event, levee failures can lead to the inundation of large areas and therefore threaten the lives as well as the property of many people in the hinterland. In order to calculate the size of the inundated zone as well as water depths that are to be expected there, hydraulic models are powerful tools, if the required data is available. Usually, the analysis of a flooding combines hydraulic models with digital surface information (elevation models) and the measures of the dike breach at its maximum extent, as it is measured after the incident.

So far, there are no terrestrial surveying or monitoring methods implemented in the emergency management measures, which would allow for a real time information supply of the breach geometry and extent over time [1]. Remote sensing techniques and satellite based data are essential tools for Disaster Risk Management especially in terms of early warning systems and providing the background

IOP Conf. Series: Earth and Environmental Science **46** (2016) 012046 doi:10.1088/1755-1315/46/1/012046

of geospatial data [2] as well as covering large areas. But due to the processing requirements, limited temporal resolution (due to revisit capability), high costs [3] as well as copyright restrictions, these systems appear to be less suitable for monitoring purposes of highly dynamic and small levee failures rather than the flooding itself.

The technical progress of unmanned aerial vehicles (UAV) or unmanned aerial systems (UAS) and a decline in acquisition costs has led to a widespread use of these platforms, in private as well as commercial sectors. Although these systems differ in some technical details, most of the commercial off-the-shelf systems are capable of producing digital high resolution aerial images virtually at no cost and without the need of extensive preparation [4].

A wide range of publications and projects demonstrated the usefulness of UAV for common mapping purposes [4] as well as rapid mapping during or shortly after catastrophic events like post-earthquake disaster surveying [5], or mapping of floodplain inundation [6].

The primary outcome of these surveys may be simple videos, single geotagged or orthorectified images or orthomosaics, depending either on the precision of the on-board GNSS and IMU systems or on the presence of reliable ground control points (GCP). Either way, ground resolution can easily be brought to a few centimetres if needed. The temporal resolution is depending on flight time as well as numbers of batteries. Therefore, the observation of a highly dynamic process like a levee failure appears to be most suitable for UAVs.

2. The Breitenhagen levee failure 2013

In 2013, prolonged and intensive rainfall caused extreme floodings all over Central Europe, concentrating in the Elbe and Danube drainage basins. Water levels recorded at the Elbe and Saale exceeded the 100-year flood (HQ100) and 200-year flood (HQ200) levels and showed the maximum values since records began. As a consequence, a number of levee failures occurred. One incident near the city of Breitenhagen was recorded using a UAV, instructed on behalf of the State Service for Flood Protection and Water Management (LHW) in Saxony-Anhalt.

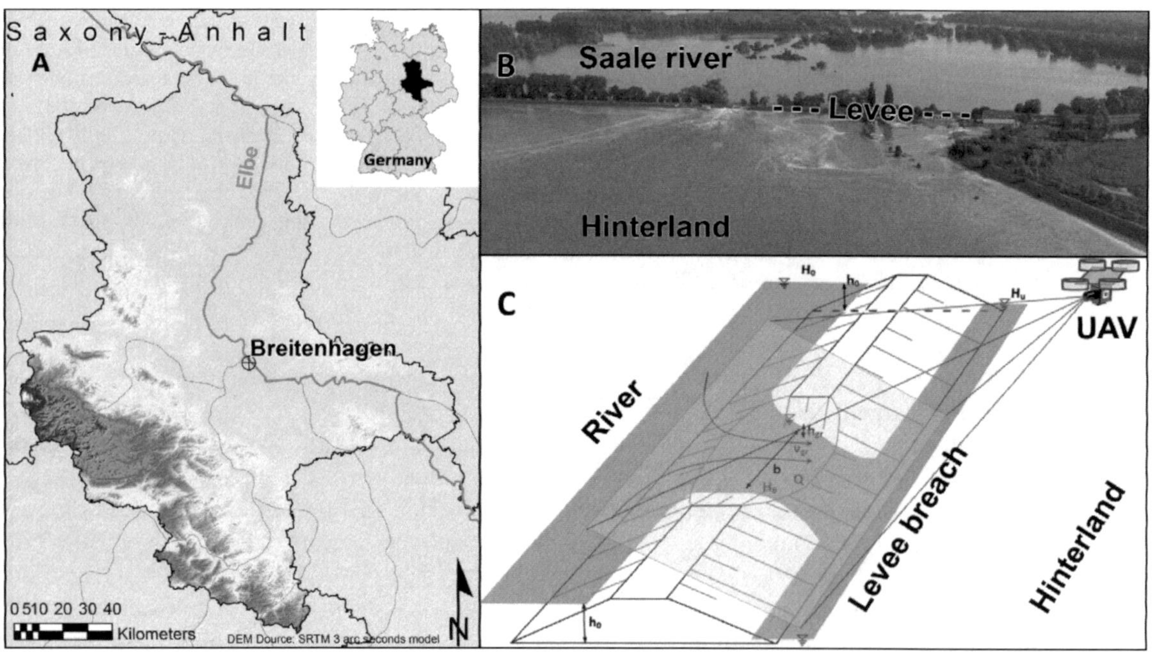

Figure 1. (A) Location of Breitenhagen (B) Picture taken from the UAV at the end of the recorded sequence (C) Scheme of the UAV survey showing the main flight pattern

The levee failure began at about 7:20 a.m. on 9th June 2016. Shortly after the beginning of the failure, the UAV took off and recorded five videos of 1:28 to 10:35 minutes length, only separated by battery replacement breaks of unknown lengths. The image resolution was 1280*720 pixel and 50 frames per second were recorded. The breach widened to about 40 m in 40 minutes and reached its maximum extend of 140 meters after nine hours. An area of 85 km² were inundated, leading to the partly evacuation of seven villages.

3. Methods

3.1. SFM modelling

Over the recent years, the increased availability of high resolution aerial footage has led to a more widespread use of photogrammetric techniques and algorithms, some of them are subsumed under the terms structure from motion or dense image matching. These software based analyses allow for a semi-automated production of 3d surface models and digital elevation models [7].

In order to process the data from Breitenhagen, the video sequences had to be converted into images using only 1 frame per second due to the slow movement of the UAV. Suitable sequences that focussed primarily the failure were arranged and analysed using the software Agisoft Photoscan.

Due to the limited resolution, four digital surface models with the maximum precision settings were created. As the number of suitable images were low for two models, additional information was added from other models. The dynamic parts, especially flowing water areas, were masked and only static areas mainly eastwards of the failure were put into the models.

The overall lack of surface information and ground control points caused some the most time consuming parts of the analysis. No geopositioning information was recorded during the flights, therefore georeferencing had to be applied indirectly by calculating an unrectified image of the surrounding area, which included locatable objects like power poles. The location of static objects could then be determined as national coordinates using a GIS. Digital surface models as well as orthoimages were combined to derive height profile lines along the levee failure. Additionally, the flow path of water within the breach was mapped to derive information about the water extent.

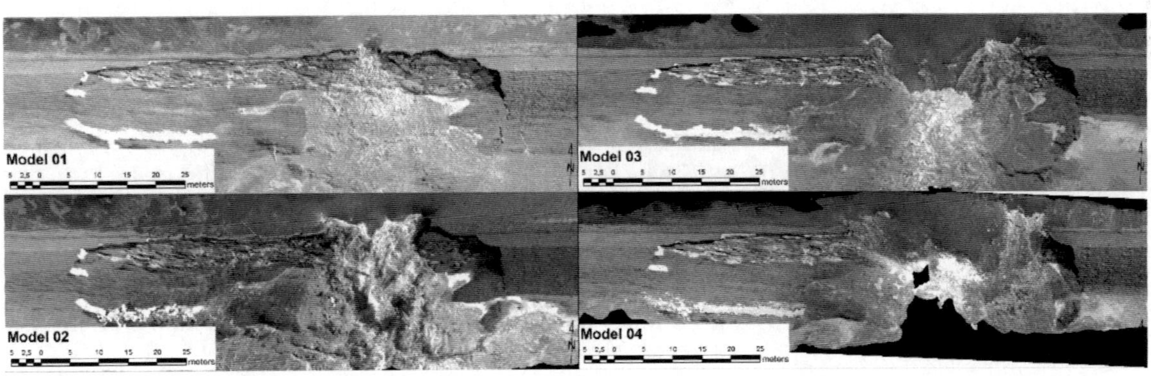

Figure 2. Orthoimages from the breach at 10, 15 21 and 27 minutes after the start of the failure

3.2. Hydrological modelling

Flow velocity and discharge are important characteristic values for the analysis of levee failures and the functional flood defence measures in the hinterland during the ongoing levee break. With these values the breach discharge, the progressive inundation of the hinterland and the inundation depth can be estimated when the terrain morphology in the hinterland and its release capacity (sluice, opening, pumping) are known. Further the size of the breach which is changing during the event as well as the flow situation upstream and downstream (river, lake, sea) should be known.

The calculation during the event will become difficult when direct measurements of the water levels, velocities and terrain points in the near of the breach are not possible. In this case it can be tried to get the ground elevation data from aerial images by means of photogrammetry to become able to run a hydro-numerical computation. Missing terrain data of the rapidly changing breach region below the water surface complicate the real time analysis. Therefore some additional assumptions have to be made as follows:

a) negligence of the approaching flow velocity far upstream of the breach in the river floodplain and downstream in the farer distance of the hinterland,
b) negligence of the change of flow direction (lateral weir effect),
c) assumption of a broad crested spillway or a Venturi flume with the critical depth and velocity respectively in the breach,
d) retarded or submerged flow when the downstream water level exceeds about ¾ of the upstream water depth,
e) no change of subcritical or supercritical flow when the downstream energy head is greater than the minimum energy head in the most constricted opening,
f) any remaining levee (if existing) acts as a overtopped threshold,
g) the breach geometry might be approximated by a trapezium,
h) the often found very steep breach flanks justify the assumption of a rectangle, which will simplify the calculation,
i) scour holes cannot be identified during the flood and can therefore only be assumed approximately.

The discharge can be calculated by multiplying the mean flow velocity with the flow area $Q = v \cdot A$ (s. a. figure 1C and [8]). Assuming the critical depth due to $Fr = v/\sqrt{g \cdot h_{gr}} = 1$ yields for an (almost) rectangular breach $Q = \sqrt{g} \cdot b \cdot h_{gr}^{3/2}$

With the minimum energy head $h_{E\,min} = 1{,}5 \cdot h_{gr} = h_{gr} + \frac{v_{gr}^2}{2g}$ the discharge with negligible approaching velocity can be estimated only from the water level difference between the upstream approach (HO) and the breach (HB)

$$\Delta h = H_O - H_B = \frac{h_{gr}}{2} + \zeta \cdot \frac{v_{gr}^2}{2g} = \frac{h_{gr}}{2} + \zeta \cdot \frac{h_{gr}}{2} = (1+\zeta) \cdot \frac{h_{gr}}{2} \qquad (1)$$

$$\text{with } h_{gr} = \frac{2 \cdot \Delta h}{(1+\zeta)} \qquad (2)$$

$$\text{and hence } Q = \left(\frac{1}{1+\zeta}\right)^{\frac{3}{2}} \cdot \sqrt{g} \cdot b \cdot (2 \cdot \Delta h)^{\frac{3}{2}} = C \cdot \sqrt{g} \cdot b \cdot (2 \cdot \Delta h)^{\frac{3}{2}} \qquad (3)$$

where C < 1 is a loss coefficient due to friction and spot losses which can also be expressed as seen above by the entrance loss coefficient ζ which takes values between $\zeta = 0{,}5$ (sharp edged breach entrance) and $\zeta \geq 0$ (well rounded breach entrance). The breach bottom, being unknown during the discharge and breach progression has the elevation

$$H_S = H_B - h_{gr} = H_B - \frac{2 \cdot \Delta h}{(1+\zeta)} \qquad (4)$$

When considering a trapezoid gap (index T) with the overflow width b and the flank slopes $1 : m_1$ und $1 : m_2$ as broad crested spillway the following equations can be applied:

$$Q_{gr} = v_{grT} \cdot A_{grT} = v_{gr} \cdot \left(b \cdot h_{grT} + \frac{1}{2}(m_1 + m_2) \cdot h_{grT}^2\right) \qquad (5)$$

$$\text{with} \quad v_{grT} = \sqrt{g \cdot h_{grT} \cdot \frac{1 + \frac{h_{grT}}{b'}}{1 + \frac{2 \cdot h_{grT}}{b'}}} \quad \text{and } b' = \frac{2 \cdot b}{m_1 + m_2} \qquad (6)$$

In the case of backwater without changes between subcritical and supercritical flow, besides the upstream water level also the downstream water level and the losses in between are needed as input data. Then the discharge can be calculated by means of the principle of energy conservation (Bernoulli equation). Depending on the operating stress a quick hydro-numerical calculation can also be carried out. But in this case the flow cross sections must be known or estimated in a pre-design stage.

4. Results

With the above equations the discharge (figure 3A) and the breach parameters were estimated (figure 3B). As there are uncertainties expected among the input data their spread was taken into account by uniformly distributed random variables in a Monte-Carlo-Calculation. The resulting confidence belts for 90 and 33 per cent respectively as well as the median are shown in figure 3A. The hydro-numerical 1-D-Calculation for the model 3 (x + 21 min) yielded similar results which represent one moment in the progressive levee breach process [9].

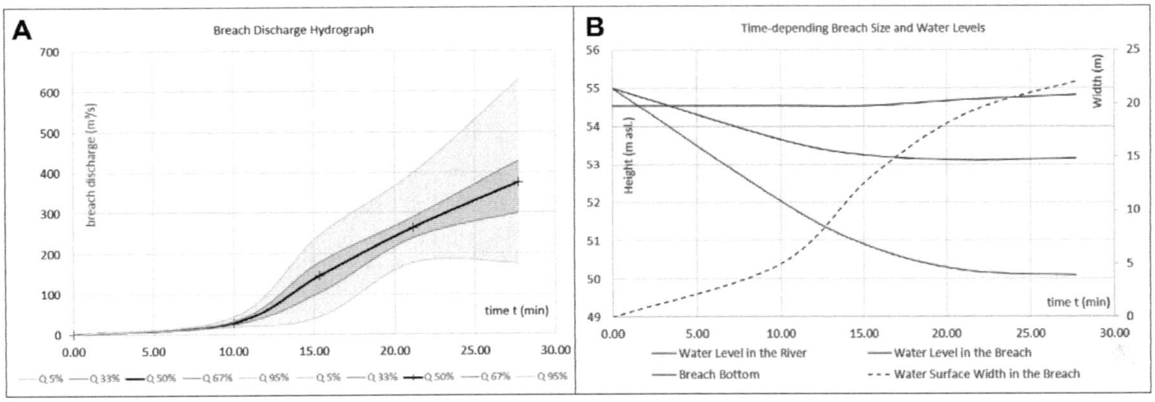

Figure 3 (A) Calculated breach discharge over time (B) Breach geometry over time

Models 02 to 04 show a distinct linear rise in breach discharge, after a steeper initial incline from model 01 to 02 (figure 3A). The temporal progression is one of the main sources of uncertainties, as the original video files were corrupted. Therefore, some assumptions about the gaps in between the video files had to be made. Due to personal expertise, the time for battery replacement breaks was set to 2 minutes (including landing and take-off).

5. Conclusions/Outlook

The results demonstrate that the implementation of UAV in emergency management measures is certainly an opportunity to use new data for emergency forces and clearly belong to the emerging technologies for monitoring levees. Real time or near real time information about the breach size and discharge would help to improve emergency management issues like evacuation lanes, etc.

Certainly, the goal of future works should be to incorporate new data sources and data management systems into emergency response strategies in order to quickly provide new information for decision support. But besides workflow improvement, these new issues also raise questions about the reliability of the spatial information in terms of precise location, fast processing and overall data quality. The most valuable information for disaster management response units is the precise location and geometry of disturbances and defective spots. Therefore, accurate georeferencing of orthoimages and 3d-models is the most crucial task as it is the most time consuming step in the processing workflow.

Some guidelines for future UAV pilots intending to survey levee failures could help to improve as well as simplify the processing workflow:

- orthogonal images are to be preferred over free camera angles
- pictures taken from an UAV usually have some coordinate information and other metadata tags stored in exchangeable image file format (Exif) and should be used instead of videos
- due to data copying issues, the correct time of the data acquisition was lost, an additional (handwritten) protocol would have been helpful
- camera objectives with a fixed focal length are preferable as the majority of SFM-programs are only capable of taking one focal length into account

References

[1] Juepner R, Assman A, Weichel T, Ettmer B, Mueller S and Bhaskar N 2015 Modelling flooding from a catastrophic dike breach along Elbe river, Germany June 2013 Proc. of the 35th Annual USSD Conference Louisville/USA 543-59

[2] Boccardo P 2013 New perspectives in emergency mapping. *Eur. J. Remote Sens.* **46** 571–82

[3] Joyce K, Belliss S, McNeill S and Glassey P J 2009 A review of the status of satellite remote sensing and image processing techniques for mapping natural hazards and disasters *Progr. in Phys. Geog.* ***33*** 1-25

[4] Nex F and Remondino F 2014 UAV for 3D mapping applications: a review *Appl. Geomat.* **Vol 6 No 1** 1-15

[5] Xu Z, Yang J, Peng C, Wu Y, Jiang X, Li R, Zheng Y, Gao Y, Liu S and Tian B 2014 Development of an UAS for post-earthquake disaster surveying and its application in Ms7.0 Lushan Earthquake, Sichuan, China *Computers & Geosciences* **68** 22-30

[6] http://www.bbc.com/news/science-environment-35353869

[7] Westoby M J, Brasington J, Glasser N F, Hambrey M J and Reynolds J M 2012 'Structure-from-Motion' photogrammetry: A low-cost, effective tool for geoscience applications *Geomorphology* **179** 300-14

[8] Bollrich G 2013 *Technische Hydromechanik* (Berlin Beuth)

[9] Brauneck J, Juepner R and Pohl R 2016 Auswertung des Deichbruchs Breitenhagen (Juni 2013) anhand von UAS-basierten Videoaufnahmen *Proc. of the 38th Dresden Conference on Hydraulic Engineering Dresden/Germany*

Radix Representation of Triangular Discrete Grid System

J Ben[1,2], Y L Li[1] and R Wang[1]

[1] Department of Remote Sensing, Zhengzhou Institute of Surveying and Mapping, 62 Science Avenue, High-Tech Zone, Zhengzhou, P.R. China
[2] State Key Laboratory of Resource and Environmental Information System, Chinese Academy of Science, 11A Datun Road, Chaoyang District, Beijing, P.R. China
benj@lreis.ac.cn

Abstract. Discrete Global Grid Systems (DGGSs) are spatial references that use a hierarchical tessellation of cells to partition and address the entire globe. It provides an organizational structure that permits fast integration between multiple sources of large and variable geospatial data. Although many endeavors have been done to describe certain discrete grid systems, there still lack of a uniform mathematical framework for them. This paper simplifies the planar class I aperture 4 triangular discrete grid system into a hierarchical lattice model which is proved to be a radix system in the complex number plane. Mathematical properties of the radix system reveal the discrete grid system is equivalent to the set of complex numbers with special form. The conclusion provides a potential way to build a uniform mathematical framework of DGGS and can be used to design efficient encoding and spatial operation scheme for DGGS.

1. Introduction

Discrete Global Grid Systems (DGGSs) are new forms of multi-resolution geospatial data fusion and geoscience modelling solution for the whole globe. It subdivides the globe uniformly in a particular way to provide a hierarchical structure without gap and overlapping, and uses cell address for data manipulation instead of traditional geographic coordinates. DGGSs are more suitable for large-scale questions and structurally support handling multiple-resolution data efficiently, compared with traditional local-regional organizational structure and application pattern of geospatial data [1].

An effective way for DGGS construction is to use regular polyhedrons to recursively subdivide the sphere, and the main steps are shown in figure 1. Firstly, choose a regular polyhedron and octahedron or icosahedron is commonly used; Secondly, unfold the regular polyhedron to the plane and obtain multiple triangular faces sharing one edge or one vertex; Thirdly, recursively subdivide all triangular faces into multi-resolution grids in the plane and triangle, quadrangle and hexagon are commonly used; Finally, by projection between polyhedral surface and spherical surface, map multi-resolution grids to the sphere and construct DGGS.

The obvious choice for a triangle is to subdivide it into smaller triangles. An equilateral triangle can be divided into $p^2(p \in \mathbb{N}, p \geqslant 1)$ smaller equilateral triangles by breaking each edge into p pieces and connecting the break points with lines parallel to the triangle edges (figure 2). This is referred to as Class I or alternative subdivision [2,3]. It generates consistent and aligned planar discrete grid systems with aperture (the ratio of the areas of a planar polygon cell at two successive levels) p^2 [2]. The smallest possible aperture of Class I subdivision is 4 ($p = 2$). The subdivision is similar to the fourfold recursive subdivision of the square grid quadtree, and therefore many algorithms developed on the square grid quadtree are portable to the triangular grid quadtree with minor modifications [3]. This subdivision approach (figure 2) was used by Goodchild and Yang[4], White et al.[5], Dutton [6],

Song et al.[7], Zhao et al.[8]. Congruent and unaligned Class I aperture 9 ($p = 3$) (figure 3) subdivision has been proposed by White et al.[5] and Song et al.[7].

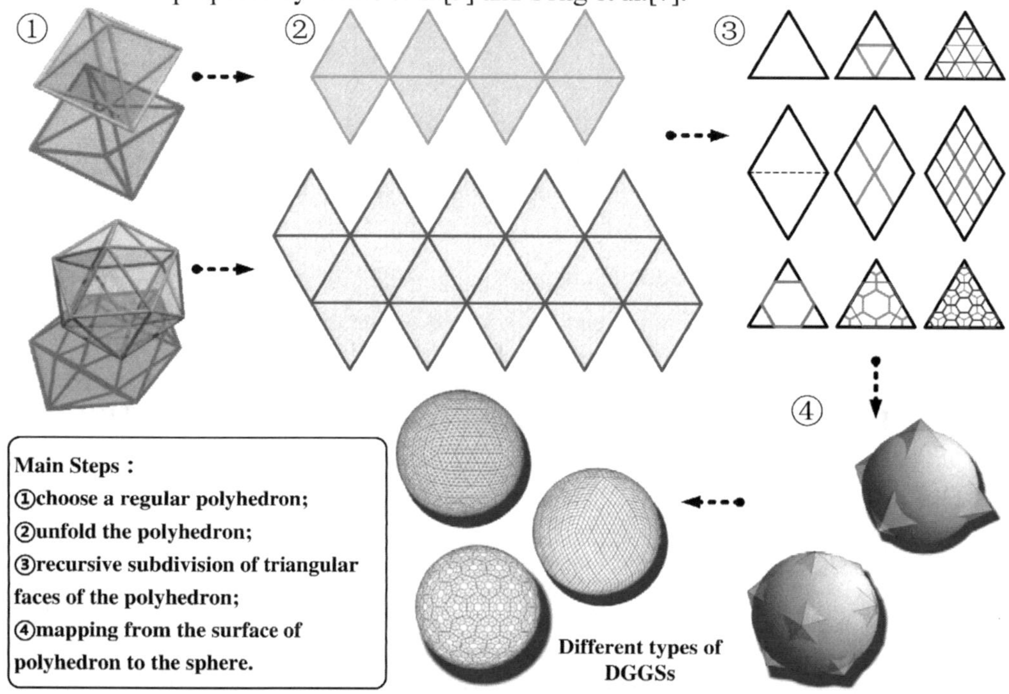

Figure 1. Main steps of polyhedron based DGGS construction.

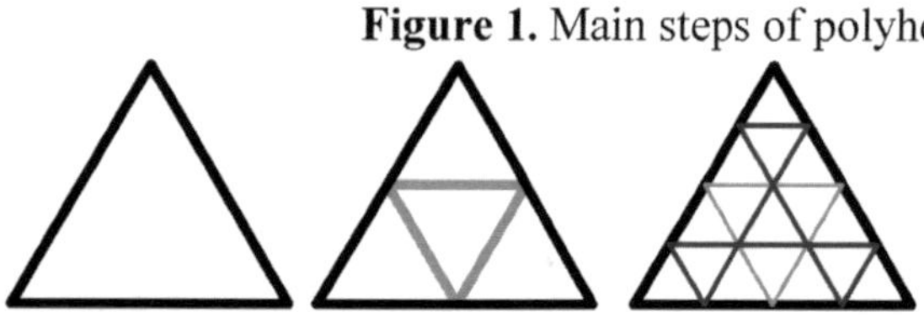

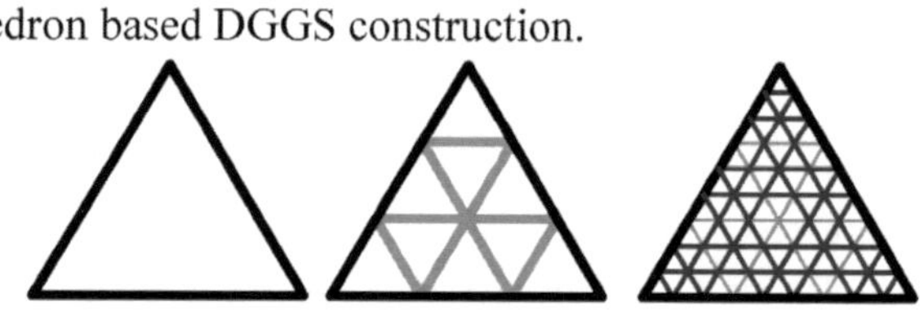

Figure 2. Aperture 4 recursive triangular subdivision

Figure 3. Aperture 9 recursive triangular subdivision

The square is the most popular cell region shape for construction of planar discrete grids. Its geometric attributes make it inappropriate on the triangular face of the octahedron or icosahedron. However, White notes that pairs of adjacent triangle faces may be combined to form a diamond or rhombus, which can be recursively subdivided as square quadtree does (figure 4)[9]. This subdivision approach yields a congruent, unaligned planar discrete grid system with aperture 4. It should be noted that, similar approach can be taken to yield a congruent, aligned planar discrete grid system with aperture 9 (figure 5).

The hexagon is consistently neighboring, which takes advantage of geospatial analysis. However, it is impossible to completely tile a sphere with hexagons. A none-hexagon polygon will be formed at each of the polyhedrons' vertexes. The number of such polygons relates to the number of polyhedrons vertexes and has nothing to do with the grid resolution. The none-hexagon polygons for octahedrons and icosahedrons are 6 squares and 12 pentagons respectively. On polyhedron's triangular faces, various hexagonal grid systems can be generated by special recursive subdivisions. The aperture 3 subdivision yields an aligned grid system, whose orientation of cell alternates with successive resolutions (figure 6). This grid system is used by Sahr et al.[3,10] and PYXIS innovation[11]. For the aperture 4 subdivision, it yields similar grid systems with fixed cell orientation (figure 7). These grid systems are used by Tong et al.[12].

Figure 4. Aperture 4 recursive rhombic subdivision

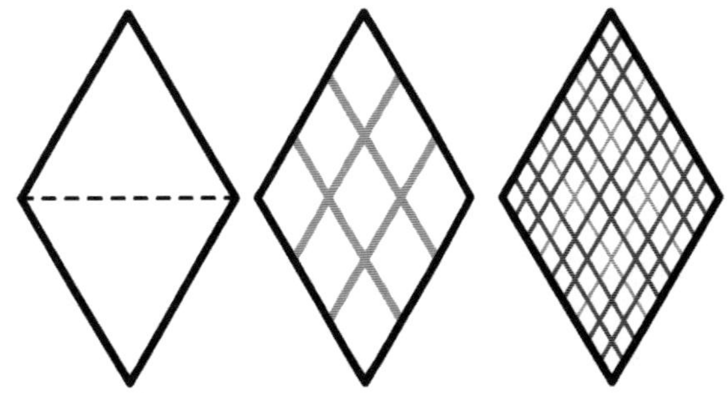

Figure 5. Aperture 9 recursive rhombic subdivision

Figure 6. Aperture 3 recursive hexagonal subdivision

Figure 7. Aperture 4 recursive hexagonal subdivision

Above all, as beneficial supplement and improvement for existing GIS theory, DGGS has raised wide academic concern. Many scholars have done research into methods of grid partition, encoding, operation, level index, grid data visualization, etc. However, in current research situations, the academia lacks a thorough understanding of the essence of grid systems, so grid expression, encoding and operation are short of necessary theoretical support, and it is difficult to improve relevant algorithms.

This paper simplifies planar triangular discrete grid system into a hierarchal lattice model which is proved to be a radix system in the complex number plane. Mathematical properties of the radix system reveal the discrete grid system is equivalent to the set of complex numbers with special form. The conclusion provides a potential way to build a uniform mathematical framework of DGGS and can be used to design efficient encoding and spatial operation scheme for DGGS.

2. Definition of planar grid system

A planar grid system consists of a set of regions that form a subdivision of the 2D plane, where each region has a single point contained in the region associated with it. Each region/point combination is a cell. Depending on the application, data objects or vectors of values may be associated with regions, points, or cells. If an application defines only the regions, the centroids of the regions form a suitable set of associated points. Conversely, if an application defines only the points, the Voronoi regions of those points form an obvious set of associated cell regions.

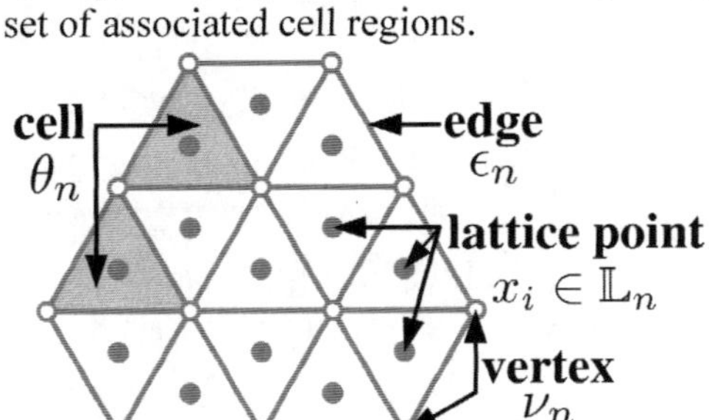

Figure 8. Definition of triangular grid

It is convenient to represent each cell in a grid by its centroid. The centroids of the cells at a given resolution $n(n \geqslant 1)$ are a finite set of points of a lattice in the plane. The elements of the lattice are called lattice points. The Voronoi cell of a lattice point x_i in a two-dimensional lattice $\mathbb{L}_n$ is a set consisting of all points of $\mathbb{R}^2$ which are at least as close to x_i as to any other lattice point of $\mathbb{L}_n$. If the Voronoi cells are regular triangles, then the lattice is called a triangular lattice (figure 8). A finite, non-empty subset of a lattice, or the corresponding set of Voronoi cells, is equivalent. Throughout this paper we often refer to the lattice point and its Voronoi cell interchangeably.

3. Mathematical model of triangular lattice

According to the analysis above, Class I triangular subdivision theory can produce grid systems with aperture p^2. But in fact, great apertures lead to severe jumping of resolution between two successive levels. Small apertures have the advantage of generating more grid resolutions, thus giving applications more resolutions from which to choose. Small apertures of triangular subdivision are 4 and 9. Subdivision of aperture 4 generates aligned grid systems while aperture 9 generates unaligned grid systems. Since aligned systems are convenient for expression of lattice and formula derivation, this article takes an aperture 4 triangular grid system as an example to establish the mathematical model of lattice.

Theorem 1 On the complex plane, letting $\omega = -\frac{1}{2} + \frac{\sqrt{3}}{2}i$, $D = \{0, \omega, \omega^2, \omega^3\}$, lattice of nth level in aperture 4 Class I triangular grid system is

$$\mathbb{L}_n = \frac{1}{2}e_1 + f(e_1)\frac{1}{4}e_2 + f(e_1)f(e_2)\frac{1}{8}e_3 + f(e_1)\cdot f(e_2)\cdots f(e_{n-1})\frac{1}{2^n}e_n = \sum_{i=1}^{n}[\prod_{j=1}^{i-1} f(e_j)]\frac{1}{2^i}e_i$$

Among this, $e_i \in D(i = 1, 2, \cdots, n)$, $f(x) = \begin{cases} -1 & x = 0 \\ 1 & x \neq 0 \end{cases}$ is the direction adjustment function, $\sum$, + are accumulation and plus operators between sets.

Proof. Since the aperture of the grid system is four, according to the law that the side ratio of two successive levels is 2, one cell in the level $i-1(i \geqslant 2)$ is divided into four cells in level i, as shown in figure 9. The lattice of one sub-cell is identical with the father cell, but the orientations are opposite. The other three sub-cells are rotational symmetry about the father cell and orientations are identical with the father cell.

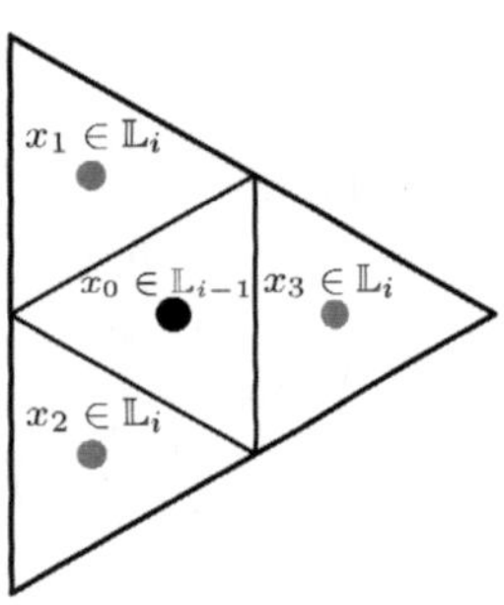

Figure 9. Lattice relationship between two successive levels

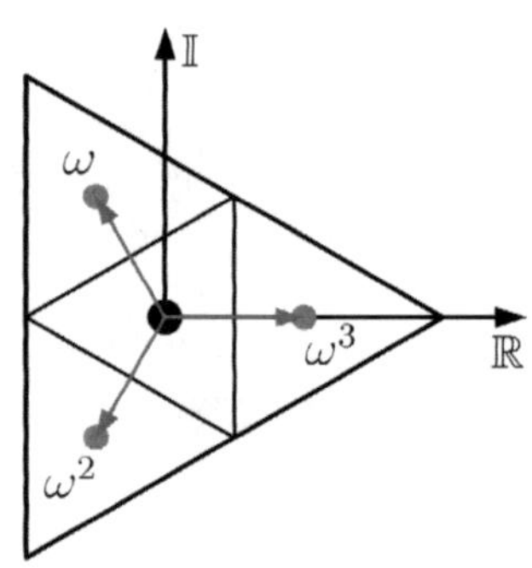

Figure 10. Lattice on the complex plane

Take the lattice in $\mathbb{L}_{i-1}$ as the origin and establish the complex plane, letting $D = \{0, \omega, \omega^2, \omega^3\}$, so lattice relationship between two successive levels is(figure 10):

$$\mathbb{L}_i = \mathbb{L}_{i-1} + \frac{1}{2^i}f(e_1)f(e_2)\ldots f(e_{i-1})e_i \quad (1)$$

Among this, $e_j \in D(j=1,2,\cdots,i)$, $f(x)=\begin{cases}-1 & x=0\\ 1 & x\neq 0\end{cases}$.

By equation (1):

$$\mathbb{L}_{i-1}=\mathbb{L}_{i-2}+\frac{1}{2^{i-1}}f(e_1)f(e_2)\dots f(e_{i-2})e_{i-1}$$

$$\mathbb{L}_{i-2}=\mathbb{L}_{i-3}+\frac{1}{2^{i-2}}f(e_1)f(e_2)\dots f(e_{i-3})e_{i-2}$$

$$\vdots$$

$$\mathbb{L}_2=\mathbb{L}_1+\frac{1}{4}f(e_1)e_2$$

So the expression of $\mathbb{L}_n$ is

$$\mathbb{L}_n=\frac{1}{2}e_1+f(e_1)\frac{1}{4}e_2+f(e_1)f(e_2)\frac{1}{8}e_3+f(e_1)\cdot f(e_2)\cdots f(e_{n-1})\frac{1}{2^n}e_n=\sum_{i=1}^{n}[\prod_{j=1}^{i-1}f(e_j)]\frac{1}{2^i}e_i.$$

Theorem 1 reveals, aperture 4 triangular lattice is essentially identical to a Radix System or a Positional Number System such as binary, octonary and decimalism, and it is a 'number' with specific form on the complex field. The system of this positional number system is 2 and the digit set is D. For $x\in\mathbb{L}_n$, its specific form is:

$$x=\frac{1}{2}e_1\pm\frac{1}{4}e_2\pm\frac{1}{8}e_3\pm\cdots\pm\frac{1}{2^n}e_n(e_1,e_2,\dots,e_n\in D)$$

If the base of power is omitted, exponents like decimalism, x is denoted by $x=.e_1e_2\cdots e_n$. The radix points show that the lattice is more and more intensive when the subdivision level is ingrown. Express elements in $\mathbb{L}_1$~$\mathbb{L}_5$ using points with different sizes and colours, their distribution on the complex plane is shown as figure 11.

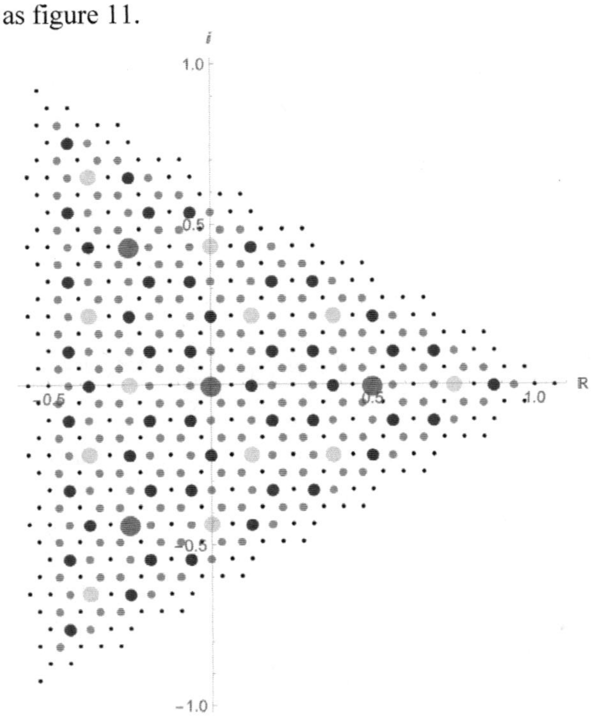

Figure 11. Lattice distribution in complex plane

Theorem 2 The aperture 4 triangular grid system $\mathbb{L}_n(n\geqslant 1)$ has the following properties:

- $\mathbb{L}_n$ has inclusion relation, that is $\mathbb{L}_1\subset\mathbb{L}_2\subset\cdots\subset\mathbb{L}_n$;
- The number of lattice in level i is 4^i.

Proof. Since $\mathbb{L}_i = \mathbb{L}_{i-1} + \frac{1}{2^i} f(e_1)f(e_2)\dots f(e_{i-1})e_i$, that is to say, lattices in level $i(i \geqslant 2)$ take lattices in level $i-1$ as centre, $\frac{1}{2^i}$ as radius and then expand three elements in the directions of nonzero elements of D, so $\mathbb{L}_{i-1}$ must be proper subset of $\mathbb{L}_i$. Combined with recursive relations, it is concluded that $\mathbb{L}_1 \subset \mathbb{L}_2 \subset \cdots \subset \mathbb{L}_n$.

Letting the lattice count of level $i(i \geqslant 2)$ is a_i, it is obvious that $a_i = 4a_{i-1}$, Combined with recursive relations, and $a_1 = 4$, so $a_i = 4^i$.

4. Conclusion

This paper employs lattice as the equivalent object of grids and reveals the mathematical essence of the planar Class I aperture 4 triangular grid system. Although only one grid system is discussed in this paper, the proof processes of relevant theorems have shown that multi-resolution regular grids are self-replication of the finite lattice set, which has nothing to do with the shape of grids. Thus, this research thought can be expanded to other regular grid systems. And as an effective tool of formal description, it perfects the theoretical basis of DGGS and provides theoretical support for design and analysis of algorithms.

References

[1] Open Geospatial Consortium 2015 *OGC Discrete Global Grid System (DGGS) Core Standard* https:// portal.opengeospatial.org/files/66643
[2] Kenner H 1976 *Geodesic math and how to use it* (California: University of California Press) p172
[3] Sahr K, White D and Kimerling A 2003 *Cartography and Geographic Information Science*, **30** (2) p121–134
[4] Goodchild M and Yang S 1992 *Graphical Models and Image Processing* **54**(1) p31–44
[5] White D, Kimerling A, Sahr K, and Song L 1998 *International Journal of Geographical Information Science* **12** p805–27
[6] Dutton G 1999 *A hierarchical coordinate system for geoprocessing and cartography*. Germany: Springer-Verlag. p231
[7] Song L, Kimerling A and Sahr K 2002 *Discrete global grids: A web book* (Santa Barbara: University of California)
[8] Zhao X, Hou M and Bai J 2007 *Spatial digital modeling of the global discrete grids* (Beijing: SinoMaps Press) p46
[9] White D 2000. *Environmental Monitoring and Assessment* **64**(1) p93–103
[10] Sahr K 2008 *Environment and Urban Systems* **32** 174–187
[11] PYXIS Innovation 2016 *Pyxis public wiki* http://www.pyxisinnovation.com/pyxwiki/index.php?title=Main_Page
[12] Tong X, Ben J, Wan Y, Zhang Y and Pei T 2013 *International Journal of Geographical Information Science* **27**(5) p898–921

Acknowledgments

Portions of this research were supported by National Natural Science Foundation of China (Grant NO. 41271391, 41671410) and by Special Financial Grant from the China Postdoctoral Science Foundation (Grant NO. 2013T60161). The research has not been subjected to the Committee of National Natural Science Foundation or the Committee of Postdoctoral Science Foundation's peer and administrative review and, hence, does not necessarily reflect the views of them.

WebGIS based community services architecture by griddization managements and crowdsourcing services

Haiyin Wang[1,2*], Jianhua Wan[3], Zhe Zeng[3] and Shengchuan Zhou[1,2]

[1] Qingdao Geotechnical Investigation and Surveying Research Institute, Qingdao, 266034, China
[2] National & Local Joint Engineering Research Center for Sea-land Geographics Information Integragtion and Application, Qingdao, 266043, China
[3] School of Geosciences, China University of Petroleum, Qingdao, 266580, China

*E-mail: why@qdkcy.com.cn

Abstract. Along with the fast economic development of cities, rapid urbanization, population surge, in China, the social community service mechanisms need to be rationalized and the policy standards need to be unified, which results in various types of conflicts and challenges for community services of government. Based on the WebGIS technology, the article provides a community service architecture by gridding management and crowdsourcing service. The WEBGIS service architecture includes two parts: the cloud part and the mobile part. The cloud part refers to community service centres, which can instantaneously response the emergency, visualize the scene of the emergency, and analyse the data from the emergency. The mobile part refers to the mobile terminal, which can call the centre, report the event, collect data and verify the feedback. This WebGIS based community service systems for Huangdao District of Qingdao, were awarded the "2015' national innovation of social governance case of typical cases".

1. Introduction

Along with the fast economic development of cites, rapid urbanization, population surge, the social community service mechanisms need to rationalize and the policy standards need to unify, which results in various types of conflicts and challenges for community services of government. To address the issue, in China, the local governments at all levels start to build community service platforms for the public.

The community service platforms are considered as a centre of collecting the social information, publishing the governance information, responding the emergency calling, and dispatching the cases. The platforms should provide all kind of the authoritative data which includes not only government service contents but also associated spatial data such as coordinates, geocoding data and etc. The data is so massive that it is a challenge for managing these data, updating the data dynamically and responding the service request immediately. Therefore, the WebGIS technology is applied into the community service platforms for the public. The paper provides an architecture of WebGIS based community services, which mainly realizes the following functions.

- A platform of addressing the social governance problem. The platform service as a centre of managing the social governance information, receiving the report of emergency events and dispatching the community services.

- A WebGIS based urban griddization management system. The WebGIS technology provides the web browser for positioning the service location and the server for responding all geographical services by gridding the management.
- A mobile application for crowdsourcing services. The mobile application can be deployed to iOS and Android smart phones. It can be used for collecting the social governance information, submitting urban management or emergency events and receiving the service tasks.

2. WebGIS based community services architecture

The WebGIS based community services architecture is based on the network infrastructure platform, hardware and software platforms, and resource information sharing platform. The digital social governance information system plays the core role in the service architecture. The government websites and mobile application give convenient access methods for the community services.

The WebGIS based community services architecture includes database clusters, standard platform, software and hardware infrastructure. There are five layers and two frameworks as following (see Figure 1):

- present layer,
- service layer,
- platform layer,
- data layer,
- hardware layer,
- standards framework,
- security framework.

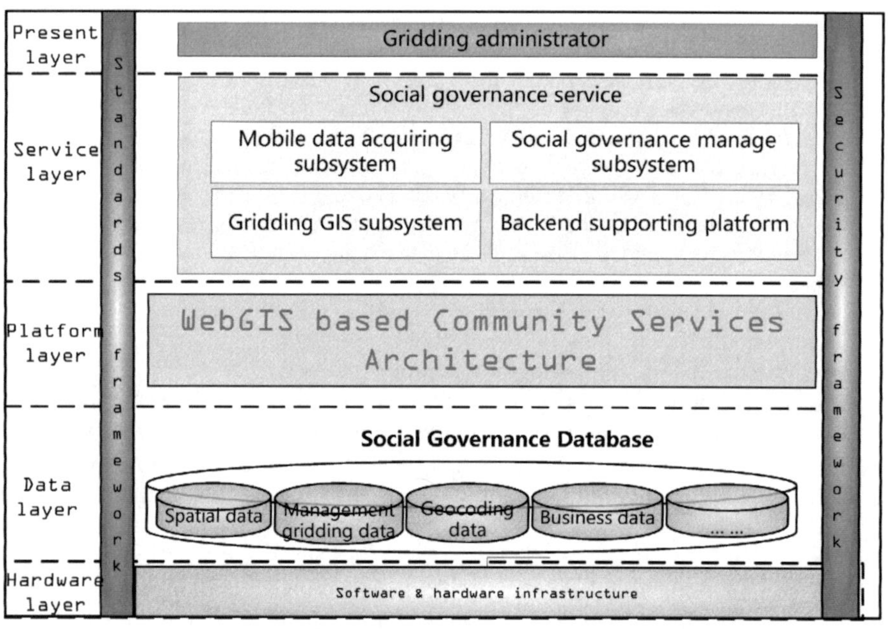

Figure 1. WebGIS based community services architecture

■ Hardwar layer

In the hardware layer, an office centre for administration should be constructed. High-speed broadband network should be established for connecting social governance centre and departments of government. A hardware platform of labs, servers, switches and terminal equipment should also be built.

■ Data & platform layer

In the data & platform layer, a unified database should be built for spatial and non-spatial data, which consist of basic topographic data, urban infrastructure, geocoding data, land information data, planning data, landscaping data, transportation data, and business data.

- Service layer

The service layer provides data exchange platform, middleware, workflow engines, third-party software, and so on, which supports system development, deployment, operations, and application.

The WebGIS based community services architecture introduces the three key technologies: WebGIS, urban griddization management, and crowdsourcing services.

2.1. WebGIS –based community services

The WebGIS based community services can publish the spatial and non-spatial information on the server side. The public can explore these information through the web browser such as IE, Firefox and etc. The database in the backend can provide the geo-data and non-spatial data. The WebGIS framework play a core role in community service architecture. The following figure depicts the WebGIS framework for the community service.

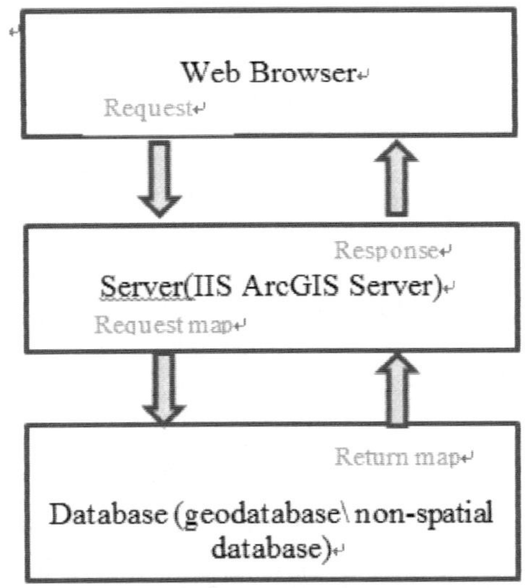

Figure 2. Framework of WebGIS

The three layers in the WebGIS framework are data layer, logic layer and presentation layer. The following figure describes the framework in details.

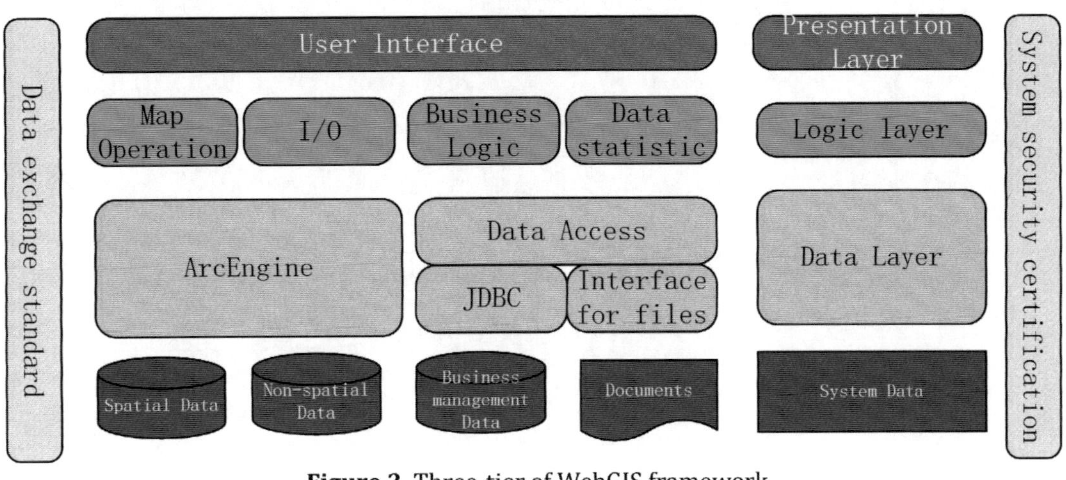

Figure 3. Three-tier of WebGIS framework

- Data Layer: storing spatial data, attribute data, text and other format data.
- Logic Layer: performing the data services and various business processes.
- Presentation Layer: presenting and visualizing the data and business operations to the user.

A standard is built for mutually accessing data in three layers, and the security control system can ensure the security of data. The three-tier structure separate the data from the logic layer. It is convenient for the extension of system and the reuse of code. The tasks are centralized into logic middle tier so that the business logic of the system is independent, that is, when the user's needs change, you can quickly build platform update business logic in the middle tier (application server), without submitting the updated applications to PC terminals.

Flexibility and scalability in the future are characteristics of the three-tier structure design for WebGIS based community services platform. The platform would not refactor the architecture when the requirement of a digital society governance changed greatly.

2.2. Griddization management

Griddization management is an administrative policy that employs divide-and-conquer strategy. It firstly divides the administrative region into grids. And then the community services are deployed into these different grids. An administrator for the community services should be dispatched into each grid. Through the gridding technology, the social governance should be better in practice. The formation of the grids follows these principles.

1) Region management principle

Each grid only belongs to one and only one community, and the boundaries cannot across to other communities.

2) Geographical distribution principle

The grid layout fits the city's geography, such as streets, alleys, courtyards, public green spaces, squares, bridges, rivers and etc.

3) On-line management principle

Courtyard managed by the self-management unit should not be split, but in accordance with specified standards for independent courtyard.

4) Convenient management principle

The division of the road and alleys should follow the travel habits of their neighbour courtyards.

5) Object management principle

The management target is the individual objects of the city which cannot be split into parts.

As an example, in community service systems for Huangdao, the district is divided into grids according to the above principles (see the following figure).

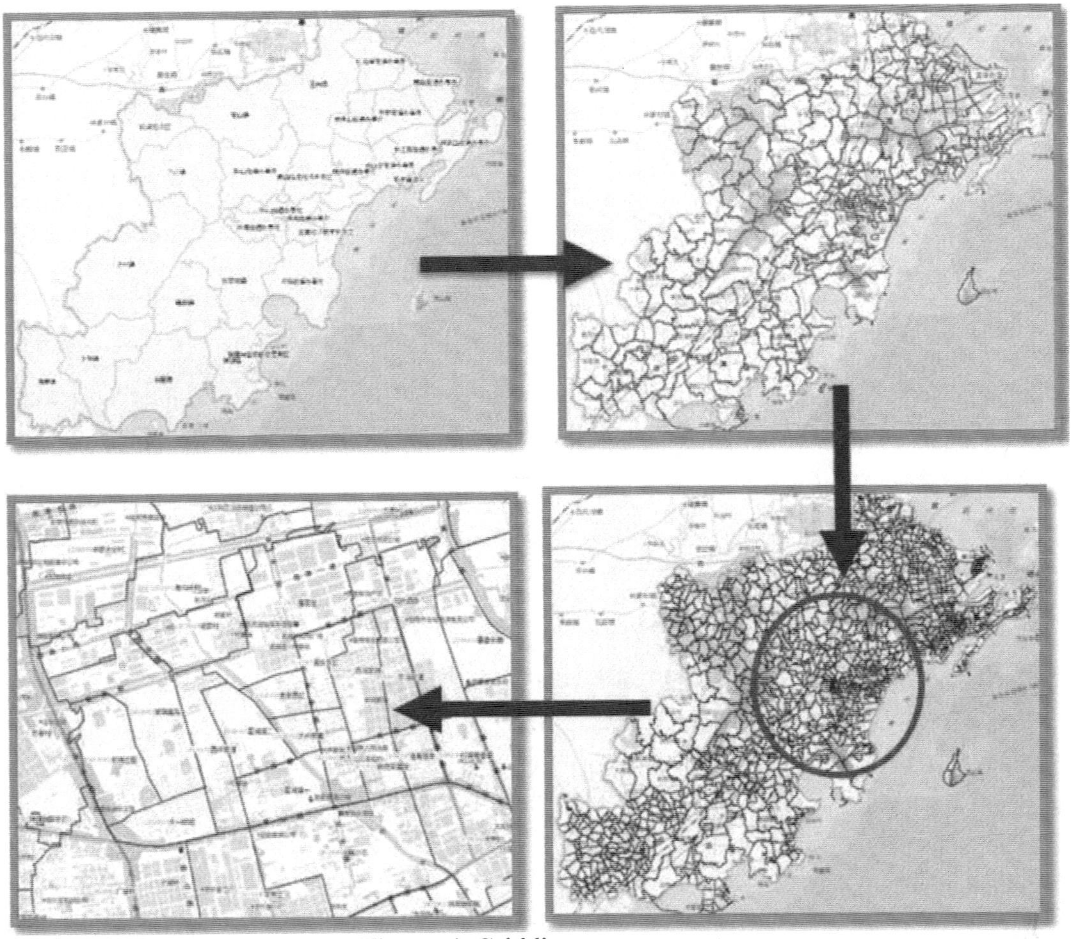

Figure 4. Gridding management

2.3. Crowdsourcing services

Crowdsourcing services refers to the service mechanism that deploy mobile application for community services to smart phones. So the mechanism crowdsources the community services into the grid administrators equipped with smart phone. The smart phones of the grid administrators are installed the community service application, which guarantee that the grid administrator carry out the service task.

The mobile application provides a condition that the grid administrator can participate in social governance anywhere and anytime. The smart phone with the community service application became a mobile terminal which could take photo, position and call on the spot when urban management event happened. By the mobile application, the grid administrator can collect data, submit a report and recheck the task, which greatly facilitates one's daily work. On the other hand, a mobile application for photo anywhere is available to the public. Everyone can take photo for helping the grid administrator through this application if he downloads the application into his smart phone. Therefore, a pattern of crowdsourcing the governance services is given. Each grid administrator as well as the individual with the smart application became the sources of the social governance services. The following figure give an illustration of the mobile applications on the smart phones.

Figure 5. Mobile application in smart phone

3. Conclusion

Based on the WebGIS technology, this article provides a community service architecture by griddization management and crowdsourcing service. The WebGIS service architecture includes two parts: the cloud part and the mobile part. The cloud part refers to community service centre, which can instantaneously response the emergency, visualize the scene of the emergency, and analyse the data from the emergency. The mobile part refers to the mobile terminal, which can call the centre, report the event, collect data and verify the feedback.

References

[1] S. Shekhar and H. Xiong, "WebGIS," in Encyclopedia of GIS, S. Shekhar and H. Xiong, Eds., ed Boston, MA: Springer US, 2008, pp. 1273-1273

[2] H. Wang, "Specifications for design of social govenance information platform (Hangdao District, Qingdao city)," ed, 2015.

Moon-based visibility analysis for the observation of "The Belt and Road"

Yuanzhen REN [1, 2], Huadong GUO [1], Guang LIU [1*], Hanlin YE [1, 2], Yixing DING [1], Zhixing RUAN [1], Mingyang LV [1, 3]

[1] Key Laboratory of Digital Earth Science, Institute of Remote Sensing and Digital Earth, Chinese Academy of Sciences, Beijing 100094, China

[2] College of Resources and Environment, University of Chinese Academy of Sciences, Beijing 100190, China

[3] School of Earth Sciences and Engineering, Nanjing University, Nanjing 210046, Jiangsu, China

* liuguang@radi.ac.cn

ABSTRACT: Aiming at promoting the economic prosperity and regional economic cooperation, the "Silk Road Economic Belt" and the "21st Century Maritime Silk Road" (hereinafter referred to as the Belt and Road) was raised. To get a better understanding of "the Belt and Road" whole region, considering the large-scale characteristic, the Moon platform is a good choice. In this paper, the ephemeris is taken as data source and the positions and attitudes of Sun, Earth and Moon are obtained based on the reference systems transformation. Then we construct a simplified observation model and calculate the spatial and angular visibility of the Moon platform for "the Belt and Road" region. It turns out that Moon-based observation of this region shows a good performance of spatial visibility and variable angular visibility, indicating the Moon being a new potential platform for large-scale Earth observation.

1. Introduction

When Chinese President Xi Jinping visited Central Asia and Southeast Asia in 2013, he raised the initiative of jointly building the Belt and Road, which has attracted close attention from all over the world. The Belt and Road Initiative aims to promote the connectivity of Asian, European and African continents and their adjacent seas, establish and strengthen partnerships among the countries along the Belt and Road, set up all-dimensional, multi-tiered and composite connectivity networks, and realize diversified, independent, balanced and sustainable development in these countries. However, the existing Earth observation systems lack the ability of long-term continuous observation at a regional-to-global scale due to the spatial and temporal coverage limitation and systematic bias of satellite observation [1]. Under this circumstance, Guo proposed to probe into ways of setting up observatory on the moon containing a group of remote sensors as a complementation for global environmental change on a global scale [2]. In order to get a better understanding of "the Belt and Road" region, considering the spatial and angular visibility, we propose the Moon platform for observation.

Taking the Moon as a platform for Earth observation has several potential advantages [3]–[5]: (1) the Moon is a satellite that never falls and its microgravity facilitates the installation, upgrading, maintenance and replacement of large telescopes, antenna arrays and other equipment. (2) The Moon's rotation period and revolution period are the same, with one side always facing the earth, and therefore, it is possible to observe a certain area of the Earth over a long period of time from a variable angle. (3) Since the geological structure of the Moon is very stable, the sensors mounted on it are much more stable than those on near-earth satellites.

In this paper, we construct a simplified observation model and calculate the spatial and angular visibility of the Moon-based observation for "the Belt and Road" region in order to quantificationally evaluate the visible ability of the Moon platform.

2. Data and processing

2.1 Study area

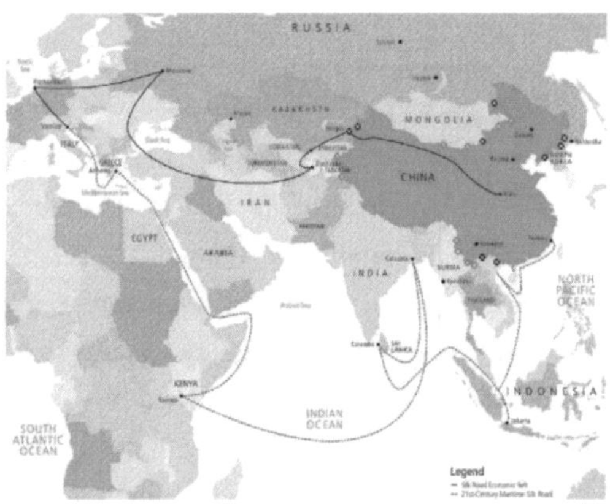

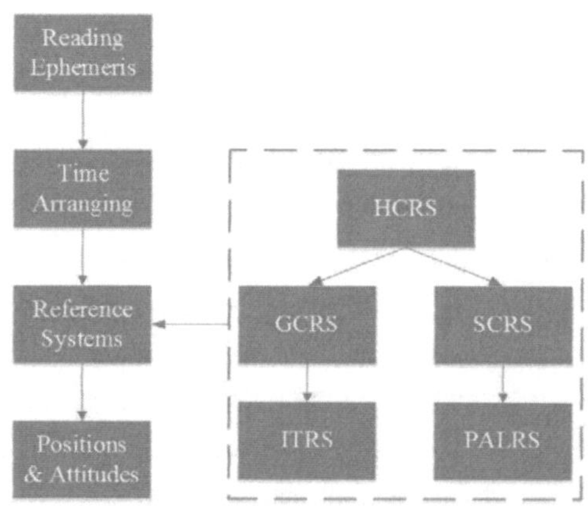

Figure 1. The routes of "the Belt and Road" with key cities

Figure 2. The flowcharts of data processing and reference systems transformation

As is shown in the fig.1, "the Belt and Road" includes the "Silk Road Economic Belt" and the "21st Century Maritime Silk Road", covering three continents: Asia, Europe and Africa. Approximately, we make use of the locations of several cities to set the boundary of this region, such as Ningbo (29°53′N 121°33′E), Djakarta (6°08′S 106°45′E), Nairobi (1°17′S 36°49′E), Rotterdam (51°55′N 4°29′E), Moscow (55°45′N 37°37′E) and Xi'an (34°16′N 108°57′E).

2.2 Data source

The ephemeris data from Jet Propulsion Laboratory (JPL) is used in this study. JPL ephemerides of DE series are the most popular ephemerides among the international astronomical community, providing the positions and velocities of certain celestial bodies at a given time, generated by fitting numerically integrated orbits of the Moon and planets to observations[6]. The positions and velocities are stored as Chebyshev interpolation polynomial coefficients fit in 32-day-long segments, which are shown in the form of planetary three-dimensional Cartesian reference of heliocentric reference system[7].

2.3 Data processing

As is shown in the fig.2, there are several data processing steps: (1) Read the ephemeris' header files and data files, and then merge each part of them chronologically. (2) Based on the given Julian time, obtain the corresponding ephemeris data, then calculate the positions and velocities of each celestial body by interpolation. (3) According to the observational perspectives of simulation, reference systems transformations are carried out.

First, time system is introduced. Julian day is the continuous count of days since the beginning of the Julian Period. The Julian date (JD) of any instant is the Julian day number for the preceding noon in Greenwich Mean Time plus the fraction of the day since that instant. Julian dates are expressed as a Julian day number with a decimal fraction added. The Julian Period is a chronological interval of 7980 years beginning 4713 BC. It has been used by historians since its introduction in 1583 to convert between different calendars.

Besides, there are a few reference systems related to the Moon-based Earth observation. It is essential to bridge the different reference systems for calculating the celestial bodies' positions from different perspectives [8]. When celestial spheres' centers are selected as the center of Sun, Earth or Moon, three different celestial reference systems are defined: heliocentric celestial reference system (HCRS), geocentric celestial reference system (GCRS) and selenocentric celestial reference system (SCRS). In addition, it is necessary to utilize the international terrestrial reference system (ITRS) when dealing with the parameters related to Earth [9]. Analogously, principal axis lunar reference system (PALRS) is introduced for the parameters related to lunar surface [10].

2.4 Calculation of visibility

As the lunar radius is much lesser than the distance between Moon and Earth, we utilize the barycentre of Moon to take the place of the observatory's position for simplification. From the preceding processing, the coordinates of lunar barycentre $L(x_L, y_L, z_L)$, Sun $H(x_H, y_H, z_H)$ and the target on Earth $T(x_T, y_T, z_T)$ could be obtained. Thus, the visibility of the target could be derived:

$$\alpha = arccos(\frac{x_L x_T + y_L y_T + z_L z_T}{\sqrt{(x_L^2 + y_L^2 + z_L^2)(x_T^2 + y_T^2 + z_T^2)}}) \tag{1}$$

$$\beta = arccos(\frac{x_T x_H + y_T y_H + z_T z_H}{\sqrt{(x_H^2 + y_H^2 + z_H^2)(x_T^2 + y_T^2 + z_T^2)}}) \tag{2}$$

$$\alpha < \frac{\pi}{2} \ \& \ \beta < \frac{\pi}{2} \tag{3}$$

If formula (3) condition is met, the target is visible. Then the observing perspective θ is:

$$x_D = x_L - x_T, \quad y_D = y_L - y_T, \quad z_D = z_L - z_T \tag{4}$$

$$\theta = arccos(\frac{x_T x_D + y_T y_D + z_T z_D}{\sqrt{(x_D^2 + y_D^2 + z_D^2)(x_T^2 + y_T^2 + z_T^2)}}) \tag{5}$$

3. Results and analysis

After a lot of experiments, we find that there is similarity between different months with little difference. So we arbitrary choose the data in May, 2016 for description. Since the Moon-based Earth observation

IOP Conf. Series: Earth and Environmental Science **46** (2016) 012049 doi:10.1088/1755-1315/46/1/012049

is continuous, Rotterdam and Ningbo, the two terminal vertexes of the "the Belt and Road", are selected for analysis.

As is shown in the fig.3, the observation period in a day of Rotterdam is a bit longer than that of Ningbo. They both show similar variation trend. The observation period of Rotterdam varies between 870 minutes and 40 minutes and that of Ningbo varies between 750 minutes and 40 minutes. The average period of these two targets are 446.77 minutes and 390.65 minutes, showing a good performance of visibility.

The fig.4 shows the observation period of the "the Belt and Road" in each day of May, 2016. The "whole" curve refers to the whole region of "the Belt and Road" could be observed and the "region" curve refers to any region of "the Belt and Road" could be observed. The average period of these two categories are 80.65 minutes and 756.77 minutes, indicating a rather good coverage of the study area.

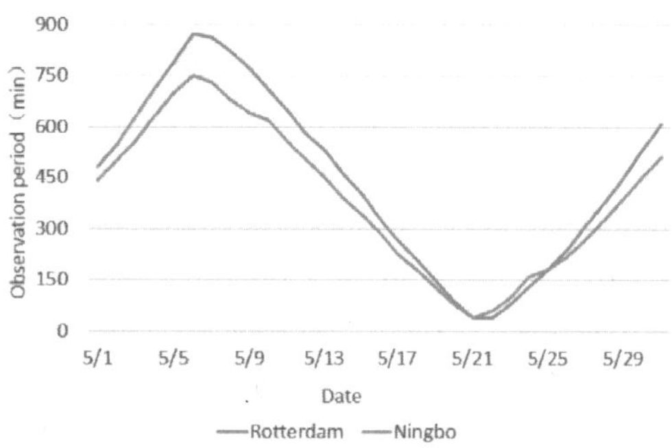

Figure 3. The observation period of each day in May 2016 of Rotterdam and Ningbo

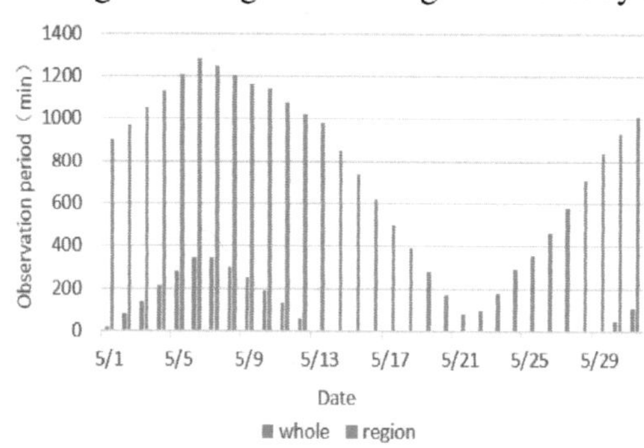

Figure 4. The observation period of each day in May 2016 of whole and region

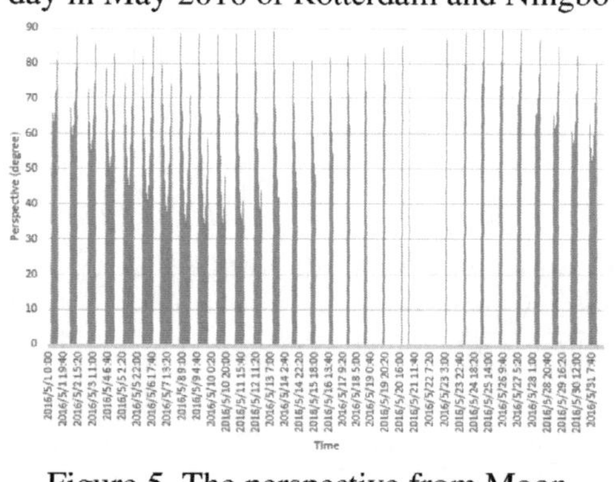

Figure 5. The perspective from Moon platform of Rotterdam in May 2016

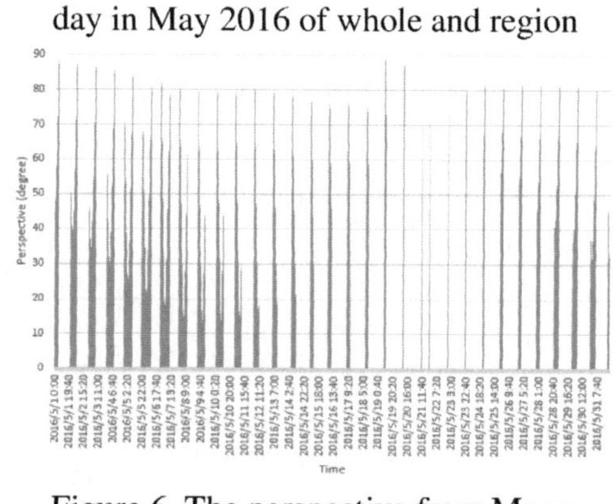

Figure 6. The perspective from Moon platform of Ningbo in May 2016

The fig.5 and fig.6 depicts the perspective from Moon platform of Rotterdam and Ningbo, showing that the perspective of Rotterdam is larger than that of Ningbo, owing to the relative positions of Moon and Earth and the latitudes of Rotterdam is larger than Ningbo. The perspective of Rotterdam varies between 34° and 90° and that of Ningbo varies between 10° and 90°. Analogously, the cities with smaller latitudes have larger perspective variation. Thus, for "the Belt and Road" region, different cities with different latitudes differ a lot at the perspective. Besides, it is obvious too large the perspective is not suitable for observation. Although at the certain time, some region is visible from the Moon platform in theory, it is possible that the perspective is too large, which is not appropriate for observation.

Actually, there is a little difference between different months for observing periods and perspective since the Moon has libration. Besides, the period of lunar rotation is one sidereal month, not synodic month. Because of the different time system, it brings the difference. In addition, the axis perpendicular to the lunar orbit's plane precess clockwise in the ecliptic, the period of which is 18.6 years. Correspondingly, the observing periods and perspective vary in different years, which adds to more diversity. Generally, the observing periods of low latitudes' locations are more stable than that of high latitudes' locations in different months and the perspective of low latitudes' locations has a larger range than that of high latitudes' locations.

4. Conclusion

This paper, facing the demand of observation of "the Belt and Road" region, proposed Moon-based platform for large-scale observation. The JPL ephemeris is selected as data source and processing procedures are introduced. The spatial and angular visibility of this region is calculated with the average period of whole region of 80.65 minutes in a day and the average period of any part of the region of 756.77 minutes, implying a good performance of overall observation. For angular visibility, the perspective vary a lot for different latitudes, showing a large range of variable angular visibility characteristic. Furthermore, the perspective of low latitudes' locations has a larger range than that of high latitudes' locations. However, for certain target, the perspective should be carefully chosen for effective observation, avoiding large distortion.

5. Acknowledgement

This research was supported by the National Science Foundation of China (41590852, 41590853), National Key Technology R&D Program of China (2012BAC16B01), Strategic Pioneer Program on Space Science, Chinese Academy of Sciences (XDA04077200), Science Foundation of the Institute of Remote Sensing and Digital Earth, Chinese Academy of Sciences (Y5ZZ14101B, Y6SJ1900CX).

6. References

[1] E. Pallé and P. R. Goode, "The Lunar Terrestrial Observatory: Observing the Earth using photometers on the Moon's surface," *Adv. Sp. Res.*, vol. 43, no. 7, pp. 1083–1089, 2009.

[2] H. Guo, "Space-based Observation for Sensitive Factors of Global Change," *Bull. Chinese Acad. Sci.*, vol. 23, no. 4, pp. 226–228, 2009.

[3] G. Fornaro, G. Franceschetti, F. Lombardini, A. Mori, and M. Calamia, "Potentials and limitations of Moon-Borne SAR imaging," *IEEE Trans. Geosci. Remote Sens.*, vol. 48, no. 7, pp. 3009–3019, Jul. 2010.

[4] A. Moccia and A. Renga, "Synthetic aperture radar for earth observation from a lunar base: Performance and potential applications," *IEEE Trans. Aerosp. Electron. Syst.*, vol. 46, no. 3, pp. 1034–1051, Jul. 2010.

[5] H. Guo, G. Liu, Y. Ding, and D. Zhang, "Conceptual Research of Lunar-Based Earth Observation for Global Environmental Change," *39th COSPAR Scientific Assembly*, vol. 39. p. 684, 2012.

[6] W. M. Folkner, J. G. Williams, and D. H. Boggs, "The planetary and lunar ephemeris DE 421," *IPN Prog. Rep. 42-178*, vol. 5, pp. 1–34, 2009.

[7] W. M. Folkner, J. G. Williams, D. H. Boggs, R. S. Park, and P. Kuchynka, *The Planetary and Lunar Ephemerides DE430 and DE431*, vol. 196. 2014.

[8] S. A. Hall, J. A. Hoffman, and D. Candidate, "Determination of Shadowing On the Lunar Surface Using a Lunar-Celestial Equatorial Coordinate System," *44th Int. Conf. Environ. Syst.*, no. July, pp. 1–6, 2014.

[9] P. Gérard and B. Luzum, "IERS Conventions (2010)," *Bur. Int. DES POIDS Mes. SEVRES*, pp. 1–179, 2010.

[10] B. A. Archinal, M. F. A'Hearn, E. Bowell, A. Conrad, G. J. Consolmagno, R. Courtin, T. Fukushima, D. Hestroffer, J. L. Hilton, G. A. Krasinsky, G. Neumann, J. Oberst, P. K. Seidelmann, P. Stooke, D. J. Tholen, P. C. Thomas, and I. P. Williams, "Report of the IAU Working Group on Cartographic Coordinates and Rotational Elements: 2009," *Celest. Mech. Dyn. Astron.*, vol. 109, no. 2, pp. 101–135, Feb. 2011.

The architecture for data, security and application in Digital Earth platform

Fang Miao[1,2], Wenhui Yang[1] , Ansheng Ye[2] , Huayue Chen[1]

[1] Institute of Spatial Information Technology, Chengdu University of Technology
[2] Research Institute of Big Data, Chengdu University

E-mail: miaofang@126.com

- **Abstract.** Digital Earth platform is an application, service and decision support system, which integrates geo-spatial data acquisition, transmission, storage, processing, analysis, statistics and visualization. It realizes comprehensive management and widely application of different kinds of huge earth data under the geo-spatial framework and in a open environment . It should consider the continues growing of data and applications, as well as data security. According to the issues of how to safely use spatial data in Digital Earth platform, an architecture for data ,security and application in Digital Earth platform was proposed in this paper. We call this architecture DOA(DOSA). In the digital earth platform on its goal is the Trusted Data. DOA(DOSA) and Trusted Data can provide data management, security grantee, application supporting for Digital Earth platform.
- **Key.** Data Oriented Architecture(DOA), Digital Earth platform, Trusted Data Data-Oriented Security Architecture(DOSA)

1. Introduce

1.1 Background

In 1998, U.S. Vice-President Al Gore articulated a vision of "Digital Earth" as a multi-resolution, three-dimensional representation of the planet that would make it possible to find, visualize, and make sense of vast amounts of dereferences information on the physical and social environment. Such a system would allow users to navigate through space and time, access to historical data as well as future predictions based for example on environmental models, and support access and use by scientists, policy-makers, and children alike [1]. Since then digital earth was largely developed globally. Huadong Guo [2] gave Digital Earth a definition in 2008. He claimed that Digital Earth is a mega-data system about the Earth in regard to the information of natural and human society, the model of earth evolvement and society development, information technology system, and the geo-spatial information obtaining system; is a development of geo- and social science in the computer technology,

Web system technology, Web sensor technology, satellite remote sensing technology, global positioning system, geography information system, virtual reality technology, data storage and database technology.

Google Earth which was produced by Google Co. in June 2005 brings digital earth technology application into a new stage. Google Earth technology made a great progress by integrates international computer technology and 3S technology in the aspects of data and application. Google fully utilizes his powerful search engine and Web server system around the world to successfully realize the mega-remote-sensing-data based functions, such as browse, query, measure, path analysis, location-based service (connected with GPS) and so on, and realize serving the public with the information by popularizing the application of digital earth science and technology for the common user on the Web.

Digital Earth platform is an application, service and decision support system, which integrates geo-spatial data acquisition, transmission, storage, processing, analysis, statistics and visualization. It realizes comprehensive management and widely application of different kinds of huge earth data under the geo-spatial framework. Digital Earth platform includes all kinds of data and information with and without geography coordinates, Including the following data (1) space-based dimensions of different regions; (2) region-based dimensions of special requirements; (3) time-based dimensions of different spans; (4) science-based dimensions of different purpose; Digital Earth platform need to adapt in and grantee and rapidly growth of data and application. It should consider the continues growing of data and applications, as well as data security.

1.2 Main Factors Considered

The era of Big Data platform for the Digital Earth is the use of mass, multi-resolution, long phase, multi-type Earth observation data and social economic data and its analysis algorithm and model building of virtual earth, its data acquisition and organization, analysis, application embodies the characteristic of the large data. A new generation of Digital Earth platform is not generally interpreted as "put the earth into a computer," and in today's era of Big Data, we can think of Digital Earth is earth's big data. We can pass on the Digital Earth platform for massive spatial data efficiently and social economic data organization, to streamline the data space for scientific information mining and analysis. We may face the following issues:

(1) Digital Earth and Smart Cities in Big Data Era

Previously we only deal with very small parts of data in some applications. The era of Big Data of Digital Earth and Smart City, is not a simple earth or city data into the computer but to pass on the Digital Earth platform for massive spatial data efficiently and social economic statistics organization, to streamline the data space for scientific information mining and analysis. Data island or data chimney become more and more serious. Sharing data across industry, cross-scrotal, cross-regional exchange is difficult. In the agricultural civilization era, the core resource is land; In the era of industrial civilization, the core is energy resources; In the Internet Big Data age, the core is data resources. Break "data island", "data chimney", to revitalize the flowing data is imminent. At the same time some fundamental problems raise, such as data transaction, data rights, data ownership, data profits belongings, etc.

(2) Big Data management

Emerging services such as Cloud Computing, Internet of Things prompted data type and size of the Digital Earth platform is growing at an unprecedented rate. We are facing data with features of huge, different structure, static & dynamic, different sources, real-time & historical, public & private, etc. Formal Big Data era, data from simple processing object into a fundamental resource. How to better manage and use the Digital Earth platform of big data has become a common topic. Big Data scale effect for data storage, management and data analysis application has brought great challenges.

(3) Data fusing and processing

The era of Big Data exist multi-source heterogeneous data; widely distributed; dynamic growth; first the characterizes of the data after the model. The extensive existence of data makes data are increasingly spreading, and there are multi-dimension, multi-phase and multi-style earth observation data and ground observation data in the Digital Earth platform. In order to carry out data integration, data analysis need to be on the Digital Earth platform widely heterogeneity is mainly reflected in the data. 1) Data types from mainly structured into structured, semi-structured, and unstructured; 2) Different types of data integration (social and economic data; 3) The diversity of the way of data changes in the data source. With the rapid development of mobile terminals, mobile phones, tablet PC, GPS and other explosive growth, the amount of data and the data produced with obvious characteristics of time and space. Data growth Apps, Apps accumulate data.

(4) Data security problem

Putting forward along with the "Internet Plus" plan of action, the mobile Internet, Cloud Computing, Big Data, such as the Internet of Things will be as an important thrust of economic development in the new period, the information system facing the environment more open, more demanding for data and information security. In open environment, safely storage of data and safely delivery, traditional methods are facing great challenges. Unauthorized or using system leaks to access data become more dangerous. Digital Earth platform requires a new generation of data security solutions in open internet environment. At present, most of the Digital Earth platform in the data center involved in data security, refers to the use of the data backup, data disaster recovery techniques to guarantee the data is not lost, don't be destroyed, but still unable to prevent dangerous behavior, such as unauthorized access to data and information leaks risk. So closed environment security method in open environment is facing great challenge, how to guarantee the Digital Earth platform of data security in open environment become an important research topic.

1.3 The New Architecture of Digital Earth Platform

Digital Earth platform needs to have a new mechanism. In the open environment, the core of information security is the security of the data, to carry out data and data as the core of the data security system research is very necessary. Facing the open Internet environment, the Digital Earth platform needs to have a lower level, directly to the architecture of data to ensure data security and the application. In this paper we consider that the Digital Earth platform based on open environment data security problem, design the application data security mechanism, introduces the technical architecture for data(DOA) [3]. And build low-level architecture, try to solve the data ownership, information sharing, data management, system function expansion, the big support data analysis and mining, software engineering, information security, and to guarantee the data owner interests, so as to build the era of Big Data Digital Earth platform security mechanism of data access.

2 Data Safely To Be Used: The Goal of Data-Oriented

Big data era in order to solve the problem of space information network service application in the G/S (General browser / Services cloud) model supported by the Cloud Computing, also known as Port/Cloud model. DOA and Port/Cloud model diagram is shown in figure 1. From the perspective of data, cloud computing is the data storage services, more is IaaS and DaaS, through distributed and virtualization technologies, the infrastructure and data (I+D, Infrastructure plus Data). For end users to provide flexibility, quantifiable, personalized data and computing services. Everything in the cloud, including all of the infrastructure, such as network, servers, storage devices, etc., more important is including all the data, the network world and physical world of data and so on.

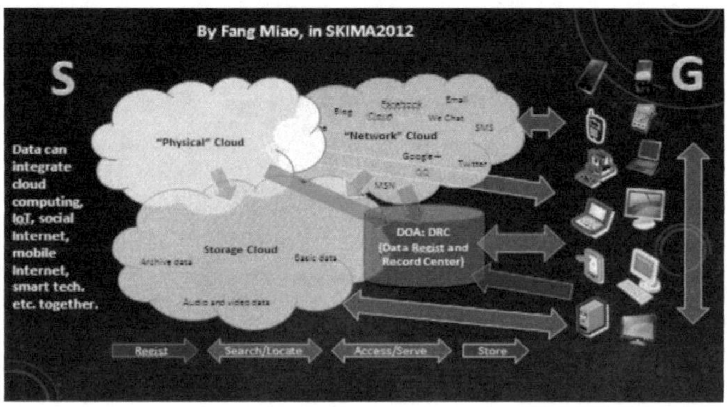

Figure 1 Data view: integrate all techniques

The architecture of the Digital Earth platform, we use the data and data as the core oriented thinking, data security problem in view of the open environment, leading into the data-oriented technical architecture, data access mechanism of building security. Figure 2 shows the data in the core position in the platform architecture.

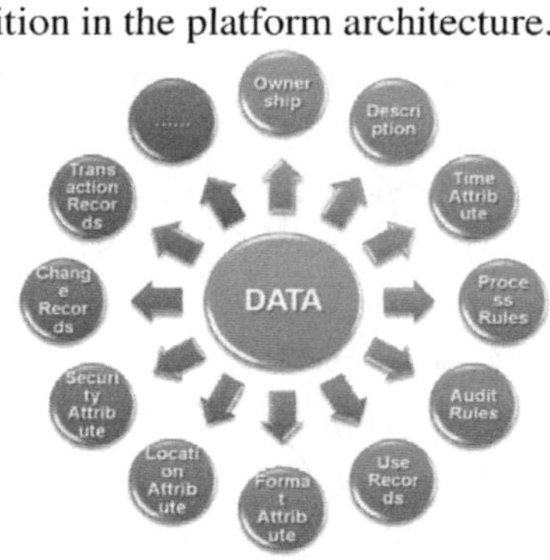

Figure2 Data As A Kernel

In under the guidance of the data and data as the core, data processing and related technologies, as shown in figure 3:

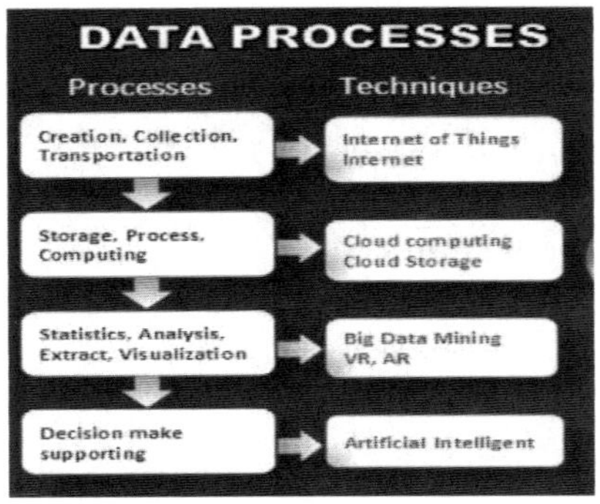

Figure3 Data Processes

Digital Earth platform with the data system is an ecosystem. It means that data and applications can be changeable, growth and sustainable development, and self-adaptive. According to the data oriented and data as core ideas to establish ecological system, various applications, including ecological data ecosystem is like as "fertile and rich data soil to grow with plentiful and luxuriant application forest". Ecosystem need to construct a logical data resource pool, support the big data platform and fragmentation and increasing the application of the support data sharing and the system can be extended. Based on the data: all measurable, all can be united, all operational, everything can be realized. Data security ecosystem as shown in figure 4.

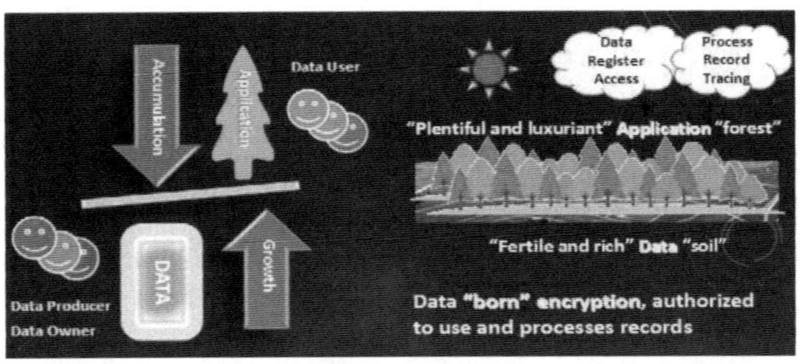

Figure4 Data Security Application Ecosystem

3 The Trusted Data and Platform——DOSA

Data-oriented security architecture based on the underlying architecture Digital Earth platform on data security system carries on the comprehensive design, including data management and application, etc. DOSA is built based on DOA, face data and based on data about the data security architecture, built up from the data protection to the mechanism of a complete set of license application. DOSA are based upon "cloud computing" and CA (Certification Authorization) / PKI (Public Key Infrastructure), as data encryption, "natural authorized the use of" for the principle, properties for registration and management of data, so as to realize the data security management and application. The purpose is to form a trusted data and its application in an open environment. DOSA is composed by DRC (Data Register and Recording Center), DAC (Data Authorization Center), DEC (Data Exception Control Center) and DAUs (Data Application Units).

3.1 Trusted Data

Emphasizes the trusted data from security protection object data itself establish credible mechanism, is the security solution of the future. Trusted data is different from trusted computing, trusted cloud computing and trusted computing environment. They all have in common is through the security infrastructure to achieve the purpose of a can trust, strengthen the sense of security. The difference is that the former is on the object, credible, the latter from the environment and the means to achieve reliable. Trusted data is to stand in the angle of the data, with the support of cloud computing, based on the CA and PKI technology system, i.e. the data-oriented security architecture. Trusted data emphasize the object of information security, that is, to protect data itself. Trusted data, or DOSA, is an open environment of "data security applications" solution, is to build a "data security applications" ecosystem, is the basis of the data in the future architecture. Data can be trusted in an open environment. Data is protected and can be used. It is one of the solutions for future information security, is also the solution for Digital Earth platform.

3.2 Data attributes

Data shall meet the following basic characteristics: the concept of generalized data, and the life and attribute such as identity, security attribute, time and spatial attribute.

(1) Generalized data: Who can be computer registration and registration of any things are called data.
(2) Identity properties: Data ownership, that is, the owner of the data (data producers and data owners), friends (data user or licensee), strangers (unauthorized and stay to the authorized person) and the enemy (not the authorized person, blacklist).
(3) Security attribute: Data has the protection function, to "wear armor", presenting an encryption, have different level of encryption and depth, the use of the data to be authorized.
(4) Time and space attribute: The generation of data, authorization and the use of time and space.

Data is the basis of the application and not dependent on the specific hardware and software environment, the same data to support different applications. For convenience of management, we to store and transfer into keeping the encrypted data and authorization in should be used when using decrypt application state. Data only when the application state is in a state of decryption, once completed or left the application environment, or by the application of the new data, data shall be immediately into the encrypted data, fully guarantee the security of the data and the use of authorization.

3.3 Data-Orient Security Mechanism

Core mechanism of data security and data security applications include data registration, data encryption, natural authorized the use of data, data record, data tracking back, etc, the relationship between them is shown in figure 5.

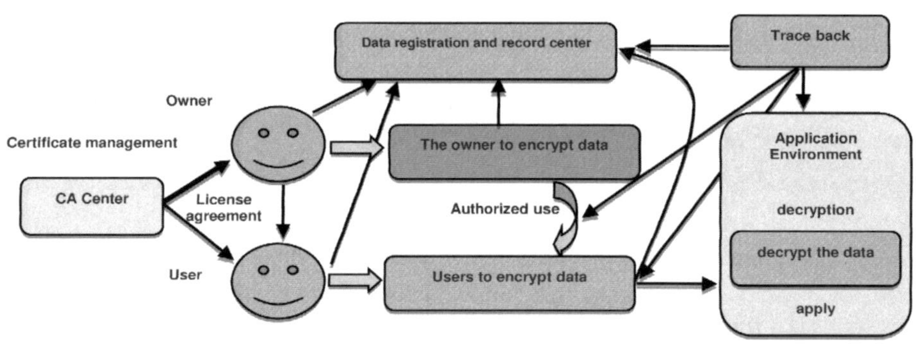

Figure 5 The relations of data ownership, security, applications

IOP Conf. Series: Earth and Environmental Science **46** (2016) 012050 doi:10.1088/1755-1315/46/1/012050

All kinds of fragmentation data platform and application can be compared to "fertile and rich data soil to grow with plentiful and luxuriant application forest". Accumulated and rich in a variety of data is defined as the soil of data. While all kinds of fragmentation, personalization and applications are defined as growing forests. The application forest can be ecology, flourishing and sustainable development. To establish an ecosystem of data platform with fragmentation applications and protection of data, is a solution of effectively deal with the growth, adaptive, the mechanism of management and sustainable development between data and applications. The Digital Earth platform with security and ecological application and service system on an open environment is built as an ecosystem, which is based on coordinating of the relationship among data, people and applications. Data security applications of ecological system consists of the following main components: CA (user authentication center), DRC (data registration center), PKI, DAC (data authority center), DEC (data exception control center), DWR (digital watermarking records center) as well as DAUS (data application units) etc., shown in figure 6.

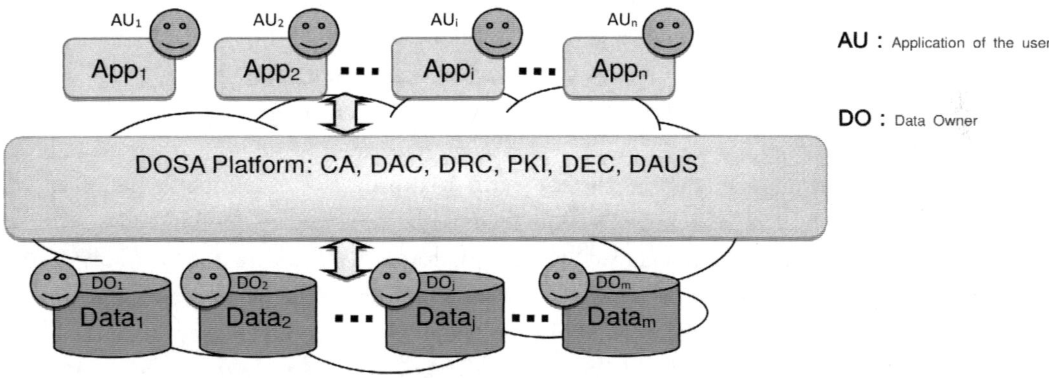

Figure 6 Data security applications of ecological system

3.3.1 People Identity Certification

Every people who do activities in cyberspace have an identity. This identity is certification authorized (CA).CA provides each person a digital certificate, that is a public key and a private key. One of the core idea of DOSA is to determine the data to the user, the relationship between the ownership of the data need to be clarified. This will require new users to participate in the activities of the network to the registration and identification. DRC to all registered users, while user identity authentication by CA. CA authentication using third-party CA authentication center, digital certificate to network users by the public and private keys. The private key in the form of a variety of safe distribution to each user, while the public key is stored in the data in the registration center.

3.2.2 Data Ownership

Data ownership is the relationship with people. Each data has its own rights ownership. The data can be divided into three kinds of relationship: data owner, data producer and data user. Other data can be set to agents in liability. Data ownership as shown in figure 7.

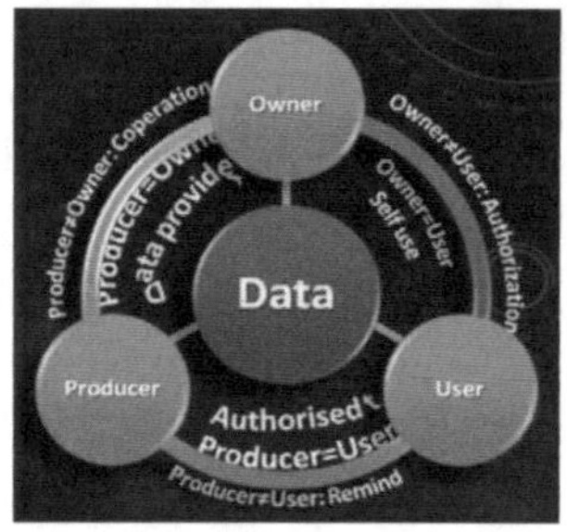

Figure 7 Data Ownership

Data owner, who owns all rights of the data. In general, who pays for the data, who is the owner of the data and purchase data only authorized users and owners.

Data producer, who is under a kind of contract with data owner to produce the data.

Data user, who is authorized by data owner to use the data.

The three relationships: they can be the same person, also can be different people and can also according to the situation changes to the data responsibility among each other.

Data agent, who is authorized by data owner to execute some functions for data owner, such as data manager, data seller, etc.

3.2.3 Data Born Encryption

All the participants on the digital earth platform has a unique network id (digital certificate), such data can be in accordance with the data owners intend to implement encryption. Once the data produced, DOSA platforms need to clear two things: First is to determine the owner of the data and the producers of data; The second is to encrypt data, to prevent others to steal. Under normal circumstances, the data of the producer is the owner of the data. No matter in storage or transport state, data is to keep the encryption. In other words, with the public key encryption, determine the owner of the data, and the data encryption present; use the private key encryption, identified the data producer, data producers in general is the data owner. Data born encryption is that data is encrypted just after data is produced. Data keep in encryption no matter it is in storage state or in transmitting state. The encryption is in concerned with data ownership. The encryption key is the public key of a data owner, or the public key of an authorized data user. Using symmetrical encryption for large data, while the symmetrical key is encrypted by the public key.

Data natural encryption use Certificate Authority and Public Key Infrastructure, With the data owner's public key to encrypt data, encrypted data can be placed in any public space, such as stored in the public cloud, Only the data owner himself by his own private key can decryption data. With the public key encryption, the owner of the data is determined, and the data encryption present; with the private key encryption, the data producer is determined, as shown in figure 8.

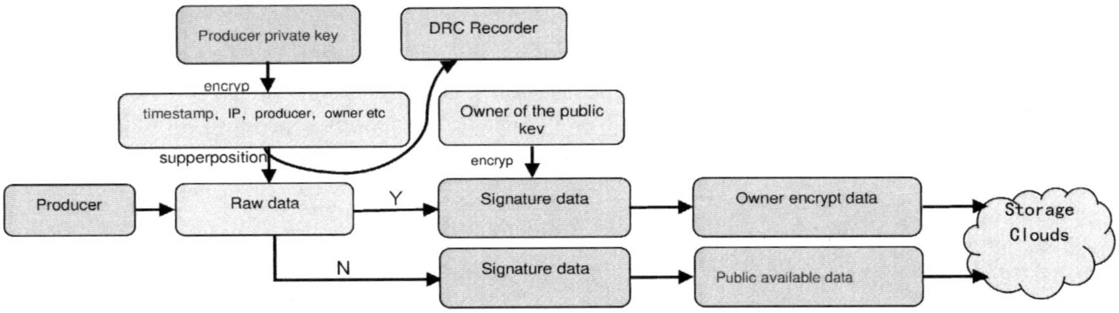

Figure8 The data flow chart of encryption by nature

3.2.4 Data Born Registered

Data is also born registered. Data resist is get unique ID and URL location. The registered data can be seen and can be searching, can be indexed, and can be analyzed. Encrypted data becomes available to use if registered. The performance of using registered data needs cloud computing. The security of registered data needs platform authorization.

DRC is the core component of DOSA, use all sorts of data attribute information of registration, including the security of the data attribute information and right of data information, etc, and the use of data process record. DRC also reserves the rights of all the data and the application of the user's public key. DRC used to construct the logic data resources pool, through the establishment of the index and search engine, implementation, implement the data and application of management and service. A DRC can be associated with other DRC, so as to realize data sharing. Through CA digital certificate and private key safe distribution to individuals with a variety of forms, the public key is stored in the data registration and record in the center. As shown in figure 9.

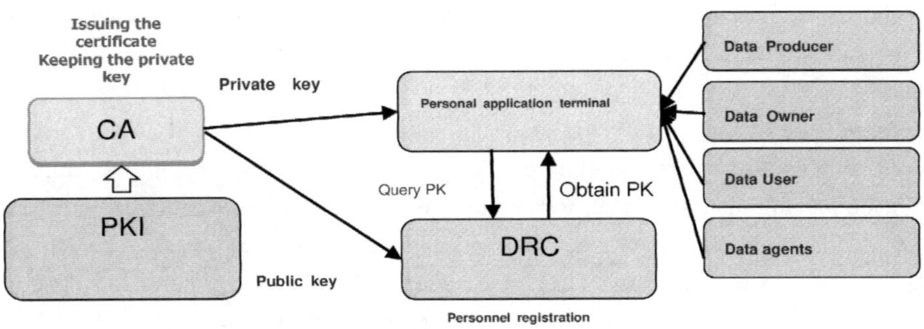

Figure9 The identity authentication In DRC

3.2.5 Authorized the use of data

DAC is the key to DOSA parts, used for authorization management data. Data in the generation, storage and transmission is encrypted and not to be used, and authorized users in the use of the data is decrypted and accessible. User needs the data owner authorization to use data. The authorization for a legal data user, needs both law contract permission and technology ownership transfer .Authorization of the data, is data ownership change, is the process of data decryption and encryption, with the data owner of the reoccupy after private

IOP Conf. Series: Earth and Environmental Science **46** (2016) 012050 doi:10.1088/1755-1315/46/1/012050

key to decrypt the data using the public key encryption. Authorization process to go through the watermark and registry data for recording and management. For larger volume data, take the symmetric key encryption method, authorization process is carried out on the symmetric key. The copy of data is necessary, the time and space consumption will need cloud computing and fast hardware support, but now it is possible because the price is very low. Authorization of data is the data of the ownership change, that is the process of data decryption and encryption, In which private key to decrypt the data owners reoccupy after data users (the licensee) the public key encryption, The data registration and authorization process records center for recording and management, In some permission, may also be authorized by superposition of watermark information. We based on the dynamic behavior of the data record and tracking, can find the data to a third party without authorization.

3.2.6 Data Security Applications

Make sure data security application environment general considerations decrypt the data in memory, to the memory data by means of a variety of security and not into stealing. Data is only decryption in an application. Open environment, to do the security of the data itself, and can safely use, the first is to encrypt data, the data should have the characteristics of "natural encryption, authorized the use of. Might as well assume that data in use is not encrypted state, then the data is not in use should be encrypted. Set data, therefore, has two states: when the storage and transmission is in a state of encryption "data" and authorized the use of the "application" in a state of decryption. DOSA as a mechanism, we must ensure that data can be in a state of the two and authorization, and encryption technology.DEC is an important part of a DOSA, adaptive management for use in data resources, ensure the uniqueness of the data and consistency, supervision and disposal of all kinds of abnormal behavior data. DOSA based on data encryption innate, authorized the use of the concept of, Data access on the basis of the data security attributes and the identity of visitors, by applying the authorization, user authorization, and then determine the authorization of the data, according to the authorization of way and the application environment, provide the decryption key or decryption algorithm, realize the safety of data is used.

3.2.7 Data Use Processes Recording

DWR will be in the form of the watermark data are recorded and authorized the use of process, together with the original data is encrypted management, facilitate tracing the source of data, accounts and data the unauthorized use of evidence. Data authorization processes and application processes should be recorded. The time stamp and space location should be recorded. Similar to Block Chain, DRC will record whole data events in its life circle. This will also take some consumption of time and space, but now we can endure it by cloud computing, etc. Once we find illegal data, we can trace and find the person who does not observe the contract. Using Big Data analysis of registered and recorded information, as well as the system log files, we can locate, and find the real illegal person.

Can be based on the analysis of the recorded information and mining, to understand the use of the data, in some application scenarios can provide valuable information. Especially the authorized the use of process data, can trace data records, can help to understand the data of normal and abnormal use.

3.2.8 Data Value Assess

Big data era, a new generation of digital earth platform under the Internet + data plays an important supporting role. Digital earth platform infrastructure, is actually the precipitation data in the platform all digital business, can we put every business use data to precipitate down. Analysis registered and recorded data information and using assessing model to calculate the data values.

Under this background, the data as an important resource, but if we don't flow will not be able to create value for sure, In data oriented architecture built digital earth platform allows data open, Shared, and financing can not only make social efficiency more effectively, and increase new blue ocean, new resources, let everybody can enjoy in the process of using the benefits of open data application data.

Data can be easily copied and transmitted, It belongs to you can copy and repeated trading assets, will become the world's most important asset; In essence, the value of the data by the extent of the circulation and has been applied. From another perspective, the data only in circulation and the collision will produce a higher value. Data is property and the owner should get paid by its value.

3.2.9 Discovery Data Violation

Data using the authorization is one of the biggest challenges facing illegal or illegal outflow data from authorized users of the data. There maybe some illegal or violation of using data. Due to the data users there, normally can't supervise data, Therefore, on the one hand, need through the establishment of rule of law and moral self-discipline to restrain illegal and irregularities, On the other hand, can only use data registration and record information to analyze and judge from whom data. According to data registered and recorded information, we can use Web spider or crawler to find data violation. To find illegal flow of data, must first determine the illegal data found the place and time, coupled with recorded data event in this paper, and the parameters of time and space, where can preliminary judgment data out. Further, you can through the large data analysis, to accurately analyze the offenders. This will become normal to patrol the network under the legal data application.

3.2.10 Data Records Tracing

Once we find illegal data, we can trace and find the person who does not observe the contract. We can according to the record contents of data dynamic behavior research data, analysis data, the dynamic behavior of the characteristics. Using Big Data analysis of registered and recorded information, as well as the system log files, we can locate and find the real illegal person. . Using big data analysis research and analysis of the data found abnormal behavior and call the police, the unauthorized data such as illegal use of data in time, judgment and tracking method, using the web crawler found illegal data and traces the offenders.

4 Applications and Prospects

DOSA, as a kind of data security concept and mechanism, is to ensure that the data in the data and application in the two aspects can be safe, reliable, convenient for management and use, can be applied in the traditional closed environment, enhance the data safety protection, and can protect the data in open environment safety and is not from unauthorized access. At present, the emergency command system in Si Chuan province people's government, geological disaster monitoring and management system and education resources public service platform and application system of Si Chuan province are applied the DOSA architecture, expected to be effective application in the digital earth platform.

The related theory and method system of information security, data security, AAA technology about network security, the CA technology, PKI technology and key system, encryption technology, the reliable technology, as well as the continuous development of network space security, system security, application environment security technology, etc., can be used in DOSA framework, But need to be further from the Angle of the oriented data and data as the core, to comb, from data security concept, theory, method and the application of the protected data mechanism, etc., adaptability and in-depth study, in order to further improve the information security guarantee. DOA & G/S model support data growth applications as shown in figure 10

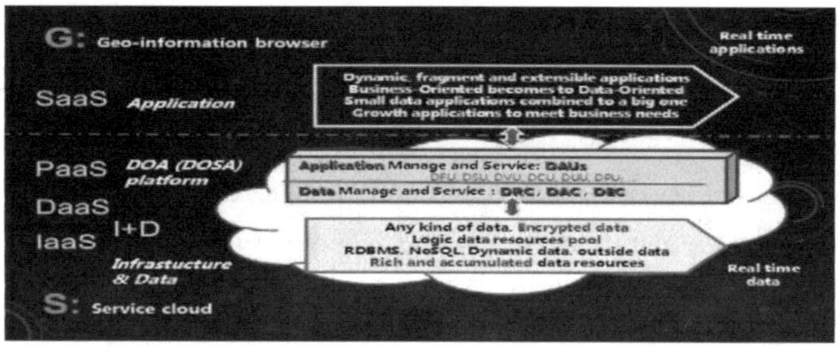

Figure10 DOA & G/S Model Support Data Growth Applications

5 Conclusion

Digital Earth platform needs an ecosystem for data growth and application growth. Open environment information security problems embodied in the data on the safety of digital earth platform of all kinds of big data security in open environment will be severely challenged. DOSA and Trusted Data can provide data management, security grantee, application supporting for Digital Earth platform. DOSA use the data-orient and the data as the core of the idea, set up data with the relations between the ownership of the user, use data "natural encryption, authorized the use of" method, By the CA, DRC, DAC, DEC, PKI, DWR, DAUS etc for data security management and security applications, Established from the data protection to the authorization mechanism of a complete set of application. Based on data-oriented security architecture of the digital earth platform may effectively solve the geospatial data security and respond to open environment, ownership, trading, sharing, management, privacy and other problems and challenges. Coordinating the relationship among data, people and applications, building a Digital Earth platform with security and ecological application and service system on an open environment, is the new architecture in the Era of Big Data.

References

[1] Max Craglia, Michael F.Goodchild, Alessandro Annoni, Gilberto Camara, Michael Gould, Werner Kuhn, David Mark, Ian Masser, David Maguire, Steve Liang, Ed Parsons.Next-Generation Digital Earth*,A position paper from the Vespucci Initiative for

the Advancement of Geographic Information Science. International Journal of Spatial Data Infrastructures Research, 2008, Vol. 3, 146-167.

[2] Guo Huadong. DIGITAL EARTH SCIENCE PLATFORM: DESP/CAS .The International Archives of the Photogrammetry, Remote Sensing and Spatial Information Sciences. Beijing, 2008, Vol. XXXVII. Part B2,1089-1091.

[3] AGRAWAL D, BERNSTEIN P, BERTINO E, et al. Challenges and Opportunities with Big Data-A Community White Paper Developed by Leading Researchers Across the United States [EB/OL].[2012-10-02]. http://wwwcra.org/ccc/files/docs/init/bigdatawhitepaper. pdf

[4]FangMiao.Data-orientArchitecturehttp://baike.baidu.com/subview/649092/12822804.htm#viewPageContent, 2014

[5] Xueqi Cheng, Xiaolong JI, Yuanzhuo Wang, etc. Big data system and analysis technology were reviewed [J]. Journal of software, 2014, 25(9): 1889-1908

[6] Xiaofeng Meng etc. Big data management: concepts, technology, and challenges [J]. Research and development of the computer, 2013, (1): 146-169

[7] Fang Miao, A preliminary study on the security architecture for data [J]. ZTE Communications, 2016, Vol.22 No.1:19-22

[8] KRESIMIR S, HRVOJE O, MARIN G. TheInformation Systems' Security LevelAssessment Model Based on An Ontologyand Evidential Reasoning Approach [J].Computers & Security, 2015, 55(6): 100-112.doi: doi:10.1016/j.cose.2015.08.004

[9] GEORGE S, VLADLENA B, JEAN N E, et al.Individual Information Security, UserBehaviour and Cyber Victimisation: AnEmpirical Study of Social Networking Users[J]. Technological Forecasting and SocialChange, 2015, (5): 320-330. doi:10.1016/j.techfore.2015.08.012

[10] GURPREET D, ROMILLA S, CRISTIANE P.Interpreting Information Security Culture:An Organizational Transformation CaseStudy [J]. Computers & Security, 2015, (4):63-69.doi: 10.1016/j.cose.2015.10.001

[11] AGRAWAL D, BERNSTEIN P, BERTINO E, etal. Challenges and Opportunities with BigData-A Community White Paper Developedby Leading Researchers Across the UnitedStates [EB/OL].[2012-10-02]. http://www cra.org/ccc/files/docs/init/bigdatawhitepaper.pdf

[12] XIAO X R, SUN X M, WANG X B, et al.DOSM: A Data-Oriented Security ModelBased on Information Hiding in WSNs [J].Information Technology Journal, 2009, 8(5):678-687

[13] Yimin Liu. Private data access control mechanism based on access to the purpose of research [D]. Shanghai: fudan university, 2012

IOP Conf. Series: Earth and Environmental Science **46** (2016) 012051 doi:10.1088/1755-1315/46/1/012051

Land use scenarios simulation based on the CLUE-S model of the Lijiang River Basin in Guilin, China

Qingwen Jin[1,2], Guang Liu[2,*], Lei Li[2,3], Chengxin He[4], Yuqing Huang[4] and Yuefeng Yao[4]

[1] Institute of Geography and Environment, Baoji University of Arts and Sciences, Baoji 721013, Shaanxi, China

[2] Key Laboratory of Digital Earth Science, Institute of Remote Sensing and Digital Earth, Chinese Academy of Sciences, Beijing 100094, China

[3] Department of Survey, School of Civil Engineering, Beijing Jiaotong University, Beijing 100044, China

[4] Guangxi Institute of Botany, Guangxi Zhuang Autonomous Region and Chinese Academy of Sciences, Guilin 541006, China

E-mail: liuguang@radi.ac.cn

Abstract. The relationship between government policy and land use change is very important, which can provide important information for understanding of land use change and for helping in development of sustainable policy. Returning Farmland to Forest Program is simulated by the CLUE-S model. Land use maps in 1993, 2006, 2010 and 2015 in Lijiang River Basin are interpreted based on remote sensing change from 1993 to 2025 under two scenarios (i.e., Natural Growth Scenarios, Government Intervention Scenarios). In the "Natural Growth Scenarios", the area of construction land and cultivated land are increased, the others are decreased. In the "Government Intervention Scenarios", the area of construction land, woodland, cultivated land, and water are increased, the others is in declined. The compared results of two scenarios provide a scientific support for the government policy in the Lijiang River Basin.

1. Introduction

Land-Use and Land-Cover Change (LUCC) is considered a critical research of global environmental change and sustainable development [1]. The government policy has large effects on land use change, which is considered to be the main driving factor [2]. Land use change models are helpful tools for simulating future scene which offer a comprehensively understand of the complex driving forces of economic, policy, physical variables that affect land use/cover[3]. The CLUE-S (Conversion of Land Use and its Effects at Small regional extent) model is developed by Peter H. Verburg [4] and is considered as a powerful tool for simulating land use change and getting scenario analysis. The CLUE-S model is suitable for different regions of the world, such as river basin, urban agglomeration, watershed, hills, the reservoir and others, the CLUE-S model makes a significant contribution to land use change[5].

The purpose of this paper is to establish a chronic record of land use change from 1993 to 2015 in the Lijiang River Basin, based on the Landsat images of 1993, 2006, 2010and 2015. The CLUE-S model is used to predict the land use change of Lijiang River Basin from 1993 to 2025 under two different scenarios and complete different of land use change under two scenarios.

The primary intention of this study were: (1) to study the tendency of the spatio-temporal distribution of land use change, analyze the degree of land use development, explore the driving factors of land use change, and identify the hotspots of land use change from 1993 to 2025; and (2) to analyze the value of the Returning Farmland to Forest (RFFP) through the difference of land use change between the NS (Natural Growth Scenarios) and GS (Government Intervention Scenarios) in Lijiang River Basin from 1993 to 2025.

2. Materials and Methods

2.1. Study Area

The Lijiang River Basin is located in the Guangxi Zhuang Autonomous Region and the southwest of the Nanling Mountains (Figure 1). The cover area of the Lijiang River Basin is 5306.03 km^2, and the length of the Lijiang River is 164 km.

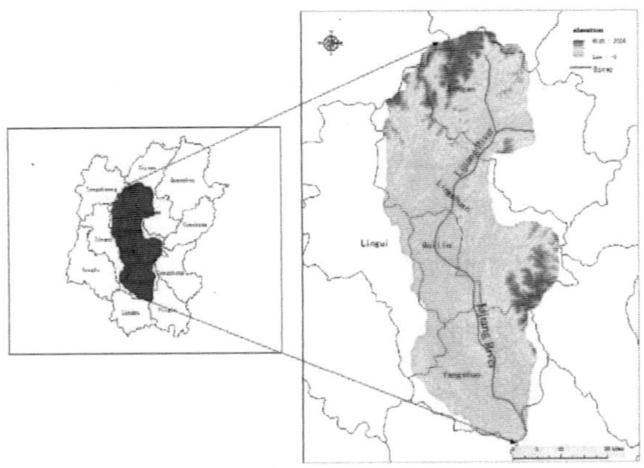

Figure 1. Study area location

2.2. Data and methods

2.2.1. Data. All Landsat images of 1993, 2006, 2010and 2015 were downloaded from the Earth Resources Observation and Science Center (EROS). In the study, all remote sensing data in WGS84 geodetic datum, Universal Transverse Mercator map projection (UTM, Zone 49N), and is processed by radiation correction and geometric distortion. According to the standards of the second national land survey of the classification system, and considering to local environment, types of land use are divided into five categories, namely construction, shrubland, water, cultivated land, woodland (Table 1). Two scenarios were simulated by CLUE-S model. The Returning Farmland to Forest was carry out in 2002 in the Guangxi Zhuang Autonomous Region. In the following, two different scenarios used for the future land use development are described in more detail (Table 2). In NS scenario, the simulated land use type in 2006 of the Lijiang River Basin that is based on land use type in 1993 by the CLUE-S model. Land use demand for this year 2025 is formulated based on the historical statistics of land use change from 1993 and 2006 by a linear interpolation of total area of each land use type. In GS scenario, land use demand for the year 2025 is formulated based on the historical statistics of land use change from 2010 and 2015 by a linear interpolation of total area of each land use type.

Table 1. The classification of the land use

Land Cover Type	Description
Construction land	Residential areas, industrial land, transportation land
Shrubland	The height of the plants is under 3m, garden plot, constructed green land

Water	Rivers, lakes and reservoirs pits
Cultivated land	Dry field, paddy field
Woodlands	The kinds of forest, such as timber stands, economic forests, etc.

Table 2. Scenario planning

Situation type	Description of scene	The reason of setting
NS[a]	Be based on the rate change of 1993-2006	The Lijiang River began to urbanization in 1993
GS[b]	Be based on the rate change of 2010-2015	The Lijiang was intervened in 2002

[a] NS is Natural growth situation.
[b] GS is Government intervention situation.

2.2.2. Methods. Logistic regression analysis is a common method for the research of land use change. The logistic regression model can be written as [6]:

$$\log\left(\frac{P_i}{1-P_i}\right) = \alpha_0 + \alpha_1 x_1 + \alpha_2 x_2 + \cdots \ldots + \alpha_n x_n \quad (1)$$

where, Pi is the probability of a cell for the occurrence of the considered land use type; x_n represents the driving factors of each variable; a_o represents the intercept; a_n represents the coefficient of each variable.
Formula 2 is represents the CLUE-S model prediction principles, and the total probability of a grid cell is calculated for land-use type [7]:

$$TPROP_{i,j} = P_{i,j} + ELAC_j + ITER_j \quad (2)$$

where, $TPROP_{i,j}$ is the probability of a grid cell i for the land use type u; $P_{i,j}$ is the space distribution probability of the answer of Logistic regression equation; $ELAC_j$ represents the elastic coefficient of land use type j; $ITER_j$ represents the iteration variable of land use type j.

3. Results

3.1. Driving forces

These driving factors be divided into two categories: natural factors and socio-economic factors, which are used in the logistic regression and the CLUE-S model. The first level is the grid unit, and the second level is the county unit. In this paper, 9 driving factors are chosen, including 6 natural factors (elevation, aspect, slope, distance to railway, distance to road, and distance to river), and 3 socio-economic factors (population, GIP, GOFFAF), Figure 2.

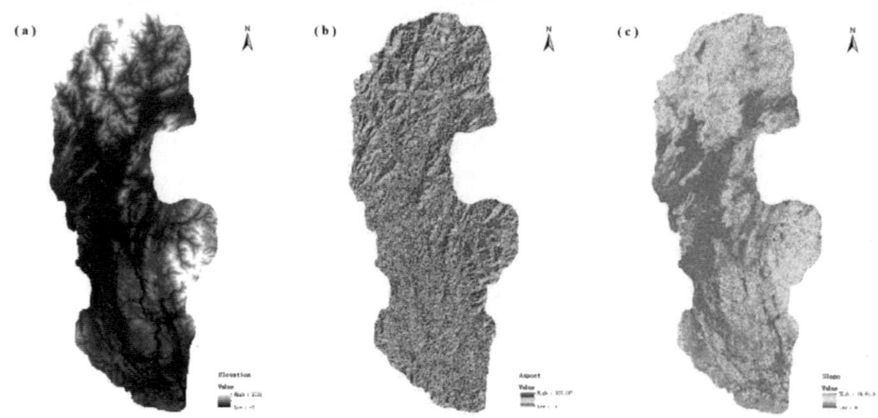

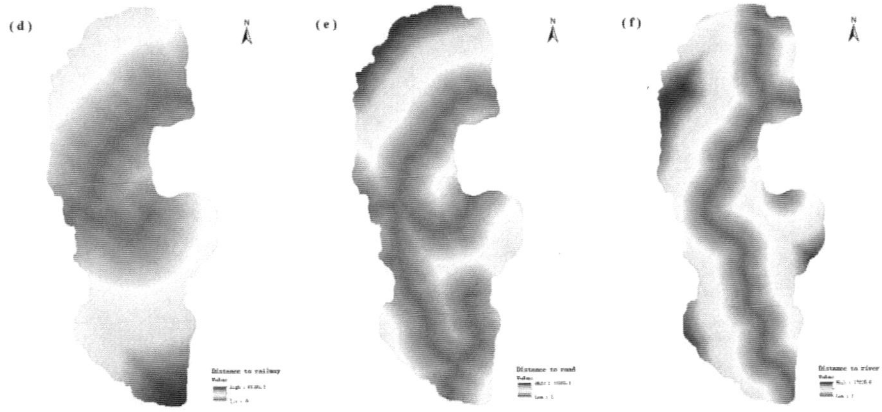

Figure 2. The spatial differentiation of the driving factors in land-use changes (a) Elevation; (b) Aspect; (c) Slope; (d) Distance to railway; (e) Distance to road; (f) Distance to river.

3.2. Regression Analysis of land use

Logistic regression is a critical step in the simulation process, which aims to check up the relationships between location factors and different land use types. In the Natural Growth Scenario (NS), construction has the maximum amount of location factors, which is mainly impacted by elevation and population. In the Government Intervention Scenario (GS), construction has the maximum amount of location factors, which is mainly impacted by elevation and slope. The ROC vales are more than 0.7 for all land use types, which suggests a high degree of spatial similarity between the simulated and real land transition.

3.3. Land use scenario simulation in Lijiang River Basin

Combining with NS and GS, the forecast of land use in 2025 is respectively simulation by CLUE-S model. Figure 3 is revealed a trend of land use map in the 2025. Under the county-level of Land use types, Natural Growth Situation is different from Government Intervention Situation (Figure 3). Under the county-level of land use types, the results under the NS are different from those obtained under the GS (Figure 4).

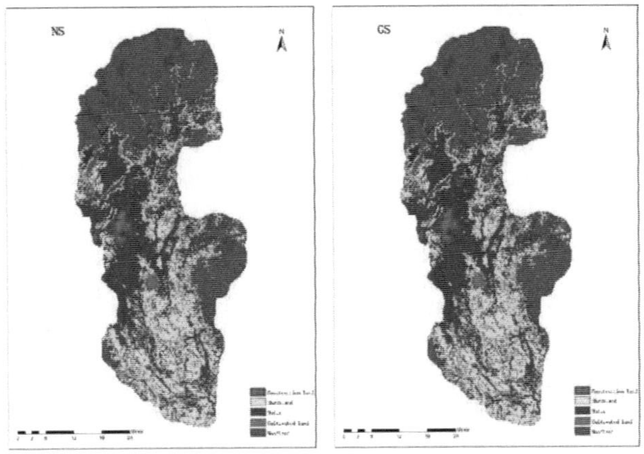

Figure 3. Simulated scenarios of land use in 2025

NS is defined as a continuation of interpolation result that linear interpolation of total area of each land use type from 1993 to 2006. The results show that during 1993-2025, the area of construction land and cultivated land have a positive trend, which is mainly at the expense of woodland, agricultural and shrubland.

GS is defined as a government intervention scenario and thus, as expected, results in the most pronounced land use pattern changes with a strong trend toward regularity during 1993-2025.

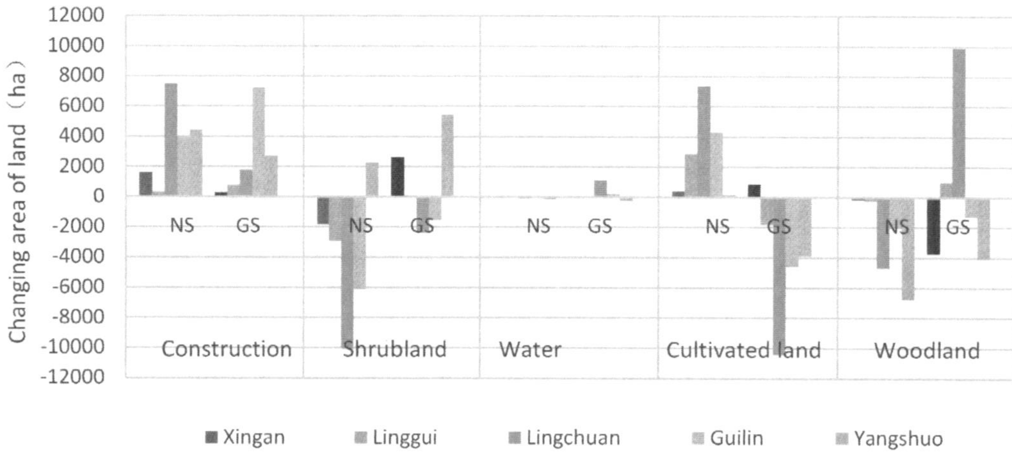

Figure 4. Land use change at the county-level in different scenarios

4. Conclusions

We used Landsat remote sensing image as data source and constructed CLUE-S model, simulated land use change from 1993 to 2025 under two scenarios (i.e., natural increase, government intervention). We explored the future characters of land use change and analyzed the driving factors of possible social. In the natural increase scenarios, by using the based data of the land use in 1993, we simulated land use change in 2025; in the government intervention scenarios, by using the based data of the land use in 2006, we simulated the trend of land use change in 2025. We analyzed the future characters of land use change by the CLUE-S model under two situations. The result of simulation produced a good Kappa coefficient above 0.85, and ROC values more than 80%. The research revealed the future possible trends of land use change and provided a sound basis for further analyses of the possible impact on the land use degree in the Lijiang River Basin. The spatial patterns of land use were obviously different during two decade. We found that woodland and water areas decreased in the NS but slightly increased in the GS, respectively. The hotspot for GS was a controlled, organized development of construction land, however, the construction of NS was ruleless, unplanned expansion. The RFFP of Lijiang River Basin program had been most successful payment. Under the correct leadership of the state forestry administration, the Lijiang River Basin was completed the various project and obtained the obvious result. By identifying the hotspots of land use change, we found that governmental intervention such as RFFP could protect ecological environment and improve the awareness of environmental protection.

Reference

[1] Turner II B L, Skole D, Sanderson S, Fischer G, Fresco L and Leemans R 1995 *Land-use and land-cover change: Science/Research Plan*

[2] Lambin E F E and Meyfroidt P 2011 Global land use change, economic globalization, and the looming land scarcity *Proc. Natl. Acad. Sci. U. S. A.* **108** 3465–72

[3] Wang J, Chen Y, Shao X, Zhang Y and Cao Y 2012 Land-use changes and policy dimension driving forces in China: Present, trend and future *Land use policy* **29** 737–49

[4] Verburg P H, Soepboer W, Veldkamp A, Limpiada R, Espaldon V and Mastura S S A 2002 Modeling the spatial dynamics of regional land use: The CLUE-S model *Environ. Manage.* **30** 391–405

[5] Jiang W, Chen Z, Lei X, Jia K and Wu Y 2015 Simulating urban land use change by incorporating an autologistic regression model into a CLUE-S model *J. Geogr. Sci.* **25** 836–50
[6] Lin Y P, Chu H J, Wu C F and Verburg P H 2011 Predictive ability of logistic regression, auto-logistic regression and neural network models in empirical land-use change modeling - a case study *Int. J. Geogr. Inf. Sci.* **25** 65–87
[7] Mao X, Meng J and Wang Q 2014 Modeling the effects of tourism and land regulation on land-use change in tourist regions: A case study of the Lijiang River Basin in Guilin, China *Land use policy* **41** 368–77

Acknowledgments
This research was supported by the the National Key Technology R&D Program of China (2012BAC16B01), International Science & Technology Cooperation Program of China (2010DFB23380), and the Dragon Cooperation Program (32365, 10606).

A Groud and Aerial BattleField Spatio-Temporal Data Unified OrganizeModel and Aplication

LI Feng, Li Qin , WAN Gang, CAO Xuefeng

Institute of Surveying and Mapping in Zhengzhou ,Zhengzhou China

E-mail:jueyun1020@126.com

Abstract. Aiming at the requirement of battlefield environment data model in joint operations, this paper proposed a groud and aerial battlefield spatio-temporal data unified organize model based on grid division. The thought of shere grid division is adopt to divide the space, and extend to the time dimension. Also, the spatio-temporal grid coding is designed. The field and object are combined to build the BSTD organization model, where the panet layer unit and panet volume unit are considered as the reference space separately to quantitatively describe the field, and the spatio-temporal grid coding is adopt to describe the space and time attributes of the objects. The relevant experiment is designed in paratroop tactics wargame simulation, and the results verify the feasibility of the model.

1. Introduction

With the rapid development of the science and technology, the range of military activities have been extended from the land surface to ocean, outer space, underground and undersea. The methods to get the environment information have made great progress in recent year.Therefore, it is possible to obtain the meteorological, electromagnetic and nuclear information in real time, which can form the massive BSTD ultimately[1].As the data origin from different departments, they have evident difference in spatio-temporal reference, data granularity, professional semantic, which greatly limit the capability. How to build the BSTD organization model, and accurately describe each element in strength, boundary,and change law, has become an urgent problem in battlefield environment science.

To simulate the comprehensive battlefield environment, the UAS develop the CTDB, define the TCDM, and make the SEDRIS to meet the requirement of the modular semiautomatic force simulation[2]. Many universities of our country have done deep research on comprehensive environment simulation and terrain database, including the National University of Defense Technology, Beijing University of Aeronautics and Astronautics, and the Armoured Engineering Insititute. Liang[3] et al. proposed the concept model of environmental factors, including natural and humanistic characteristics,which effectively describe the data of environmental factorsbased on SEDRIS,considering the classification principle in EDCS, Xie[4] et al. built the object framework of marine and atmospheric environment based on SEDRIS, Liu [5] et al. combined the object and field model to build the battlefield data model, according to which the database is carried out. Aiming at the requirement of battlefield environment data model in CGF decision, Dong[6] et al. combined the grid and vector to build the battlefield environment database. Zhong[7] et al. proposed a battlefield space modeling method based on space division, which can effectively describe the battlefield space.

This paper adopts the earth sphere grid division method to construct the groud and aerial BSTD organization model. The spatio-temporal grid division system is constructed based on space division and the extension to time dimension. The BSTD is effectively organized by combined the object and field. And the method is adopted in airborne flight path planning and site selection.

2. Battlefield space grid division

2.1. The thought of earth sphere grid division

The purpose of the sphere layer space grid is to describe the whole earth, and the sphere space is the abstraction of the earth space. According to certain rules, the sphere space is discretized into grid unit, which has unique code, than the process and mutual spatial relation of the geographic entities can be effectively described by several grid units, based on which the environment data can be effectively integrated and organized.

In the method, the geocentric coordinate system is the space reference, the earth center is the sphere center, the earth sphere is the starting surface, a series of concentric spheres are constructed by zooming the sphere in different scale. The concentric sphere is the sphere layer reference, the three-dimensional volume is formed by close the neighboring reference layer, which is also called shere volume reference.According to the sphere grid division model, the reference circle layer is divided into sphere grid unit, which is called as sphere grid. The neighboring sphere grid is closed into starting sphere volume, which can be divided into sphere volume with layer detail according to the octree rules in each layer.

The reference sphere layer division method is corresponding to the sphere grid division, where the hybrid sphere grid can be adopted.The basic thought is dividing the sphere into 6 basic surface with the same size, then dividing each surface according to the quadtree, to get the sphere grid unit with the same size. The sphere is divided into 2 spherical crowns and the middle zone by cutting using a planet, whose latitudes is 48.2. Then the middle zone is divided into 4 parts to get 6 basic surface with same size, which is coded using the number 0-5.

According to the octree rules, the sphere layer is divided into 4 equal parts, the circle layer radium is divided into 2 parts by the same distance, and the whole circle volume is divided into 8 parts, each of which is divided according to octree division, the results is shown in Fig. 1.

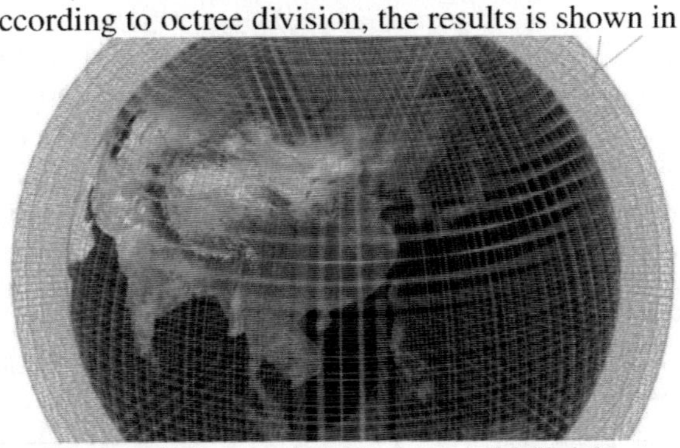

Figure 1. the results of sphere grid division.

2.2. Extension to the time dimension

The battlefield elements is changing all the time, and the sample space description just reflect the statue at certain time, therefore it can not effectively organize the BSTD. The sphere space grid is just the discretization of the three-dimension space, and the space attribution is described by the discrete grids.Time is also the one dimensional data, same as the space. In order to effectively describe the time attribution, it is essential to discretize and code the time. Thus, the division is extended to time dimension based on space division, and the space grid and the space coding is combined to construct the spatio-temporal coding system.

Since 1912, China adopted the Christian era to represent the time. The sidereal time, ephemeris time, and atomic time are adopted in astronomy, weapons manufacture, and the christian era is widely adopted to describe the other military activities and environment elements. The start of the the christian era is the birthday of Jesus, before the start time is called as the "bc", and after that is "ad".In the Christian era, the "year, month, day, hour, minute, second" is considered as the time unit to discretize the time, which is same with the space grid division. Different time units describe the layers in different level.

2.3. Spatio-temporal grid coding

In order to divide earth space into the sphere volume grid with different detail level, each grid must have the unique coding number, in which case, the space grid unit can be effectively described. The grid coding mechanism is the core of the space grid system, and it is the foundation of describing the attribution and mutual relation of grid units, and the premise of the code operation.

In the method of sphere grid division, there need 5 parts to identify a sphere volume unit, including the reference sphere volume, the division level, the starting sphere volume, the sphere radium, and the sphere grid.The sphere volume coding is consisted of the reference sphere volume coding, the sphere radium coding, the sphere grid coding, and the sphere grid coding is consisted of the division level, the row number and the column number. Therefore, the whole coding system is consisted of the reference sphere volume coding, the division level, the sphere radium coding, the basic surface number, the basic column number, and the row number. And the coding system can be represented as :

$$SolidCellCode = (DatumSphereID, RootSolidID, \ , Level, RadialSegID, SphericalRowID, SphericalColID)$$

Where $SolidCellCode$ is the sphere volume coding, $DatumSphereID$ represents the reference sphere volume coding, $RootSolidID$ denotes the starting sphere volume coding, $Level$ represents the division level, $RadialSegID$ is the sphere radium coding, the $SphericalRowID$ denotes the sphere grid row number, and the sphere grid column number is represented by $SphericalColID$.

Similar with the space grid, the time grid can be identified by the unique coding. The level coding, such as the year + month + day + hour +minute + second, can represent the time in different precision.Considering the habits in battlefield, the coding is like "YYYYMMDDHHM`M`SSIII", where "YYYY" is the year coding, "MM" is the month coding, "HH" is the hour coding, "M`M`" is the minute coding, "SS" is the second coding, and "III" is the millisecond coding. The time coding is decimal coding with 17 numbers.

Combined with the sphere grid space coding, the spatio-temporal coding system can be designed as : spatio-temporal coding = space coding + time coding, the whole structure of the spatio-temporal coding is depicted in Fig. 2, where the yellow part is the space coding, the blue ones represent the time coding. The time coding can be set as 0, when the high precision of the time description is no need, for example, there is no need to record the second and millisecond coding when describe the thunderstorm areas, because the high precision of the calculation model is minute.

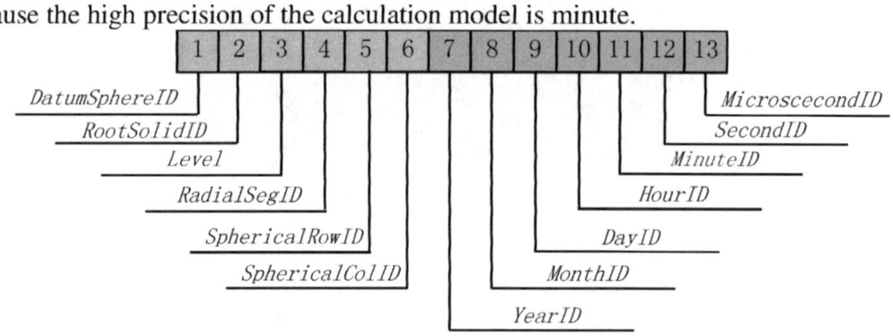

Figure 2. structure of the spatio-temporal coding

3. The battlefield data organization model based on grid division

3.1. data organization strategy of combining the field and object

Different elements in battlefield have different distribution characteristic and evolvement rules, according to which, the battlefield elements can be divided into field elements and object elements.The field elements have the continuous distribution, such as the elevation, soil type, the atmospheric temperature, seawater density, and the seawater velocity. This kind of data is the spatio-temporal distribution characteristics of certain attribution in battlefield, and it can be further divided into two-dimension sphere elements and the three-dimension sphere volume elements according to the difference in time dimension.The object elements have the clear boundary, complete function and discrete distribution in space, such as the building, the three, and the fortifications. This kind of data is an independent individual in space, which can be divided into point, line, surface and volume object.

The sphere grid model is a method of spatial description, and the spatial attribution of object can be described by the grid unit coding and the subsidiary spatial coordinates. It can answer "where is?", for example, the grid coding can describe the location and the range of the spatial points, lines, surfaces, and volume objects. The sphere grid model is also a spatial reference, and the battlefield elements can be effectively recorded and quantified based on sphere grid unit. Thus, it is convenient to adopt the sphere grid model to manage both the field elements and the object elements, the thought of organizing the battlefield environment data based on sphere grid model, is shown in Fig. 3.

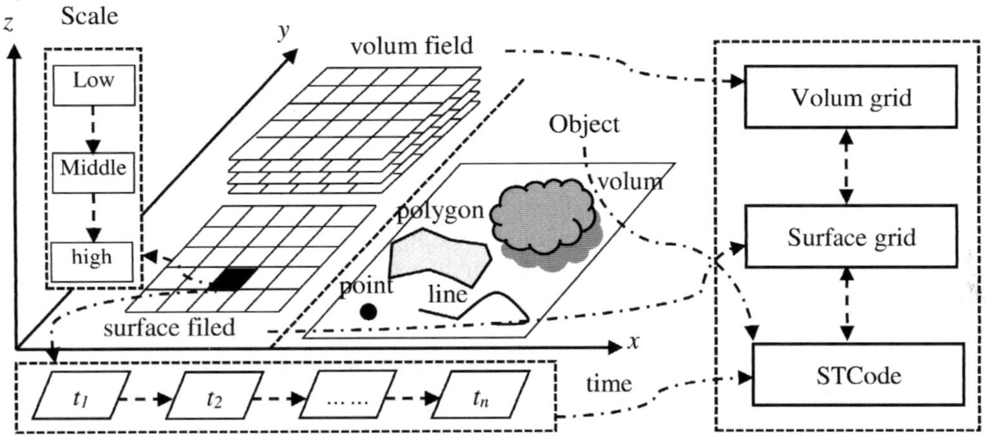

Figure 3. the battlefield environment data organize model

The point, line, surface, and the volume elements can be described by the spatio-temporal coding, where the spatial position and range is described by the space coding, the time attribution is described by the time coding, and the level and neighboring relation of the grid unit can be the index of the object elements. The field elements in two dimension sphere can be quantified with the reference of the sphere grid, and those in three dimension sphere can referring the sphere volume grid. The scale attribution of the battlefield elements can be described by the level relation of the sphere grid model.

3.2. the grid coding description of the object elements

There are two purposes to adopt the sphere grid model to manage the object elements. One is to describe the spatio-temporal attribution of the environment object as a whole, the other is to describe the spatial distribution and the evolution characteristics of the environment objects in detail. In order to achieve the goal, the object coding structure, consisted of the spatio-temporal coding and the object coding, is essential to be designed. The whole spatio-temporal coding is described by the whole spatio-temporal coding, and the internal spatio-temporal attribution of the environment objects is described by the spatio-temporal coding set. The object structure based on the circle grid model is shown in Fig. 4 , where the target ID is the identical mark of the environment object, which is formed according to other information system.

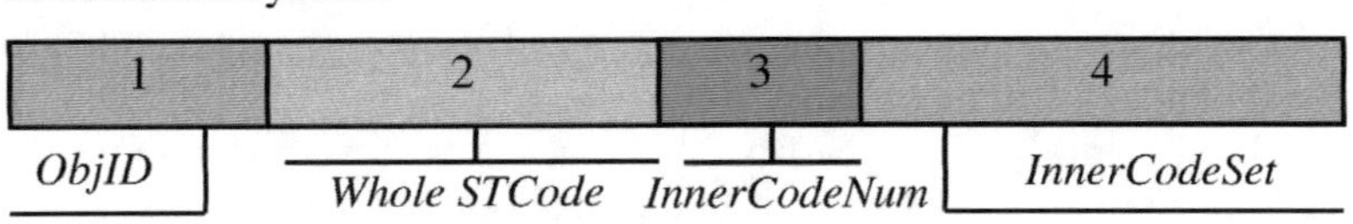

Figure 4. object coding structure

The whole spatio-temporal coding record the space coding of the object' s grid location and the time coding of the data collecting and updating.It can loosely locates the environment objects, which can be used in the spatial index and search. The internal spatio-temporal coding record the spatial distribution and the time attribution of the environment objects in detail, it is just the spatio-temporal coding collection. And it is designed mainly for the line, surface, and the volume elements, and it can describe the spatio-temporal information, such as the spatial distribution and the time evolution. Also, it can be adopted to search the internal elements and make a judgement on spatio-temporal relation of the environment elements.

3.3. the grid organization of the field elements

The field elements can be effectively managed by adopting the spatial grid unit to record the attribution. The two-dimension and three-dimension sphere field elements can be organized separately by the sphere grid and the sphere volume grid.

The two-dimension sphere elements, such as the elevation, slope, soil, the land type which is distributed in the earth surface, can be quantified based on the sphere grid unit. In the process, each unit records the element value, which is also called as the data record with time coding, representing the element value in different time. Many battlefield elements can be described based on the sphere grid unit, such as the elevation, the remote sensing image, the slope, the soil, the land type, the earth surface weather, the visibility, and the shaded degree.

Some elements are spread in certain sphere volume in earth, such as the air temperature, humidity, visibility, seawater temperature, salinity, density and the sound velocity, which ca be considered as the three dimension sphere volume field data. And the data can be quantified based on the sphere volume grid unit, each grid unit record the element value.

The whole space is divided into underground, the undersea, in the air, the mix of ground and air, the mix of water and air, and the mix of the ground and water, according to the material in the volume grid unit, and the material is also the basic attribution of the grid. With the refinement of partial levels, the sphere grid model can effectively simulate the real battlefield.

Some sphere volume units are filled with gas, such as the air, the mix of the ground and air, and the mix of the water and air, many attributions of this elements need to be quantified, including the atmospheric temperature, humidity, visibility, wind speed, wind direction, cloud, special weather phenomena.

Some sphere volume grid units are filled with liquid, such as the water volume, the mix of the ground and water, and the mix of the air and the water. The attribution of the field elements, need to be quantified, include the water temperature, salinity, density, the sound velocity, fluid velocity and the fluid direction.

4. Application and analyse

In order to verify the validity and the practicability of the model, the proposed model is applied in paratroop tactics wargame simulation. The BSTD of the system can be divided into 7 parts, include the elevation data of the task area, with the 2 meter resolution produced by the digital map, on a scale of 1:10000, the slope and the slope direction information produced by the elevation data, the shaded degree, soil, and the land type extracted from the digital, the three dimension building data by manual model, the aerocraft information received in real time, the electromagnetic environment information produced by the electronic countermeasure model, and the meteorological data received from the meteorological model.

Different battlefield elements are organized in different methods, for example, the elevation, sensing image, slope, soil, landform, and the shaded degree can be organized by the sphere grid model, and the sphere grid unit is considered as the reference to quantify the environment elements, the atmosphere, and the electromagnetism can be organized by the sphere volume grid model, and the sphere volume grid unit is considered as the reference to quantify the elements, the building and the aerocraft in low altitude can be described by the spatio-temporal grid coding. Experimental results demonstrate that the proposed model can effectively support the flight planning of the helicopter, the selection of the airborne area, and the construction of the battle efficiency evaluation model.Application results are shown in figure 5. The figure 5-a is helicopter landing terrain location analysis, figure 5-b is helicopter penetration path planning.

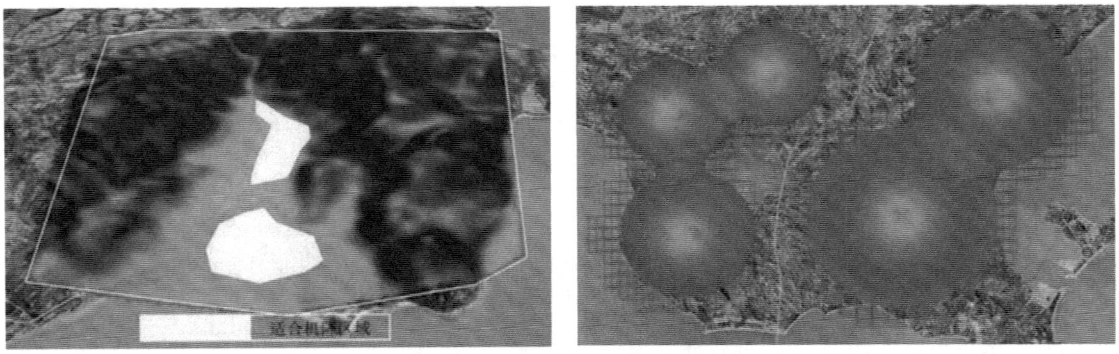

a helicopter landing location analysis **b** helicopter penetration path planning

Figure 5. object coding structure

5. Conclusion

Aiming at the multi-dimensional battlefield data support of the paratroops in joint operation, a ground and aerial BSTD organization model is proposed. The space and time are both divided into grid, and the spatio-temporal grid coding model is designed, based on which, the environment elements is organized by combining the field and the object. The model is applied in the paratroop tactics wargame simulation, the experimental results demonstrate that the model can effectively describe different kinds of data in battlefield with high efficiency. Further research on the multi description of the BATD and the conversion of field elements and object ones is still needed .

References

[1] Yu Xiong. Battlefield environment simulation [M]. Beijing: PLA press, 2012.
[2] Paul A. Birkel. SEDRIS Data Coding Standard [A].Proceedings of the 1999 Spring Simulation Interoperability Workshop Conference [C].99S-SIW-011, March 1999.
[3] Liang HaoZhe, Zhang Jun, Lian Lin, et. Based on SEDRIS synthetic battlefield environment data representation model [J]. Computer and modernization for (7): 18-23.
[4] Xie Kongshu , Zhao Guorong, Jiang Xiaorui, etc.A SEDRIS sea battlefield environment based on description method [J]. Computer applications and software, 2012,29 (6): 196-201.
[5] Liu Weihua, Wang Hangren. Design and implementation of integrated natural environment data model [J]. Journal of system simulation, 2004,16 (9): 1978-1981.
[6] Dong Zhiming. Study on the battlefield environment database suitable for computer generated forces [D]. Beijing: The Academy of Armored Forces Engineering.2004.03:2-9.
[7] Zhong Hui, Li Qun, Wang Weiping. Modeling and command and control of the battlefield space modeling based on space division 2007,32 (8): 20-24.
[8] Wu Lixin Jie Qing Yu. Based on sphere degenerated octree global 3D grid and deformation characteristics [J]. Geography and geo information science, 2009, 25 (1): 1-4.
[9] Cheng et al. Introduction to spatial information section organization [M]. Beijing: Science Press,.2012.
[10] Cao Xuefeng. Earth space mesh theory and algorithm research [D]. Zhengzhou: The PLA Information Engineering University, 2012.

Acknowledgments

This thesis is Supported by the National Natural Science Foundation of China（41301428）and State Key Laboratory of Geo-information Engineering,NO SKLGIE2015-Z-4-1.

BEMD-based high resolution image fusion for land cover classification: A case study in Guilin

Lei Li[1], Guang Liu[2]*, Qingwen Jin[3], Chengxin He[4], Yuqing Huang[4] and Yuefeng Yao[4]

[1] Department of Survey, School of Civil Engineering, Beijing Jiaotong University, Beijing 100044, China;
[2] Key Laboratory of Digital Earth Science, Institute of Remote Sensing and Digital Earth, Chinese Academy of Sciences, Beijing 100094, China;
[3] Institute of Geography and Environment, Baoji University of Arts and Sciences, Baoji 721013, Shaanxi, China
[4] Guangxi Institute of Botany, Guangxi Zhuang Autonomous Region and Chinese Academy of Sciences, Guilin 541006, China.
* E-mail: liuguang@radi.ac.cn

ABSTRACT: Analysis of image texture feature can help to reduce the adverse effects of the condition that same object but different band or same band but different object. Therefore, if it can add and highlight the texture information to the remote sensing image, it will be very helpful in the classification of ground objects. In this paper we consider to add SAR image information in classification. Bidimensional empirical mode decomposition (BEMD) has been widely applied to the analysis of non-stationary and non-linear signals. This paper proposes a new method for fusing high resolution SAR and optical image in Guilin area, based on Bidimensional empirical mode decomposition (BEMD) method.

1. Introduction

SAR image contains information such as surface roughness, object shape, orientation, and moisture content pertaining to the target location. Texture information is more stable than spectral characteristics, and therefore holds special significance in the classification of SAR images [1]. Multi-source image data fusion has been shown to be preferable to single image for automated land cover classification.

A number of methods for texture analysis have been developed over the years. However, the textural information extracted by GLCM depends on direction of object features, adjacent distance between two different surface features, and sub-image size. The mean value of Gray characteristic in four directions, at 0°, 45°, 90°, and 135°, is typically used to obtain the final textural features of an image. It has been shown that an adjacent distance of no greater than 1/3rd of the size of the small window should be chosen [2]. Thus, determining the window size is the crucial step.

Certain factors—e.g., foreshortening, range and azimuth shift, layover, speckle noise, etc.—make SAR images difficult to interpret [3]. A panchromatic image with high spatial resolution can reflect the edge information of the relevant objects in the image [4]. Combining the characteristics of SAR and panchromatic images can yield a high-quality image that can benefit image processing tasks, such as land cover classification [5]. BEMD has been widely applied to many fields, such as signal detection [6] and noise reduction [7]. Because of its superiority over other decomposition methods in non-stationary and non-linear signal processing, it has been used in multispectral image fusion and panchromatic image fusion [8]. Since the local area characteristics of an image should be represented by pixels that are strongly correlated, an area-based fusion scheme may yield improved fusion results.

2. Study area and dataset

2.1. Study area

Our chosen area for our experiment is Guilin, a world-renowned tourist destination located in the northeast of Guangxi Zhuang Autonomous Region. It is located between 24°18′ ~ 25°41′north latitude and 109°45′~110°40′east longitude, with an estimated area of 2.78×10^4 km^2 (Fig.1).

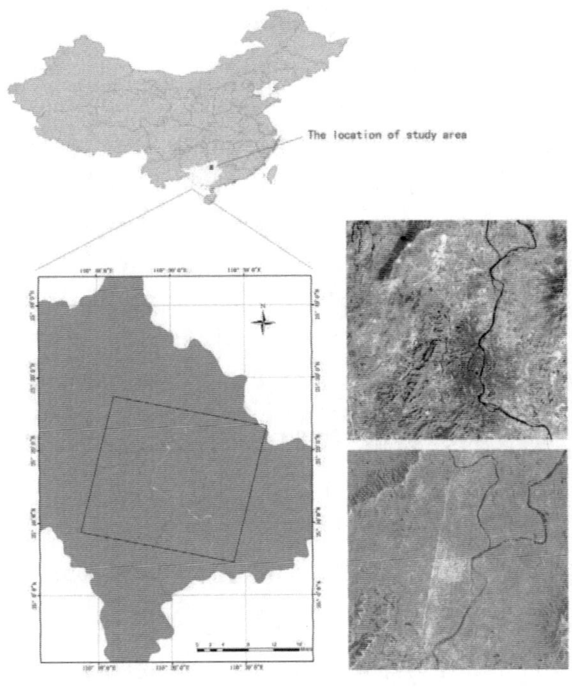

Figure 1. Location of the study area.

2.2. Dataset

In our experiments, we used TerraSAR-X images acquired on 14 October 2013, at a spatial resolution of 3m with a single polarisation in the HH mode. The Gaofen-1(GF-1) is a high-resolution Earth observation satellite developed by the Institute of Space Technology at the China Aerospace Science and Technology Corporation. The GF-1 remote sensing data used in this study consisted of high resolution panchromatic images which acquired on 14 October 2013 at 03:34:40 and the spatial resolution of panchromatic waveband is 2 m × 2m.

3. Data Processing

In this section, we explain our work in the following steps: data pre-processing, panchromatic and SAR image fusion, extraction and selection of texture features, and land cover classification.

3.1. Pre-processing

Speckle noise reduction is necessary for SAR images. It can be reduced by multi-look processing or spatial filtering. For SAR image processing, the Lee filter can be expressed as follows:

$$\hat{x}_{\text{Lee}}(i) = \bar{z}(i) + k(i)(z(i) - \bar{z}(i)) \quad (1)$$

where $\hat{x}_{\text{Lee}}(i)$ denotes the minimum mean square error estimate of pixel i, and z and x represent the SAR image and the speckle-free image, respectively. $k(i)$is a weighting function, and can be defined as $k(i) = \sigma_x^2(i)/\sigma_z^2(i)$. $\sigma_x^2(i)$ and $\sigma_z^2(i)$ represent the local statistics of $x(i)$ and $z(i)$, respectively, $\bar{z}(i)$ is estimated over a small neighbourhood. The filtered image can preserve the inherent characteristics of the SAR image while minimizing speckle noise.

3.2. The fusion of SAR and panchromatic image based on BEMD

BEMD is an integral transform technique which is highly efficient and adaptive, and can be adapted to analyse on-linear and non-stationary datasets [9].Through BEMD, any complicated signal can be decomposed into multiple scales of spatial frequencies, called intrinsic mode functions (IMFs). The components are extracted from signals using a sifting algorithm. BEMD picks out the highest remaining frequency oscillation in the signal. This means that each IMF contains components of lower frequency than the preceding one. Considering its superiority, BEMD offers considerable improvement for the result of image fusion. For a given SAR image (D^{SAR}) and a panchromatic image (D^{PAN}), the following steps can be adopted to obtain the fused image based on BEMD[10]:

(1) Decompose images D^{SAR} and D^{PAN} into modes of frequencies $\{I_i^{\text{SAR}}, R^{\text{SAR}}\}$ ($1 \le i \le n$)and $\{I_j^{\text{PAN}}, R^{\text{PAN}}\}(1 \le j \le n)$,respectively, by using BEMD. I is the IMF and R is the residual.

(2) Suppose that $T(T = \min\{m, n\})$is the number of the IMFs of each of the SAR and the panchromatic images. The left IMFs (m-T or n-T) of the image are incorporated into the corresponding residual. Then images can be expressed as the sum of the IMFs and the residual as follows:

$$D^{\text{SAR}} = \sum_{k=1}^{T} I_k^{\text{SAR}} + R_T^{\text{SAR}}, \quad D^{\text{PAN}} = \sum_{k=1}^{T} I_k^{\text{PAN}} + R_T^{\text{PAN}}(1 \le k \le T) \quad (2)$$

(3) Fuse the IMFs $\{I_k^{\text{SAR}},\ I_k^{\text{PAN}}\}$and residual$\{R_T^{\text{SAR}}, R_T^{\text{PAN}}\}$with an adaptive area-based weighting scheme. In order to preserve as much information as possible regarding the SAR and the panchromatic images, we use a weighted algorithm instead of simple replacement. The weights from an adaptive area-based scheme of IMFs and residuals can be calculated as follows: If the window is smooth, the weight of each pixel in the same window can share the same weight. The weight is the average weight of pixels obtained from the pixel-based scheme. Otherwise, the weight from the pixel-based scheme is retained. This step can be represented as

$$\gamma_{\text{area\_based}} = \begin{cases} \overline{\gamma_{\text{pixel\_based}}}, & if\ the\ window\ is\ smooth \\ \gamma_{\text{pixel\_based}}, & otherwise \end{cases} \quad (3)$$

Once the weight of the *k*th IMF, denoted by α_k^{SAR}, and the residual of the SAR image, β_T^{SAR}, are obtained.The fused *k*th IMF and the residual can be calculated as follows:

$$I_k^{\text{fuse}} = \alpha_k^{\text{SAR}} I_k^{\text{SAR}} + (1 - \alpha_k^{\text{SAR}}) I_k^{\text{PAN}} \quad (4)$$

$$R_T^{\text{fuse}} = \beta_T^{\text{SAR}} R^{\text{SAR}} + (1 - \beta_T^{\text{SAR}}) R^{\text{PAN}} \quad (5)$$

(4) Obtain the fused image using inverse BEMD:

$$D^{\text{fuse}} = \sum_{k=1}^{T} I_k^{\text{fuse}} + R_T^{\text{fuse}} \quad (6)$$

3.3. Result of fusion of SAR and panchromatic images based on BEMD

Lee filter was chosen for speckle noise reduction in the SAR image. Following registration, the speckle-free SAR image and the panchromatic image were fused using BEMD. To prove the superiority of our proposed fusion method, we compared it with the method based on wavelets. The basis of the wavelets was Daubechies, and their decomposition level was 3 (Fig.2). The details of original images and fusion images obtained using different fusion methods show in Fig.3.

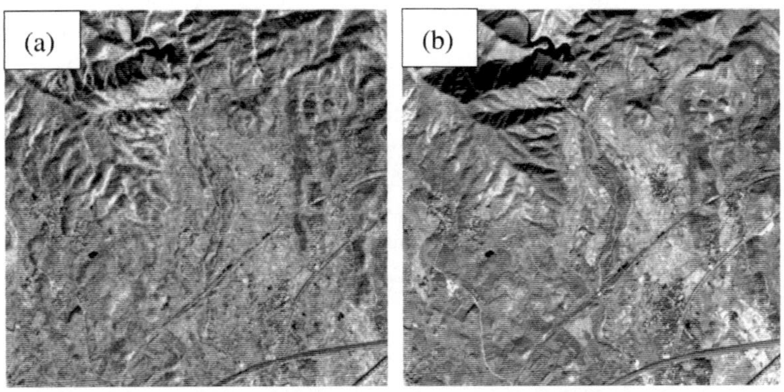

Figure 2. (*a*) Fusion image based on the BEMD method. (*b*) Fusion image based on the DWT method.

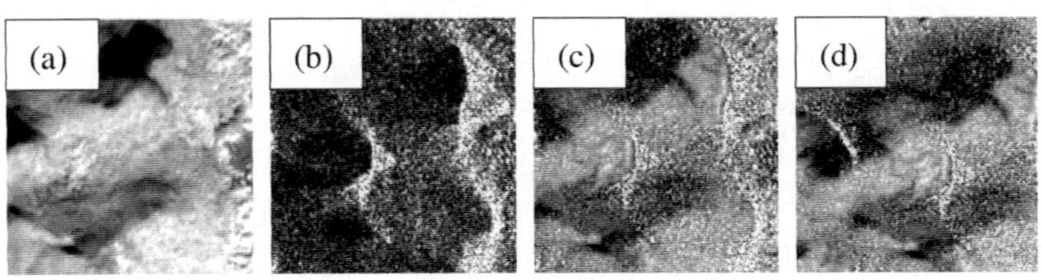

Figure 3. (a) Details of panchromatic image (b) Details of SAR image (c) Details of fusion image by DWT method (d) Details of fusion image by BEMD.

By visual inspection, we can see that the fusion images preserved the character of the SAR image and combined the contour information of the panchromatic image. In order to prove the superiority of our proposed BEMD fusion method, we calculated entropy, spatial frequency, and average gradient to quantitatively analyse the performance of each method. These quantitative indices are listed in Table 1.

Table1. Evaluation results of fusion methods

Fusion method	Entropy	Average gradient	Spatial frequency
BEMD	7.8818	35.3146	64.4228
Wavelet	7.8825	34.5929	62.9135

According to the definition of entropy, spatial frequency, and average gradient, the larger the values of the indices, the more information a given image contains. From the table, we can see that the entropies of the fused images obtained using the two methods were nearly identical in value, whereas the average gradients and spatial frequencies of the images fused using our BEMD-based method were significantly greater than those of the images obtained using the wavelet-based method. This means that the fusion image obtained using our method contained more useful information.

3.4. Result of land cover classification based on fusion image and texture analysis

Along with the six eigenvectors of the BEMD fusion image obtained by GLCM, PCA is employed to reduce their dimensionality. The top two components, which contained the greatest amount of information, were joined with the original image to generate a new image. The new image was divided into four classes—water, building, vegetation, and farmland—using the SVM classifier. Evaluation criteria for classification results, such as overall classification accuracy, mapping accuracy, user's accuracy, and the kappa coefficient, are shown in Table 2.

Table 2. Confusion matrix of classification of BEMD fusion image.

	Building	Water	Vegetation	Farmland	Total	User's accuracy
Building	423	0	1	25	471	95.2%
Water	0	381	0	0	381	100%
Vegetation	0	0	306	70	376	81.4%
Farmland	12	0	44	299	355	84.2%
Total	457	381	351	383	1572	
Producer's Accuracy	97.4%	100%	87.2%	78.1%		

*Note: Overall accuracy=91% Kappa coefficient=0.8800

We conducted three additional experimental studies to evaluate the classification results based on different texture analysis methods and images. These covered the following cases: (1) A SAR image with texture obtained by adaptive window used for classification; (2) A wavelet-based fusion image with texture obtained by adaptive window used for classification; (3) A BEMD-based fusion image with texture obtained by a window of a certain size used for classification

Table 3 lists the accuracy values of these methods.

Table 3. Classification accuracy values of different approaches.

Classification accuracy	Proposed approach	Case 1	Case 2	Case 3
Overall accuracy	91.0%	86.2%	89.4%	87.7%
Kappa coefficient	0.8800	0.8153	0.8588	0.8359

Our proposed classification approach yielded the best classification accuracy in table 3. According to these results, we see that the semivariogram can be used to assess the best parameters of the GLCM, and texture features obtained using the adaptive window is useful.

4. Conclusion

In this paper, we proposed an approach for land cover classification involving the fusion of SAR and panchromatic images based on BEMD. An improved GLCM using semivariogram was then used to determine the best parameters for texture analysis and obtain the eigenvectors. Experiments' result showed that our proposed classification approach can improve land cover classification accuracy.

5. Acknowledgements

This research was supported by the National Key Technology R&D Program of China (2012BAC16B01), National Natural Science Foundation of China (41590852, 41001264), and the Dragon Cooperation Program (32365, 10606).

6. References

[1] Z. Liu, "Minimum distance texture classification of SAR images in contourlet domain," in *Proceedings - International Conference on Computer Science and Software Engineering, CSSE 2008*, 2008, vol. 1, pp. 834–837.

[2] Q. Chen and P. Gong, "Automatic variogram parameter extraction for textural classification of the panchromatic IKONOS imagery," *IEEE Trans. Geosci. Remote Sens.*, vol. 42, no. 5, pp. 1106–1115, 2004.

[3] M. Shimada, "Ortho-rectification and slope correction of SAR data using DEM and its accuracy evaluation," *IEEE J. Sel. Top. Appl. Earth Obs. Remote Sens.*, vol. 3, no. 4 PART 2, pp. 657–671, 2010.

[4] C. Pohl and J. L. Van Genderen, *Multisensor image fusion in remote sensing: Concepts, methods and applications*, vol. 19, no. 5. 1998.

[5] A. Vidal and M. R. Moreno, "Change detection of isolated housing using a new hybrid approach based on object classification with optical and TerraSAR-X data," *Int. J. Remote Sens.*, vol. 32, no. 24, pp. 9621–9635, 2011.

[6] E. M. Decomposition, "Harmonic Detection of Active Power Filter Based on Empirical Mode Decomposition," *Society*, vol. 24, pp. 1–6, 2009.

[7] R. Song, H. Guo, G. Liu, Z. Perski, H. Yue, C. Han, and J. Fan, "Improved goldstein SAR interferogram filter based on adaptive-neighborhood technique," *IEEE Geosci. Remote Sens. Lett.*, vol. 12, no. 1, pp. 140–144, 2015.

[8] X. Zhang, W. Wang, D. Wang, and Y. Zhang, "A fusion algorithm for remote sensing images based on nonsub-sampled pyramids and bidimensional empirical decomposition," *Sci. China Technol. Sci.*, vol. 53, no. 1 SUPPL., pp. 196–204, 2010.

[9] Z. Yang, "Signal Period Analysis Based on Hilbert-Huang Transform and Its Application to Texture Analysis," *Third Int. Conf. Image Graph.*, no. 2003, pp. 430–433, 2004.

[10] Z. Liu, P. Song, J. Zhang, and J. Wang, "Bidimensional Empirical Mode Decomposition for the fusion of multispectral and panchromatic images," *Int. J. Remote Sens.*, vol. 28, no. 18, pp. 4081–4093, 2007.

IOP Conf. Series: Earth and Environmental Science **46** (2016) 012054 doi:10.1088/1755-1315/46/1/012054

Research on Geo-information Data Model for Pre-selected Areas of Geological Disposal of High-level Radioactive Waste

M Gao[1], S T Huang[2], P Wang[1], Y A Zhao[1], H B Wang[1]

[1]Engineer,CNNC Key Laboratory on Geological Disposal of High-level Radioactive Waste, Beijing Research Institute of Uranium Geology, Beijing 100029, China

[2]Senior Engineer, National Key Laboratory of Remote Sensing Information and Image Analysis Technology, Beijing Research Institute of Uranium Geology, Beijing 100029, China

E-mail:gaomin820@163.com hshutao@hotmail.com

Abstract. The geological disposal of high-level radioactive waste (hereinafter referred to "geological disposal") is a long-term, complex, and systematic scientific project, whose data and information resources in the research and development ((hereinafter referred to "R&D") process provide the significant support for R&D of geological disposal system, and lay a foundation for the long-term stability and safety assessment of repository site. However, the data related to the research and engineering in the sitting of the geological disposal repositories is more complicated (including multi-source, multi-dimension and changeable), the requirements for the data accuracy and comprehensive application has become much higher than before, which lead to the fact that the data model design of geo-information database for the disposal repository are facing more serious challenges.

In the essay, data resources of the pre-selected areas of the repository has been comprehensive controlled and systematic analyzed. According to deeply understanding of the application requirements, the research work has made a solution for the key technical problems including reasonable classification system of multi-source data entity, complex logic relations and effective physical storage structures. The new solution has broken through data classification and conventional spatial data the organization model applied in the traditional industry, realized the data organization and integration with the unit of data entities and spatial relationship, which were independent, holonomic and with application significant features in HLW geological disposal. The reasonable, feasible and flexible data conceptual models, logical models and physical models have been established so as to ensure the effective integration and facilitate application development of multi-source data in pre-selected areas for geological disposal.

1. Introduction

With the development of spatial information technology, many important achievements have been have made in the techniques establishment and application of the spatial data model both at home and abroad. In the field of environmental resources, geological data is already one of the most complex spatial data, however, the data involved in the process of research and engineer of high-level radioactive waste geological disposal repository is much more complicated (multi-source, multi dimension and variability), and the requirements for data precision and comprehensive application is even much higher. Therefore, data model design for pre-selected area of repository is facing more serious challenges. How to fully control and systematic analyse data resources, understand application requirements of geological disposal of high-level radioactive waste deeply, and solve the key problems of multi-source data entity and reasonable classification system, complex logical relationships, effective physical storage structure have become the hot and difficult problems in the research field.

2. Objective of data model design

The objective of geo-information data model design goal is to fully meet the specific requirements of environmental data storage layer, development layer and application layer in repository preselected

area of repository, to classify and name the data fit in with the relevant codes and standards, to high efficiency integrate and organize multi-source geo-information data. The geo-information data can meet the following requirements:
(1) the data model can comprehensively describe the multi-source data characteristics of the environmental objects in the area of the repository, and fully meet the requirements of data resources needed by the guidelines of the site selection;
(2) the data model can fully reflect the multidimensional characteristics of data entities in time dimension and space dimension;
(3) the data model has accurate data entity classification mechanism and classification coding system, and can be classified according to the different attributes of the data entity;
(4) the logical relationship of the data model is consistent with the spatial and non-spatial attributes, which is convenient for the system integration of multi-source data;
(5) data model can ensure that the physical storage structure of data is clear and simple, and easy to compatible and migrate in the database management platform (such as MS, SQL Sever, Oracle, PostgreSQL, etc.).
(6) the data model can effectively guarantee the data exchange between different application platforms (such as ArcGIS,QGis, ERSI, etc.);
(7) the data model has sufficient flexibility and scalability to facilitate the expansion of a new data type and compatible with the external data in the subsequent stages of waste disposal;
(8) the data model can make the development layer easy to control all data objects, and can make the development of automatic data entities to establish effective data association;
(9) the data model can be conducive to the representation of multi-source, multi-dimensional data to achieve fast and accurate, and easy to realize the mutual representation of different forms (graphics, images, text table);
(10) the data model has the relative independence and integrity, and the special data model can be used independently, which is not restricted by the data model associated with.

3. Contents of data model design

Based on the mastery of data object features and the understanding of application requirements, the contents of data model design to provide comprehensive concept model and logical model, and then to carry out thematic data model design on the basis of the comprehensive data model. The specific contents are as follows:
(1) master data objects and their features: data list was established on the existing data resources, and non-node database was formed. The basic characteristics of the entity data was described in the data list, including data source, accuracy, existing form (data format), application mode and etc. The characteristics of the analysis of new data generated in the follow-up phase was carried out the investigation of relevant fields in foreign countries.
(2) understand the application demand for data: according to ten factors of site selection criteria, a detailed understanding of the data acquisition, analysis and application in the whole process of the data entity characteristics is carried out, and then the specific requirements are finally determined. (3) classify data entity: the entity data are classified according to the spatial and non-spatial features. For spatial features, the data entity are classified according to the horizontal and vertical space classification, and in the coordinates of the spatial points are to three-dimensional coordinate of space graphics expression. For non-spatial properties, the data entity are classified according to the classification of the subject topics, and then multi-level classification and coding are established.
(4) propose conceptual model: the conceptual data model is put forward in accordance with data object and its characteristics, the understanding of data application requirements and the classification system of the data cntities.
(5) confirm the logical relationship: the hierarchical relation in the special subject classification and code table of entity data is the most important data logical relationship. All data from preselected area are processed as spatial data, and fuzzy space range positioning is used for no clear spatial location information documents, pictures and other data entity. Then, the logical relationship of measuring points and the sample in the data model is determine at last.

(6) establish physical structure: special database storage structure and the corresponding elements are organized according to the subject classification of data entities and existing form and attribute description table, including basic geographic data, basic geology, hydrogeology, geophysics, geochemistry, rock mechanics drilling data, changes in the natural environment, humanistic economic, etc.
(7) analyse the rationality and test feasibility: the actual data are used to test the rationality of the physical storage of data entity and its relationship, including the spatial graph, the data package, the picture, the picture, the document and so on.

4. Data model design ideas

The research describes the data model of pre-selected areas in details. First, the design of multi-source data conceptual model are carried out to comprehensively reflect the characteristics and embody relations of the data required by the sitting criteria. Then, the thematic database are set up with various subjects, as well as the description of the thematic data, in particular with the thematic conceptual model, logical model and physical model, are established according to the specific characteristics and application model of the thematic data in multiple areas. The design of the thematic data model not only consider the comprehensive integrity, but also take the independent use of thematic data into account. Thematic data model is the main unit composed geo-information the data model for pre-selected areas of geological disposal, and drilling, samples, documents and pictures and other data objects content involves many disciplines and belong to various disciplines in the logical structure. But the physical structure and storage and other thematic data are in the same layer from the point view of the simplicity of database management system, therefore, these data are associated with other thematic data through two or three level classification encoding.
The design method is to create directory based on all the selected data, establish unstructured database, make specific advice to the professional staff according to the nuclear safety guidelines HAD 401/06-2013 and make clear to the access to all the data entity form, physical meaning, dependence and application mode etc..The data model is preliminary designed in MS Access, and then all kinds of thematic data model are put forward to experts in various fields to for review. The experts are centralized organized to demonstrate after corrections and make actual tests for the modified data model. Each thematic database in MS Access are transported to PostgreSQL database platform by PLSQL programming, and each subject database corresponds to each special topic. Eventually, SQL is compiled to a comprehensive test to the data storage structure and data relations.

5. Implementation method

There are 10 factors in the sitting criteria for repository, and each factor have a cross overlap with other factors(as shown in figure 1). In order to effectively and systematically integrate these data, data acquisition method and convenience should be taken into account in the design of database structure to set up conceptual model and logical model. Therefore, data storage with different subjects is conducive to the expansion of database, specific organization and application.

5.1. space and time domain

SRID defined in ESPG is used for reference coordinate system in 6 pre selected area, and there is only one spatial reference system in principle. the million years (MA) and second (s) time units are used to describe the data associated with timing respectively. The time of geological age and geological events occurred is described in the attribute table, as well as time description of the field measurement and real-time monitoring data entities (including data packets, data table).

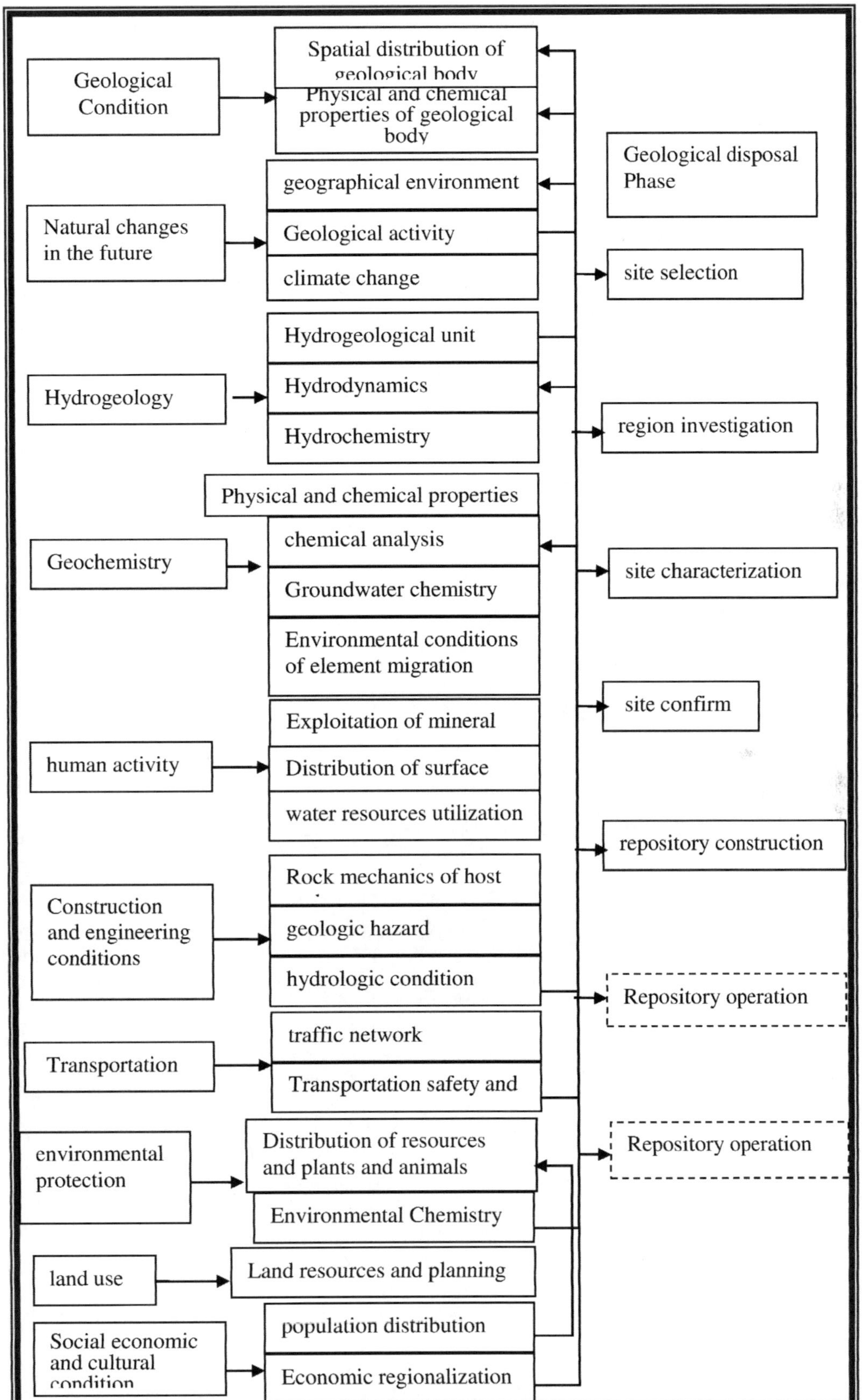

Figure1. Data requirements and relationship diagram in Sitting Criteria

5.2. data entity storage and data interface

Spatial vector data is stored as WKB and WKT, and pictures, documents and data packet is stored as BLOB. The BLOB can save all the information of original data package. A pre pyramid image structure is set up to store large amounts of remote sensing data(1GB or more),at the meantime metadata is recorded in the database.

Analysis of data entity is carried out through the description of three levels of metadata. First of all, special database and data elements of the table is getting from the meta database, and then all information of data entities are obtained on the basis of thematic classification code, spatial index, attribute index. All spatial data entities (vector and raster) make good use of Gdal/Ogr data interface, and the special file data package is stored using BLOB data storage (such as logging, spectral and multi-data format).

Base on the above research, the conceptual model(as shown in Figure 2) and logical model of geo-information data model is designed in details. Each thematic data model for preselected area of repository has been taken into account. Based on simplified model of logical relationships of data entities for HLW repository, various logic relation of data entities is set up according to the concrete forms respectively, including two-dimensional horizontal surface spatial graph (as shown in Figure 3), vertical section surface spatial graph (as shown in Figure 4), a one-dimensional curve graph (as shown in Figure 5), point data entity (as shown in Figure 6), documents, pictures and other multimedia data entity (as shown in Figure 7).

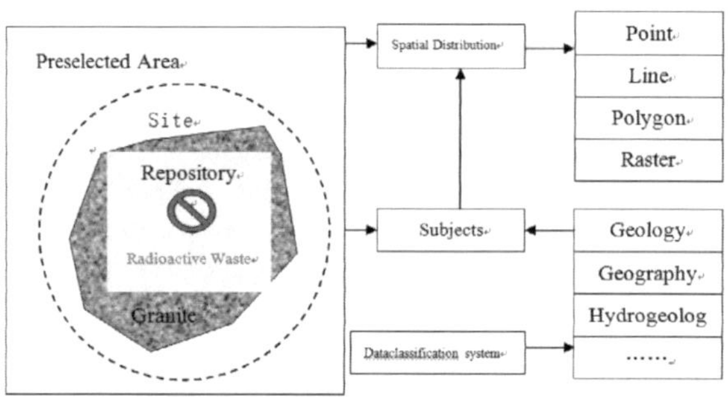

Figure 2. Conceptual data model for Preselected area of HLW repository

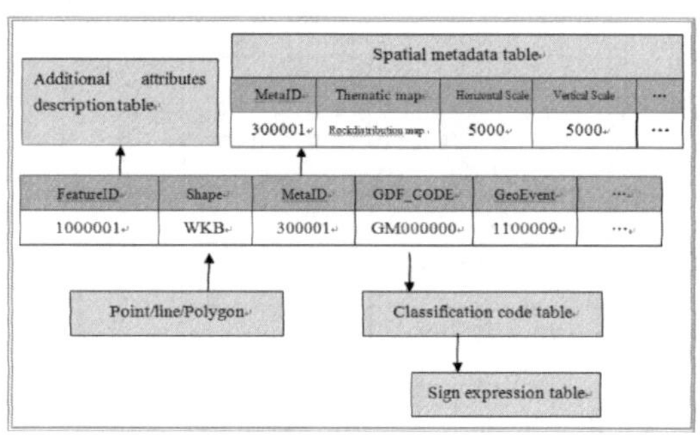

Figure 3 Logic relationship diagram of spatial data entity description for plane thematic map of pre-selected area

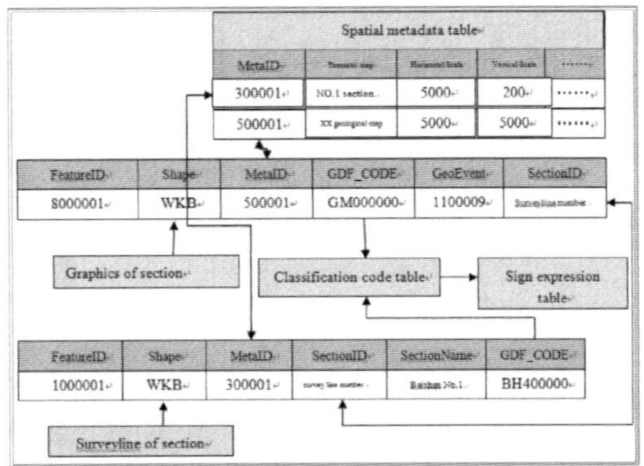

Figure 4 Logic relationship diagram of spatial data entity description for section thematic map of pre-selected area

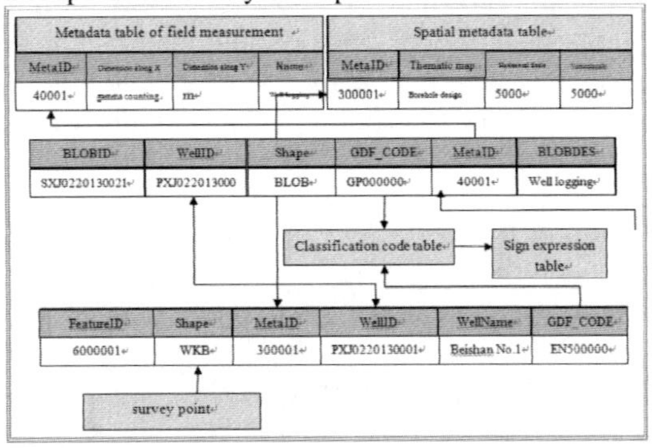

Figure 5 Logic relationship diagram of curve data for preselected area of repository

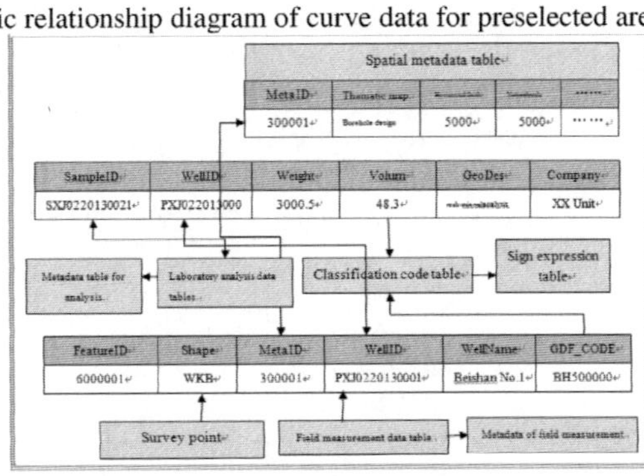

Figure 6 Logic relationship diagram of survey point, sample and analysis datafor preselected area of repository

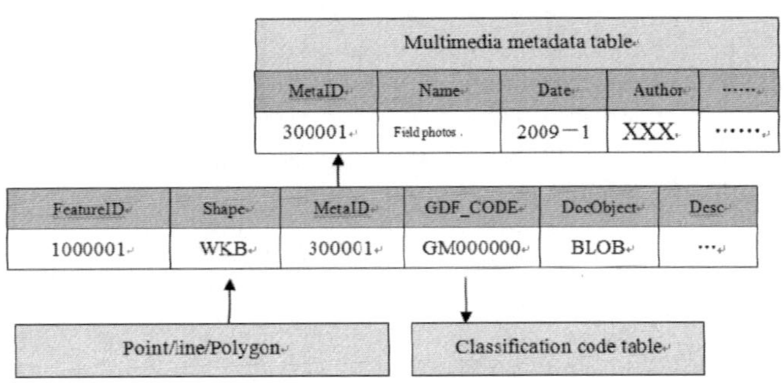

Figure 7 Logic relationship diagram of multimedia data for preselected area of repository

IOP Conf. Series: Earth and Environmental Science **46** (2016) 012054 doi:10.1088/1755-1315/46/1/012054

6. Conclusion

The research has made a solution for a series of key technologies, including the rational classification system, complex logical relation, effective physical storage structure ,etc.. The research has broken through the traditional industry application of data classification and conventional spatial data organization model, realized in the independent, complete data organization and integration by data entities and spatial relations for the basic as the basic unit with characteristics applied significance in the field of geological disposal, established a reasonable, feasible and flexible multisource data concept model, logical model and physical model, which guarantee the effective integration and convenient application of multi-source data of pre-selected areas for repository.

The integrated data model can easily be extended, and provides an open interface development of new data model for subsequent geological disposal system. It is proved that the established data model can completely meet ten data resources, storage and integrated management needs through the test and practical application, and give a strong support to multi-source data integration and management from storage layer in pre-selected areas for geological disposal repository. Geo-information data model is a relatively comprehensive, system data model in the field of geoscience, even there is no similar precedent both at home and abroad, and also can be extended to other nuclear facilities decommissioning and treatment, monitoring or other research fields.

7. Reference

[1] National Standard of People's Republic of China: Metadata Standard of geo-information data ISO 19115: 2003;

[2] Geological survey technology standard of China Geological Survey Bureau: Metadata Standard of geology data DD2006-05;

[3] Core information metadata standard: TD/T 1016-2003;

[4] Geological survey technology standard of China Geological Survey Bureau: Spatial database of Digital geological map DD2006-06;

[5] Geological survey work standard of China Geological Survey Bureau: Standard for the evaluation of regional groundwater resources survey and evaluation DD2010-03;

[6] Content Standard for Digital Geospatial Metadata Workbook(For use with FGDC-STD-001-1998)Version 2.0Federal Geographic Data CommitteeMay 1, 2000

[7] Content Standard for Digital Geospatial Metadata，FGDC-STD-001-1998

Beyond Moore's Law: Harnessing spatial-digital disruptive technologies for Digital Earth

Timothy W. Foresman

6219 Rockburn Hill Rd, Elkridge, MD 21075 USA

foresman@earthparty.org

Abstract. Moore's law will reach its plateau by 2020. Big data, however, will continue to increase as the Internet of Things and social media converge into the new era of 'huge data'. Disruptive technologies, including big data and cloud computing are forces impacting business and government communities. The truth of our collective future is suggested to align with the Digital Earth (DE) vision. Benefits of technological advances will be manifested from business performance improvements based on capitalizing the locational attributes of corporate and government assets - the foundation of big data. Better governance and better business represents a key foundation for sustainability and therefore should be explicit DE guiding principles.

Moore's law, the acclaimed barometer for Information Technology's pace of progress, is predicted to reach its plateau by 2020 according to no better an authority than Gordon Moore himself [9]. The exact date for the downward turn of this phenomenal growth curve remains unknown; however, contemplative minds are considering the consequences of this technology-era shift. Up to the present, market and technology experts have relied on the predictable advances of computer chips for their economic projections and business applications. The Digital Earth (DE) vision was borne from within the IT's industry's rapid expansion and therefore projections for the DE future are influenced by the digital technology development [5]. If the demise of Moore's law affects the future of exponential increases in transistors with associated lower cost for computer chips and memory, the question can be posed as to what this technology shift will mean for the Digital Earth community and its aspirations. This paper places a focus on this question within the context of the dynamics of the spatial-digital economy and the Digital Earth international community.

The impact of cramming more and more electronic components into integrated circuits' transistor density on computer chips has enabled the explosive growth of automation and computational complexity for society's benefits ranging from GPS and world transportation systems to global financial transactions and robotics. Members of the Digital Earth community have benefitted from these communication and computing advances as underpinning technologies for the development of creative applications. The collective impact of the Digital Earth community on society has been overshadowed, however, by the advances from commercial ventures by Google, Facebook, DigitalGlobe, Microsoft, and Apple in the USA initially and now with Uber, AirBNB, HERE (Europe), with Didi Kuaidi (China). A plethora of visualization, web-based globes provide access for citizens to ever increasing amounts of satellite (EOS) data and increasing volumes from active ground-based fleets of data collection vehicles (laser/camera). Interconnecting these increasing components of the Internet of Things (IoT) is the Holy Grail for tens of thousands of businesses in the consumer electronics market.

Where is the Digital Earth role in this rapidly expanding, commercially driven, new world? Craglia and colleagues [1] offered suggestions for research agendas appropriate for the next-generation Digital Earth. A decade since of observation at the annual symposiums and summits, and in the published literature, however, shows no evidence that the academic or government members of DE have impacted society. More often than not, members point to Google Earth as the Digital Earth (DE) analogy. Perhaps we should examine the DE members' appropriate role and build on our academics' capacity to actively monitor developments within myriad components of the Digital Earth vision and contemplate the best value propositions for our community. These questions are worth considering during this time of rapidly expanding disruptive spatial-digital technologies - technologies and business applications that capture society's attention from one market domain to another. The market success of these DE oriented tech companies should hopefully foster a deliberate conversation regarding issues of sustainability and improving the welfare of billions of our most needy citizens. It is worthy for the DE community to raise salient concerns about the global value of these disruptive technologies while recognizing that these innovations will continue to garner profound public receptivity and market share. This profound receptivity arrives in the guise of addressing insatiable entertainment opportunities for society and in facilitating customer transactions with enhanced communications for a new generation of digital citizens. Is the Digital Earth community positioned to documenting history, or are there other active roles for the skills and talents from this community?

What are the big disruptions that impact a Digital Earth? Big data, that collective buzzword, will continue to increase as the IoT and social media converge into the new era of 'huge data'. Disruptive technologies over the past couple decades include RFIDs, smartphones, smart buildings, smart meters, broadband network and communications, cloud computing, analytics for complex big data, visualization, drones and the expansion of platforms for social media. The media hype around these disruptive technologies states these are explosive and immediate forces that rock the foundations of the business and government [7].

Perhaps we are amidst a multitude of advancing and new enterprises that will ride out Moore's law technical capacity in two directions. First, is the path based upon the technical advantages from lower cost and increased performance of computer chips. Second, is the multitude of pathways based upon the capacity of innovators to enhance performance using existing technology for practical business and commercial application through integration of spatial-digital data, models, and visualization. Uber innovation exemplifies the second path model based on using existing technology. Uber does not require the next generation of supercomputers to continue succeeding in the market. Uber only needs to ride on the spatial-digital capacity of smartphones and act as a competitive broker between customer and service provider. If this model of innovation continues, through innovative horizontal expansion of spatial-digital applications, then the demise of Moore's law will not represent a doomsday spectre for growth of business creation and capitalism. Is DE the big data platform for visualisation and analytics?

The truth of our collective future is possibly aligned with the Digital Earth vision (along with updates such as, the 2009 Beijing Declaration) regarding aspirations for significant economic and societal benefits and increased utility available to those who harness integration of the spatial-digital disruptive revolution. To set a distinctive framework, a suggested second law of geography is introduced that states "Objects that know where they are can capitalize on their locational knowledge. Whether the object knows, or an external agent knows, spatially enabled entities have potential for financial and functional utility." Spatial enablement of the IoT can be expected to have a deep influence on further disruption to the disruptive agents. A list of disruptions is offered to anticipate the value-added role of Digital Earth within major societal and market shifts. This is not an exhaustive list and no such list exists.

- Virtual Reality (VR) and Augmented Reality (AR) – Once people experience VR its engaging properties are addictive. VR is so powerful on the human mind that health and trauma specialists are investigating its therapeutic potential. VR brings the next generation of entertainment, tourism, and community participation, as well as engineering and design applications. AR promises to upgrade the performance for legions of technicians by at least threefold while lowering risks associated with ignorance and fatigue.
- Electronic Banking and Financing -Bitcoin and block chain encryption are slow to take over the world finance system, albeit, these advances can stymy the conventional methods of corruption long held in institutions labelled too big to fail. Business to business and person to person financial transactions and loans threaten to undermine the world's economic system. Corruption and fraud may decrease as cash transactions for all purchase and services become more transparent and are a documented part of history. The trail left behind is but a harbinger of the trail ahead. Fintech's potential is also recognised for financial disruption, building off of digital payments and automated wealth management.
- Citizen Science – Volunteer Geographic Information (VGI) [4] and citizen upwelling of information for science and public safety represents a force multiplier for governments and society [3]. Gruntify (Brisbane based) represents a recent smartphone technology disrupter that enables citizens to capture and submit video, photos, and audio from anywhere to upload to the cloud useful for safety, VGI, and a new wave of applications.
- Drones – The market growth of drones for hobbies is matched by the growth of business and government use. The skies will be filled with robots that can recognize you and interact with you; even delivering your mail and parcels.
- Facial Recognition – Anonymity is no longer viable for anyone operating in the public sphere. For over a decade, Iceland has applied facial recognition technology for all inbound travellers. Ubiquitous urban cameras leave nowhere to hide.
- Health Bits – As health data is accrued through personal tracking devices, health statistics can enter a new era of populations and census data. These trends will include epidemics and pandemics in our increasing mobile world. Connection to field maps increases the efficacy of tracking and monitoring health conditions for individuals and regions.
- Carbon Accounting – From the 2015 Paris Climate Agreements on forward, nations and business will be more closely tracking their use of fossil fuels. This will influence energy trends; from electric cars to solar cells to batteries.
- 3D Printing – A range of disruptions has been identified from this innovation, most of which have been assessed as positive. Manufacture of non-metallic handguns provides a distinguishing exemption. Advances are offering significant advantages to aerospace industry while shortening supply chains for precision manufacturing.
- Self-driving Cars – Major corporations have billion dollar investments in this disruptive arena where the future of hundreds of thousands of transport drivers' is threatened. Success will require supreme geo-spatial feature recognition and navigation automation.
- IoT – The internet of things represents a phenomenal expanding universe of objects that can be linked and/or controlled with wireless and wired networks. Current estimates of 30 billion 'things' in the IoT are expect to reach 200 billion by 2020. Harnessing segments of the IoT, for profit or better asset management, is driving a last-frontier mentality across many markets [6].
- Cloud Computing – The arrival of wireless connections and cloud computing services has emancipated operations from tethered connections and the complex and costly overhead of data storage and administration. The disruptive forces of cloud computing provides unprecedented computational resources to individuals and small businesses that were previously limited to major universities. This will foster a force multiplier effect for creativity.

The spatial-digital revolution, from which the Digital Earth vision was borne, gained its prowess in accordance with Moore's law, but also with the newly posited second law of geography (SLG). The

SLG acknowledges the potential for financial and functional utility through spatial enablement. This law helps explain, in part, the phenomenal expansion of applications and enterprises over the past decade, often labelled disruptive. Again, Uber is a classic example. Myriad other examples are evident across the commercial world as the idea of business intelligence and locational intelligence gains currency in the leading business management circles. Empowering clients with purchasing knowledge at the front door of commercial shops and sharing price comparisons on customers' smartphones is also feeding the retail industry's individual and aggregate customer profiles for increased sales efficiencies through big data analytics. A recent innovation with spatial enablement of the IoT can be found, for example, at COEX mall in south Seoul. The Lucky Chouett outlet uses 'smart mirror' with coin-size beacons on each hanger with information, accessories, color options, and inventory data displayed to the potential customer who is standing in front of the magic mirror at the store front [8].

Capitalization has not been at the forefront of the Digital Earth community's conversation. Capitalization is, however, what is driving the phenomenal growth in the more visible manifestation of the Digital Earth in society (e.g., Google Earth, HERE, IoT, et cetera). When attending to the benefits of spatially-enabled enterprises for local government and business applications, the DE community can provide evidence of the significant return on investments (ROIs) and expanded business opportunities. Benefits of DE technological advances can be communicated as business performance improvements based on capitalising or improved utilisation of the locational attributes for corporate and government assets. Spatial enablement and integration of assets is the context for big data value. Better governance and better business represents a key foundation for sustainability and therefore should be explicitly stated guiding principles. While the DE community of academics and government representatives should not expect to compete in the public sense with the market leaders, we might consider what enhancements our collective efforts might possibly yield. As an example, a call for focus on ethics was the final topic raised at the ISDE 9 Halifax symposium.

Desha and her colleagues [2] recommended an elegant approach to incorporate the benefits of DE within a policy and management framework. The Pivotal Principles proposed by Desha et al, could illuminate the potential in the spatial-digital era for a broad audience of international stakeholders across a range of applications. These Principles provide for a comprehensive management philosophy regarding harnessing data policies, operations, and engagement with key stakeholders. Members of the DE community will gain an innovative perspective from these Principles requisite for strategic decision making in this disruptive age. Can the DE community consider a new approach to our collective promise by offering pragmatic policy and management solutions based upon the DE society's comprehensive knowledge of society and technology's trajectory? As the ISDE reflects upon a decade of progress, it is appropriate to assess the value proposition of the Society relative to the wave of commercial and market drivers that are most influencing the public.

References

[1] *Craglia,* M., M.F. Goodchild, A. Annoni, G. Camara, M. Gould, W. Kuhn, D. Mark, I. Masser, D. Maguire, S. Liang, and E. Parsons *(2008). Next-Generation Digital Earth. International Journal of Spatial Data Infrastructures Research, (3):146-167.*

[2] Desha, C., T. Foresman, A. Vancheswaran, A. Reeve, and J. Hayes (2016) Pivotal Principles for Digital Earth Development in the 21st Century. International Journal for Digital Earth, DOI: 10.1080/17538947.2016.1161090

[3] Elwood, S, M. F. Goodchild & Daniel Z. Sui (2012) Researching

[4] Volunteered Geographic Information: Spatial Data, Geographic Research, and New Social Practice, Annals of the Association of American Geographers, 102:3, 571-590

[5] *Foresman, T. W. (2008).Evolution and implementation of the Digital Earth vision, technology and society.. International Journal of Digital Earth **1** (1): 4–16.*

[6] Greengard, S. (2015) The Internet of Things. MIT Press, Cambridge, Massachusetts.

[7] Hersh, E.D. (2015) Hacking the Electorate: How Campaigns Perceive Voters. Cambridge

University Press.
[8] Korea Joongang Daily (2016) Article of 22 Feb 16, pg 5.
[9] The Economist (2016) After Moore's Law. March 12, 2016, pg 3-10.

Digital disaster evaluation and its application to 2015 M_s8.1 Nepal Earthquake

Xiaoqing WANG[1], Jinxia LV[2], Xiang DING[1] and Aixia DOU[1]

[1] Institute of Earthquake Science, CEA, Beijing, 100036, China
[2] China Earthquake Networks Center, Beijing, 100045, China

E-mail: wangxiaoq517@163.com

Abstract. The purpose of the article is to probe the technique resolution of disaster information extraction and evaluation from the digital RS images based on the internet environment and aided by the social and geographic information. The solution is composed with such methods that the fast post-disaster assessment system will assess automatically the disaster area and grade, the multi-phase satellite and airborne high resolution digital RS images will provide the basis to extract the disaster areas or spots, assisted by the fast position of potential serious damage risk targets according to the geographic, administrative, population, buildings and other information in the estimated disaster region, the 2D digital map system or 3D digital earth system will provide platforms to interpret cooperatively the damage information in the internet environment, and further to estimate the spatial distribution of damage index or intensity, casualties or economic losses, which are very useful for the decision-making of emergency rescue and disaster relief, resettlement and reconstruction. The spatial seismic damage distribution of 2015 M_s8.1 Nepal earthquake, as an example of the above solution, is evaluated by using the high resolution digital RS images, auxiliary geographic information and ground survey. The results are compared with the statistical disaster information issued by the ground truth by field surveying, and show good consistency.

1. Introduction

The satellite and airborne high resolution digital RS (HRS) technologies have been developed fast since 1990s. The new technologies have been applied widely to the earthquake disaster monitoring and assessment. The typical and successful application evens are those of 1999 M_s7.6 Taiwan earthquake of China, 2003 M_s6.3 Bam earthquake of Iran, 2004 M_s9.0 earthquake and tsunami of Indonesia, 2008 M_s8.0 Wenchuan earthquake of China, 2010 M_s7.0 earthquake of Haiti, and 2011 M_s9.0 earthquake of the Pacific coast of Tohoku of Japan, etc. The high resolution image based methods have been widely studied for the extraction of building damage, earthquake induced geological disasters such as landslide, and road damage. While, as the mass RS data with high resolution and complexity of seismic damage features in image, The extraction of seismic damage are usually interpreted visually from images up to now. The effect of automatic extraction of seismic damage from HRS is remain in improvement. By the way, the conventional image pre-procession, seismic damage extraction and assessment in personal computer or intranet environment need also to be improved. At present, as the fast development of internet technology, it is available to interpret the seismic damage through resource sharing and collaboration over Wide Area Network. The RS data process and seismic damage analysis platform has changed from 2D map system to the 3D digital earth system with the development of digital earth techniques [1,2]. This enhances greatly the ability of seismic damage interpretation. The paper probes the digital earth based seismic damage extraction and assessment techniques with HRS images, discussed the effect of the application mode through the case study of 2015 M_s8.1 Nepal earthquake.

2. The basic idea of collaborating interpretation of seismic damage based on Digital Earth System (DES) on internet environment

As the same RS images are similar in spectrum and spatial resolution in whole disaster area, the RS based seismic damage interpretation is monotonous repetitive work. The precision of computer based automatic interpretation need to be improved. Although higher precision seismic damage distribution will be obtained by visual interpretation, the efficiency is relatively low and cannot meet the needs of timeliness requirement of decision-making for emergency and rescue. The development of seismic damage interpretation system based on DES solves the problem in a certain degree. On a digital earth platform, the RS data will be pre-processed, enhanced and shared (published) in the server-side, then a large number of users can view simultaneously the damage information on digital earth windows (3D) in multi-angles and the details in the client-side, in which the accuracy and timelines will be improved obviously through the parallel extraction of seismic damage by a large number of users in internet environment (figure 1).

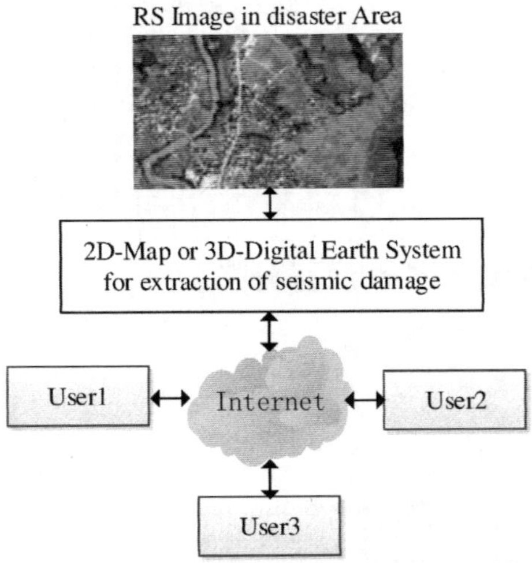

Figure 1. Work mode for seismic damage extraction based on Digital Earth System in internet environment.

The examples of internet based disaster information extraction system will be found in the Internet environment. Such as the "DigitalGlobe FirstLook" developed by the DigitalGlobe Corporate [3]. The system provides to the volunteers the latest block images of a catastrophe events, for example, the Hurricane, the aircraft accident such as Malaysia Airlines (MH370, figure 2a), or the event such as 2015 M_s8.1 earthquake of Nepal; The volunteers will be allocated different block images for viewing, finding and locate the suspected damaged objects, and then reporting them (providing the information) to servers of DigitalGlobe.

In order to improve the application ability of Chinese earth observation infrastructure for disaster reduction, the "National coordination system & data sharing service platform for spatial data acquisition, application & emergency response", also known as the "Reservoir emergency response platform", was developed by the Academy of Opto-Electronics (AOE), Chinese Academy of Sciences (CAS),under the grant of the Ministry of Science and Technology (MOST) of China [4]. The platform provides functions to coordinate various civil aerospace resources to acquire RS images of the disaster stricken area, and disseminate the data to relevant institutions involved in disaster reduction as soon as possible. The platform has been playing an important role in spatial data sharing of emergency response for Yushu earthquake in Qinghai, China, Lushan earthquake in Sichuan, China, Lushan earthquake in Sichuan, China, etc. by the authors.

The system is only a collecting and sharing centre of the earth observation data for emergency response without functions of damage extraction.

Other example is the information system of rescue and relief of 2015 M_S8.1 Nepal earthquake (figure 2c) by The International Centre for Integrated Mountain Development (ICIMOD), in which all the information associated the administrative, population and economic, casualties and geohazards, rescue as well as geographic data can be accessed. Although there are no interpretation functions in the system, the abundant information is very helpful for the disaster identification, decision-making of rescue and relief etc.
The last example is a digital earth system for the seismic damage evaluation (figure 2b) developed by the Institute of Earthquake Science (IES), CEA (Wang et al, 2015). The system has functions of fast estimation of seismic intensity, seriously damaged area identification, or the disaster information such as building damage, road damage, as well as earthquake induced landslides etc. based on the post-earthquake RS images. The system runs in the internet environment and thus provide a fast and 3D visual damage interpretation by multi-users, which plays a very important role in damage identification in the emergency stage of cases such as 2013 Lushan, 2014 Ludian and 2015 Nepal earthquakes.

(a) DigitalGlobe FirstLook[3]

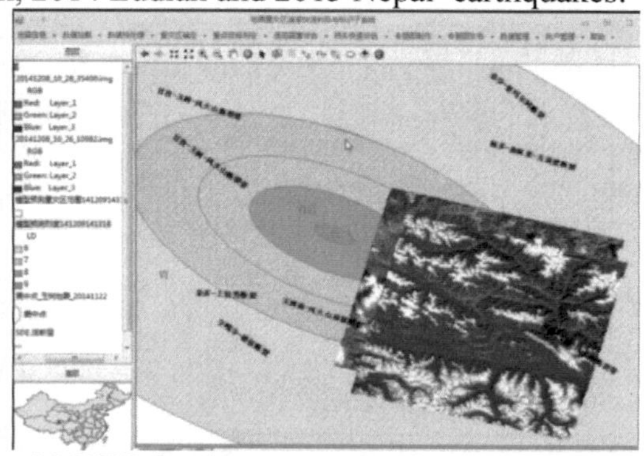

(b) DES for damage evaluation by IES,CEA[2]

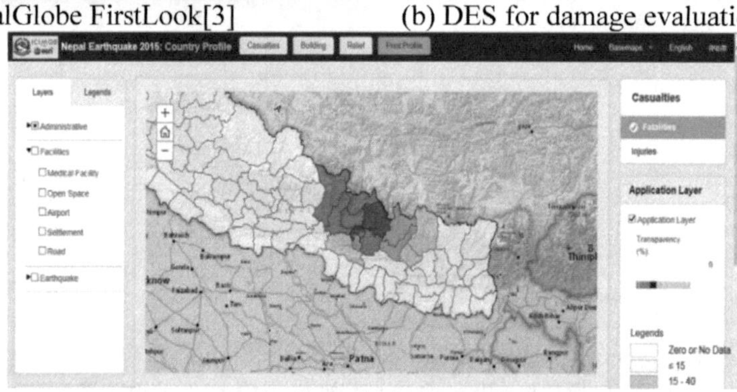

(c) Information system of rescue and relief of 2015 M_S8.1 Nepal earthquake by ICIMOD [5]

Figure 2. Examples of damage interpretation system or information platform based on 2D map (block image) or 3D digital earth in internet environment.

3. General functions of DES-based quantitative evaluation of seismic damages

The seismic intensity is usually applied to present the disaster scale and extent, essential for the emergency rescue and relief. According to "Seismic Intensity Scale of China" (GB/T 2008), the seismic intensity can be determined by the seismic damage index, the latter will be estimated quantitatively by field survey. based on the individual building damage, but it usually takes too long time to meet the need of emergency rescue. Wang et al (2003,2008) developed a RS-based seismic damage estimation method [7,8]. It estimates quantitatively the seismic damage index based on the individual building damage grades extracted from satellite or aerial borne very high resolution remote sensing images (VHR) acquired immediately after an earthquake. The damage index in a settlement, as statistical unit, is the sum of proportion of each building damage grade weighted by the mean damage index (1.0, 0.5, 0.0 respectively to collapsed, partial collapsed and un-collapsed grades). The seismic damage index equivalent to the ground truth on each residential unit will be obtained through the conversion relationship between damage indices by the field survey and by extraction from RS images of historical earthquakes [9].

The conventional method extracts the damage grade of individual building from 2D RS images by visual identification or automatic classification of damage grades. But the precision and timelines is usually low. The digital earth system provides a new tool to interpret the building damage, landslide or other earthquake induced geological disasters with stereo and more accurate images in 3D visual environment developed by IES, CEA [1,2], as shown in figure 3. Besides, the DES runs as a client on the Internet, which implicates that many users have possibility to join the interpretation work at the same time and will meet the needs of timelines in the emergency stage.

The DES for damage evaluation developed by IES,CEA has functions in the earthquake emergency stage as following [1]:

(1) To provide RS image data management functions such as upload and publish of RS data.
(2) To provide maps of background in disaster area, such as the administrative zoning map, earthquake epicentre distribution, population distribution, building distribution, chorography map etc.
(3) To view the pre- and post-RS images in earthquake-stricken areas.
(4) To extract damages of buildings and lifeline systems, landslides and other disasters from RS images on DES by internet users.
(5) To provide services of real-time collection of damage information interpreted by internet users and publish at the server-side. Thus the users can understand the process and take collaboration of interpretation, then the decision-makers will get the damage information in time.

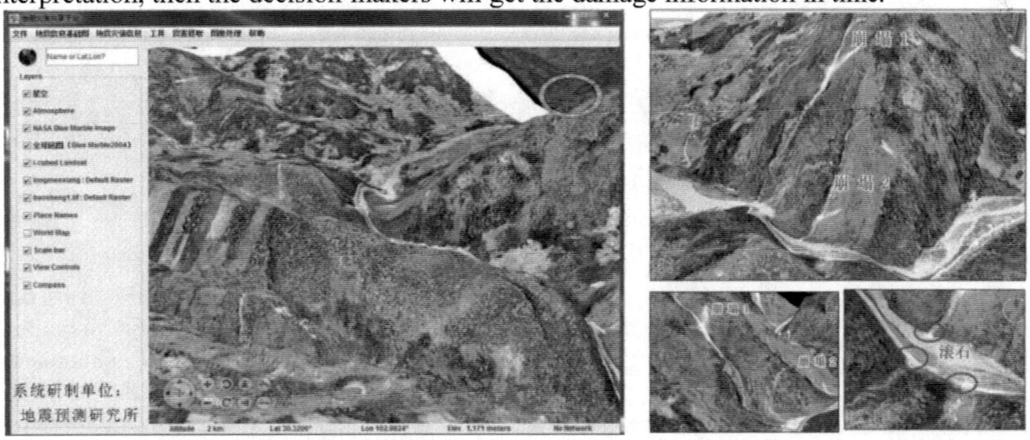

(a) The main interface of DES (b) The collapse hazard and rock shed

Figure 3. The application of DES, developed by IES,CEA, to interpretation of geological disasters induced by 2013 M_S7.0 Lushan earthquake. The text shows collapse; Circle shows rock shed.

4. The DES-based quantitative evaluation of building damages due to 2015 earthquake, Nepal.

A strong earthquake with M_S8.1 occurred near Gorkha, Nepal at 11:56 am (6:11am, UTC) on April 25, 2015 (84.7°E, 28.2°N) with a depth of 20 km. 34 minutes later, a strong aftershock of M_S7.0 occurred with a depth of 30 km in the area near Pokhara (84.8°E, 28.3°N) to the northeast of and about 5 km away from the epicentre of the main event. About 12 hours later, another strong aftershock of M_S7.1 occurred to the southeast of the main earthquake area (85.9°E, 27.8°N) at 12:54 pm on April 26 (UTC +5:45), with a depth of 10 km. 17 days later, still another huge earthquake of M_S7.5 struck Nepal (86.1°E, 27.8°N) at 12:50 on May 12, with a depth of 10 km. The earthquakes caused extensive damage to buildings and thousands of deaths and injuries It is reported that, by the end of May, the death toll is about 8953, among which 218 caused by the event occurred on May 12.

After the earthquake occurred, many satellites immediately acquired high remote sensing images, such as WorldView, GF1/2, ZY2/3, et al. in disaster areas during the emergency stage. We used images of GF and ZY for disaster estimation, which were applied to support the emergency and rescue. After that the RS images, including WorldView (0.5m), were used to support the extraction of building damages, lifeline system damages and landslide et al., which is applied by China team of field investigation of damages and losses in the disaster areas in Nepal during from June 6th to 20th, 2015. The disaster information was extracted mainly with DES before and during the field work stage (figure 4).

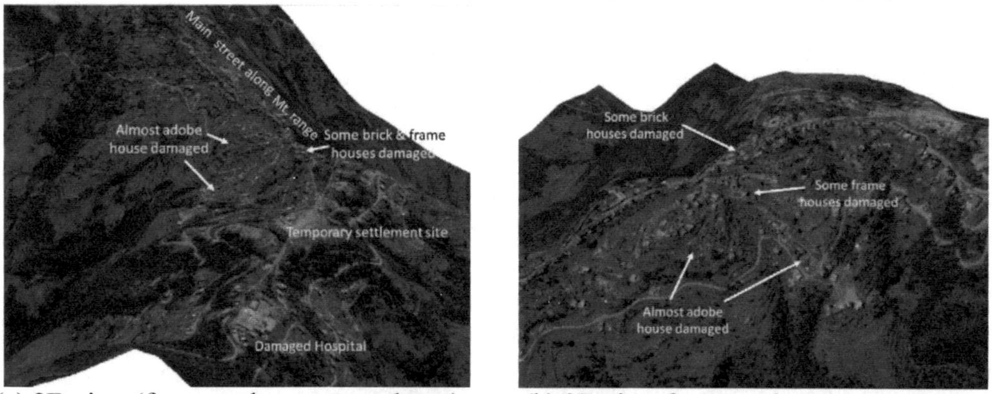

(a) 3D view (from northwest to southeast) (b) 3D view from southwest to northeast

(c) The collapsed frame house (d) The hazard and rock shed of

Figure 4. 3D view (a & b, from Worldview image) and photos (c & d, by field survey) of building damages in Chautara, Sindhupalchok, induced by 2015 Nepal earthquakes.

Compared to 2D view, images on DES will provide more clear disaster information and its possible causes which might be observed from multiple orientations. The RS image features of building damages, landslides and other disasters will be established by comparison and analysis of RS images and field survey as examples shown in figure 4.

Based on the processing and analysing of the pre- and post-earthquake high resolution RS images, combined with the field survey, the building structure types and their damages were interpreted. Then the distribution of damage level of buildings due to Nepal earthquake was obtained and shown in figure 5 (Wang et al, 2015). The results show that, the spatial distribution of building damage exists obvious influence of the combination of different structures such as stone, masonry and RC frame buildings, the locations of main event and strong aftershocks, as well as the distribution of active faults in the disaster area. The seriously collapsed areas are mainly located surrounding the epicentres of main event (M_s8.1) and strong aftershock (M_s7.5). But the two regions are separated in space. Those residential areas with very few buildings collapsed are large enough into one region. The distribution is identical to the damage distribution determined by ground survey [11].

5. Discussion and Conclusion

The authors in the article reviewed the advance of RS-based methods and techniques of disaster information extraction and evaluation, probed the technique resolution and examples of collaborating interpretation of seismic damage with 2D map system and 3D digital earth system (DES) based on the internet environment, introduced the general functions of DES-based quantitative evaluation of seismic damages developed by the Institute of Earthquake Science, China Earthquake Administration. Then described the application of DES to the seismic damage extraction of 2013 Lushan, China earthquake and 2015 Nepal earthquake.

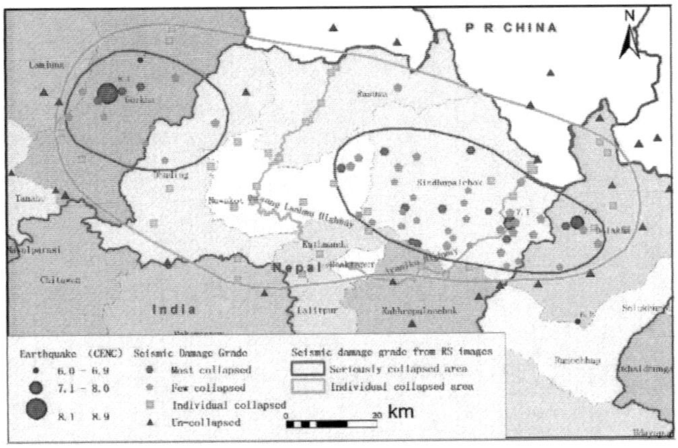

Figure 5. Distribution of damage grade of buildings due to M_S8.1 Nepal earthquake interpreted from RS images [10].

The primary application of DES in the earthquake emergency and later field survey and loss estimation of latest damage earthquake events, such as Lushan earthquake and Nepal earthquake et al, shows great advantage in the seismic damage information extraction. It featured faster, more precise which are crucial to the decision-making for emergency and rescue. While more effort should be put to realize a good service-oriented, closely collaborative on internet, digital earth system for emergency seismic damage information interpretation.

References

[1] Guo J X, Wang X Q, Dou A X, Yuan X X and Zhang W Remote sensing data dissemination system for earthquake based on OpenGIS and digital earthquake platform 2013 *Earthquake* **33**(2) 123-31 (in Chinese)

[2] Guo J X, *Study on the Visulization and Extraction of Seismic damage based on digital earthquake platform* [D] 2013 (Institute of Earthquake Science, China Earthquake Administration) pp 1-116

[3] http://www.digitalglobe.com/products/firstlook.

[4] http://english.aoe.cas.cn/rh/rps/201412/t20141219_133444.html

[5] http://www.icimod.org/?q=17851

[6] General Administration of Quality Supervision, Inspection and Quarantine of P R China, Standardization Administration of P R China *National Standard of the People's Republic of China, GB/T 17742-2008, Chinese seismic intensity scale* (Beijing: Standards Press of China) (in Chinese) pp 1-3 (in Chinese)

[7] Wang X Q, Wang L, Zhang X H, Ding X, Qiu H, Dong Y F, Dou A X, Zhang F Y and Liu Z T. Primary quantitative study on the earthquake damage extracted from remote sensing imagery-A case study of Doujiangyan due to the Wenchuan earthquake with magnitude 8.0. 2009. *Earthquake* **29**(1) pp 174-81 (in Chinese)

[8] Wang X Q, Wei C J, Miao C G, Zhang J F, Shan X J and Ma Q Z Extraction of seismic damage from remote sensing images-A case study of Bachu-Jiashi earthquake with Ms=6.8 occurred on Feb. 24, 2003 2003 *Earth Science Frontiers*, **10**(z1) pp 285-91 (in Chinese)

[9] Wang X Q, Dou A X, Wang L, Yuan X X, Ding X and Zhang W RS-based assessment of seismic intensity of the 2013 Ms7.0 Lushan earthquake, Sichuan, China 2015 *Chinese J. Geophys.* **58**(1) 163-71 (in Chinese)

[10] Wang X Q, Huang S S, Ding X, Cui L P, Dou A X, Li Y W Extraction and analysis of building damage due to Nepal Ms8.1 earthquake from remote sensing images 2015 *Technology for Earthquake Disaster Prevention* 10(3) 481-90

[11] Sun B, Yan P L Damage characteristics and seismic capacity of buildings during Nepal Ms 8.1 earthquake. 2015 *Earthquake Engineering and Engineering Vibration*, 14(3) 571-578

Acknowledgments

The research was supported by the Fundamental Research Funds of Institute of Earthquake Science, China Earthquake Administration (Grant No. 2013IES0204).

Mobile 3D laser scanning technology application in the surveying of urban underground rail transit

Youmei Han [1,2] **Bogang Yang** [1,2] **Yinan Zhen** [1]

[1] Beijing Institute of Surveying And Mapping, China
[2] Beijing Key Laboratory of Urban Spatial Information Engineering, China

E-mail: hanyoumei@126.com

Abstract. Mobile 3D laser scanning technology is one hot kind of digital earth technology. 3D completion surveying is relative new concept in surveying and mapping. A kind of mobile 3D laser scanning system was developed for the urban underground rail 3D completion surveying. According to the characteristics of underground rail environment and the characters of the mobile laser scanning system, it designed a suitable test scheme to improving the accuracy of this kind of mobile laser scanning system when it worked under no GPS signal environment. Then it completed the application of this technology in the No.15 rail 3D completion surveying. Meanwhile a set of production process was made for the 3D completion surveying based on this kind of mobile 3D laser scanning technology. These products were also proved the efficiency of the new technology in the rail 3D completion surveying. Using mobile 3D laser scanning technology to complete underground rail completion surveying has been the first time in China until now. It can provide a reference for 3D measurement of rail completion surveying or the 3D completion surveying of other areas.

1 Introduction

The main task of traditional underground rail transit completion surveying is to survey the underground structures, and compare the measurement data with the original planning and designing data. Then the results would be provided to the relevant departments to carry out a comprehensive analysis [1]. However, with the requirements, the rail transportation management department also hopes to make the 3D completion surveying to support the rail transportation operation and management [2]. In recent years, the static ground 3D laser scanner has been used to the 3D completion surveying [3,4]. However, the efficiency of this kind of equipment is relatively low compared to the rail transit tight and heavy task. So it set up a kind of vehicle borne 3D laser scanning system, which consists of one 360 degree 3D laser scanner, one GPS, one IMU, one distance measuring instrument and some cameras. All the sensors are fixed on the slide rail car with the technology of the time and space synchronization. And it finished 1.5km rail 3D surveying data collection and modelled this data.

2 The vehicle borne laser scanning system and its principle

The main sensors of this vehicle borne system are one 360 degree laser scanner, one high accuracy IMU, one GPS and a wheel distance measuring instrument. (Figure 1), the parameters of these sensors are shown in Table 1.

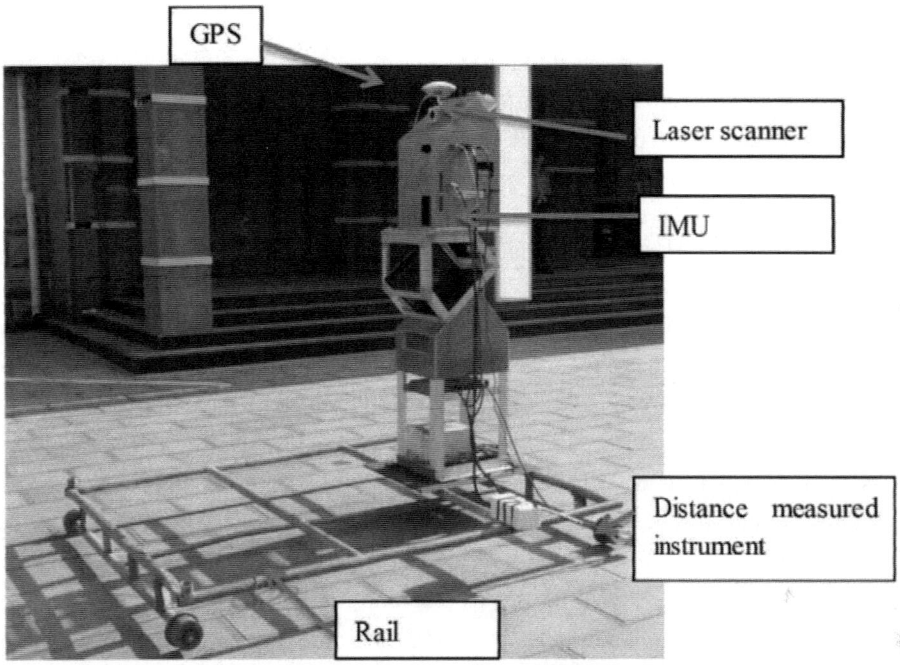

Figure 1 rail vehicle borne laser scanning system

Table 1 The parameters of these main sensors

Sensors	Style	Main technic parameters
Laser scanner	RA-360 III	Range resolution: 5mm Range scope: 0.6m-300m Range precision: 2cm(100m) Divergence angle: 0.3 mrad(0.017°) Angle accuracy: 0.1 mrad(0.0057°) Angle scope: 0-360° Acquisition frequency: 200kHZ
IMU	POS50	Heading angle accuracy: 0.03(°)/hr(RMS) Roll angle accuracy: 0.03(°)/hr(RMS) Pitch angle accuracy: 0.05(°)/hr(RMS) Acquisition frequency: 200Hz
GPS	Trimble5700	Acquisition frequency: 20Hz Post processing differential positioning accuracy:

		5mm+1ppm CEP Time resolution: 20ns
Distance measuring instrument	sick	Frequency: 10000Hz， Resolution: ≤5mm

All these sensors are fixed on the rail car rigidly, with the time and space synchronization technology to all devices. After a comprehensive calibration obtained sensors relative spatial relationship [5]. It acquires space 3D information with the vehicle moving forward. GPS, IMU and wheel encoder together constitute the system positioning and attitude determination system. With the inertial Explorer processing software, the three navigation data are computed combined to get the center vector corresponding to the attitude and position data [6]. It can get the absolute coordinates by these three position sensors.

3 The system precision analysis

The precision of vehicle borne 3D laser scanning technology is affected by the range and angle accuracy of the laser scanner, the positioning accuracy of the vehicle, and other factors. The range and angle accuracy of the laser scanner have been removed by the system calibration, which can't been optimized during working [7,8]. So the system accuracy depends on the positioning accuracy of the carrier during navigation. However there is no GPS signal during underground mobile measurement, And the location accuracy got by IMU itself will be getting worse with the extension of the time, so it add a distance measuring instrument to correct the navigation accuracy [9], which provides a direction of the distance information for IMU to help integrated navigation. But this kind of integrated navigation technology will be out of operation when the GPS signal missing in a long time, it also needed other means to improve the navigation precision, in order to meet the rail traffic completion surveying requirements.

Based on the principle of the control surveying, it added some control points along the rail to improve the accuracy of the rail transition completion surveying based on the vehicle borne 3D laser scanning system. The distribution of the points should be even in some extend and must be surveyed by higher level methods. Then the control points can be used to correct the whole laser scanning data.

4 Experiment analyses

4.1 Experiment profile

This experimentation area is the Beijing subway Line 15, Guan Zhuang station to An Li road station, which is a typical underground rail transit new line. Before the laser scanning data collection, some planning control points were set up along the line, which were designed by high reflection film (shown in Figure 2). After some test, it found this kind of control points were easily to be recognized in the scanning data (also named point cloud). In this experiment these control points were put up on the wall of the rail left and right alternately by 25m interval. Some of these points are for control, the others are for the accuracy detection.

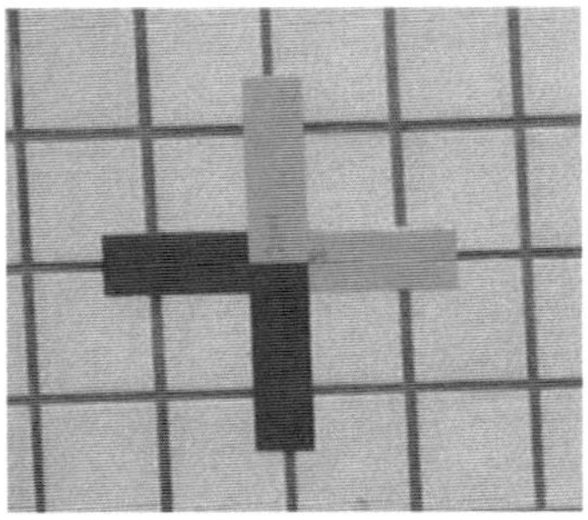

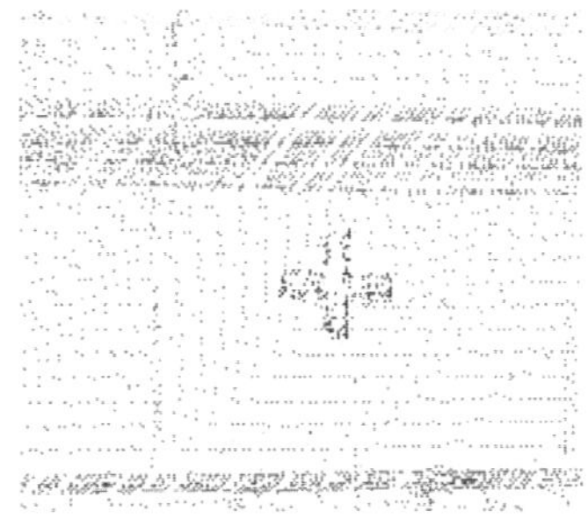

Figure 2 Control point design picture （left）、 the point cloud capture of the control point（right）

The position of this 3D laser scanning system is differential positioning, so it needed a reference station on the road surface in the surveying area, which is relatively open. This experiment reference was set up on the Guan Zhuang Station. At the beginning of this experiment, it needed initial the system in about 15 minutes on the ground, and then it opened the laser scanner and moved this system from ground to underground rail vehicle to obtain 3D point cloud of this section rail. At the end of the experiment, it also needed another initial period. This experiment need finished as soon as possible to minish the GPS lock time and improve the position accuracy.

4.2 The technic framework

It designed a technic framework for this experiment (shown in figure 3). First it would capture the laser data and the position data. Then it computed the position data and tested the result. After the real coordinate point cloud would be computed. The core step was the point cloud correction. It needed correct the plane firstly then correct the elevation. Each step need to be test before go ahead. Aim to some rough data it needed to add more control points to get the right data which can meet the requirements.

The result was the point cloud which was culled noise.

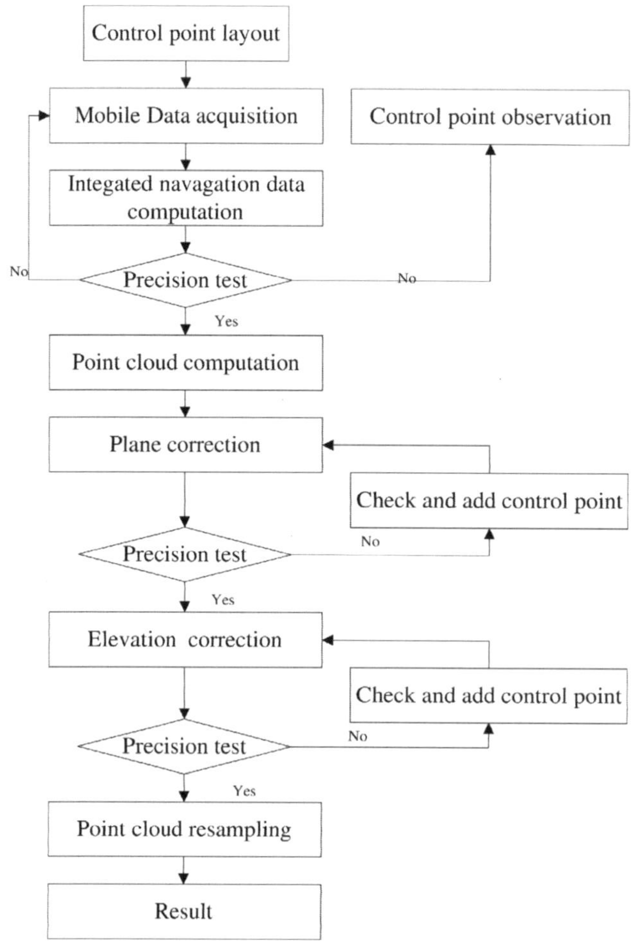

Figure 3 A technic framework

The final point cloud was shown in figure 4.

Figure 4 The point cloud for the underground rail

4.3 Point cloud correction

The correction principle is the bilinear interpolation. The software is programmed by us named as SWDY.

I) Plane correction

The detailed steps of this section were shown blow.

a. Control point projection (shown in figure 5).

Figure 5 SWDY point cloud software

b. Mapped each point at the corresponding area which was the real position of each control point. Map one by one with the planning 50m interval.

c. Checked the plane accuracy of the correct point cloud, and added necessary control point at the area which accuracy was out of requirements.

d. Corrected the whole point cloud.

e. Checked the plane accuracy of the whole correct point cloud.

II) Elevation correction

After the plane correction, the elevation correction would be done. The detail steps were following.

a. The elevation control point projection.

b. Projected the point cloud, map each control point at the real area of each control point on the point cloud with 50m interval.

c. Check the elevation accuracy of the correct point cloud, and adds necessary control point at the area which accuracy was out of requirements.

d. Correct the whole point cloud.

e. Check the elevation accuracy of the correct point cloud.

4.4 Accuracy analysis

I) Plane accuracy

There were 39 plane points to test the plane accuracy with about 50m interval. The adjustment model is formula (1):

$$m = \pm\sqrt{\frac{vv}{2n}} \tag{1}$$

The plane error of mean squares was m=±0.039m.

2）Elevation accuracy

There were also 39 elevation points to test the elevation accuracy with about 50m interval. The adjustment model is:

$$m_h = \pm\sqrt{\frac{vv}{2n}} \tag{2}$$

The elevation error of mean squares was m_h=±0.026m.

The accuracy of the correct and uncorrected point cloud was shown in table 2 and table 3.

Table 2 The accuracy of the uncorrected point cloud

ID	Coordinate of the point cloud for each teest point/m			High precision coordinate for each test point/m			Difference value /m		
	x	y	z	x	y	z	dx	dy	dz
k1	450820.863	4429665.474	9.289	450815.330	4429676.415	18.157	5.533	-10.941	-8.868
k2	450784.417	4429670.743	9.645	450778.865	4429681.748	18.307	5.552	-11.005	-8.662
k3	450753.897	4429680.675	8.641	450748.387	4429691.816	17.216	5.510	-11.141	-8.575
k4	450708.795	4429684.214	9.896	450703.440	4429695.139	18.570	5.355	-10.925	-8.674
k5	450662.131	4429696.363	9.853	450657.103	4429707.251	18.653	5.028	-10.888	-8.800
k6	450615.771	4429711.979	9.841	450611.198	4429722.941	18.838	4.573	-10.962	-8.997
k7	450549.091	4429738.812	9.716	450545.315	4429749.953	19.070	3.776	-11.141	-9.354
k8	450503.595	4429757.874	9.590	450500.334	4429769.121	19.257	3.261	-11.247	-9.667
k9	450457.285	4429777.331	9.400	450454.414	4429788.612	19.400	2.871	-11.281	-10.000
k10	450411.241	4429796.436	9.186	450408.724	4429807.907	19.553	2.517	-11.471	-10.367
k11	450391.484	4429809.837	8.051	450388.483	4429819.865	17.223	3.001	-10.028	-9.172
k12	450367.333	4429813.657	9.163	450365.095	4429825.402	19.844	2.238	-11.745	-10.681
k13	450323.067	4429828.739	9.820	450320.990	4429840.764	20.622	2.077	-12.025	-10.802
k14	450278.926	4429841.146	10.499	450276.860	4429853.337	21.431	2.066	-12.191	-10.932
k15	450259.061	4429851.228	9.784	450257.122	4429863.563	20.809	1.939	-12.335	-11.025
k16	450212.676	4429860.306	10.522	450210.584	4429872.660	21.580	2.092	-12.354	-11.058
k17	450188.104	4429858.719	11.724	450185.948	4429870.962	22.903	2.156	-12.243	-11.179
k18	450165.241	4429866.626	11.212	450163.289	4429878.977	25.074	1.952	-12.351	-13.862
k19	450142.611	4429863.514	12.565	450140.307	4429875.761	23.727	2.304	-12.247	-11.162
k20	450095.636	4429865.792	13.504	450093.120	4429877.915	24.563	2.516	-12.123	-11.059
k21	450048.922	4429865.973	14.659	450046.161	4429877.818	25.557	2.761	-11.845	-10.898
k22	449999.201	4429864.831	15.787	449996.243	4429876.461	26.337	2.958	-11.630	-10.550
k23	449926.547	4429868.170	16.311	449923.415	4429879.707	26.337	3.132	-11.537	-10.026

k24	449905.570	4429863.257	15.479	449902.477	4429874.561	25.122	3.093	-11.304	-9.643
k25	449862.476	4429862.292	15.866	449859.297	4429873.469	24.982	3.179	-11.177	-9.116
k26	449779.359	4429859.690	17.321	449776.131	4429870.826	24.893	3.228	-11.136	-7.572
k27	449757.617	4429863.218	18.557	449754.365	4429874.420	25.705	3.252	-11.202	-7.148
k28	449736.731	4429857.756	19.549	449733.451	4429868.690	26.463	3.280	-10.934	-6.914
k29	449579.518	4429848.067	21.226	449578.664	4429863.998	26.135	0.854	-15.931	-4.909
k30	449546.673	4429846.533	21.370	449539.136	4429862.532	26.091	7.537	-15.999	-4.721
k31	449505.237	4429845.483	20.680	449496.519	4429860.412	26.032	8.718	-14.929	-5.352
k32	449480.256	4429844.559	20.375	449476.059	4429864.290	25.135	4.197	-19.731	-4.760
k33	449454.078	4429843.427	19.532	449453.676	4429857.797	25.963	0.402	-14.370	-6.431
k34	449389.851	4429829.565	14.560	449380.964	4429856.355	23.206	8.887	-26.790	-8.646
						REMS		±13.623	±9.343

Table 3 The accuracy of the corrected point cloud

ID	Coordinate of the point cloud for each test point/m			High precision coordinate for each test point/m			Difference value /m		
	x	y	z	x	y	z	dx	dy	dz
t1	449817.692	4429872.170	24.904	449817.686	4429872.251	24.941	0.006	-0.081	-0.037
t2	449880.687	4429877.544	26.091	449880.691	4429877.738	26.092	-0.004	-0.194	-0.001
t3	449796.974	4429875.011	25.924	449796.998	4429875.225	25.903	-0.024	-0.214	0.021
t4	449712.447	4429873.752	25.799	449712.370	4429873.384	25.562	0.077	0.368	0.237
t5	450792.757	4429684.836	17.049	450792.754	4429685.013	17.001	0.003	-0.177	0.048
t6	450770.932	4429688.175	17.135	450770.966	4429688.289	17.098	-0.034	-0.114	0.037
t7	450739.322	4429688.148	18.380	450739.288	4429688.031	18.445	0.034	0.117	-0.065
t8	450724.880	4429695.888	17.324	450724.891	4429696.011	17.234	-0.011	-0.123	0.090
t9	450681.258	4429705.781	17.485	450681.266	4429705.902	17.379	-0.008	-0.121	0.106
t10	450635.333	4429719.663	17.624	450635.428	4429719.790	17.517	-0.095	-0.127	0.107
t11	450589.739	4429736.914	17.850	450589.846	4429737.008	17.739	-0.107	-0.094	0.111
t12	450524.262	4429764.478	18.098	450524.308	4429764.618	18.010	-0.046	-0.140	0.088
t13	450477.980	4429784.120	18.228	450478.038	4429784.274	18.122	-0.058	-0.154	0.106
t14	450433.458	4429803.091	18.388	450433.499	4429803.201	18.281	-0.041	-0.110	0.107
t15	450345.098	4429838.124	19.282	450345.146	4429838.297	19.157	-0.048	-0.173	0.125
t16	450299.980	4429852.399	20.086	450300.053	4429852.584	19.914	-0.073	-0.185	0.172
t17	450233.420	4429863.307	21.966	450233.374	4429863.147	22.019	0.046	0.160	-0.053
t18	450117.107	4429882.367	23.253	450117.168	4429882.505	23.229	-0.061	-0.138	0.024
t19	450069.377	4429883.235	24.068	450069.448	4429883.409	24.026	-0.071	-0.174	0.042
t20	449970.592	4429880.922	25.431	449970.594	4429880.935	25.427	-0.002	-0.013	0.004
t21	449946.172	4429875.051	26.740	449946.178	4429874.922	26.691	-0.006	0.129	0.049
						REMS		±0.039	±0.026

It got that, the plane accuracy of point cloud, which got only by combinative navigation, would be ±13.623m and the elevation one would be ±9.343m. This result would never meet the need of the completion surveying. While some control points were used to correct the point cloud, the results were improved greatly. And the point accuracy of the corrected point cloud was 0.039m, and the elevation one was 0.026m, which can meet the accuracy requirements of rail traffic completion surveying.

5 Conclusion

It developed a set of underground vehicle borne rail traffic laser scanning system and used it to finished one section rail transit completion surveying. One set of completion surveying technic rout based on this kind of vehicle borne scanning system was summarized. When there were no enough GPS signals in underground rail transit area, some control points can improve the point cloud accuracy greatly. This experiment area was one typical trail transit, and there were no spatial terrain which would influence the result. Another conclusion was the control point interval should be about 50m which could improve the point cloud accuracy well.

In order to verify the efficiency of the mobile measurement technology, the traditional surveying method was also used to complete this section rail transit. The practice showed that the vehicle borne time was 16 days. In the same environment, the traditional surveying method need about 20 days, the vehicle borne scanning technology compared with the traditional surveying method has obvious advantages. With a high precision integrated navigation technology advance in the future, the cost of this new technology would decline. The point cloud result can also be used to build the 3D scene model, to improve the data support for subsequent trail transit operations and management. So the vehicle borne scanning technology would have more advantages in the area of the underground rail transit construction completion surveying. It was the first experiment of the vehicle born laser scanning technology in China in the area of the underground rail transit, which can provide a reference for the other areas completion surveying and even 3D surveying.

References

[1] Liu Zhichao. Research on Completion and Acceptance Control Measurement Techniques of Guangzhou Metro Line [J]. Geomatics and Spatial Information Technology, 2014（07）

[2] Gao Zhiguo, Li Changhui. Research on 3D Completion Surveying Method Based on Terrestrial LiDAR[J]. Urban Geotechnical Investigation and Surveying, 2014(03)

[3] Xie Xiongyao, Lu Xiaozhi, Tian Haiyang, Ji Qianqian, Li Pan. Development of a modeling method for monitoring tunnel deformation based on terrestrial 3d laser scanning, 2012（11）

[4] Xing Hanfa, Gao Zhiguo, Lyu Lei. Application of 3D terrestrial laser scanner technology in building completion surveying [J]. Geotechnical Investigation and Surveying, 2014(05)

[5] Han Youmei, Yang Bogang. Calibration theory and Methods for Vehicle-borne Mobile Surveying system [M]. Surveying and Mapping Press, 2014.06

[6] Zhang Tao. The Key Techniques of Ultra-Tightly Integrated GPS/SINS Navigation System[D]. A Dissertation for the Doctor Degree, Harbin Engineering University, 2010.06 2011（2）

[7] Wang Liuzhao, Han Youmei, Zhong Ruofei. The taper scanning Angle calibration of 360° Laser scanner[J]. BULLETIN OF SURVEYING AND MAPPING，2010(9)

[8] Han Youmei1, Wang Liuzhao, Lu Xiushan1. Angle error calibration of 360° laser scanner[J]. Geotechnical Investigation and Surveying, 2011.2

[9] Cai Haiyong. Research on the combined navigation based on the distance measuring instrument [D], Capital Normal University, 2015.05

Acknowledgments

Authors wishing to acknowledge assistance or encouragement from colleagues of the Beijing Key Laboratory of Urban Spatial Information Engineering.

Considerations on Geospatial Big Data

Zhen LIU, Huadong GUO and Changlin WANG

Institute of Remote Sensing and Digital Earth, Chinese Academy of Sciences, Beijing 100094, China

E-mail: liuzhen@radi.ac.cn

Abstract. Geospatial data, as a significant portion of big data, has recently gained the full attention of researchers. However, few researchers focus on the evolution of geospatial data and its scientific research methodologies. When entering into the big data era, fully understanding the changing research paradigm associated with geospatial data will definitely benefit future research on big data. In this paper, we look deep into these issues by examining the components and features of geospatial big data, reviewing relevant scientific research methodologies, and examining the evolving pattern of geospatial data in the scope of the four 'science paradigms'. This paper proposes that geospatial big data has significantly shifted the scientific research methodology from 'hypothesis to data' to 'data to questions' and it is important to explore the generality of growing geospatial data 'from bottom to top'. Particularly, four research areas that mostly reflect data-driven geospatial research are proposed: spatial correlation, spatial analytics, spatial visualization, and scientific knowledge discovery. It is also pointed out that privacy and quality issues of geospatial data may require more attention in the future. Also, some challenges and thoughts are raised for future discussion.

1. Introduction

Currently, big data has been a hot topic worldwide covering academic, governmental and commercial communities. Typically, about 80% of datasets relate to a spatial location [1][2][3]. Scientists usually call this kind of data geospatial data. According to statistics, Google generates 25 PB of data per day, and geospatial data takes up a significant position among it [4]. By 2014, ESA alone had exceeded 1.5 PB of Earth observation data [5].

Researchers have been devoted to complicated technologies, architectures and applications in the big data landscape. Pioneering technologies, such as Apache Hadoop [1] and MapReduce [6], data infrastructures, and analytic tools, have been fully developed. However, few studies have been conducted to look into the evolution of geospatial big data with scientific research methodologies and the capabilities of geospatial data in the overarching (big data) landscape, which are the real driving components for the development of the technologies and architectures mentioned above. Miller and Goodchild [7] argued that we have entered into a data-driven era. It might be more useful and helpful to review and examine the nature of these data than simply rush to advance the relevant technologies in terms of extracting knowledge from huge geospatial data.

By reviewing current research findings, this paper proposes that geospatial big data has been experiencing a huge evolution in terms of its scientific research methodology and the potential rules of

[1] http://hadoop.apache.org

dealing with these geospatial data. Particularly, in the data-driven context, key research focuses are pointed out with regard to the future of research on geospatial big data.

We start by summarizing the components and features of geospatial big data (not general big data). Then, scientific research methodologies of geospatial big data are reviewed to examine the evolving pattern of geospatial data in the scope of the four Science Paradigms. Following that, some visions are discussed in terms of potential key research on geospatial data. At the end, we summarize some challenges and thoughts that might be instructive for future studies.

2. Definition of geospatial big data

Traditionally, geospatial data refers to geo-referenced data that correlates to Earth's environmental components and processes and further to the interaction between humans and Earth by using spatial technology assisted with ground station systems. The original generation of geospatial data and its outbreak into big data later on greatly benefited from the rapid development of remote sensing, computing, and information communication technologies, among others. It has been noted that the geospatial data increase has been such an explosive trend that it has outpaced the existing capacities and growth rates of storage, computing and analysis systems [8][9]. For example, the amount of remote sensing images produced by advanced airborne, satellite and ground-based remote sensing systems is increasing at the rate of one terabyte per day, and even a single image set can reach tens of gigabytes. Statistically, remote sensing data on a global scale is approaching the exabyte (EB) level (1 EB = 1024 PB, 1 PB = 1024 TB, 1 TB = 1024 GB) [5]. These geo-related, spatiotemporal data are almost always scientifically oriented and mainly controlled by governmental or commercial agencies (although current policies allow much open access to these data), and are normally labeled as 'authoritative' or 'official' [10][11].

If the explosive growth of spatiotemporal satellite data is attributed to the development of spatial technology, computing, communication, and other relevant technology, then the advancement of social networks, Web 2.0 and mobile devices and the policy of free public access to satellite images has really promoted the collection of public-contributed data [12][13]. During the last decades, some emerging data sources enabled the appearance of other forms of location-related data. These emerging geospatial data encompass *in situ* sensor network data (e.g., OpenStreetMap), GPS trace data from mobile devices, geo-social media data (e.g., Twitter), and crowdsourcing/VGI data [9][12][14][15][16][17][18]. These emerging geo-related data are mostly contributed by the volunteered public and largely correlate to the creator's motivation, behavior and circumstance. They are called 'user-generated' or 'volunteered' data [11][16][19].

With the emergence of 'user-generated' data, the concept of geospatial data has been stretched to a broader scope (see the differences in Table 1), and the amount of these data are in turn growing explosively. Along with the increasing amount of data, the analysis and computing technologies and rapid processing capability are accordingly required to constantly improve in order to match the rapid generation of huge geospatial data. In view of this, some researchers define geospatial big data as spatial datasets exceeding the existing capability of computing systems [9].

Table 1. The definition of geospatial data

	Label	Data scope	Receiving methods
Traditional geospatial data	Authoritative or official data	Geo-referenced spatiotemporal data	Spatial technology, e.g., satellite remote sensing
Emerging geospatial data	User-generated or volunteered data	Geo-related behavior-based data	Social networks, Web 2.0 and mobile devices

3. Features of geospatial big data

Many studies have been conducted on geospatial big data. However, most of them simply point out that geospatial data qualifies as big data because of the '3Vs' (or more) [7][8][9][10][20] with few descriptions of what the Vs of geospatial big data are. In this section, the features of geospatial big data are systematically discussed.

3.1. The Vs

As mentioned, geospatial big data has the features of big data, falling well within the '3Vs', namely *volume*, *velocity* and *variety* [21].

Volume – The images collected from Earth observing satellites contain rich information with high spatial, temporal and radiometric resolution, high acquisition rate, and short observing period. The higher the resolution is, the bigger the size of the image is. Considering that there are more than 500 satellites globally and some satellites have worked for decades (e.g., the Landsat satellites have served almost 30 years), the received satellite image data have become quantitatively enormous [22].

Velocity – The real-time or near real-time monitoring of satellites means a constant flow of image data, which demands high computing and analysis capabilities. Data from new sources, such as VGI, are based on users' interactivity, so that the value of these data can only be found and used when they are provided, processed and shared dynamically, almost in real-time [7][9].

Variety – (i) Geospatial data has three basic models: raster (grid, e.g., satellites images), vector (encompassing points, lines, and polygons), and graph (spatial network) [17][18]. (ii) Multiple sources of and approaches to data collection, such as the global Earth observation systems based on remote sensing technology and the *in situ* measurements of sensor networks, mobile devices, geo-related social media, and citizen sensors, have generated various types of geospatial data [22][23][24][25][26]. (iii) Some data are originally derived from sensors or software, but some are generated from complex operational/modeling systems. (iv) Heterogeneous data are produced in various formats, encompassing structured (tables and relations), unstructured (text and imagery), and semi-structured (auxiliary) data [7][22][27].

In addition, other 'Vs' have been proposed to define geospatial big data, such as *value*, *veracity*, and *visualization* [20][28].

Value – Scientific technologies have largely advanced to manage and process geospatial data by extracting the essential information (the valuable part) from redundant noise, so as to discover new insights and scientific knowledge [8][29]. However, it is hard to make each dataset valuable because of the huge volume and complexity. As for geospatial data, its big volumes have a very low value density, described as 'abundant data with scarce information' [29]. It is impossible and unnecessary to extract all information from huge amounts of data; to find the most valuable data, Turner *et al.* [30] proposed five criteria to define 'target-rich' data. Target-rich data: are easy to access, are real-time, have a large footprint, are transformative, and have 'intersection synergy'.

Veracity – Data accuracy has always been given attention for various uses in different fields. Satellite images and the processing procedure encompass many uncontrollable factors, and the credibility of social media data is concerning in terms of the measuring accuracy and certainty of the data [31]. The uncertainty of data is a result of it coming from many sources and including noise, deletions, inconsistencies and ambiguities. As Goodchild [32] argued, 'all location references are subject to uncertainty', and maybe the veracity issue can only be overcome through modeling and analyzing the huge collection of geospatial data.

Visualization – The term 'big data' was originally produced in the context of computer systems being challenged by visualization [33]. The spatial characteristic of geospatial data makes it possible and reasonable to transform and display the data onto the screen to enhance interactive processing with users. With the increase in size and dimension, and the demand for quick display shifting from 2D to 3D presentation, geospatial data also challenges the capacity of computer processing.

3.2. Beyond Vs

Besides the 'Vs', the unique characteristics of geospatial data should also be highlighted in terms of space and geography, as it is these characteristics that support the 'V' features of geospatial big data and are more important for people to discover the value and relations behind them.

Most basically, geospatial data has the properties of spatial auto-correlation (Tobler's First Law) and spatial heterogeneity [17]. Spatial auto-correlation means that the attribute values of geographic targets are correlated and more importantly there exists a neighborhood effect. Spatial heterogeneity refers to inconsistent observing results caused by spatial location differences when observing [34]. These features of geospatial big data are much more related to the differences in spatial location of observed targets themselves, rather than the aforementioned *variety* related to various sources or formats of the observing methods and tools.

Meanwhile, geospatial data have '3H' features: high dimensionality (including spatial, spectral, and temporal dimensions), high complexity (complex modeling and computing methods and systems), and high uncertainty (the uncertain sensing and empirical process) [22][35]. Compared to traditional spatial data, which are mostly static maps and regular or irregular survey data that statically describe Earth's surface, geospatial big data cover a more refined space and time granularity and are qualified with the typical liquid spatiotemporality. With these characteristics, the emergence of large amounts of data in a short period may lead to unlimited possibilities.

4. Evolving pattern

No matter how big the geospatial data are, the unprecedentedly advanced capabilities of acquiring, storing, processing, computing and analyzing geospatial data make the scarce-data era part of the past in humankind's understanding of Earth. We have entered into a data-rich era. While some scientists make progress on processing/computing technology, others are starting to think about how people should react to the 'exaflood' of data and learn to 'drink from a fire hose' [36][37]. This idea stirs up deep thinking about the evolving pattern of geospatial big data. It may be a good time for people to re-evaluate the role of data and make some changes when using data to solve problems.

Data, since the concept originally appeared, have always accompanied the evolution of science and technology. Looking at the four Science Paradigms makes us clearly aware of the relationship between the improvement of science and technology and data [38]. In the Empirical Paradigm, attempts were made to explain natural phenomena—it was a technology-driven phase seeking the accumulation of original data. Moving into the Theoretical Paradigm, models were used to generalize the empirical data—it was a theory-driven phase of verifying data. When entering into the Computational Paradigm, computers were applied to simulate complex phenomena to find any possible rules—it was a model-driven phase of describing and predicting trends. In 2007, the concept of data-intensive scientific discovery was proposed as the Fourth Paradigm—it is a data-driven phase of knowledge discovery (See Table 2).

Table 2. The role of data in the four Science Paradigms

Science Paradigm	Driving Pattern	Aims
Empirical	Technology-driven	Seeking data to explain natural phenomena

Theoretical	Theory-driven	Verifying data
Computational	Model-driven	Simulating to find rules
Data intensive	Data-driven	Discovering knowledge

The four Science Paradigms show, just as Miller and Goodchild [7] argued, that it is reasonable to describe the development of data as an evolution rather than a revolution. Below we shall look deeply at the evolving pattern of geospatial data from a data-seeking to a data-driven phase. Specifically, the data-intensive paradigm will be examined to find the role of data in knowledge discovery.

- *From hypothesis to data*

According to classical scientific methodologies, questions or hypotheses (and maybe predictions) were created before executing experiments [39]. Data, which were produced through methods or tools, were used to support or explain what had been queried, hypothesized or predicted. In this sense, the motivation of people seeking data was largely attributed to them having something in mind to test or explain. In this paradigm, the role of data in scientific research is passive.

Because traditional measuring methods might have been time-consuming, especially prior to the satellite era, the obtained geographic data would have been very limited and scarce in terms of efficiency, scale, and comprehensiveness, compared to that in the big data era. Considering the limitations of data acquired by traditional methods, the usage of data was strongly target-oriented. These ‘sampling’ data were collected to solve certain problems [40], which could lead to the trend of data being typical, not in generalization [7]. Maybe it is necessary to argue about how valid problems are solved by using these data and about the generality of them [7].

- *From data to questions*

Since the launch of the first Earth observing satellite, *Nimbus*, in 1964, there have been 514 Earth observing satellites as of 2011[5]. Many countries have proposed initiatives for Earth observation based on remote sensing satellite technology. With the establishment of global Earth observing satellite systems and the advanced capability of multi-scale, real-time dynamic monitoring, satellite data has exploded. With these abundant data on hand, users are allowed to examine datasets for themselves to find the useful ones for their particular research purposes.

As for location data from emerging sources (e.g., social media and mobile devices), they are mostly voluntarily contributed by the public. Any users of social media or mobile devices can be a geo-located data generator and provider. None of them can tell the possible future uses of their contributions in advance. They just generate and share their data through geo-social media or sensor networks individually. However, these individual volunteered location data are compiled into a big data pool and only a set of the data from the pool can possibly assist in solving problems through new technologies, e.g., OpenStreetMap or eco-routing systems.

It has become a trend that authoritative data and emerging volunteered data might be generated and shared without precise, pre-set hypotheses or problems that need to be tested or to solved. They are quite often collected and stored with a general purpose. For example, some satellite images are aimed at monitoring environmental changes, but to monitor what kind of environmental changes (e.g., water resources, soil use, or vegetation cover), and where, is pendent. Environmentalists and biologists are likely to use data with different spatiotemporal characteristics when conducting their specific research. Taking another example, VGI data is generated and circulated by individuals with few or no exact purposes. But these data may help governmental agencies efficiently make decisions on traffic issues, or assist the public with travel plans.

Currently, big populations of data are not solely generated for usage in a certain project. Although some thematic and scientific satellites are designed based on pre-set tasks, these data are automatically received and could be used for other purposes or in other fields. Users are free to access data and to extract the information they are concerned about. In this sense, compared to 'sampling' data's passive mode in scientific research, the 'population' property of geo-related data is more active and powerful in driving people to discover knowledge [40].

- *Challenges*

The shift from 'hypothesis to data' to 'data to questions' coincides so well with geographical thinking that it may be much more reasonable to explore the world in a 'bottom to top' view [41]. Instead of talking about how to collect data to solve existing problems, people increasingly think about what kinds of problems could be addressed by using the collected data. This helps to find the generality of big data, and also leads to its broader usage in the future.

Normally, with excessive focus on computing capacity, models, and algorithms in specific fields, scientists pay little attention to the generality of explosively growing geospatial data. This can result in scientific hypotheses or questions being left to those specific scientists to think about and solve. With the increasing accumulation of scientific hypotheses and questions, the common scientific questions (or generality) of big data can be extracted and proposed in hopes of overcoming scientific barriers [42].

When faced with big spatial data, it is easy to imagine a scenario where researchers are overwhelmed with choices, like a hungry man at a buffet. How do they select the right data from the big data pool either to address a certain problem or to make a decision? How do they acquire effective information or knowledge from such a big data pool? Maybe these are issues that only data scientists can address. Maybe we need to establish auxiliary information classification or filtering mechanisms to assist with information retrieval and knowledge discovery.

5. Future research on geospatial big data

In the scarce-data era, the use of computers made processing and analyzing data convenient, and somehow needed a growing amount of data to satisfy and verify advancements. This changed dramatically when entering into the big data era due to the ability to collect data no longer being a problem. The computer loses control of the data to some extent when data comes like an 'exaflood' [36][37]. Moreover, scientific and technical methodologies strongly improved to meet the demands caused by the growing amount of data. No longer hindered by the capabilities of collecting and computing, geospatial data enabled people to exploit and discover knowledge. In this sense, big data became a driving force, and plays a pivotal role together with technologies in helping human beings meet various challenges.

Undoubtedly, we have entered into a data-driven phase in geographic research. Given the circumstance that data collection and relevant processing technologies are not the terminal goal of big data research, how to effectively use these data has become the core of attention. In this section, four potential research areas for geospatial data are proposed.

- *Spatial Correlation*

It has been discussed that the correlation among data is more valuable in scientific knowledge discovery than the data themselves or even models that describe these data [35]. Differing from traditional logical inference, it is more suitable to employ analytic induction to find correlation or relevance in huge volumes of data by using searching, comparing, clustering, and classifying technologies. Through correlation analytics, association networks that might be hidden in the data can be figured out [42].

In terms of geospatial big data, it is argued that research on geographic science has developed beyond the phase of explaining phenomena or findings (knowing *why*), and rather focuses more on correlations of the data collected from *in situ* experiments, sensor monitoring, computer generation, and even individual observations. It is the correlation that enables geographic scientists to produce the detection and representation patterns of data and to make possible predictions [7]. As discussed before, more than 80% of data are related to location [1][2][3], making it meaningful to examine the spatial correlation of geospatial data based on their location information.

- *Spatial Analytics*

Traditional spatial analytics refers to using appropriate statistical analysis and artificial intelligence algorithms to analyze geospatial data, extract useful information, and summarize the general process. With the increase in magnitude of geospatial data, traditional methods for spatiotemporal analysis of geospatial big data have been unable to meet demands, and must be improved. Spatial partitioning, multi-dimensional data structures, static and dynamic load balancing, multiple iterations, and modeling algorithms should be taken into account [17].

Given that geospatial data consists of both location and time information, integrating and analyzing timely data (e.g., VGI data) makes it possible to detect the data generator's mobility habits and model their routes. This capability benefits greatly from advanced trajectory modeling technologies [43][44][45][46]. Along with this behavioral information, geospatial data also contain a wealth of other social information. Some researchers argue that existing models cannot meet the requirements of processing non-spatiotemporal data that accompany the raw geospatial data [47]. With the involvement of social network activities, human behavioral patterns and the context within which data are generated, high-dimensional spatial analysis models are needed. Such advanced models may be complex, but could be more precise in representing the reality of the data and help people extract useful information and form basic knowledge from the raw geospatial data.

- *Spatial Visualization*

Geospatial data is characterized by high dimensionality, including high spatial, spectral and temporal dimensions. With the continuous development of computing graphics, imaging techniques and data acquisition techniques (e.g., LiDAR), the ability to collect spatial-topological information has been gradually strengthened. The increase in size and dimension of geospatial data demands the enhancement of spatial data visualization. Although 3D and 4D technology has been developed, traditional representation and visualization methods are still largely restricted by the two-dimensional display of computers. In addition to computing technology, key technologies, such as multi-dimensional spatial data models, spatiotemporal integration management technology, and 3D spatiotemporal integration modeling methods, are needed to visualize geospatial big data [48].

- *Scientific Knowledge Discovery*

The fourth paradigm of data-intensive scientific discovery raises new concerns about analyzing and mining big data in terms of relevant correlation and rules in order to find, new knowledge and even new rules that the previous scientific methodologies could not discover [35]. As for geographic knowledge, it was initially derived from ground-based observations and measurements. These data are limited mostly at temporal and spatial scales. Geospatial data, achieved by space-air-ground monitoring technology and systems, provides long-term, multi-scale (from local, regional to global), study area-oriented data for geographic research. Based on rich, multi-source data, advanced computing and analysis technology entail the task of value exploration.

From big data, to data-intensive scientific data, and then to geospatial big data, a variety of models and algorithms have been generated for analysis and interpretation. But rather than only practicing computing capability, the ultimate target of acquiring data is to describe reality, discover knowledge, support decision making, and understand the real world.

6. Challenges

Although the collection of massive geo-related data provides unprecedented opportunities for human beings with abundant information and knowledge to understand Earth, we have to admit that geospatial big data also brings challenges for privacy and data quality issues.

As discussed in the Introduction, authoritative geospatial data are largely controlled by governmental and commercial agencies. The large-scale commercial mode of utilizing these geospatial data runs the risk of reducing necessary, proper applications. In spite of the increasingly open policies for access to satellite data and free mechanisms for sharing data to reduce such risks to some extent, there still exists a challenge caused by open data policies in terms of privacy. Furthermore, because data are contributed from crowdsources or individual volunteers, the reproducibility and privacy of data also create challenges [17].

In addition, multi-source geospatial data raise a critical problem with data quality, which turns out to be a prominent challenge when analyzing and using these data [49]. The massive Earth observation data from aerial and satellite remote sensing technology are quite varied in terms of their technical parameters, storage formats, image resolutions, and observation scales (in both space and time). This leads to some problems in using multi-source heterogeneous geospatial data. Similarly, user-generated geospatial data, such as VGI data, are based on public input. They are timely, but challenge the balance between efficiency and quality. At the same time, public-contributed data are a necessary supplement to scientific data. The availability of such timely, low-cost data brings significant changes to research in social sciences as well as natural sciences [50][51]. However, considering the identities of data contributors and the environment in which data are collected, the data become suspicious in terms of authenticity, credibility and reliability [52][53][54]. Taking OpenStreetMap as an example, discussions have continued about the position accuracy, completeness, and currency [55][56][57][58][59][60].

7. Conclusion

Geospatial data, as an important portion of big data, has gained widespread attention. Depending on the different collection sources and methods, geospatial data can be defined in different scopes. Aside from the '3Vs' (and other 'Vs'), geospatial big data has its own unique features.

In the big data era, data processing and analysis technologies have been strongly driven by massive data. Compared to the acquisition of data, the capability of extracting information and discovering knowledge from data seems to have greater requirements. Accordingly, the role of geospatial big data in scientific methodology has changed from being produced to test hypotheses to being exploited to discover knowledge, i.e., from 'hypothesis to data' to 'data to questions'.

We pointed out four main future research areas of geospatial data, covering spatial correlation, analytics, visualization and scientific knowledge discovery. Finally, there are privacy and quality issues caused by geospatial big data and a new set of challenges to meet in the future.

8. References

[1] Densham P J and Goodchild M F 1989 Spatial decision support systems: a research agenda *Proc. GIS /LIS'89 (Orlando FL)*

[2] Shekar S and Xiong H 2007 *Encyclopedia of GIS* (New York: Springer)

[3] Li Q and Li D 2014 Big Data GIS *Geomatics and Information Science of Wuhan University* **39** 6

[4] Vatsavai R R, Ganguly A, Chandola V, Stefanidis A, Klasky S, Shekhar S 2012 Spatiotemporal data mining in the era of big spatial data: algorithms and applications *Proc. 2nd ACM*

SIGSPATIAL Int. Workshop on Analytics for Big Geospatial Data (Redondo Beach CA) pp 1–10

[5] He G J *et al* 2015 Processing of earth observation big data: challenges and countermeasures *Chin. Sci. Bull.* **60** pp 470–478

[6] Dean J and Ghemawat S 2008 MapReduce: simplified data processing on large clusters *Commun. ACM* **51** 1 pp 107-113

[7] Miller H J and Goodchild M F 2014 Data-driven geography *GeoJournal*

[8] Xu C and Yang C 2014 Introduction to big geospatial data research *Ann. GIS* **20** 4 pp 227-232

[9] Lee J G and Kang M 2015 Geospatial big data: challenges and opportunities *Big Data Res.* http://dx.doi.org/10.1016/j.bdr.2015.01.003

[10] Schade S 2015 Big data breaking barriers – first steps on a long trail *Proc. 36th Int. Symp. on Remote Sensing of Environ*ment *(Berlin, Germany)*

[11] Bakillah M, Lauer J, Liang S H L, Zipf A, Arsanjani J J, Mobasheri A and Loos L 2014 Exploiting big VGI to improve routing and navigation services *Big Data Techniques and Technologies in Geoinformatics* ed. H A Karimi (CRC Press)

[12] Heipke C 2010 Crowdsourcing geospatial data *ISPRS J. Photogramm. Remote Sens.* **65** pp 550–570

[13] Salk C, Sturn T, See L, Fritz S and Perger C 2016 Assessing quality of volunteer crowdsourcing contributions: lessons from the cropland capture game *Int. J. Digital Earth* **9** 4 pp 410–426

[14] Ramm F, Topf J and Chilton S 2011 *OpenstreetMap: using and enhancing the free map of the world* (England: UIT Cambridge)

[15] Howe J 2006 The rise of crowdsouring *Wired Magazine* **14** 6

[16] Goodchild M F 2007 Citizens as sensors: the world of volunteered geography *GeoJournal* **69** 4 pp 211–221

[17] Cugler D C, Oliver D, Evans M R, Shekhar E S and Medeiros C B 2015 Spatial big data: platforms, analytics, and science *GeoJournal*

[18] Evans M R, Oliver D, Zhou X and Shekhar S 2014 Spatial Big Data Case Studies on Volume, Velocity, and Variety *Big Data Techniques and Technologies in Geoinformatics* ed H A Karimi (CRC Press)

[19] Coleman D, Georgiadou Y and Labonté J 2009 Volunteered geographic information: the nature and motivation of producers *Int. J. Spatial Data Infrastructures Res.* **4** pp 332–358

[20] Nativi S, Mazzetti P, Santoro M, Papeschi F, Craglia M and Ochiai O 2015 Big Data challenges in building the Global Earth Observation System of Systems *Environmental Modelling & Software* **68** pp 1–26

[21] Laney D 2001 3D Data Management: Controlling Data Volume, Velocity and Variety (Gartner) Available at: http://blogs.gartner.com/doug-laney/files/

[22] Song W, Liu P, Wang L and Lv K 2014 Intelligent Processing of Remote Sensing Big Data: Status and Challenges, *J. Eng. Stud.* **6** 3 pp 259–265

[23] Miller H J 2007 Place-based versus people-based geographic information science *Geography Compass* **1** 3 pp 503–535

[24] Miller H J 2010 The data avalanche is here. Shouldn't we be digging? *J. Regional Sci.* **50** 1 pp 181–201

[25] Sui D and Goodchild M F 2011 The convergence of GIS and social media: Challenges for GIScience *Int. J. Geographical Inf. Sci.* **25** 11 pp 1737–1748

[26] Townsend A 2013 *Smart cities: Big data, civic hackers, and the quest for a new utopia* (New York: Norton)

[27] Dumbill E 2012 What is big data? An introduction to the big data landscape, http://strata.oreilly.com/2012/01/what-isbig-data.html

[28] Fromm H and Bloehdorn S 2014 Big Data-Technologies and Potential *Enterprise-Integration* ed G Schuh and V Stich (Berlin: Springer) pp 107–124

[29] Tao X, Hu X and Liu Y 2013 Review on Big Data Research *J. System Simulation* **25** (Supplement) pp142–146.

[30] Turner V, Gantz J F, Reinsel D and Minton S 2014 The Digital Universe of Opportunities: Rich Data and the Increasing Value of the Internet of Things *IDC Analyze the Future* http://www.emc.com/leadership/digital-universe/2014iview/high-value-data.htm

[31] Castillo C, Mendoza M and Poblete B 2011 Information Credibility on Twitter *WWW 2011, Hyderabad*, March 28–April 1

[32] Goodchild M F 2010 Twenty years of progress: GIScience in 2010 *J. Spatial Inf. Sci.* **1** pp 3–20

[33] Cox M and Ellsworth D 1997 Application-controlled demand paging for out-of-core visualization *Proc. IEEE 8th Conf. on Visualization*

[34] Shekhar S, Evans M R, Kang J M and Mohan P 2011 Identifying patterns in spatial information: A survey of methods *Wiley Interdisciplinary Reviews: Data Mining and Knowledge Discovery* **1** 3 pp 193–214

[35] Guo H D *et al* 2014 Scientific big data and digital Earth *Chin. Sci. Bull.* **59** pp 1047–1054

[36] Sui D, Goodchild M F and Elwood S 2013 Volunteered geographic information, the exaflood, and the growing digital divide *Crowdsourcing geographic knowledge* ed D Sui, S Elwood and M F Goodchild (New York: Springer) pp 1–12

[37] Waldrop M M 1990 Learning to drink from a fire hose *Science* **248** 4956 pp 674–675

[38] Gray J 2009 eScience: a transformed scientific method *The fourth paradigm: Data-intensive scientific discovery* ed T Hey, S Tansley and K Tolle (Microsoft Research)

[39] Gauch H G 2003 *Scientific method in practice* (Cambridge University Press)

[40] Mayer-Schonberger V and Cukier K 2013 *Big Data* (Houghton Mifflin Harcourt Publishing Company)

[41] Goodchild M F 2012 Invigorating GIScience www.geog.ucsb.edu/~good/presentations/ucgismay12.pptx

[42] Li G and Cheng X 2012 Big data research: the significant strategic areas of the future development of the science, technology, economy and society - the research status of big data and scientific thinking *Proc. Chin. Acad. Sci.* **27** pp 647–657

[43] Ashbrook D and Starner T 2003 Using GPS to learn significant locations and predict movement across multiple users *Personal and Ubiquitous Computing* **7** 5 pp 275–286

[44] Quannan L, Zheng Y, Xing X, Yukun C, Wenyu L and Ma W Y 2008 Mining user similarity based on location history *Proc.17th ACM SIGSPATIAL Int. Conf. on Advances in Geographic Information Systems* **34** 1–10

[45] Giannotti F, Nanni M, Pinelli F and Pedreschi D 2007 Trajectory pattern mining *Proc. 13th ACM SIGKDD Int. Conf. on Knowledge Discovery and Data Mining (San Jose, CA)* pp 330–339

[46] Yu Z, Zhang L Z, Xie X and Ma W Y 2009 Mining correlation between locations using human location history *Proc.17th ACM SIGSPATIAL Int. Conf. on Advances in Geographic Information Systems (Seattle, WA)* pp 472–475

[47] Assam R and Seidl T 2014 Insights and Knowledge Discovery from Big Geospatial Data Using TMC-Pattern *Big Data Techniques and Technologies in Geoinformatics* ed H A Karimi (CRC Press)

[48] Li J 2014. Research on key techniques of 3D visualization of complex spatial data *J. Scientific and Technical Entrepreneurship*

[49] Goodchild M F 2013 The Quality of Big (Geo) Data *Dialogues in Human Geography* **3** 3 pp 280–284

[50] King G 2011 Ensuring the Data-Rich Future of the Social Sciences *Science* **331** pp 719–721

[51] Goodchild M F and Li L 2012 Assuring the quality of volunteered geographic information *Spatial Statistics* **1** pp 110–120

[52] Elwood S, Goodchild M F and Sui D 2013 Prospects for VGI Research and the Emerging Fourth Paradigm *Crowdsourcing Geographic Knowledge: Volunteered Geographic Information (VGI) in Theory and Practice* ed D Sui, S Elwood and M F Goodchild (New York: Springer) pp 361–375

[53] Tang J C, Cebrian M, Giacobe N A, Kim H W, Kim T and Wickert D B 2011 Reflecting on the Darpa Red Balloon Challenge *Communications of the ACM* **54** 4 pp 78–85

[54] Flanagin A J and Metzger M J 2008 The credibility of volunteered geographic information, *GeoJournal* **72** pp 137–148

[55] Haklay M 2010 How good is volunteered geographical information? A comparative study of OpenStreetMap and Ordnance Survey datasets *Environment and Planning B: Planning and Design* **37** pp 682–703

[56] Haklay M, Basiouka S, Antoniou V and Ather A 2010 How many volunteers does it take to map an area well? The validity of Linus' Law to volunteered geographic information *The Cartographic J.* **47** 4 pp 315–322

[57] Neis P, Zielstra D and Zipf A 2011 The street network evolution of crowdsourced maps: OpenStreetMap in Germany 2007–2011 *Future Internet* **4** 4 pp 1–21

[58] Canavosio-Zuzelski R, Agouris P and Doucette P 2013 A photogrammetric approach for assessing positional accuracy of OpenStreetMap roads *ISPRS Int. J. Geo-Information* **2** 2 pp 276–301

[59] Hecht R, Kunze C and Hahmann S 2013 Measuring completeness of building footprints in OpenStreetMap over space and time *ISPRS Int. J. Geo-Information* **2** 4 pp 1066–1091

[60] Asahara A, Maruyama K, Sato A and Seto K 2011 Pedestrian-movement prediction based on mixed Markov-chain model *Proc. 19th ACM SIGSPATIAL Int. Conf. on Advances in Geographic Information Systems* (*Chicago, IL*) pp 25–33

Acknowledgments
This study is supported by the Open Foundation of the Key Laboratory of Digital Earth Sciences, Chinese Academy of Sciences (project number 2014LDE001).

Wave height possibility distribution characteristics of significant wave height in China Sea based on multi-satellite grid data

W Han and J Yang

The First Institute of Oceanography, SOA, Qingdao, China
E-mail: yangjg@fio.org.cn

Abstract. This paper discusses the group of wave height possibility distribution characteristics of significant wave height in China Sea based on multi-satellite grid data, the grid SWH data merges six satellites (TOPEX/Poseidon, Jason-1/2, ENVISAT, Cryosat-2, HY-2A) corrected satellite altimeter data into the global SWH grid data in 2000~2015 using Inverse Distance Weighting Method. Comparing the difference of wave height possibility distribution of two schemes that scheme two includes all of 6 satellite data and scheme one includes all of other 5 satellite data except HY-2A in two wave height interval, the first interval is [0,25) m, the second interval is [4,25) m, finding that two schemes have close wave height probability distribution and the probability change trend, there are difference only in interval [0.4, 1.8) m and the possibility in this interval occupies over 70%; then mainly discussing scheme two, finding that the interval of greatest wave height possibility is [0.6, 3) m, and the wave height possibility that the SWH is greater than 4m is less than 0.18%.

Keywords: Significant wave height; China Sea; Satellite grid data; Possibility distribution

1. Introduction

Significant wave height (SWH), defined as the crest-to-trough height of the 1/3 highest waves in the sea wave field of view, is an important parameter of ocean research and marine environment forecast. SWH measurements are used for a variety of purposes, these studies improve the understanding of the physical processes responsible for wind-wave evolution to the validation and calibration of models for waves and other ocean processes as well as wave climate investigations, which have implications for shipping and offshore engineering projects[1].

Wave data over periods of many years are necessary for estimating the wave climate all over the world, traditionally, data coverage over the ocean has been poor, and wave observations have come from ships and moored buoys clustered around coastlines. During the 1970s, global monitoring of SWH became a reality and resulted in the first global atlas of SWH data derived from satellite observation [6]. Since then, the SWH data from several different satellites have become available, including SEASAT (1978), GEOSAT (1985-1990), TOPEX/POSEIDON (1992-2006), ERS-1 (1991-2000), ERS-2 (1995-2011), ENVISAT (2002-2012), Jason-1 (2002-2013)[7], Jason-2 (2008-present), Cryosat-2 (2010-present), HY-2 (2011-present), Saral/Altika (2013-present), Jason-3(2016.01.17), SENTINEL-3A (2016.02.16)[2].

In this paper, first, introducing the satellite altimeter data and multi-satellite grid data, next, representing wave height possibility statistics analysis method, then, comparing the difference of wave height possibility distribution of two schemes and mainly analyzing the wave height possibility of scheme two, finally, the last part states some conclusions.

2. Data

Merging SWH data comes from 6 satellite altimeter data that are distributed by the various space agencies, the detailed information of satellite altimeter is given in Table1[1]. SWH data from these satellite altimeters is corrected through regression equations, merging corrected SWH data into global grid SWH data using Inverse Distance Weighting(IDW) method[3,4], the grid data starts Jan. 1, 2000, ends to Sep. 1, 2015, the spatial resolution is 0.25° × 0.25° [5], temporal resolution is one day. Selecting the grid data in China Sea that is 100~130° E, 0~40° N.

Table 1. Basic Information for satellite radar altimeter[2]

Satellite	T/P	Jason-1	ENVISAT	Jason-2	Cryosat-2	HY-2A
Operational Period	1992.8.10-2005.10.18	2001.12.7-2013.07.03	2002.03.01-2012.04.08	2008.06.20-	2010.04.08-	2011.08.16-
Data Period	1992.09.25-2005.0924	2002.01.15-2012.12.11	2002.05.17-2012.04.08	2008.07.04-2015.09.01	2010.07.16-2014.09.12	2011.10.01-2015.08.16
Orbital Altitude/km	1336	1336	799.8/782.4	1336	717	971/973
Inclination /°	66	66	98.5	66	92.03	99.34
Spacing at equator /km	315	315	80	315	7.5	207.6/18.3
Repeat /days	9.9156	9.9156	35/30	9.9156	369	14/168
Institute	NASA,CNES	NASA,CNES	ESA	NASA,CNES	ESA	NSOAS

3. Methods

In order to possibility statistics analysis[8], first defining statistics parameters, assume M_h is the total number that SWH belongs to [h, 25) m, [$SWH_{h,i-1}$, $SWH_{h,i}$) , i=1,2,..., I_h is SWH the interval [h,25) m(the maximum measurement value of satellite altimeter SWH is 25m), I_h is the total number of interval [h, 25) m, $SWH_{h,i-1}$ and $SWH_{h,i}$ are the left and right SWH threshold value in the group i in the interval [h, 25) m, $N_{h,i}$ is the number of SWH in the group i in interval [h, 25) m, $P_{h,i}$ is the frequency of SWH in the group i in interval [h, 25) m , $F_{h.,i}$ is the accumulative frequency of SWH in the group i in interval [h, 25) m, the expressions are as follows:

$$N_{hi} = 0+\sum_{j=1}^{M_{ii}} 1, SWH_{hi-1} \le SWH_{hj} < SWH_{hi} \tag{1}$$

$$P_{hi} = \frac{N_{hi}}{M_{hi}} \tag{2}$$

$$F_{hi} = \sum_{i=1}^{I_h} P_{hi} \tag{3}$$

using above statistical models discuss probability distribution of merging grid SWH, the internal [0,25) m and [4,25) m are selected, that is $h_1=0$, $h_2=4$.

4. Results and discussions

Using IDW method merges 6 satellite altimeters according to two schemes, one scheme is foreign satellite altimeter data including T/P, Jason-1/2, ENVISAT and Cryosat-2, other scheme adds HY-2A base on scheme one.

Selecting grid SWH data that starts Oct.1, 2011 due to HY-2A data starts from Oct. 1, 2011. Here h=0, the interval is [0, 25) m, grouping is that using interval range 0.2m divide the interval [0,4.8) m into 24 groups, the interval [4.8,25) m is the 25th group. Calculating the wave height possibility of two schemes and its possibility difference, it is showed in figure 1, it reveal two schemes have the close wave height probability distribution and the probability change trend, there are difference only in interval [0.4,1.8) m, the possibility in this interval occupies over its' total 70%, the wave height possibility of scheme two is less than the wave height possibility of scheme one in this interval, the result is opposite in interval [1.0,1.8) m, this phenomenon demonstrates that the wave height probability in interval [0.4,1.0) m is becoming lower and is becoming higher in interval [1.0,1.8) m.

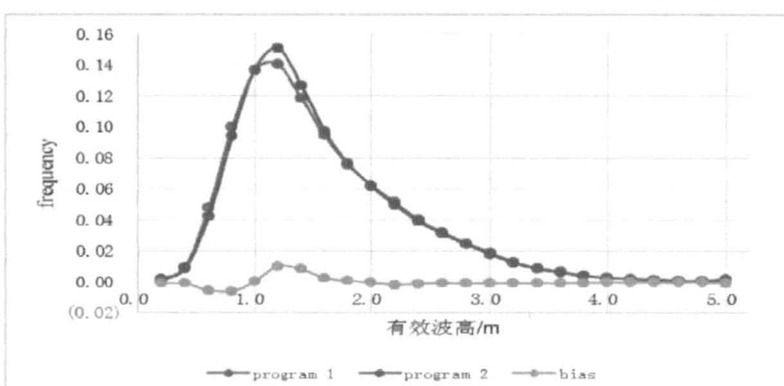

Figure 1. wave height possibility distribution of two schemes

Analyzing the SWH that the wave height is greater than 4m, here h=4, the group interval is [4, 25) m, the grouping is that using interval range 1.0m divide the interval [4, 25) m into 21 groups, calculating the wave height possibility of two schemes and its possibility difference as show in Figure 2, there are difference in the interval [4, 7) m between two schemes, this interval is the interval of higher SWH, the wave height possibility of scheme two is greater than it of scheme one in interval [4, 5) m, whereas, scheme two is becoming smaller in interval [5, 7) m. Above all, the wave height possibility is greater in interval [4, 5) m after adding HY-2A and is becoming smaller in other interval.

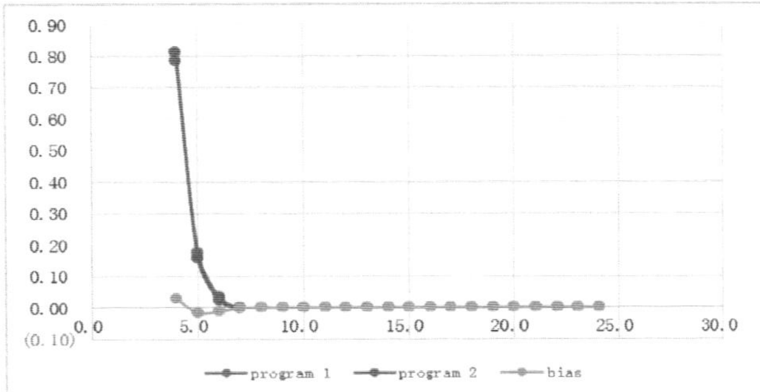

Figure 2. wave height possibility distribution of two schemes in interval [4,25)m

Above the discussion between two schemes could find that the difference of SWH in the interval [0, 2) m after adding HY-2A, the change trend of bias like Sine function. New HY-2A data could illustrate wave height possibility, so, it is essential to analyze the long term wave height possibility distribution of scheme two. The wave height possibility of SWH shows in Figure 3, data starts from Oct. 1, 2000 to Sep. 12, 2015 is close 16 years, the SWH mainly distributes the interval [0.6, 3) m, and the sub-peak distribution interval is [0, 0.2) m, the greatest wave height possibility is greater than 14%, the interval is [1,1.2) m .

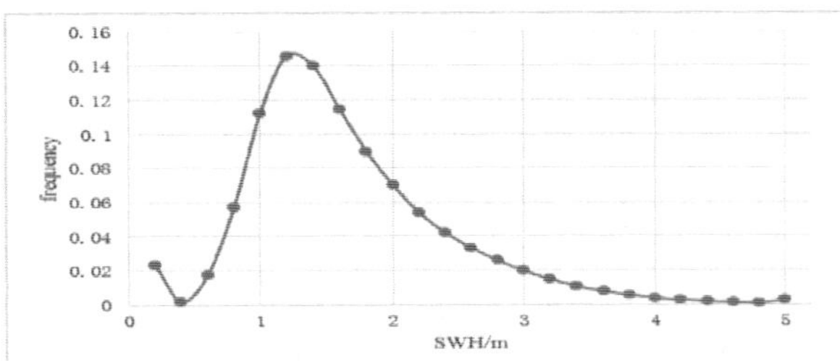

Figure 3. wave height possibility distribution in the interval [0,5)m in 2000~2015

Figure 4 mainly discusses that the wave height possibility of every interval is less than 0.18% while SWH is greater than 4m, and is almost 0 while SWH is greater than 8m, so, in general, we always focus on the SWH that is lower than 8m, above is the reason.

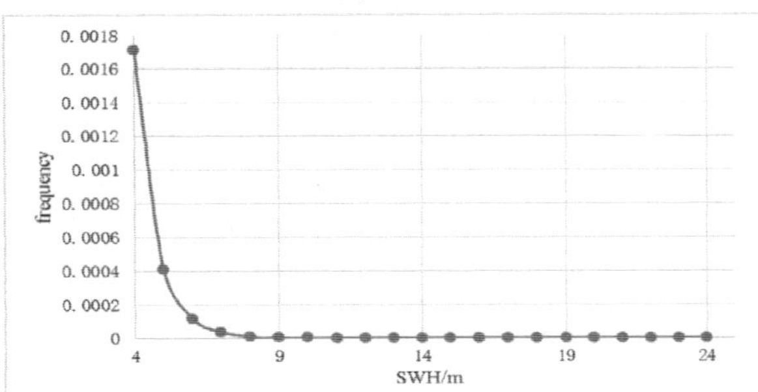

Figure 4. wave height possibility distribution in the interval [4,25) m in 2000~2015

5. Conclusions

This paper discusses the wave height possibility of two SWH grid schemes in two intervals, finding that the interval of greatest wave height possibility is [0.6, 3） m, and the wave height possibility that the SWH is greater than 4m is less than 0.18%. Above two schemes have the close probability distribution and the probability change trend, there are difference only in interval [0.4, 1.8） m, the possibility in this interval occupies over 70%; the possibility is greater in interval [4, 5） m and smaller in other interval after adding HY-2A; main distribution interval of SWH is [0.6, 3） m and possibility is less than 0.18% in interval [4, 7） m.

References

[1] Birol F Roblou L Lyard F et al 2006 Towards using satellite altimetry for the observation of coastal dynamics *Proceedings of Symposium on 15 Years of Progress in Radar altimetry* 614.

[2] Li J and Jin T 2013 On the main progress of satellite altimetry and its applications *Journal of Geomatics* **38(4)** 1-8

[3] Queffeulou P Croizé-Fillon D Queffeulou P et al 2010 Global altimeter swh data set Version 7 *Biochemistry & Oral Biology* **100(1)** 1-4

[4] Li X Yuan F Jia C et al 2012 Contrastive study of s-a multifractal filtering method based on inverse distance weighted and kriging interpolation *Science of Surveying & Mapping*

[5] Shanas P Kumar V Hithin N 2014 Comparison of gridded multi-mission and along-track mono-mission satellite altimetry wave heights with in situ near-shore buoy data *Ocean Engineering* **83(2)** 24-35

[6] Mcmillan J 1981 A global atlas of geos-3 significant wave height data and comparison of the data with national buoy data

[7] Durrant T Greenslade D Simmonds I 2009 Validation of jason-1 and envisat remotely sensed wave heights *Journal of Atmospheric & Oceanic Technology* **26(1)** 123-134

[8] Shanas P Kumar V Hithin N 2014 Comparison of gridded multi-mission and along-track mono-mission satellite altimetry wave heights with in situ near-shore buoy data *Ocean Engineering* **83(2)** 24-35

Acknowledgements

The authors thank the European Space Agency for providing ENVISAT RA-2 altimeter data, National Aeronautics and Space Administration (NASA) for providing T/P, Jason-1/2 data, National Data Buoy Center (NDBC) in the National Oceanic and Atmospheric Administration (NOAA) for providing NDBC buoys and National Satellite Ocean Application Service of State Oceanic Administration for providing HY-2 data. This study is supported by National 863 Program (No.2013AA122803), the National Marine Public Welfare Industry Special Funding Scientific Research Program (No.201305032) and Chinese National Natural Science Foundation (No.41406007).

A classification algorithm based on Cloude decomposition model for fully polarimetric SAR image

Hongmao Xiang[1], Shanwei Liu[2], Ziqi Zhuang[2] and Naixin Zhang[2]

[1]Shandong Province Institude of Land Surveying and Mapping, 2301 Jingshi Road, Jinan 250102, China
[2]China University of Petroleum (East China),66 Changjiang West Road , Qingdao 266580, China

E-mail:shanweiliu@163.com

Abstract. Remote sensing is an important technology for monitoring coastal zone, but it is difficult to get effective optical data in cloudy or rainy weather. SAR is an important data source for monitoring the coastal zone because it cannot be restricted in all-weather. Fully polarimetric SAR data is more abundant than single polarization and multi-polarization SAR data. The experiment selected the fully polarimetric SAR image of Radarsat-2, which covered the Yellow River Estuary. In view of the features of the study area, we carried out the H/α unsupervised classification, the H/α -Wishart unsupervised classification and the H/α -Wishart unsupervised classification based on the results of Cloude decomposition. A new classification method is proposed which used the Wishart supervised classification based on the result of H/α -Wishart unsupervised classification. The experimental results showed that the new method effectively overcome the shortcoming of unsupervised classification and improved the classification accuracy significantly. It was also shown that the classification result of SAR image had the similar precision with that of Landsat-7 image by the same classification method, SAR image had a better precision of water classification due to its sensitivity for water, and Landsat-7 image had a better precision of vegetation types.

Key words: polarimetric MAR; Cloude decomposition; coastal zone; classification

1. Introduction

Coastal zone is the area between the sea and land. It is rich in natural resources and has always been in the center of people's attention. Remote sensing technology has always been playing an important role on coastal zone research. However, the climate of coastal zone is cloudy and rainy, optical data is difficult to obtain in this area so that inconvenience comes to extraction and supervision. SAR is a kind of active microwave remote sensing sensor and it can work all-weather and all-day. Therefore, SAR data can be an effective data to replace optical data in the coastal zone.

With the development of SAR technology, the working mode of SAR has been gradually turned into the multi-polar, multi-band imaging mode with more information. Applications of SAR are more extensive, including military, agriculture etc. Image classification is an important application of SAR data[1]. Foreign research about SAR image classification has achieved remarkable results. Fukuda(2001) proposed the SVM classification method[2]. Lee(2004) proposed a classification method based on Freeman decomposition and Wishart classifier [3]. Singh (2011) used PCA to analyse spatial texture, and extracted the most effective texture feature for SAR image classification[4]. Ataollah(2011) combined ANN and genetic algorithm for PolSAR image classification, and achieved higher classification accuracy than MLC and Wishart method[5]. Kayabol(2012) proposed an unsupervised classification algorithm based on hierarchical clustering and statistical expectation maximization for high resolution SAR images[6].

SAR technology started late in China, the present study focuses on the improvement of foreign advanced algorithm. Chen Jin-song(2004) proposed a NN classification method based on the target decomposition for full polarization SAR image[7]. Liu Xiu-qing(2004) proposed an iterative classification method based on unsupervised classification[8]. Zhang Xiang(2013) combined different decomposition methods to extract classification features, then through SVM classification experiment concluded that information of coherent and non-coherent decomposition are complementary[9]. Chen Jun(2014) proposed a supervision classification method based on the Pauli decomposition and SVM[10]. This paper chose a Radarsat-2 image that covers Yellow River estuary area to research on classification method of SAR image with characteristics' information of coastal zone features. We performed four kinds of classification methods on Cloude decomposition result image, including the H/α classification, H/α-Wishart classification, H/A/α-Wishart classification, and a new Wishart classification which based on the H/A/α-Wishart classification result, and evaluated these methods. We also classified a Landsat-7 image covering the same region to analyse the respective advantages and causes of optical and SAR images.

2. Data and Pretreatment

2.1. Data

Radarsat-2 data is used in this study. The work mode is fine full polarization mode, and the working band is C band, while the incident angle is from 25.79°~27.62° with the distance to the resolution of 12 meters, azimuth resolution to 8 meters. The data type is single look complex data (SLC) with the range of 37.59°N~37.88°N and 118.92°E~121.79°E, imaging time on July 3, 2009.

2.2. Data Pretreatment

2.2.1 Filtering processing

The existence of speckle noise is the most basic feature of SAR image, which is the inevitable principle defect of SAR system. The existence of noise seriously affects the quality of the image and causes difficulties on image interpretation, so this article chosen enhance Lee method[11] of "3 * 3" window to filter the SAR image.

2.2.2 Polarization target decomposition

The features of PolSAR data is mainly random scatterer, so incoherent polarization target decomposition has a wider application, so this article used Cloude decomposition (incoherent decomposition) method for the SAR image after filtering processing to get three components - entropy, scattering angle and anti-entropy, which showed as in figure 1.

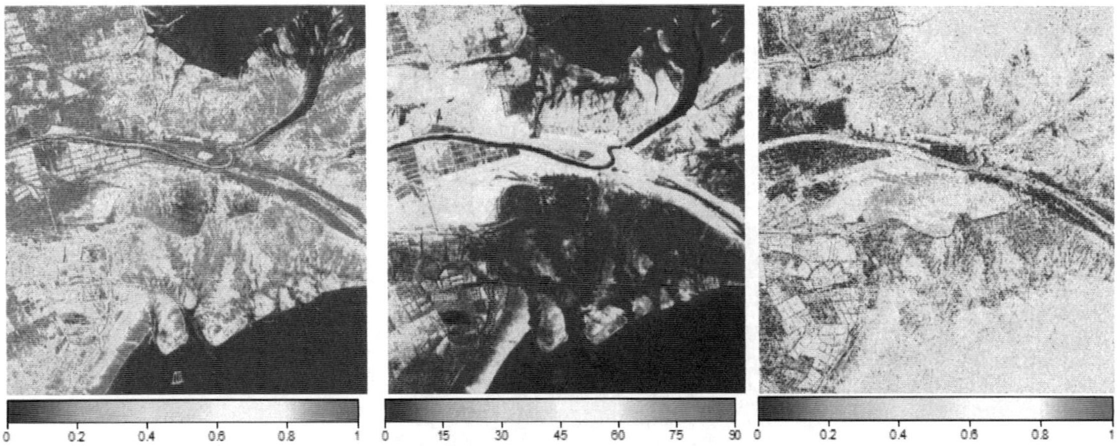

(a) Scatter entropy H (b) Average alpha angle (c) Anti-entropy A

Figure 1. the feature quantity from Cloude decomposition

From figure 1 (a), it can be seen that the scattering entropy of water body and water body-related features is small, and the scattering entropy of different vegetation types is larger. From the physical meaning of scattering entropy, it can be concluded that the water body can be regarded as a point target. The scattering process is more stable and the scattering type is single, while the different types of vegetation can be regarded as random scatterers. The scattering process consists of many pure target linear combinations and more chaotic, scattering type is more complex. From figure1 (b), it can be seen that the average scattering angle of different types of vegetation is larger, and the average scattering angle of water body and water body-related feature type is smaller. This is mainly because the scattering type of the water body is isotropic surface scattering, and the scattering type of vegetation is body scattering or even scattering. From figure1 (c), it can be seen that the entropy of water body and water body-related features is relatively large, while the anti-entropy of different types of vegetation is small.

3. Experiment

SAR image feature classification can be divided into two categories - supervised classification and unsupervised classification. Supervised classification is mainly anthropogenic selection of object categories and unsupervised classification is mainly clustering categories or specific types of scattering. At present, the mainstream polarization SAR remote sensing classification method is based on the classification of polarization target decomposition. In this paper, H / α -Wishart classification and $H / A / \alpha$ -Wishart classification are used to classify Cloude decomposing images based on $H / A / \alpha$ -Wishart classification. Unsupervised classification results were then subjected to Wishart supervised classification. And then by visual interpretation and quantitative evaluation to compare the advantages and disadvantages of these four different classification methods. Finally, in order to compare the advantages and disadvantages of SAR image and optical image in feature extraction, this paper chooses an optical image to classify and contrast the experiment.

3.1. Unsupervised classification based on Cloude decomposition image

(1) H/α unsupervised classification

The experimental data were subjected to H / α unsupervised classification. The classification results are shown in Figure 2.

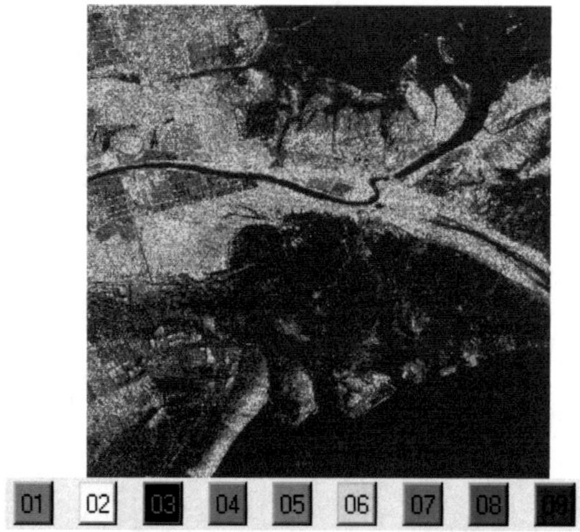

(a) H / α unsupervised classification images

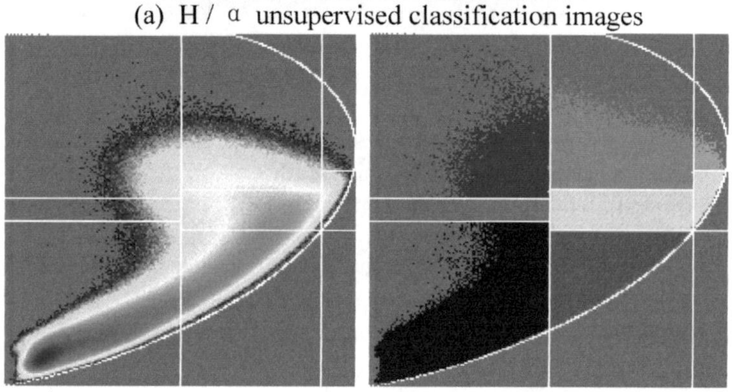

b) H_alpha_occurence (c) H_alpha_segmented

Figure 2. H / α unsupervised classification

Figure 2 (b) (c) is the distribution of the features in the H / α plane and the coloring of the images in the classification. From Figure2 (b), it can be seen that the features of this area are mainly concentrated in low scattering entropy and medium scattering entropy. From the classification results in Figure 2 (a), the overall classification of this classification method is not effective, and the obvious obscuration phenomenon is found in the seawater, river, pond, tidal flat mud and Suaeda salsa. Other vegetation is simply divided into two categories. The classification result is determined by the classification principle of H / α unsupervised classification. The H/ α plane classification is classified according to the target scattering mechanism. The backward scatter characteristics of the water bodies such as seawater and rivers are similar in this area, and the wetland of Suaeda salsa is characterized by high water content in the soil and water Characteristics, and therefore also be divided into this category. The other vegetation types had similar scattering characteristics. The reed marshes were classified as one species because of their high plant height and scattering type, which were distinguished from the dwarf plant species such as arable land.

(2) H/α-Wishart unsupervised classification

The H / a-Wishart unsupervised classification of the test data is shown in Figure 3.

It can be seen from Figure3 ming area is clearer. However, it is still not good to classify the types of water bodies, such as sea water, rivers, ponds and aquaculture areas, which are closely related to water bodies.

(3) H/A/α-Wishart unsupervised classification

The H / A / α -Wishart unsupervised classification of the test data is shown in Figure4.

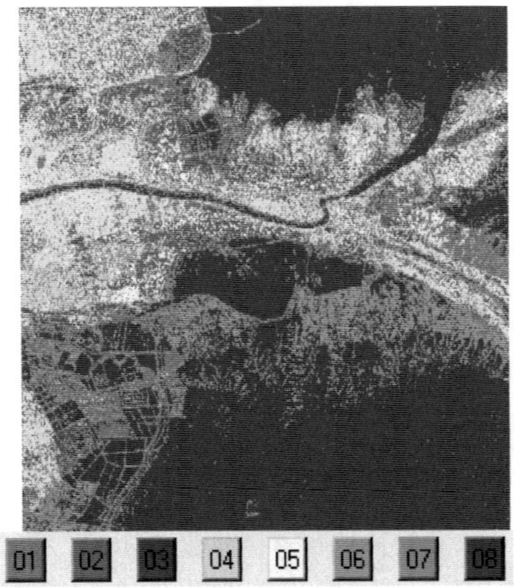

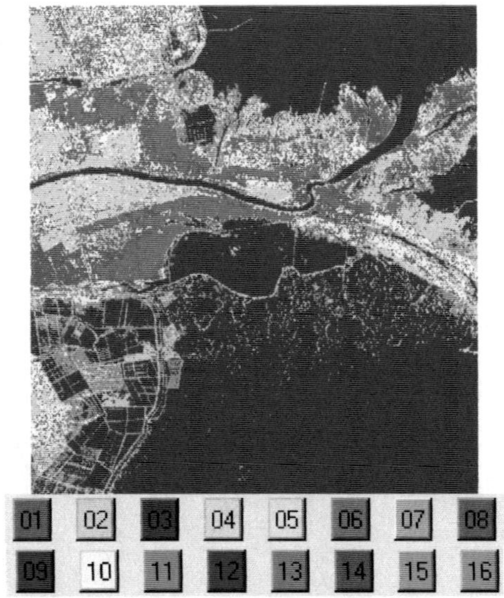

Figure 3. H / a-Wishart unsupervised classification image **Figure 4.** H / A / α -Wishart unsupervised classification image

It can be seen from Figure4 that different types of vegetation have serious misclassification phenomenon. The introduction of anti-entropy A scattering component makes the information increase, resulting in information redundancy, and the types of ground objects are divided into too many types. Water, rivers, aquaculture areas and other water-related features or is divided into a class. Suaeda salsa can be clearly separated from adjacent seawater and ponds.

3.2. Supervised classification on unsupervised classification result

Three kinds of unsupervised classification above are all based on the scattering characteristics of ground features, proning to produce erroneous classification phenomenon. To solve this problem, this paper improved classification methods above, proposed a Wishart supervised classification method that based on the H/A/α-Wishart unsupervised classification result. The main idea of this method is to perform the Wishart supervised classification on the H/A/α-Wishart unsupervised classification result, combining it with the prior knowledge. The classification results are as shown in Figure 1, in which nine kinds of ground feature have better classified effect.

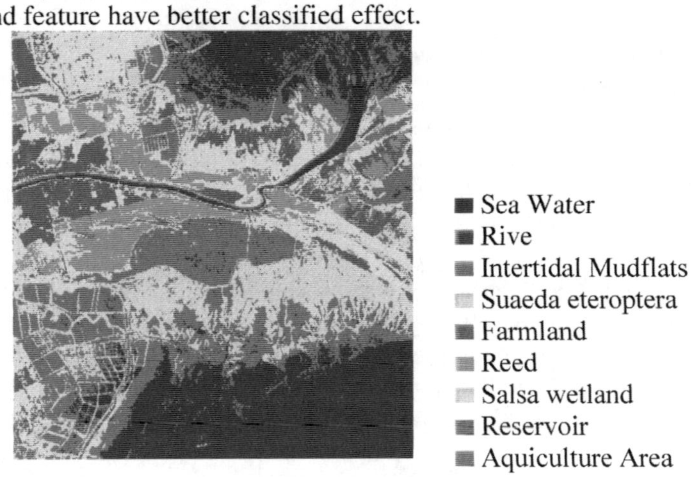

Figure 5. SAR image classification result
(Wishart classification based on H/A/α-Wishart classification).

3.3. Classification of Landsat7 image

In the optical contrast experiment, a Landsat7 image of August 31, 2009 was selected. The imaging data is similar to the SAR data imaging time, and the variation of the surface features is small, and the water level of the Yellow River Estuary is high. The classification result is comparatively high. Because of the vegetation coverage in the study area, the 4, 3, 2-band combination of vegetation sensitivity was selected in the Landsat7 image, and the Wishart monitoring based on the H / A / α - Wishart unsupervised classification. The classification results are shown in figure6.

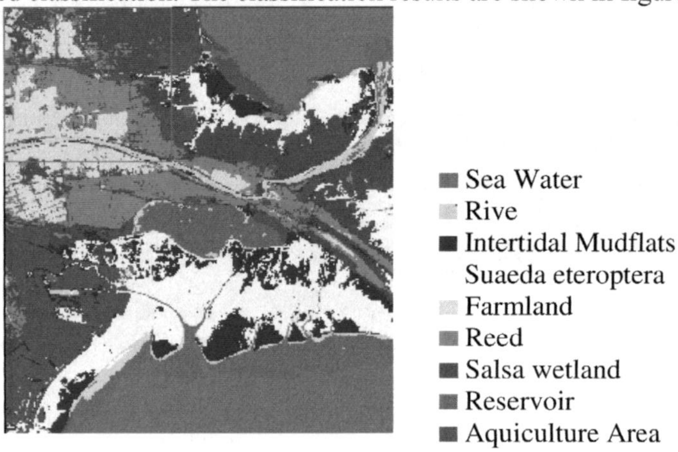

Figure 6. Landsat7 image SVM supervised classification result Results and analysis

It can be seen from Figure6 that 4, 2, 2-band pseudo-color synthetic images of the overall classification results are better, reed marsh, Suaeda wetlands, cultivated land, breeding and other objects have been more accurately divided. However, there is still some misclassification in the ground surface. Part of the sap was broken up into tidal flat mud, and some seawater and river were mistakenly divided into ponds.

4. Results and analysis

4.1. Comparison and analysis of SAR image classification results

For the three unsupervised classification methods, the H/ α -Wishart classification was significantly improved compared to the H / α classification. The different types of vegetation were well classified. The advantage of this method over the former was vegetation type and the Wishart measure is more sensitive to vegetation. The overall effect of H/A/ α -Wishart classification is between H / α classification and H / α -Wishart classification. The biggest advantage of this method is that the classification effect of Suaeda salsa is improved.

On the whole, the supervised classification based on unsupervised classification results can effectively overcome the shortcomings of the above three kinds of unsupervised classification, and the nine types of objects are well classified. According to the classification accuracy (Table 1), it can be seen that the accuracy of classification of other landform types is more than 90%, and the cultivated land is more than 85%. The average precision and accuracy are more than 97% ,which fully demonstrates the validity of Wishart supervised classification method based on H / A / α -Wishart unsupervised classification results.

Table 1. SAR image classification result (Wishart classification based on H/A/α-Wishart classification).

object	Accuracy(%)	average accuracy(%)	weighted accuracy(%)
Sea Water	95.79		
River	99.74		
Intertidal Mudflats	96.71		
Suaeda Heteroptera	100.00		
Farmland	87.52		
Reed	100.00		
Salsa wetland	99.67		
Reservoir	100.00		
Aquiculture Area	99.22	97.63	97.13

4.2. Comparison and analysis of SAR image and optical image classification results

Both of optical image and SAR images have achieved good classification results. According to the classification results (Table 2),we can found the classification accuracy of all classes reached more than 90%, and the average accuracy and weighted accuracy reached more than 96%. Comparing with the classification accuracy of SAR image and optical image,we can see there points.
(1) For water-related objects' classification, SAR image is better than optical image, such as Sea Water, River, Reservoir, Intertidal Mudflats, Suaeda Heteroptera.
(2) For vegetation objects' classification, optical image is better than SAR image, such as Aquiculture Area, Suaeda Wetland, Reed, Farmland.
(3) The average accuracy and weighted accuracy of SAR image reached more than 97%, the average accuracy and weighted accuracy of optical image reached more than 96%, both of them have achieved satisfactory classification accuracy.

Table 2. optical image classification result.

	Accuracy(%)	average accuracy(%)	weighted accuracy(%)
Sea Water	90.54		
River	93.73		
Intertidal Mudflats	90.23		
Suaeda Heteroptera	100.00		
Farmland	95.64		
Reed	97.13		
Salsa wetland	100.00		
Reservoir	99.84		
Aquiculture Area	99.74	96.32%	96.22%

5. Conclusions

(1) Four types of Radarsat-2 fully polarized images have been classified by Cloude decomposition using four different methods, including H / α classification, H / α -Wishart classification and H / A / α -Wishart classification. Unsupervised classification method, and a Wishart supervised classification based on H / A / α -Wishart unsupervised classification results. Among them, Wishart supervised classification accuracy based on H / A / α -Wishart unsupervised classification is more than 97%, which is the best classification method.
(2) By comparing the effects of polarimetric SAR images and optical images on the classification of coastal landforms, it is found that the classification accuracy of the two types of images is similar, but the SAR images are more accurate for classification of water-related features Optical images are more accurate for vegetation classification. This is because SAR images are sensitive to water and can be

used to classify water-related features. And the optical image uses 4,3,2 band combination, and it can highlight the performance of vegetation characteristics, therefore, it has more accurate classification of vegetation types.

Reference

[1] WuYong-hui, Ji Ke-feng, Yu Wen-xian. Comparition of Classification Performance od Full-, Dual- and Single-Polarization SAR Images Using SVM[J]. Journal of Remote Sencing,2008, 12(1): 46-52.

[2] Zhou Xiao-guang, Kuang Gang-yao, Wan Jian-wei. A Review of Polarimetric SAR Image Classification[J].Signal Processing, 2008, 24(5): 806-812.

[3] J.S.Lee, M.R.Grunes, E.Pottier.Unsupervised terrain classifi, cation preserving Polarmetric scattering characteristics[J]. IEEE Trans.Geosci.Remote Sensing, 2004, 42(4):722-731.

[4] Mandeep Singh , Gunjit Kaur.SAR Image Classification Using PCA and Texture Analysis[J].Information Technology and Mobile Communication , 2011, vol.147:435-439

[5] Ataollah Haddadi G., Mahmodreza Sahebi.Combination of GA and ANN to High Accuracy of Polarimetric SAR Data Classification[J].Advances in Computational Intelligence, 2011, vol.6691:207-214

[6] Koray Kayabol, Vladimir A. Krylov, Josiane Zerubia.Unsupervised Classification of SAR Images Using Hierarchical Agglomeration and EM[J].Computational Intelligence for Multimedia Understanding, 2012, vol.7252:54-65

[7] Chen Jin-song, Shao Yun, Li Zheng. Neural Networks Classification of Quad-polarization SAR Data Based on Target Decomposition ABSTRACT[J]. Journal of Image and Graphics, 2004, 9(5): 552-556.

[8] Liu Xiu-qing, Yang Ru-liang. The Iteration Classification Method and Experiment Study Based on Unsupervised Classification of Fully Polarimetric SAR Image[J]. Acta Electronica Sinica, 2004, 12(32): 1982-1986

[9] ZHANG Xiang, DENG Ka-zhong, FAN Hong-dong, ZHAO Hui. PolSAR SVM supervised classification method combining with polarimetric target decomposition[J]. Application Research of Computers, 2013, 30(1): 295-298.

[10] CHEN Jun, DU Pei-jun, TAN Kun. A Supervised Classification Algorithm Based on Pauli Decomposition and SVM on Polarimetric SA R Image[J]. Science Technology and Engineering, 2014, 14(17): 104-108.

[11] Sun Nan, Wang Yan-fei, Zhang Bing-chen. Multi-look Polarimetric SAR Speckle Filtering Based on Unsupervised Classification[J]. Journal of Electronics & Information Technology, 2008, 1(30): 220-223.

Acknowledgments

This research has been supported by the high resolution special project for remote sensing monitoring and application demonstration in coastal zone and the project of global change and air-sea interaction.

Comprehensive evaluation and analysis of ecological environment quality ofLaoshan Natural Reservebased on Remote Sensing

Wenlian Cui[1], Shanwei Liu[2], Yan Liu[2], Yanling Wang[1] and Naixin Zhang[2]

[1]Qingdao Environmental Monitoring Center, No.41Yan'an One Road, Qingdao 266003, China
[2]China University of Petroleum (East China), No.66Changjiang West Road , Qingdao 266580, China

E-mail:shanweiliu@163.com

Abstract. We used the remote sensing images of 2001, 2005 and 2010, statistics of reservoir water quality, air quality data, precipitation data and population data to evaluate and analyze the ecological environment quality of Laoshan Natural Reserve. In this decade, the ecological environment of tourism scenic area in Laoshan Natural Reserve becomes significantly better than that of the surrounding area, and it is the urban sprawl and increase of cultivated land area that resulted in the reduction of the scenic plants; Reservoir water quality was stable, but PH value and total nitrogen content still did not meet the standards because of the use of the sewage and pesticide fertilizer in the neighborhood; Air quality decreased slightly, however, the situation of acid rain had improved; Residential population continued to grow in Laoshan district and scenic tourists have increased, so human activity has become the main impacting factor of ecological environment of Laoshan Natural Reserve.

1. Introduction

Laoshan Natural Reserve is located in Qingdao, and has the most complex plant species in the same latitude of China. This area possesses a high ecological conservation and scientific research value. In recent years, with the exploitation of the natural reserve and its surroundings, the disturbance by human activities is becoming more and more serious.

In the early 90's of the last century, Strobel etc. evaluated the estuary eco-environment of Virginia[1]. Some researchers combined the human ecological footprint model with the ecological system evolution model for evaluation of urban ecosystem[2]. Some used GIS technology to evaluate the effects of agricultural production on the environment of Clay Mona province in Italy[3]. In China, with the development of remote sensing and GIS technology, the ability of getting, processing, and analysing information can provide information services to monitor the human environment. Liu Zhuang, Xie Zhiren etc. established the evaluation system of eco-environment in Jiangsu province[4]; Ma Ronghua used TM remote sensing data in 1998to evaluate the ecological types in Hainan province[5]; Zhang Xiuying and Zhao Chuanyan used MCA method for potential eco-environment evaluation research of Loess Plateau[6]; Du Suojun, Yin Yimin etc. evaluated the eco-environment of Zhang Jiagang[7]; Zheng Hongping, Bai Liang etc. evaluated the eco-environment quality of Fujian province and analysed the dynamic changes[8]. Jiang Zhenlan, Sha Jinming etc. built the eco-environment evaluation model in Fuzhou[9]. Chen Tao, Xu Yao etc. evaluated the

IOP Conf. Series: Earth and Environmental Science **46** (2016) 012061 doi:10.1088/1755-1315/46/1/012061

eco-environment quality of Sichuan province and established the local eco-environment comprehensive evaluation system[10].

In conclusion, remote sensing technology has become an important means of study on ecological landscape pattern. In this paper, we used three Landsat-TM images to monitor the Laoshan Natural Reserve from 2001 to 2011, and integrated multi-source data to evaluate the environmental quality synthetically.

2. Study area and data

2.1. Natural environment conditions

Laoshan is located in the northeast of Qingdao city, the south of Shandong peninsula. Its eastern side is close to the yellow sea. It is located in the middle latitudes, which have the continental monsoon climate. Both sides of Laoshan are close to the sea, and the length of the coastline is 87.3 kilometres long. The coastal landscape is almost marine abrasion landform. There are many plant species in the region, such as forest vegetation, economic crops, economic trees, crops, etc.

2.2. Remote sensing data

The remote sensing data used in this paper is Landsat-7 image in 2001, Landsat-5 image in 2005 and Landsat-5 image in 2010.They are as shown in figure 1.

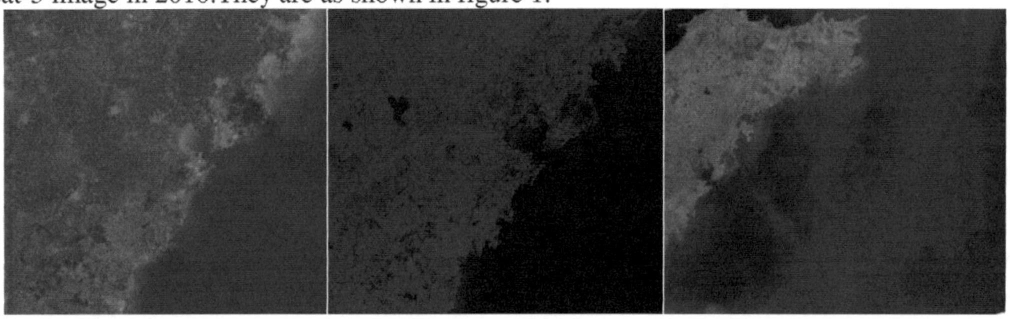

a. Landsat-7 (2001) b.Landsat-5(2005) c.Landsat-5 (2010)

Figure 1. Remote sensing data

2.3. Other data

We also collected the monitoring data of water quality from four reservoirs in the natural reserve (table 1), the air monitoring data, the statistical data of precipitation and demographic (table 2).

Table 1. Monitoring data of water quality

	2001		2005		2010	
Reservoir	water quality classification	Standard components	water quality classification	Standard components	water quality classification	Standard components
Dahedong	III	PH, total nitrogen	II	PH, total nitrogen	II	total nitrogen
Dashicun	III	null	II	null	II	null
Quanxinhe	II	null	I	null	I	null
Liuqinghe	II	total nitrogen	II	total nitrogen	II	total nitrogen

Table 2. Monitoring data of air, precipitation, demographic

year	air quality				precipitation data		demographic
	SO_2	CO	NO_2	PM_{10}	pH value in rain	acid rain（%）	quantity
2001	0.34	0.90	0.013	0.095	5.19	29.79	191881
2005	0.053	1.3	0.025	0.090	6.27	0.0	214634
2010	0.045	1.3	0.025	0.090	6.43	0.0	241818
standard value	0.06	4	0.04	0.1	/	/	/

3. Extraction and analysis on the types of landscape ecology

3.1. Landscape ecological types extraction and characteristic analysis

We pre-processed remote sensing image of the three years (2001, 2005, 2010), and used GIS software to make the graph of landscape ecological classification system (figure 2). The ecological classification system is shown in table 3.

Table 3. Ecosystem classification system

	Level 1 class	Level 2 class
1	Forest	Mixed broadleaf-conifer Forest
2	Water	Reservoir Canal
3	Plough	Farmland
4	Artificial surface	Residence Industrial land Transportation land
5	Bare soil	Bare mountain

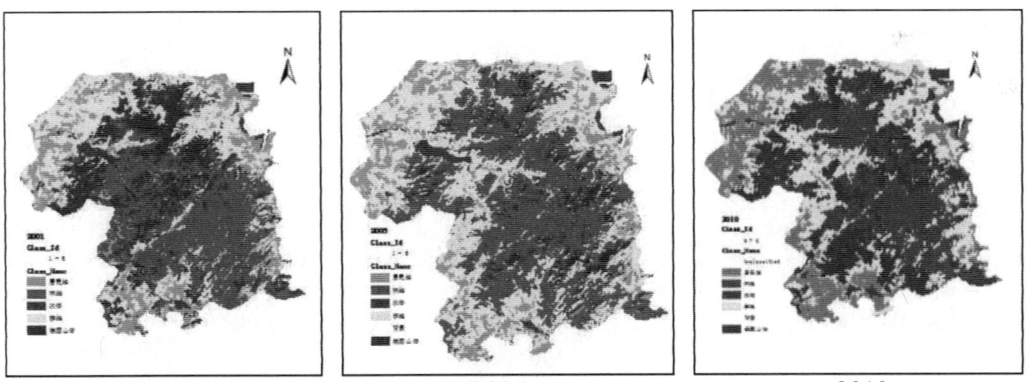

a.2001 b.2005 c.2010

Figure 2. Extraction results of landscape ecological types

3.2. Analysis of characteristics of landscape pattern

Table 4 is the table of ecosystem landscape area change for ten years in Laoshan Natural Reserve. From the proportion of landscape, we found that forest accounted for maximum, and farmland followed by, while others accounted for smaller and water accounted for minimum. From the situation of area change, we found the area of forest has been reduced during ten years, and the area of residence (including industrial land and transportation land) has increased obviously. The area of farmland is constantly increasing, and the area of water had no change. Setting 2000 to 2005 as the first phase and 2000 to 2005 as the second phase, the area of forest reduced in both phases, but the forest reduction amplitude in the first phase is less than the second phase; the farmland increase amplitude in the second phase is obvious higher than that of the first phase; the increased area of residence is the same as farmland, and the increase amplitude in the second phase is higher.

Table 4. Decade change table of ecosystem landscape area

	2001		2005		2010	
Type	Area(ha)	Percentage(%)	Area (ha)	Percentage(%)	Area(ha)	Percentage(%)
Forest	26316.28	56.10	21605.05	51.87	23524.12	50.01
Plough	11749..77	26.04	12618.78	27.96	13239.73	28.15
Water	1132.47	2.51	1184．03	2.62	1179.97	2.51
Residence	5926.22	15.35	7920．48	17.55	9093.42	19.33

3.3. Analysis of change driving mechanism

The area of forest in the Laoshan Natural Reserve trended to reduce in this decade, but the decreasing rate has been declining, which shows that people are paying attention to the survival condition of the natural vegetation in the Laoshan Natural Reserve and have already begun to take some measures to protect the vegetation. As can be seen from the table, the effect of forest protection turned out to be good enough after 2005.

The surrounding areas of Laoshan scenic spot had a certain degree of urban sprawl. Residence, industrial land, transportation land, and farmland increased in a certain extent. The increase inevitably reduced the area of natural vegetation, but the area of water reduced little.

The main factor that influenced the ecological landscape change has always been the human activities. From the population data of Laoshan Natural Reserve, we can found the population of Laoshan Natural Reserve rendering out the trend of continuous growth during ten years. The population density growth results in the needs of more human housing and food, and the area of residence, industrial land, transportation land, and farmland are constantly increasing. All this leads to the natural vegetation in natural reserve under threat. In the 10 years from 2000 to 2010, the increase of the area of residence and farmland is at the cost of large areas reducing of vegetation resources. Therefore, the most important factor that effects ecological landscape conversion of Laoshan Natural Reserve is human's driving force.

4. Comprehensive assessment of eco-environment quality

We comprehensively evaluated and analysed the eco-environmental quality of Laoshan Natural Reserve with the information of precipitation, air quality, reservoir water quality, and population change.

(1)Analysis of reservoir water quality

We trimed the reservoir water quality data of Laoshan Natural Reserve -- the water quality of Dashicun reservoir is standard, but the PH value and total nitrogen are excessive; the water quality of Dahedong reservoir belongs to a good level; the water quality of Liuqinghe reservoir is good, but the total nitrogen is excessive; the water quality of Quanxinhe reservoir achieves class Ⅰ standard.

Overall the water quality in study area is good and all achieve class Ⅱ standard of surface water (GB3838-2002). But some overweight indicators reflect the local reservoir water has a certain level of pollution. Among them, the exceeding PH value is the reason for the surrounding dam and reservoir construction; the exceeding total nitrogen shows that the reservoir water quality has been the situation of the eutrophication in a certain extent, which is mainly caused by human factors. From population variation of Laoshan Natural Reserve from 2000 to 2010, we can find that due to the increase of the population, the pollution around the reservoir increased. The wastewater emissions, the use of nitrogen fertilizer and pesticide have caused a certain degree of pollution. In addition, the fish and shrimp fed by farmers in the reservoir can also make the nitrogen content of water over top. The rapid development of tourism leads to the increase of restaurant wastewater, and rainfall will also carry the total nitrogen.

(2)Analysis of air quality

According to the datum, the air quality of Laoshan Natural Reserve in 2001 is good. The content of SO_2, CO, NO_2, and PM_{10} were not overweight. Four kinds of air-monitoring index in 2005 and 2010 were not overweight too, and they have decreased compared with those of 2001. Overall, the air quality of Laoshan Natural Reserve is better, and the local vegetation contributes a lot. In 2001, the impact of PM_{10} in the air became serious for air environmental. Through the survey we found it's because of the circulating fluidized boiler of the nearby high-tech company that was put into operation in November. That indicates the human activities have a great influence on the composition of air.
(3)Analysis of precipitation quality
The precipitation monitoring data of Laoshan Nature Reserve be obtained in this study mainly used to test the PH value and the rate of acid rain. We found that the PH value of total precipitation is less than the standard of acid rain (5.60) in 2001, and the PH value is greater than 5.19 in 2005 and 2010, which not belong to the scope of the acid rain.

5. Conclusions

(1)In this study, we processed remote sensing images of three years (2001, 2005, 2010) in the Laoshan Natural Reserve, and classified the landscape ecosystem, then we analysed ecological landscape characteristics. Research shows that the eco-environment of tourism scenic spots in Laoshan Natural Reserve is much better than that of the surrounding areas, but the urban expansion of surrounding areas and the increase of cultivated land area have a harmful effect on the eco-environment of scenic spots to a certain extent, such as the decrease of vegetation in scenic spots, and more importantly, the influence for air quality.
(2)The PH value of reservoir water and total nitrogen in the Laoshan Natural Reserve are overweight. The main reason is the sewage discharge and excessive use of pesticide nitrogen fertilizer by nearby residents. In addition, feeding fish and shrimp in the reservoir has a certain impact as well. Through analysed factors such as air quality, population changes, we concluded that the most important factor of eco-environment changes in Laoshan Natural Reserve is the producing practice of human.

Reference

[1] Strobel C. J., et al. Environmental Monotoring and Assessment Program: Current Status of Virginian Province (U.S.).Estuaries [J]. Environmental Monitoring and Assessment.1999, 56(1):1-25.
[2] Matthew A, Luck G.Darreljendrdtte the Urban Funnel Model and the Spatially Heterogeneous Ecological FootPrint [J]. Eeosystems, 2001.4(8) :782-796.
[3] MareoTrevisan. NonPoint-Source Agricultural Hazard Index:ACasestudy of the Province of Cremona [J]. Environmental Management, 2000, 26(5):577-584.
[4] Liu Zhuang, Xie Zhiren, Shen Weishou. A New Method that can Improve Regional Eco-environmental Evaluation -- combining GIS with AHP [J]. Resources and Environment in the Yangtze Basin, 2003, 12(2): 163-168.
[5] Ma Ronghua , Hu Mengchun. Assessment of the Ecological Environment in Hainan Island Based on RS &GIS[J]. Tropical Geography, 2001, 21(3): 198-201.
[6] Zhang Xiuying, Zhao Chuanyan. Study on a Potential Eco-environment Evaluation Based on GIS in Longzhong Loess Plateau[J]. Journal of Lanzhou University (Natural Sciences), 2003, 39(3): 73-76.
[7] Du Suojun, Yin Yimin, Xie Dongjun. Ecological Environmental Assessment of Zhang Jiagang[J]. Jiangsu Environmental Science and Technology, 2006,19(2): 74-75.
[8] Zheng Hongping, Bai Liang, Lin Feng. Eco-environmental quality Assessment and Dynamic Analysis of Fujian province[J]. Energy and Environment, 2006, 3: 6-8.
[9] Jiang Zhenlan, Sha Jinming , Yang Wunian. Comprehensively Commented Model Based on Remote Sensing Evaluation of Ecological Environment of Fuzhou [J]. Remote sensing for land & resources, 2004, 3: 46-48.

[10] Chen Tao, Xu Yao. Eco-Environmental Synthetic Analyses Based on RS and GIS Technology of Sichuan Province[J]. Journal of China West Normal University (Natural Sciences), 2006, 27(2): 153-157.

Acknowledgments
This research has been supported by the high resolution special project for remote sensing monitoring and application demonstration in coastal zone and the project of global change and air-sea interaction.

Calculating the Marine Gravity Anomaly of the South China Sea based on the Inverse Stokes Formula

Liang Liu[1,2,3], Xiaoguang Jiang[1], Shanwei Liu[4*] , Lei Zheng[4], Jinxia Zang[4], Xuehua Zhang[2,3] and Longfei Liu[1,2,3]

[1] University of Chinese Academy of Sciences, Beijing 100049, China
[2] National Disaster Reduction Center of China, MCA, Beijing 100124, China
[3] Satellite Application Center for Disaster Reduction, MCA, Beijing 100124, China
[4] China University of Petroleum, Qingdao 266580, China

E-mail: shanweiliu@163.com

Abstract Marine gravity field information has a great significance for the resource, environment and military affairs. As a new way to get marine gravity data, the satellite altimetry technique makes up for what ship measuring means lack. The paper carries out the researches on how altimeter data applied for calculating marine gravity anomaly based on inverse Stokes formula. In the article, the editing of 14-track Jason-1 data over South China Sea for 7 years is for collinear processing and cross-point adjustment. The inverse Stokes formula and fast Flourier transform technique are applied to calculate marine gravity anomaly of the region (0°~23°N, 103°~120°E), and to draw gravity anomaly map. Compared with the gravity anomaly by ship observation, RMS is 12.6mGal, and single altimetry satellite has a good precision.

1.Introduction

Satellite altimetry data to calculate marine gravity anomaly can provide abundant marine gravity and seabed geophysical information. It is of great significance in building the high-precision global gravity model, in the study of climate, and in providing information for further exploitation and utilization of marine resources.

So far, the altimetry satellites launched by countries are mainly as follows: Geosat Satellite by the United States, the T/P, Jason-1 and Jason-2 Satellite by the United States and France jointly, and ERS-1, ERS-2, Envisat Satellite by the European Space Agency. Besides, China launched the first radar altimetry satellite HY-2 in 2011. After the successful launching of the altimetry satellites, how to use satellite altimetry data to calculate marine gravity anomaly becomes a hot topic in the field physical geodesy. Hereby, experts both inside and outside of China carry out the related research. Inverse Stokes formula proceeds inverse operation based on the principle of Stokes formula, which resumes the geoid into a gravity anomaly. This method uses Fast Fourier Transform (FFT) algorithm to get grid-type gravity anomaly rapidly without the global integration. Houze Xu[1] used inverse Stokes formula to calculate the 30'×30' marine gravity anomaly in China offshore waters by T/P, ERS satellite altimetry data, with the accuracy at 35mgal. Haiying Wang[2] inverted 30'×30' gravity anomaly in the China Sea using the FFT inversion method of gravity anomaly, on the basis of inverse Stokes formula with T/P altimetry satellite data for 4 years and ERS 1/35 days of satellite altimetry data for 1.5 years, with the accuracy at 10 mgal. Xiaoyun Wan[3] discussed the accuracy analysis of the remove-restore method in the calculation of inverse Stoke formula[4-8].

2.Methods

2.1.Inverse Stokes formula

If g_0 represents the gravity at a point on the geoid, and γ_0 represents the normal gravity at this point on the average projection on ellipsoid surface, the difference between g_0 and γ_0 is marine gravity anomaly.

We can push to a spherical approximation of the geoid height by the fundamental formula of Geodesy and Bruns formula, Stokes integral formulas:

$$N = \frac{R}{4\pi\gamma}\iint \Delta g S(\psi) d\sigma \tag{1}$$

In the formula (1), *N*, *R*, γ, Δg, ψ respectively represent the geoid height, the earth average radius, the normal gravity value, the gravity anomaly value, and the spherical angle range between Point *P* and the flow point *Q*. $S(\psi)$ is the Stokes formula (2) and γ is the formula (3):

$$S(\psi) = \frac{1}{\sin\frac{\psi}{2}} - 6\sin\frac{\psi}{2} + 1 - 5\cos\psi - 3\cos\psi \ln(\sin\frac{\psi}{2} + \sin^2\frac{\psi}{2}) \tag{2}$$

$$\gamma = 978.0327(1 + 0.0053024\sin^2 B - 0.0000058\sin^2 2B) \tag{3}$$

In the formula (3), *B* is the latitude.

The formula (2) can be rewritten as:

$$S(\psi) = \frac{1}{s} - 6s - 4 + 10s^2 - (3 - 6s^2)\ln(s + s^2) \tag{4}$$

In the formula (4), $s = \sin\frac{\psi}{2}$

Converting Stokes formula (4) into the formula with latitude and longitude as the independent variables:

$$\sin^2\frac{\psi}{2} = \sin^2\frac{1}{2}(\varphi_P - \varphi_Q) + \sin^2\frac{1}{2}(\lambda_P - \lambda_Q) * [\cos^2\varphi_M - \sin^2\frac{1}{2}(\varphi_P - \varphi_Q)] \tag{5}$$

In the formula (5), λ_P and φ_P are respectively the latitude and longitude of calculated gravity anomaly points. λ_Q、φ_Q are respectively the latitude and longitude of Integral floating point, φ_M is the average latitude of Point *P* and *Q*, it can approximate by an average of integral area of approximate latitude, so the Stokes formula (4) turns into the following form:

$$S(\psi) = S(\varphi_P - \varphi_Q, \lambda_P - \lambda_Q) \tag{6}$$

Taking the formula (6) into formula (1), geoid high integral form can be received:

$$N(\varphi_{P,}\lambda_P) = \frac{R}{4\pi\gamma}\iint[\Delta g(\varphi_{Q,}\lambda_Q) * \cos\varphi_Q]S(\varphi_P - \varphi_{Q,}\lambda_P - \lambda_Q)d\varphi_Q d\lambda_Q \tag{7}$$

Be discrete and the convolution formula is as follows:

$$N(\varphi,\lambda) = \frac{R}{4\pi\gamma}[\Delta g\cos\varphi] * [S(\varphi,\lambda)] \tag{8}$$

In the formula (8), φ is the latitude difference between Point *P* and Point *Q*, so its spectrum calculation formula is as follows:

$$F[N(\varphi,\lambda)] = \frac{R}{4\pi\gamma}F[\Delta g\cos\varphi] * F[S(\varphi,\lambda)] \tag{9}$$

If the geoid high *N* is known, the formula (9) becomes:

$$\Delta g(\varphi,\lambda) = \frac{4\pi\gamma}{R\cos\varphi_M}F^{-1}\{\frac{F[N(\varphi,\lambda)]}{F[S(\varphi,\lambda)]}\} \tag{10}$$

In the formula (10), F and F^{-1} represent respectively two-dimension FFT and inverse transform operator, φ_M is the average latitude of integral area.

2.2. *Inner circle zone effect*

Due to the Stokes formula will encounter integral formula singular problem in the process of the integral, it must be considered the influence of calculation data points in the square, so it need to calculate the inner loop effects:

$$N_{ib}(P) = \frac{\sqrt{ab}}{\gamma\sqrt{\pi}}\Delta\bar{g}_{ib} \tag{11}$$

In the formula (11), $N_{ib}(P)$ is the inner loop effects, $\Delta\bar{g}_{ib}$ is the average inner loop gravity anomaly, $a = R\Delta\varphi$, $b = R\Delta\lambda\cos\varphi$.

3.Experiments

3.1. *Study Area and Data*

The paper studies on the area of South China Sea (0°~23°N，103°~120°E), as shown in figure 1, using 7-year Jason-1 satellite altimetry data of 14-rail. Jason-1 satellite was launched successfully in December, 2001 E.S.T. Its observation precision in offshore waters is up to 2-3 cm, the orbit height at 1336 km, orbital tilt angle at 66°, orbital repetition period of 9.9 days, with operating range covering 90% of the global ocean.

IOP Conf. Series: Earth and Environmental Science **46** (2016) 012062 doi:10.1088/1755-1315/46/1/012062

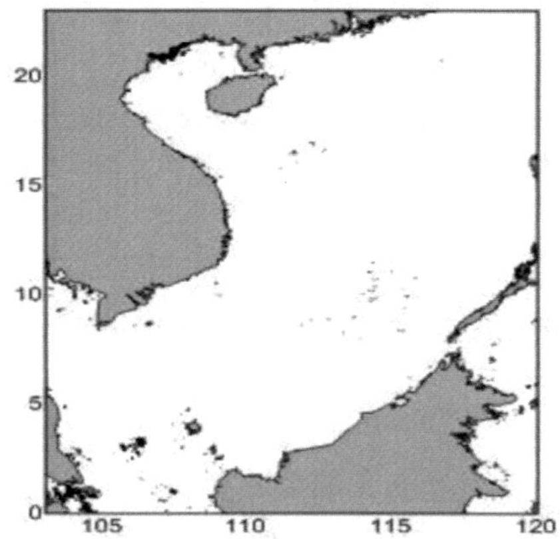

Figure 1. The South China Sea Waters

3.2. Technical Procedures

Data pre-processing includes data editing, collinear processing, cross-over point determining and adjustment, aimed at getting the processed data of the height of sea level.

The pre-processing sea surface height taking off the surface terrain model values is the geoid height. Using geoid height minus the corresponding EGM2008 geoid model values, and then making use of the residual geoid height calculate the residual gravity anomaly based on inverse Stokes formula, and combining with the corresponding EGM2008 gravity anomaly model values, the gravity anomaly will be restored. Finally, ship gravity anomaly is applied to evaluate the accuracy of calculated value. The technical procedures are shown in figure 2.

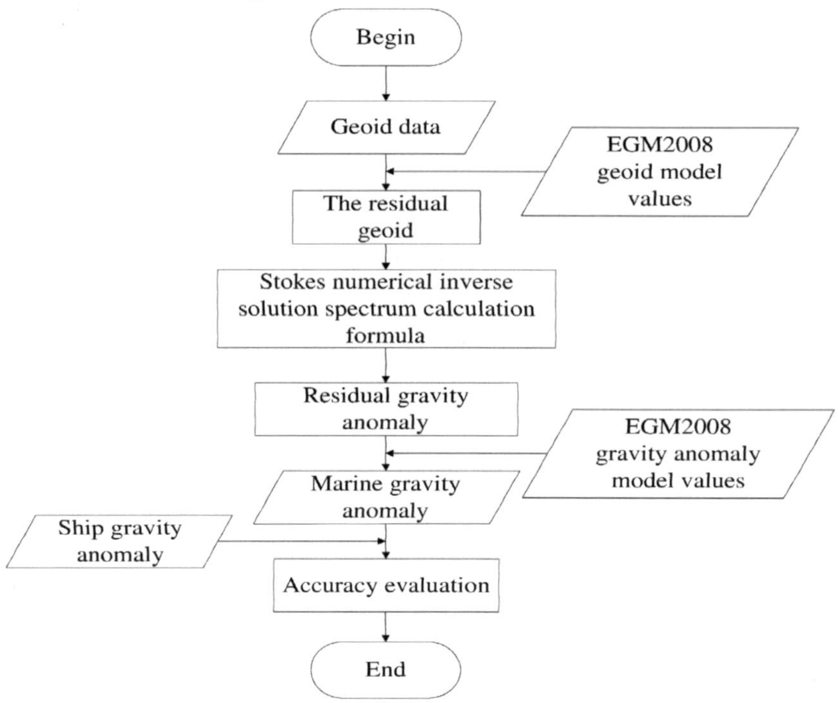

Figure 2. Technical procedures

3.3 Result and Analysis

Figure 3 is a 30′×30′ gravity anomaly chart drawn by Surfer software. The gravity anomaly inversion measurement precision is validated through ship gravity data of the South China Sea area. The data is preceded with accuracy control, gross error elimination and grid processing, at the accuracy of 1-3mgal in figure 4. The red dot represents 30′×30′grid of the ship gravity anomaly with the measuring points well-distributed and typical.

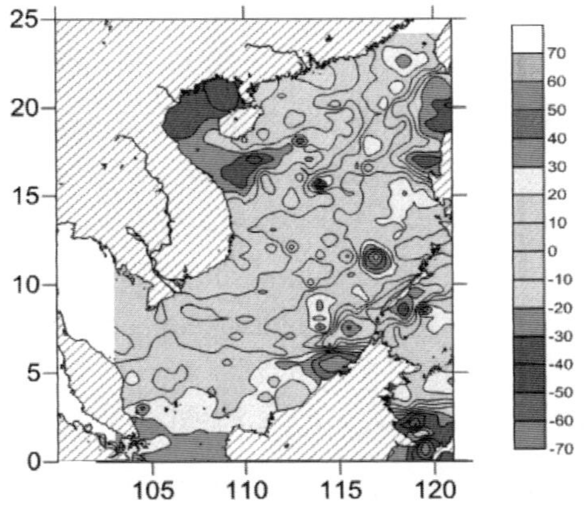

Figure 3. Gravity anomaly in the South China Sea

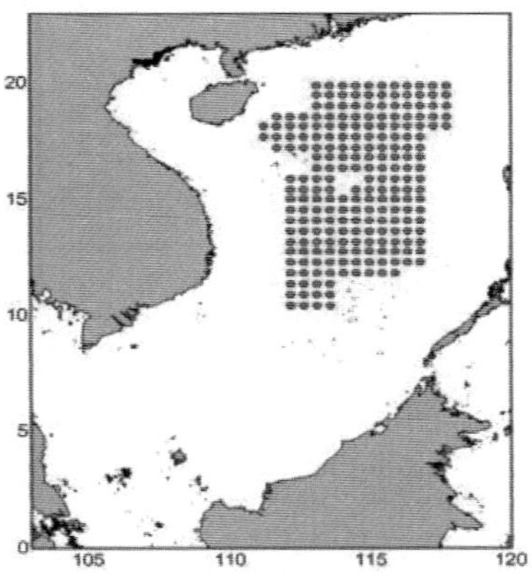

Figure 4. The ship gravity anomaly data distribution

As a result of the ship gravity not exactly corresponding to the calculation result, it needs interpolation fitting out their corresponding gravity anomaly values, by using measuring points around the ship within 4′ of the calculated value to interpolate, with 54 efficient points. Using the difference between the calculated value and measured value of the maximum and minimum value, the average difference and RMS four precision indexes to evaluate the precision, the results are shown in table 1.

Table 1. The comparison of calculated value and measured value

	Points	maximum value	minimum value	The average difference	RMS
Calculated value - ship measured values	54	31.21	-21.79	-1.15	12.63

In this table, the mgal is the unit of maximum value, minimum value, the average difference and RMS. Table1 indicates that the calculation accuracy of gravity anomaly is 12.63mgal, higher than the results from Houze Xu. Besides, Haiying Wang used method of gravity anomaly along the path to invert the 2'x2' gravity anomaly of the South China Sea along, at the accuracy up to 10mgal, corresponding with the result in this article.

4.Conclusion

This paper uses inverse Stokes formula to carry out the study on gravity anomaly calculation, by using the 7 year's 14-rail Jason-1 satellite altimetry data to calculate gravity anomaly in the South China Sea, with the precision of 12.63mgal. The huge track interval based on the Jason-1 satellite altimetry data only will have effect on the accuracy of gravity anomaly to some extent. By contract, the calculation accuracy of gravity anomaly from multi-altimetry satellite data will be higher.

References

[1] HSU HOU-Tse, et al. Geoid undulations and gravity anomalies from T/P and ERS altimeter data in the China sea and vicinity [J]. *Chinese Journal of Geophysics*, 1999, **42(4)**:465-470

[2] Wang Haiying, Lu Yang, Wang Guangyun. Recovery of Gravity Anomalies from Satellite Altimeter Data in China Seas and Vicinity--I .Method of Inverse Stokes Formula with 2D FFT [J].*Chinese High Technology Letters*, 2000, **(10)**:48-51

[3] Wan Xiaoyun, Yu Jinhai. Accuracy Analysis of the Remove-Restore Process in Inverse Stokes Formula [J]. *Geomatics and Information Science of Wuhan University*, 2012, **37(1)**:77-80

[4] Wang Haiying, Wang Guangyun. Inversion of Gravity Anomalies from Along-track Vertical Deflections with Satellite Altimeter Data and Its Applications [J]. *Acta Geodaetica et Cartographica Sinica*, 2001, **30(1)**:21-26

[5] Li Jiancheng, Ning Jinsheng, Chen Junyong, et al. Determination of Gravity Anomalies over the South China Sea by Combinationof TOPEX /Poseidon,ERS2 and Geosat Altimeter Data [J]. *Acta Geodaetica et Cartographica Sinica*, 2001, **30(3)**:197-202

[6] Huang Motao, Zhai Guojun, Guan Zheng, et al. On the Recovery of Gravity Anomalies from Altimeter Data [J]. *Acta Geodaetica et Cartographica Sinica*, 2001, **30(2)**: 179-184.

[7] Huang Motao, Wang Rui, Zhai Guojun, et al. Integrated Data Processing for Multi-satellite Missions and Recovery of Marine Gravity Field [J]. *Geomatics and Information Science of Wuhan University*, 2007, **32(11)**:988-993

[8] Hu Shumei, Wen Hanjiang, Li Hongchao, et al. Inversion of Gravity Anomalies over South China Sea by Use of Combination of Multi-satellite Altmeter Data [J]. *Journal of Geodesy and Geodynamics*, 2011, **31(4)**:56-59

Acknowledgments

This research has been supported by NSFC (61571009, 41571352, 1471297), a program of Qingdao applied basic research (15-9-1-50-jch), Marine public welfare industry research (201305032) and a project cooperation between China and the EU (ID. 10466).

Institute of Physics
Dirac House, Temple Back
Bristol BS1 6BE UK

CURRAN ASSOCIATES INC.
proceedings.com

ISSN: 1755-1307
ISBN 978-1-5108-3271-8